Thermal Conversion of Solid Wastes and Biomass

Thermal Conversion of Solid Wastes and Biomass

**Jerry L. Jones and
Shirley B. Radding,** EDITORS

SRI International

ASSOCIATE EDITORS

**Shigeyoshi Takaoka
A. G. Buekens
Masakatsu Hiraoka
Ralph Overend**

Based on a symposium sponsored by
the Division of Environmental Chemistry
at the 178th Meeting of the
American Chemical Society,
Washington, D.C.,
September 10–14, 1979.

ACS SYMPOSIUM SERIES 130

AMERICAN CHEMICAL SOCIETY
WASHINGTON, D. C.   1980

Library of Congress CIP Data

Main entry under title:
Thermal conversion of solid wastes and biomass.
(ACS symposium series; 130 ISSN 0097-6156)

Includes bibliographies and index.

1. Waste products as fuel—Congresses. 2. Biomass
energy—Congresses.
I. Jones, Jerry Latham, 1946-   . II. Radding,
Shirley B., 1922-   . III. American Chemical Society.
IV. American Chemical Society. Division of Environmental Chemical. V. Series: American Chemical Society. ACS symposium series; 130.

TP360.T46   662'.8   80-14754
ISBN 0-8412-0565-5 ASCMC8 130 1-747 1980

ACS Symposium Series

M. Joan Comstock, *Series Editor*

FOREWORD

The ACS Symposium Series was founded in 1974 to provide
a medium for publishing symposia quickly in book form. The
format of the Series parallels that of the continuing Advances
in Chemistry Series except that in order to save time the
papers are not typeset but are reproduced as they are submitted by the authors in camera-ready form. Papers are reviewed under the supervision of the Editors with the assistance
of the Series Advisory Board and are selected to maintain the
integrity of the symposia; however, verbatim reproductions of
previously published papers are not accepted. Both reviews
and reports of research are acceptable since symposia may
embrace both types of presentation.

CONTENTS

Preface xi

U.S. DEPARTMENT OF ENERGY SPONSORED PROGRAMS
CONCERNED WITH URBAN WASTES AND BIOMASS

1. **An Overview of the Department of Energy Program for the Recovery of Energy and Materials from Urban Solid Waste**   3
   D. K. Walter

2. **Thermochemical Conversion of Biomass to Fuels and Feedstocks: An Overview of R&D Activities Funded by the Department of Energy** 13
   G. F. Schiefelbein, L. J. Sealock, Jr., and S. Ergun

AN OVERVIEW OF TECHNOLOGIES FOR THERMAL CONVERSION OF WOOD

3. **Preliminary Economic Overview of Large-Scale Thermal Conversion Systems Using Wood Feedstocks** 29
   S. M. Kohan

4. **Wood Gasification: An Old Technology with New Potential for the Eighties** 43
   J. J. Fritz, A. Talib, and J. J. Gordon

COMBUSTION TECHNOLOGY

5. **Mass Burning of Municipal Solid Waste—Worldwide** 59
   P. R. Beltz

6. **An Evaluation of Modular Incinerators for Energy Recovery from Solid Wastes** 67
   D. C. Bomberger and J. L. Jones

7. **Energy Recovery from Wood Residues by Fluidized-Bed Combustion** 85
   J. W. Stallings and M. Oswald

8. **Fluid-Bed Combustion of Solid Wastes** 93
   R. H. Vander Molen

9. **Energy Recovery from Municipal Solid Waste and Sewage Sludge Using Multisolid Fluidized-Bed Combustion Technology** 109
   W. E. Ballantyne, W. J. Huffman, L. M. Curran, and
   D. H. Stewart

DENSIFICATION TECHNOLOGY

10. **The Preparation and Properties of Densified Refuse-Derived Fuel** .. **127**
    H. Alter and J. A. Campbell

11. **The Eco-Fuel II Process: Producing a Storable, Transportable
    Fuel for Co-Combustion with Oil** **143**
    F. Hasselreiis

12. **Small-Scale Source Densification of Navy Solid Waste** **151**
    D. Brunner

13. **Biomass Densification Energy Requirements** **169**
    T. B. Reed, G. Trezek, and L. Diaz

14. **Densification Systems for Agricultural Residues** **179**
    T. R. Miles and T. R. Miles, Jr.

15. **Replenishable Organic Energy for the 21st Century** **195**
    J. L. Stafford and A. D. Livingston

PYROLYSIS, GASIFICATION, AND LIQUEFACTION PROGRAMS IN THE UNITED STATES

16. **Gasoline from Solid Wastes by a Noncatalytic, Thermal Process** ... **209**
    J. P. Diebold

17. **Fuel Production from Wastes Using Molten Salts** **227**
    R. L. Gay, K. M. Barclay, L. F. Grantham, and S. J. Yosim

18. **Biomass Gasification at the Focus of the Odeillo (France)
    1-Mw (Thermal) Solar Furnace** **237**
    M. J. Antal, Jr., C. Royere, and A. Vialaron

19. **Operation of a Downdraft Gasifier Fueled with Source-Separated
    Solid Waste** **257**
    S. A. Vigil, D. A. Bartley, R. Healy, and G. Tchobanoglous

20. **The Enerco Pyrolysis System** **275**
    E. W. White and M. J. Thomson

21. **Development of Modular Biomass Gasification and
    Combustion Systems** **285**
    J. Fletcher and R. Wright

22. **Solid Waste Gasification and Energy Utilization** **291**
    F. G. Rinker

23. **Thermodynamics of Pyrolysis and Activation in the Production of
    Waste- or Biomass-Derived Activated Carbon** **301**
    F. M. Lewis and C. M. Ablow

PYROLYSIS, GASIFICATION, AND LIQUEFACTION PROGRAMS IN CANADA

24. **Canada's Biomass Conversion Technology R&D Program** **317**
    R. Overend

25. **Pyrolysis of Agricultural Residues in a Rotary Kiln** **337**
    R. H. Arsenault, M. A. Grandbois, E. Chornet, and G. E. Timbers

from 4615 N Scott

26. **Fluidized-Bed Gasification of Solid Wastes and Biomass: the CIL Program** 351
J. W. Black, K. G. Bircher, and K. A. Chisholm

27. **The Direct Liquefaction of Wood Using Nickel Catalysts** 363
D. G. B. Boocock, D. Mackay, and H. Franco

28. **The Role of Catalysis in Wood Gasification** 369
D. P. C. Fung and R. Graham

4945/1

29. **Wood Gasification for Gas and Power Generation** 379
A. Verma and G. A. Weisgerber

PYROLYSIS, GASIFICATION, AND LIQUEFACTION PROGRAMS IN EUROPE

30. **Basic Principles of Waste Pyrolysis and Review of European Processes** 397
A. G. Buekens and J. G. Schoeters

4940

31. **Pyrolysis of Plastic Waste and Scrap Tires Using a Fluidized-Bed Process** 423
W. Kaminsky and H. Sinn

32. **Two-Stage Solid Waste Pyrolysis with Drum Reactor and Gas Converter** 441
O. Tabasaran

33. **The Design of Co-Current Moving-Bed Gasifiers Fueled by Biomass** 463
M. J. Groeneveld and W. P. M. van Swaaij

34. **Pyrolytic Recovery of Raw Materials from Special Wastes** 479
G. Collin

35. **Gasification of Solid Waste in Accordance with the SFW–FUNK Process** 485
H. Funk and H. Hummelsiep

PYROLYSIS, GASIFICATION, AND LIQUEFACTION PROGRAMS IN JAPAN

36. **Overview of Pyrolysis, Thermal Gasification, and Liquefaction Processes in Japan** 493
M. Hiraoka

37. **Pilot Plant Study on Sewage Sludge Pyrolysis** 509
T. Kasakura and M. Hiraoka

4935

38. **Gasification of Solid Waste in Dual Fluidized-Bed Reactors** 525
M. Kagayama, M. Igarashi, M. Hasegawa, J. Fukuda, and D. Kunii

39. **Disposal of Municipal Refuse by the Two-Bed Pyrolysis System** 541
N. Andoh, Y. Ishii, Y. Hirayama, and K. Ito

40. **Pyrolysis Process for Scrap Tires** 557
S. Kawakami, K. Inoue, H. Tanaka, and T. Sakai

41. **PUROX System Demonstration Test on Simulated Japanese Refuse** . 573
T. Masuda and T. F. Fisher

42. Integrated System for Solid Waste Disposal with Energy Recovery and Volumetric Reduction by a New Pyrolysis Furnace 587
  T. Fujii

43. Development of a Solid Waste Disposal System with Pyrolysis and Melting 603
  M. Onozawa

BIOMASS ENERGY FOR DEVELOPING NATIONS

44. The Potential of Biomass Conversion in Meeting the Energy Needs of the Rural Populations of Developing Countries—An Overview . . . 617
  V. Mubayi, J. Lee, and R. Chatterjee

45. Power Generation from Biomass Residues Using the Gasifier/ Dual-Fuel Engine Technique 635
  A. A. Dennetiere, F. Leorat, and G. F. Bonnici

46. Studies on the Practical Application of Producer Gas from Agricultural Residues as Supplementary Fuel for Diesel Engines . . . 649
  I. E. Cruz

47. Third World Applications of Pyrolysis of Agricultural and Forestry Wastes 671
  J. W. Tatom, H. W. Wellborn, F. Harahap, and S. Sasmojo

48. A Preliminary Analysis of the Potential for Developing Nations to Use Biomass Fuels in Vehicle Engines 689
  J. L. Jones and A. K. Chatterjee

49. Social and Economic Aspects of the Introduction of Gasification Technology in Rural Areas of Developing Countries (Tanzania) . . . 705
  M. J. Groeneveld and K. R. Westerterp

Index 723

PREFACE

The conversion to useful energy of solid wastes, agricultural residues, and forestry residues by thermal processes has long been practiced to varying extents in many industrialized and developing areas of the world. Within the last decade, however, environmental concerns and rising fossil fuel prices have led to increased efforts to develop and implement efficient and economical methods for recovering energy from solid wastes and biomass. During 1977, researchers within the Chemical Engineering Laboratory at SRI International reviewed the worldwide status of pyrolysis, gasification, and liquefaction methods to identify unique processes and topics for future research. At that time very few books were available that presented comprehensive reviews of on-going research and development activities related to thermal conversion of solid wastes and biomass. Consequently, a symposium was planned under the auspices of the American Chemical Society (ACS) to discuss major research and development programs in the United States. Twenty-one papers from this symposium— held at the ACS National Meeting in Anaheim, California, during March 1978—were published as ACS Symposium Series No. 76 during the summer of 1978. Because of the success of the first symposium and the interest expressed by participants in the status of research being done in other countries, planning began in August 1978 for a second symposium with an expanded scope. The symposium was to include speakers from North America, Western Europe, and Asia. Financial support for the symposium was obtained from: the Urban Waste and Municipal Systems Branch of the U.S. Department of Energy; the Municipal Environmental Research Laboratory and the Industrial Environmental Research Laboratory of the U.S. Environmental Protection Agency; and the Division of Environmental Chemistry of the American Chemical Society.

The symposium, held at the September 1979 National Meeting of the American Chemical Society in Washington, D.C., was jointly sponsored by the ACS Divisions of Environmental Chemistry, Fuel Chemistry, and Cellulose, Paper and Textiles. It was also planned in cooperation with other professional societies including: the American Institute of Chemical Engineers; the American Society of Agricultural Engineers; the American Society of Civil Engineers; and the Forest Products Research Society.

J. L. Jones and S. B. Radding of SRI International were the chief editors of this book and were responsible for the overall planning of the symposium. The following individuals assisted in the planning and organization: A. G. Buekens, University of Brussels, Belgium; M. Hiraoka, Kyoto University, Japan; R. Overend, National Research Council, Canada; T. B. Reed, Solar Energy Research Institute, United States; and S. Takoka, SRI International, Japan.

A. G. Buekens, M. Hiraoka, R. Overend, and S. Takaoka also served as associate editors and were responsible for editing many of the contributions from their respective regions or countries.

The book contains 49 chapters, 47 of which were presented at the symposium. The remaining two chapters were contributed by authors who were interested in presenting a paper at the symposium but were unable to do so because of a lack of available time on the program.

The first section describes the many research, development, and demonstration programs funded under the U.S. Department of Energy's Urban Waste and Municipal Systems Branch and the Biomass Energy Systems Branch.

Conversion of wood to fuels, chemicals, and energy in the form of steam or electrical power is the subject of the second section. Chapter 3 deals with large wood conversion systems and presents a very comprehensive economic analysis of the different systems. Chapter 4 discusses the production of fuel gas from wood. A listing of process developers is provided, along with estimates of investment and operating costs.

The third section deals with combustion processes, which currently represent the major thermal conversion route for biomass and solid waste fuels. Chapter 5 presents the findings of a survey of the current technological state of the art and system economics for combustion of municipal solid wastes (MSW) in Western Europe. Again, the units described are large field-erected boilers, some of which have the capacity to burn more than 1000 metric tons/day of solid wastes. The use of modular or shop-fabricated combustion equipment by small municipalities or by commercial users is the topic of Chapter 6. Such small capacity systems are not yet widely used in the United States or elsewhere for energy production. An economic analysis is given for hypothetical systems with capacities to fire up to approximately 100 metric tons/day of wastes. Chapters 7, 8, and 9 deal with fluidized-bed combustion systems. Chapter 7 describes a commercially available combustor now used at about 30 installations in the forest products industry for burning wood residues; Chapter 8 describes another commercially available fluidized-bed combustor for wood residues and also contains information concerning experiments with burning processed MSW. A multisolid fluidized-bed unit now under development in pilot-scale equipment for solid waste and sludge combustion is the topic of Chapter 9.

Densification of solid wastes and biomass to produce solid fuels with more desirable physical properties is the subject of the fourth section.

Densification produces solid fuels that are free flowing and easily handled as opposed to shredded wastes or agricultural residues. Chapter 10 considers the mechanical processing of MSW and describes individual unit processes that typically precede densification. The densification process (pelletizing) also is described, as are the physical properties of the product. Chapter 11 presents another approach to producing a densified fuel. After various mechanical processing steps, the cellulosic fraction is embrittled with acid and then ball-milled to produce a powdered solid fuel. This process currently is being demonstrated in a large commercial facility. Chapters 10 and 11 deal with processes to be used principally at large capacity facilities. Chapter 12 describes the development of densification equipment for solid wastes for use at naval bases. Such equipment also may have applications for commercial and industrial users as well as some small communities. Chapter 13 presents results of experiments conducted to determine the energy required for densification under laboratory conditions, and compares these results with the actual energy use of commercial scale equipment. The effects of temperature on energy consumption are shown to be very significant. Densification of agricultural residues and wood is the subject of Chapters 14 and 15. Chapter 14 reviews the types of equipment now available, product forms, and power requirements, and cites specific examples of commercial practice. Chapter 15 discusses a current business venture in the biomass densification field and characterizes emissions from burning the product fuel pellets.

The fifth section is the first of four parts dealing with specific pyrolysis, gasification, and liquefaction projects in various regions. It includes chapters describing U.S. process development and commercialization activities, most of which were not described in the ACS Symposium Series No. 76 published in 1978. Chapter 16 concerns pyrolysis work being conducted at the laboratory scale by U.S. Navy researchers, where pyrolysis of solid wastes or biomass is accomplished in an indirectly heated transport reactor. By various separation steps, an olefin-rich gas is produced and then polymerized under high temperature and pressure conditions. The conditions under which formation of olefins is maximized rather than formation of char, tar, or water-soluble organics are discussed. Chapter 17 describes the use of a molten salt bath reactor for gasifying solid wastes such as refinery acid pit sludge, rubber tire scraps, wood, leather scraps, and waste photographic film. Experiments have been conducted in both bench-scale and pilot plant-scale equipment. Chapter 18 concerns pyrolysis research work conducted by a U.S. researcher in cooperation with French scientists at the Odeillo (France) solar furnace. The solar furnace was selected as a heat source so as to obtain an extremely high energy flux through the quartz window reactor. At the extremely high flux rates, char yields were very low and condensible product yields were high. Chapter 19 describes experiments with a downdraft gasifier (previously described in the ACS Symposium Series No. 76) that uses densified fuel pellets produced from processing MSW. Chapter 20 contains information on a mobile pyrolysis reactor now being tested in commercial applications. Such a unit can be moved to the biomass source to produce charcoal for later use as a fuel. Chapter 21 deals with the subject of retrofitting existing oil- or gas-fired industrial boilers with a crossflow biomass gasifier that would use densified fuel pellets. Prototype units currently are undergoing field testing. A description of a vertical shaft countercurrent-flow (or fixed-bed) gasifier is given in Chapter 22. Two such units are being installed to gasify waste cigarette papers and filters, packaging paper, cardboard, and other wastes. The product fuel gas will be burned in a boiler. Chapter 23 discusses the energy requirements (in the form of steam) for activating carbon char produced from biomass or wastes. Efficient use of steam for activation can be very important to the overall system economics.

Biomass utilization and conversion research in Canada are the topics addressed in the sixth section. Chapter 24 contains a review of the analysis procedure used to aid in the formulation of the government-supported biomass R&D program in Canada. The major goal is to produce a liquid fuel or a liquid fuel substitution product. A short-term goal is associated with synthesis gas production from biomass. The reader should note the Dulong diagrams used by the author for describing biomass conversion routes. Chapter 25 concerns pyrolysis of agricultural residues in a pilot plant-scale, continuous-flow, indirectly fired rotary retort. The principal products are char and a fuel gas of intermediate heating value. Conversion of MSW and wood to a low heating value fuel gas, using an air-blown fluidized-bed gasifier, is discussed in Chapter 26. The process has been demonstrated for wood gasification in a unit capable of processing approximately 25 metric tons/day of wood. A mechanical processing system and synthesis gas production are also discussed. Chapter 27 describes a direct wood liquefaction process being developed in a laboratory-scale batch feed system. The reported results indicate the ability to liquify wood at elevated temperatures in the presence of a nickel catalyst and in the absence of hydrogen. Pyrolysis of wood in a bench-scale fluidized-bed reactor is discussed in Chapter 28. Experiments with and without alkali carbonate catalysts were conducted. Chapter 29 describes an air-blown, vertical shaft wood gasifier now being demonstrated. Preliminary performance data are given for the gasification system, which is coupled to a diesel engine power generation system.

European technology for pyrolysis, gasification, and liquefaction is described in the seventh section. Chapter 30 includes an interesting discussion of why the development of such processes for MSW has not been as rapid in Europe as in the United States or Japan. During the period from 1960 to 1975, many large mass-burning incineration plants were constructed in Western Europe to recover energy from MSW. In general, these installations have proved to be an acceptable solution to the problem of disposal of MSW. A very comprehensive review of the process development activities is presented as well as an excellent review of the fundamentals of pyrolysis and gasification. A laboratory research program concerned with pyrolysis in a fluidized-bed reactor is also described. Chapter 31 describes the testing of a relatively large indirectly heated fluidized-bed pyrolysis reactor that is capable of accepting whole scrap tires on a semicontinuous feeding basis. Most other pyrolysis processes designed for scrap tires require that the tires be shredded prior to being fed to the reactor. Chapter 32 describes a pyrolysis process that uses an indirectly heated rotary retort and is now being demonstrated for commercial use. If the product fuel gas is to be used without tar or pyrolysis oil recovery, a second-stage autothermic cracking reactor is added. Use of a cocurrent-flow vertical shaft reactor for wood gasification is the subject of Chapter 33. The process has been tested and modeled, so this chapter should prove to be quite useful to designers of cocurrent-flow biomass gasifiers. Chapter 34 describes another process that uses an indirectly heated rotary retort for pyrolysis of solid waste and presents the results of tests in a pilot plant-scale reactor. Chapter 35 describes the use of a countercurrent-flow, oxygen-blown vertical shaft reactor being developed for gasification of MSW. A pilot plant system is described.

The eighth section contains eight excellent chapters concerning the development and demonstration of pyrolysis and gasification technology in Japan. Chapter 36 presents an overview of the technologies being used and developed. The installation of large mass-burning incinerators based on European technology for disposal of MSW has been under way in Japan since the early 1960s. However, heat recovery is not as widely practiced in Japan as in Europe. Since 1973, more than a dozen major process development programs have been under way to develop new thermal processes that can recover energy more efficiently from the very wet MSW in Japan and thus minimize pollution problems. These processes are reviewed and an economic analysis is presented to illustrate why advanced processes are being developed. Chapter 37 describes a drying-pyrolysis process that incorporates a multiple hearth reactor and that was developed specifically for municipal sewage sludge. The new process is compared with two alternative thermal processes. Chapters 38 and 39 describe the use of dual fluidized-bed reactors (with sand circulating between the reactors) for pyrolysis of processed MSW. Pyrolysis of the MSW occurs in a fluidized bed of hot sand. The char and sand from the first bed move to a second fluidized bed where the char is burned to heat the sand. The hot sand is then returned to the first bed. The basic concept in terms of use of a dual reactor system is the same for the two processes, although differences exist in the mechanical details of how the sand circulates between the two beds. The fluidizing gases are also different. In the first process, steam is the fluidizing gas for the first reactor while in the second process, product gas is recirculated. Both processes have been under development for some time. A 450-metric-ton/day commercial plant is scheduled to begin operation in 1980 using the first process. A large 100-metric-ton/day prototype plant using the second process is to be fully operational before the end of 1979. Pyrolysis of scrap tires in an indirectly fired rotary retort is the subject of Chapter 40. The process now is being used on a commercial scale in a demonstration plant with a capacity to process 7000 tons/year of scrap tires. Shredded tires are pyrolyzed to produce an oil and a carbon black product. Chapter 41 describes the use of the Purox Process, originally developed in the United States, for thermal conversion of Japanese refuse. The slagging gasification reactor is a vertical shaft that is oxygen blown with the gases flowing countercurrent to the solids. A comparison of typical MSW samples from the United States and from Japan is presented to illustrate why special design features are necessary for the process. Product gas is burned immediately so as to avoid production of a large wastewater stream. An economic comparison of the process is made with a conventional incineration process. Chapter 42 describes a process that uses a vertical shaft air-blown reactor and a second-stage ash melting furnace. The product gas is burned directly in a boiler for power generation. The process has been tested in a 20-metric-ton/day pilot plant. Chapter 43 describes yet another process that uses a vertical shaft furnace, blown with heated air or heated oxygen-enriched air, that operates as a slagging gasifier. The process has been developed with a pilot plant-scale reactor, and a facility with a 150-metric-ton/day capacity is now under construction.

The final section deals with the use of thermal processes for biomass conversion in developing countries. Chapter 44 discusses the current extent of use of biomass for fuel among rural populations in several developing countries. Direct combustion is currently the most widely used means for energy production from biomass. The types of biomass available in various countries and the estimated current use are discussed, as well as difficulties in implementing large biomass conversion programs. Chapter 45 describes a commercially available biomass gasification unit coupled to a diesel engine electric power generation system. The gasifier and diesel engine are produced by a French company. The vertical shaft reactor is air-blown and operates in a cocurrent flow mode (or downdraft). The units can be used industrially as well as for rural electrification. Up to 90% of the normal diesel fuel use may be substituted by using the low heating value fuel gas. Chapters 46 and 47 also are concerned with the gasification of biomass to produce a low heating value gas. Chapter 46 describes work being done in the Philippines on the operation of diesel engines using a fuel gas produced from air-blown gasifiers. The chapter presents a great deal of data and empirical correlations that will be useful to gasification system designers. Applications for gasifier/engine sets include electric power generation and irrigation pumping. Chapter 47 presents information on a gasifier that is being developed for use in rural areas. The unit is very simple to fabricate and requires a relatively low investment cost. The use of alcohol fuels produced from biomass and the use of vehicle-mounted biomass gasifiers are discussed in Chapter 48. These options offer the potential to reduce the amount of petroleum fuels currently being used for vehicles. While the use of alcohol fuels has been receiving a great deal of attention worldwide, the use of vehicle-mounted gasifiers apparently has not been considered in many cases where they would prove beneficial. Further study is required to determine if vehicle-mounted gasifiers offer a reasonable solution to petroleum fuel shortages in developing countries. Chapter 49 has been prepared in part by the author of another chapter in the book concerned with the design of a cocurrent-flow biomass gasifier. In this chapter the authors present an analysis of how the introduction of biomass gasification into a developing country will benefit the owners of the gasification units as well as the national economy. The effects on employment, rural development, and balance of trade are examined. Tanzania is considered as a case study.

This book represents the efforts of many individuals including those within the two U.S. agencies and ACS who arranged for financial support for the symposium, the representatives of cooperating technical societies who arranged for publicity, the session chairmen who invited many of the speakers, and the co-editors. Special recognition and appreciation is given to the many authors whose diligent efforts made the timely publication of this book possible.

SRI International   Jerry L. Jones
333 Ravenswood Avenue
Menlo Park, CA 94025   Shirley B. Radding
January 21, 1980

U.S. DEPARTMENT OF ENERGY SPONSORED PROGRAMS CONCERNED WITH URBAN WASTES AND BIOMASS

An Overview of the Department of Energy Program for the Recovery of Energy and Materials from Urban Solid Waste

DONALD K. WALTER

Urban Waste and Municipal Systems Branch, U.S. Department of Energy,
Mail Stop 2221-C, 20 Massachusetts Ave., Washington, DC 20545

The primary mission of the Department of Energy Urban Waste Program is to promote energy conservation through the widespread use of urban solid waste as a source of energy and materials. Specifically, the Urban Waste and Municipal Systems (UWMS) Branch seeks to replace conventional fuels with the energy recovered from urban solid wastes; replace virgin materials with materials recycled from urban solid waste; and reduce the amount of energy used in urban waste treatment and disposal. By recovering energy and materials from urban wastes, the word "waste" becomes a misnomer. Garbage, refuse, certain sludges, and certain solids and liquids which are discarded by households, institutions, and businesses are more aptly defined as "urban byproducts" which are awaiting conversion to a useful form.

Development of these previously neglected resources reinforces several national policies including energy, resource conservation, and the environment. For example, a typical 900 tonne (1,000 ton) per day energy and materials plant:

o Produces approximately two quadrillion joules (two trillion Btu's) of low-sulfur fuel per year -- the equivalent of roughly 300,000 barrels of oil;

o Produces in a year as much as 18,000 metric tons of ferrous, 1,100 metric tons of aluminum and other nonferrous metals, and 14,000 metric tons of glass for use by industry in various manufacturing processes;

o Reduces the amount of material for land disposal from over 270,000 metric tons of mixed wastes (including organic wastes and metals) per year to about 54,000 metric tons per year of relatively inert material.

The use of energy and materials recovery technologies conserves fossil fuels directly by replacing them as sources of energy for combustion, and indirectly by replacing virgin materials with intermediate products which require less energy to reconvert into a final form. However, central plant processing of urban solid waste to recover energy and materials is

still in the early stages of development. In pursuing its
mission to achieve maximum development of the resource potential
of urban solid wastes, the UWMS Branch undertakes activities in
three areas: technical processes, institutional impediments to
waste use, and program (monetary) support. These areas, although
logically separable, are highly interrelated and underline the
Branch strategy to demonstrate promising energy recovery technologies at a commercial scale in a variety of institutional settings.
This strategy is aimed at assisting public and private entities
to conserve energy by utilizing the resources contained in their
urban solid wastes; it focuses on existing impediments to the
implementation of energy recovery projects.

The program activities -- in technical processes, institutional impediments to waste use, and program support -- address
the problems that hinder progress in demonstrating resource
recovery technologies. The following discussion attempts to
describe the breadth of Branch activities in these areas.

Technical Processes

Figure 1 shows the range of technical process options in
resource recovery. Three broad technologies are available to
recover energy and energy intensive materials from urban wastes:
mechanical, thermal, and biological.

Mechanical. Mechanical processing separates wastes into
various components including metals, glass, and a refuse derived
fuel (RDF). Generally, a mechanical process is a preliminary step
to the thermal and biological technologies. The metallic components, glass, and paper fibers are recycled to displace virgin
materials.

In general, mechanical processes are employed for size reduction and separation by size, weight, shape, density and other
physical properties. A typical processing line would utilize
shredding for size reduction of raw refuse, followed by some form
of air classification to separate the particles into light
(organics) and a heavy (inorganics) material stream. The light
fraction, without further processing, has come to be known as
fluff RDF.

A demonstration unit sponsored by the Environemtal Protection
Agency (EPA) to produce RDF at St. Louis proved the basic feasibility of mechanical separation processes, transport and storage
techniques, and combustion of fluff RDF to replace 5 to 27 percent of the pulverized coal used in suspension-fired utility
boilers. However, the refinement of equipment components and the
technical and economic optimization of the basic technology still
require a great deal of work.

There is one operating facility at Ames, Iowa which recovers
and uses fluff RDF on a daily basis. This facility has encountered a series of economic and technical problems. A second
facility at Milwaukee, Wisconsin is entering its first year of

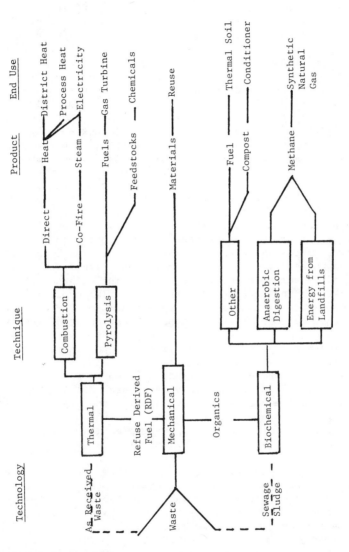

Figure 1. Technical processes

production evaluation prior to finalizing fuel sales contracts.
Several other facilities are in a shakedown phase or under construction at Chicago; Monroe County, New York; Bridgeport, Connecticut; etc.

The preparation of densified RDF (d/RDF) by pelletizing,
briquetting, or extruding is being explored and evaluated. It is
particularly adapted for stoker and spreader-stoker furnaces.
However, it has not been demonstrated commercially, and the costs,
handling characteristics, and firing characteristics are yet to
be evaluated. The anticipated advantages of densified RDF are
greatly improved storage and transportation characteristics.

Production of powder RDF (particles smaller than 0.15 milli-
meter) is being developed in a proprietary pilot-plant process by
embrittling waste with organic acid. After adding an embrittling
chemical, coarsely shredded waste is pulverized. The resulting
powder has a higher Btu content than fluff RDF, as well as a
greater density, homogeneity and decreased moisture content. In
addition, powder RDF may be capable of direct co-firing with fuel
oils. However, the dust-like composition necessitates special
handling to minimize the danger of explosion.

A "wet" mechanical separation process utilizes hydropulping
technology adapted from the pulp and paper industry to reduce raw
waste to more uniform size and consistency, followed by a series
of processes to separate the pulped mass into light and heavy
fractions and to remove some of the water. The original EPA
sponsored pilot plant at Franklin, Ohio, is still operating,
although it no longer produces low grade fiber for a roofing plant
as originally intended. After successful test burns of the 50
percent organics, 50 percent moisture pulp, a full-scale facility
designed to burn the pulp to recover steam is now in start-up at
Hempstead, New York.

Ferrous metal recovery systems are the most advanced material
recovery systems. Paper fiber recovery by both pulping and dry
processes has been demonstrated successfully, and aluminum and
glass recovery has been demonstrated with limited success.
Efforts are planned to improve system efficiencies in terms of
energy use, quantity, and quality of material recovered.

Thermal. Combustion techniques burn waste for the recovery
of heat energy. Waterwall combustors are the most technically
developed energy recovery systems and employ special grates to
burn "as received" urban waste and recover steam either at saturated or superheated conditions. Over 250 plants are operating
worldwide; seven of them in the United States. Three of the
seven were originally built as incinerators. Worldwide there have
been a number of technical problems, with the control of corrosion
and erosion being the most serious. The most recent European designs have solved these problems but at an increased capital cost.

The more popular U.S. development seems to be the recovery
of RDF for sale to coal-using facilities although this may be
changing. With modifications, existing boilers can use RDF as a

supplemental fuel. Most development has been aimed at the large
suspension coal fired utility boiler. While test burns have been
encouraging, technical problems have developed related to burning
characteristics, slagging, and environmental control equipment
performance. All can be solved. However, upgrading control
devices such as electrostatic precipitators may require significant increases in the capital costs of the system.

Another variation being demonstrated is the combustion of
RDF in a dedicated boiler as a principal fuel. Normally the
boiler is of spreader-stoker design with some consideration given
to the use of fossil fuels such as high sulfur coal as a load
leveler and steam production stabilizer.

The only available small-scale system is a packaged two
chamber incinerator with waste heat recovery. This technique is
practical at the 25 to 100 tons per day (TPD) scale. In these
units, partial oxidation occurs in the first section of the unit
and causes a portion of the waste material to degrade and give off
combustible gases. These gases, as well as products of combustion
and particulate from the first chamber, flow to a second chamber
where they are combusted with excess air and a natural gas or oil
pilot flame. The combustion products then flow through appropriate heat transfer equipment to produce steam, hot water, or hot
air. Today, four small cities and more than sixty industrial
plants use the technique with heat recovery equipment.

Thermal gasification and pyrolysis systems are also under
development with several systems approaching the full scale demonstration stage. Specific discussion is not included on individual
techniques since they are still in the developmental stages.
Fuels from these processes include gases, oils and chars.

Biological. Biological techniques use living organisms to
convert organics into useful energy forms. These processes are
in the developmental stages. The Department of Energy (DOE) is
sponsoring an anerobic digestion process that converts the
organics in urban waste to methane under controlled conditions.
This process is at the proof-of-concept stage and is not expected
to be commercialized until the late 1980s.

Another bioconversion process to recover methane from existing landfills is near commercialization. Because of the explosive
nature of the gas, increasing attention has been paid to controlling its migration. Today there is one operating site, two sites
in development, one site under construction and ten sites in an
advanced planning stage to utilize the recovered gas. It is
estimated that 28 million cubic meters (one trillion cubic feet)
of methane is potentially recoverable from existing landfills,
with 1.5 billion cubic meters (55 billion cubic feet) of pipeline
quality methane available yearly, just from the 100 largest landfills.

Generally, the mechanical, thermal and biological resource
recovery technologies can be ranked in order of their states of
development as shown in Table 1. The DOE research, development

Table I. Developmental Stage of Resource Recovery
Technologies

Level of Development	Technology
High	• Ferrous Metal Recovery • Anaerobic Digestion of Municipal Solid Wastes • Waterwall and Modular Controlled Air Combustors • Coarse, Fluff, and Wet Pulped RDF • Paper Fiber Recovery • Landfill Gas Recovery • Glass, Aluminum and Other Nonferrous Metal Recovery • Dust and Densified RDF • Pyrolysis and Gasification • Anaerobic Digestion of Solid Waste
Low	• Enzymatic and Fungal Synthesis

and demonstration (RD&D) program addresses all of these technologies and is active in promoting their development. However, true development requires a good deal more than proving the scientific or engineering viability of alternative resource recovery options. History demonstrates that the major difficulties of putting a new technology on a sound commercial footing occur after the successful operation of the first prototype. The Branch's program is fashioned to deal with these problems.

Nontechnical Issues

Figure 2 illustrates the range of nontechnical issues in resource recovery. The three broad categories -- institutional, socioeconomic, and legal -- are interrelated and often raise obstacles more difficult to overcome then technical problems.

Energy and material recovery facilities, for instance, are similar to other capital intensive manufacturing operations in that they cannot operate economically unless they receive regularly sufficient feedstocks to utilize production capacity. A guaranteed supply of waste is the foundation upon which an economically feasible energy recovery project is built. However, municipalities frequently control only a fraction of the wastes collected within their jurisdictions.

Most urban waste is collected by private haulers who determine the least cost place of disposal. Attempts to mandate that privately-collected wastes be delivered to a recovery system have met with considerable opposition from both private haulers and landfill operators. One such attempt precipitated litigation that is now awaiting a decision in the U.S. District Court for the Northern District of Ohio.

The waste control problem is further aggravated by the need to aggregate sufficient quantities of waste for the recovery facility. Because many recovery technologies currently are economically feasible only at a large scale, they frequently require regionalization of the waste management system. The participation in a recovery project of numerous communities -- each with different perceptions, needs, and expectations -- is difficult to achieve because of the perceived risks and costs associated with recovery systems. Thus, intermunicipal agreements to achieve regionalization are difficult to develop and negotiate. The complexities of this process are often a major cause of delay in implementing promising projects.

Another major impediment is that resource recovery facilities must compete with landfills in most communities. Most of these landfills fall far short of satisfying even reasonable protection against pollution of ground waters and threats to public health and safety. Further, few communities account for the full cost of land disposal. The value of land used as a landfill disposal facility will diminish greatly as it reaches capacity and must be closed. Few municipalities reflect this decrease in value in calculating disposal costs. Sometimes

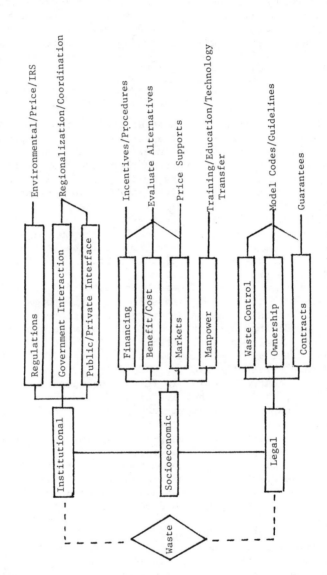

Figure 2. Nontechnical issues and their interrelationships

capital equipment costs have not been considered in the determination of disposal costs. Consequently, the cost of landfilling
is cheap in areas that do not enforce reasonable environmental
and public health standards. In addition, it is perceived as
cheap in areas which have good enforcement but fail to fully
account for the true costs.

This situation severely constrains the ability of energy and
materials recovery systems to compete with land disposal operations for the acquisition of urban wastes. Economy in transportation is one potential advantage that energy recovery facilities
enjoy relative to land disposal. Because landfill sites are
increasingly available only at considerable distance from the
urban centers which generate the waste, the costs of transporting
the waste is greatly increased.

Unfortunately, this advantage is largely reduced because
certain types of energy recovery facilities cannot be located in
urban areas that cannot meet EPA ambient standards for particulates. These nonattainment areas may not add a new stationary
source of particulate emissions without an equal or greater offset in emissions from other sources. Current Federal environmental regulations do not recognize the other offsetting regional
environmental benefits of an energy and materials recovery plant.
In order to increase the ability of energy recovery systems to
compete with land disposal, DOE is urging the Congress and the
Environmental Protection Agency to revise environmental regulations so as to encourage demonstration of environmentally sound
recovery systems.

The Branch program addresses not only these pressing issues
but also provides assistance regarding financing of resource
systems, market development for recovered resources, advice on
negotiating equipment contracts and for developing procurement
documents. Branch activities in these nontechnical areas as well
as in technical areas are coordinated through its program support
efforts.

Program Support

DOE's Urban Waste Technology funds are primarily used to
support RD&D. Most projects are funded by grants, contracts and
cooperative agreements. The objectives of our RD&D efforts are
two-fold: to conduct research and development that will provide
planners with sufficient options for choosing a waste-to-energy
system that fits specific local needs, and to provide the
financial support to demonstrate a wide range of these technologies. Major efforts concentrate on developing technologies that
can have wide application. DOE also provides training and
technical assistance to maximize the effectiveness of its RD&D
investment.

Demonstration of effective energy technologies will reduce
many of the barriers that cause municipalities and the private
sector to hesitate in undertaking energy recovery projects. In

the iterim, DOE is developing techniques to provide financial and economic incentives for moving desirable projects forward. These incentives -- loans, loan guarantees and price supports -- will be used to shift much of the financial risk from the local partici- pants to the Federal government. When in place, a limited pro- gram of price supports will help develop markets for recovered fuel, steam and electricity. Loan guarantees will be used to guarantee 75 percent of total project costs, or 90 percent of construction costs. DOE will, however, limit its loan guarantee support to those projects for which there is a reasonable assur- ance of repayment. As of writing, loan guarantee and price sup- port regulations are being developed. Appropriations are not yet available for these programs.

An active program of training, technical assistance and information transfer is also being initiated. This includes preparation of case studies, conducting workshops and seminars and developing university and apprenticeship programs. And lastly, the program is being coordinated among the Federal agencies so that the expertise of each is used to the fullest.

To summarize, the DOE Urban Waste Program aims at helping the public and private sectors conserve energy by utilizing the energy resources contained in urban wastes. The strategy is to reduce or eliminate both the technical and nontechnical barriers to using resource recovery on a large scale. The initial goal is to reduce the technical barriers by supporting R&D efforts that will lead to commercial scale demonstration of promising technolo- gies. Successful demonstrations will reduce many of the current risks which inhibit municipalities and the private sector from adopting resource recovery. In the interim, the Federal govern- ment will assume major financial responsibility for project risks.

Ultimately, it is hoped that all of these efforts will make resource recovery a common and popular method of waste disposal and energy production.

REFERENCES

1. CSI Resource Systems Group, Inc., Federal Assistance by the Department of Energy in the Demonstration of Energy and Materials Recovery Systems. Contract No. 31-109-38-4493, Argonne National Laboratory (August 1978).

2. Cohen, Alan S., An Overview of the Department of Energy's Research, Development and Demonstration Program for the Recovery of Energy and Materials from Urban Waste.

RECEIVED December 21, 1979.

Thermochemical Conversion of Biomass to Fuels and Feedstocks: An Overview of R&D Activities Funded by the Department of Energy

G. F. SCHIEFELBEIN and L. J. SEALOCK, JR.

Pacific Northwest Laboratory[1], P.O. Box 999, Richland, WA 99352

S. ERGUN

Lawrence Berkeley Laboratory[2], Building 77F, Berkeley, CA 94720

This paper is intended to provide an overview of thermochemical conversion technology development activities within the Biomass Energy Systems Branch of the U.S. Department of Energy (DOE). The Biomass Energy Systems Branch (BESB) is a part of DOE's Division of Distributed Solar Technology. The biomass thermochemical conversion technology development activities sponsored by the Biomass Energy Systems Program can be catagorized into four main areas; direct combustion, direct liquefaction, gasification, and indirect liquefaction via synthesis gas.

Pacific Northwest Laboratory (PNL) and Lawrence Berkeley Laboratory (LBL) have been selected to provide program management services to the Biomass Energy Systems Program. PNL is responsible for the technical management of development activities directed toward the thermochemical conversion of biomass feedstocks by direct combustion, gasification and indirect liquefaction via synthesis gas. LBL is responsible for the technical management of development activities on the direct liquefaction of biomass feedstocks.

Biomass comprises all plant growth, both terrestrial and aquatic and includes renewable resources such as forests and forest residues, agricultural crop residues, animal manures, and crops grown on energy farms specifically for their energy content. Biomass production and conversion is considered a solar technology because living plants absorb solar energy and convert it to biomass through photosynthesis.

Program Objective

The objective of the thermochemical conversion technology development activities of the Biomass Energy Systems Program is to

[1] Operated for the U.S. Department of Energy by Battelle Memorial Institute

[2] Operated for the U.S. Department of Energy by The University of California

develop cost competative processes for the conversion of biomass feedstocks to fuels and other energy intensive products. This objective can be accomplished by the direct combustion of biomass materials and the substitution of biomass derived fuels and chemical feedstocks for those produced from conventional sources.

Thermochemical conversion processes employ elevated temperatures to convert biomass materials to more useful energy forms. Examples include:

• Combustion to produce heat, steam, electricity, or combinations of these;

• Pyrolysis to produce gases (low or intermediate BTU), pyrolytic liquids and char;

• Gasification to produce low or intermediate BTU fuel gas;

• Gasification to produce synthesis gas for the production of synthetic natural gas (SNG), ammonia, methanol, alcohol fuels, or Fischer-Tropsch liquids and gasoline via catalytic processes; and

• Direct liquefaction to produce heavy oils, or with upgrading, lighter boiling liquid products such as distillates, light fuel oils and gasoline.

Program Organization and Implementation

Thermochemical conversion technology development activities sponsored by the Biomass Energy Systems Program can be divided into the following four categories:

• Direct Liquefaction

• Direct Combustion

• Gasification

• Indirect Liquefaction Via Synthesis Gas

In the remainder of this paper we will briefly discuss individual projects in each of these catagories.

Direct Combustion Systems. The direct combustion of biomass feedstocks is already widely practiced by several industries, especially the forest products industry. Many types of direct combustion equipment are commercially available for this purpose. New developments in direct combustion technology are expected to have a near term impact on energy supplies through the utilization forest residues and other readily available biomass feedstocks. Therefore, direct combustion technology development projects being funded by the Biomass Energy Systems Program are categorized as near term systems development activities.

Two projects are currently being funded by the Biomass Energy Systems Program in the area of direct combustion technology. These projects are shown in the organization chart illustrated in Figure 1. The Aerospace Research Corporation project is developing a wood fueled combustor which can be directly retrofitted to existing oil or gas fired boilers. Direct retrofit requires that heat release rates equivalent to those obtained when firing oil or gas be obtained in the wood fired combustor. Heat release rates on this order have been achieved when firing wood by preheating the combustion air to 800-1000°F. The Wheelabrator Cleanfuel Corporation project is a demonstration of large scale co-generation based on wood feedstock. The scope of this project includes the design of the plant plus additional tasks such as preparation of an environmental impact statement, demonstration of large tree harvesting equipment and determination of feedstock availability for a large facility. The draft final report for this project has been completed and is currently being reviewed.

Gasification-Indirect Liquefaction Systems. Development of biomass direct liquefaction, medium BTU gasification and indirect liquefaction technologies are catagorized as mid term development activities because these technologies are not expected to have a substantial impact on U. S. energy supplies for 10 to 20 years. Biomass gasification technologies can be divided into processes which produce a low BTU gas and those which produce a medium BTU gas.

Low BTU gasification technology is commercially available for most types of biomass feedstocks and can be expected to have an impact on energy supplies by 1985. Many of these commercial processes are based on low BTU coal gasification technologies and the gas produced can best be used as fuel for supplying process heat, process steam or for electrical power generation.

The versatility of low BTU gas is limited and its use is subject to the following limitations:

- Substitution of low BTU gas for natural gas as a boiler fuel usually requires boiler derating and/or extensive retrofit modifications.

- The low heating value of the gas usually requires that it be consumed on or near the production site in a close coupled process.

- The high nitrogen content of low BTU gas precludes its use as a synthesis gas for most chemical commodities which can be produced from synthesis gas.

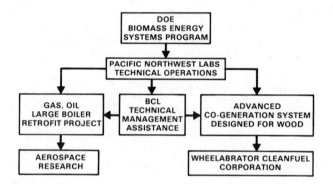

Figure 1.   Direct combustion development activities (near term)

Medium BTU gas (MBG) offers the following advantages over low BTU gas:

- Boiler derating is usually less severe when substituting MBG for natural gas than when substituting low BTU gas for natural gas and may not even be required in some cases;

- MBG can be transported moderate distances by pipeline at a reasonable cost;

- MBG is required for the synthesis of derived fuels and most chemical feedstocks and commodities which can be produced from synthesis gas.

The versatility of MBG is illustrated in Figure 2. The major disadvantage of MBG is that its production by conventional means requires the use of an oxygen blown gasifier which is expensive to operate due to the cost of the oxygen.

If the thermochemical conversion of biomass is to achieve its maximum potential for supplementing existing U. S. energy supplies in the mid term, the following two points will have to be addressed.

- Barring serious coal production constraints, biomass conversion will have to be economically and environmentally competitive with synthetic fuels produced from coal.

- Thermochemical biomass conversion must have an impact on the availability of liquid fuels and chemical feedstock supplies as well as supplementing gas for heating purposes.

Biomass has two potential advantages over coal. First, biomass is a renewable resource and coal is not. Second, and more important from a thermochemical conversion standpoint, biomass is more reactive than coal. It has the potential for gasification at lower temperatures, without the addition of oxygen, to produce medium BTU gas. Several of the gasification process development activities sponsored by the Biomass Energy Systems Program are attempting to exploit this advantage. These development activities are also directed toward improving the competitiveness of biomass gasification through the use of catalysts and unique gasification reactors to produce, directly, specific synthesis gases for the production of SNG, methanol or methyl fuels, ammonia and hydrogen. Success in these efforts could eliminate the necessity for external water gas shift or methanation reactors when producing these commodities. The potential elimination of the oxygen requirement and the water gas shift step are indicated by the dashed lines in Figure 2.

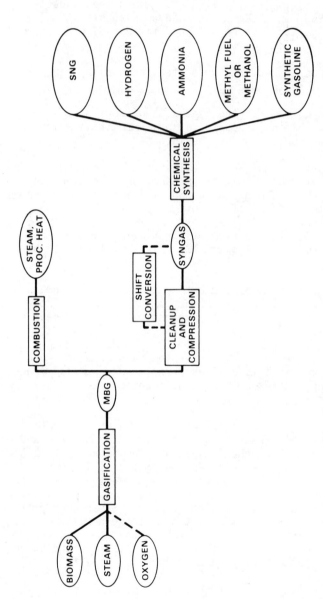

Figure 2.   Medium Btu gasification (MBG)—indirect liquefaction technology

The gasification technology development activities of the Biomass Energy Systems Program include processes for the production of medium BTU heating and synthesis gases. This appears to be the area where additional biomass gasification research and development could have the greatest impact on supplementing U. S. energy requirements.

Potential advantages of the thermochemical conversion of biomass to fuels and chemical feedstocks over competing coal conversion technologies have been addressed above. It would be inappropriate not to point out some disadvantages of biomass feedstocks. Biomass feedstocks are usually disperse and may contain 50 to 60% moisture. Therefore, there is a substantial cost associated with the collection and transportation of biomass to a central processing point. One concept which may offer a partial solution to these economic disadvantages would be to locate biomass gasification units in a disperse manner within a large forest or energy farm. The medium BTU synthesis gases produced by these units could be transported 25 to 50 miles by pipeline to a large, centrally located, chemical or fuels synthesis plant. This concept offers the advantages of shorter hauling distances, eliminates transporting the moisture long distances and offers the opportunity for a large scale central synthesis plant. However, this concept requires more economic evaluation before its merit can be determined.

Gasification and indirect liquefaction technology development projects currently sponsored by the Biomass Energy Systems Program are dipicted on the organization chart shown in Figure 3. A project breakdown includes operational process development units (PDUs) at Garrett Energy Research, the University of Arkansas, Texas Tech University, Pacific Northwest Laboratory, West Virginia University, and a much larger fluidized bed gasification PDU at the University of Missouri at Rolla. PDUs are in the design stage at Wright-Malta and Battelle Memorial Institute. Brief resumes for each of these projects, including PDU sizes, are shown in Table 1.

Supporting systems studies are being conducted by Gilbert/ Commonwealth, Inc., Gorham International, Inc., Science Applications, Inc., and Catalytica Associates, Inc.

The Gilbert/Commonwealth project is aimed at the development of a biomass resource allocation model based on linear programming. The model, which will integrate regional/seasonal biomass feedstock availability, conversion process efficiencies and economics and regional/seasonal end product demand projections, is intended for use as a planning tool by the Department of Energy. It will provide an organized structure for evaluating the potential economic and regional impact of new biomass conversion technologies. Experimental data on biomass feedstock characteristics and conversion efficiencies is being supplied by West Virginia University and Environmental Energy Engineering Company under subcontract to Gilbert/Commonwealth.

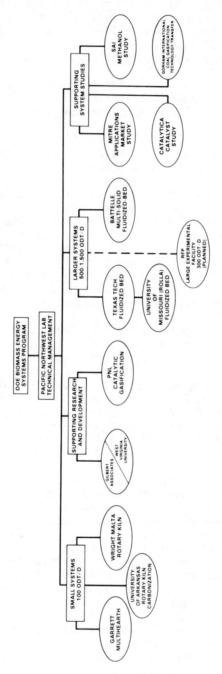

Figure 3.   Gasification—indirect liquefaction systems (midterm)

TABLE 1. Biomass Gasification-Indirect Liquefaction Process Development Units (PDUs)

Project	Reactor Type	PDU Size	Potential Feedstocks	Product(s)	Potential Applications	Status
Garrett Energy Research and Engineering Hanford, CA	• Multihearth • Non-Catalytic • Low Pressure • No Oxygen Required	3 ODT/D*	Manure Cotton Gin Thrash Wood Corn Stover	MBG (~400 BTU/scf)	Small industrial applications (under 100 ODT/D)	Operational
Wright-Malta Corporation Ballston Spa, NY	• Rotary Kiln • Soda Ash Catalyst • High Pressure • No Oxygen Required	3 ODT/D	Wood Chips	MBG (~450 BTU/scf) Methanol synthesis gas	Small industrial plants, pulp/paper mills	PDU designed startup mid 1980
University of Arkansas Fayetteville, AK	• Rotary Kiln • Non-Catalytic • Low Pressure • No Oxygen Required	0.5 ODT/D laboratory unit at Fayetteville 28 ODT/D PDU at Jonesboro Arkansas	Wood Chips, Saw Dust, Rice Hulls, Cotton Gin Thrash, Nut Hulls, and Other Agricultural Residues	Low BTU gas and char	Small industrial and agricultural applications	Operational
Pacific Northwest Laboratory Richland, WA	• Fluidized Bed • Catalytic-Various Catalysts • Pressure (1-10 atm) • No Oxygen Required	0.5 ODT/D	Wood Various others	Methane, hydrogen, and synthesis gases for ammonia, methanol and liquid hydro-carbons	Current PDU for catalyst screening studies - concept applicable to various gasifiers	Operational
Battelle Memorial Institute Columbus, OH	• Multisolids Fluidized Bed • Catalytic • High Pressure • No Oxygen Required	2 ODT/D	Wood-whole tree-Chips and Residues	'MBG (~350 BTU/scf) or Methane	Larger units, energy farms (300-1,500 ODT/D)	PDU designed startup mid 1980
Texas Tech University Lubbock, TX	• Counter Current Fluidized Bed (top fued) • Non-Catalytic (Steam-Air Gasification) • Low Pressure • No Oxygen Required	0.5 ODT/D	Manure, Wood Cotton Gin Thrash, Corn Stover, and Other Agri-cultural Stalks	MBG (270-360 BTU/scf as is basis) NH₃ synthesis gas	Larger units 300-1,500 ODT/D ammonia production from feedlot wastes, other industrial uses	Operational
University of Missouri Rolla, MO	• Fluidized Bed • With or Without Catalyst • Low Pressure • No Oxygen Required	2.4 to 24 ODT/D	Sawdust, Hogged Wood Chips and Others	Low BTU gas, MBG and synthesis gas	Larger units 300-1,500 ODT/D PDU will be used to verify Texas Tech projections and develop scale up data for commercial size units	Operational

*oven dried tons/day

Gorham International is conducting a study to determine the potential for retrofitting coal gasifiers to operate with wood feedstocks.

Science Applications, Inc., has recently completed a comprehensive technical and economic assessment of producing methanol from biomass feedstocks employing developed gasification technology. This study includes an assessment of biomass availability and the distribution and markets for methanol fuels as well as thermochemical conversion technology.

Catalytica Associates, Inc. is conducting a study to produce a systematic assessment of the role of catalysis in thermochemical conversion via gasification and liquefaction. This study is also examining the potential impact of catalytic concepts under development in other areas, such as coal conversion, and new reactor technology on biomass conversion.

Included on the organization chart is a proposal for a large experimental facility (LEF) for the gasification of biomass. This facility, which would have a projected capacity of 300 oven dried tons per day of biomass, is still in the planning stage and has not been formally approved.

The LEF would serve as a demonstration unit for the gasification processes currently being developed at the PDU stage and would provide process information that could be used for the design of a commercial sized facility.

Direct Liquefaction Systems. The Biomass Energy Systems branch is sponsoring efforts to develop a direct liquefaction process for the thermochemical conversion of biomass to liquid fuels. Lawrence Berkeley Laboratory (LBL) is responsible for the technical management of this program. An organization chart for the direct liquefaction effort is shown in Figure 4. The main thrust of the effort is centered on the operation of a direct liquefaction PDU located at Albany, Oregon. Various subcontractors provide program support.

The original process flowsheet for the Albany PDU was based on a series of bench scale batchwise biomass liquefaction experiments conducted by the Pittsburgh Energy Research Center (PERC) of the U. S. Bureau of Mines in the 1960s and early 1970s. In this flowsheet biomass flour (30 parts) is mixed with a vehicle oil (70 parts) and 20% sodium carbonate solution (7.5 parts) and injected into a high pressure vessel (3,000 psi) along with synthesis gas. The slurry is heated to about 700°F at a rate of about 12°F/min. The product stream is then cooled and flashed in a pressure let-down vessel. The bottoms are diverted into a three phase centrifuge to separate the solid residue (as a sludge) and the aqueous phase containing the sodium salts from the oil phase. Part of the oil (about 15 parts) is withdrawn as a product and most of it (70 parts) is recycled to serve as a vehicle oil.

The Albany PDU, designed to process three tons of wood per day, was constructed during 1975-1976 and commissioned by Bechtel

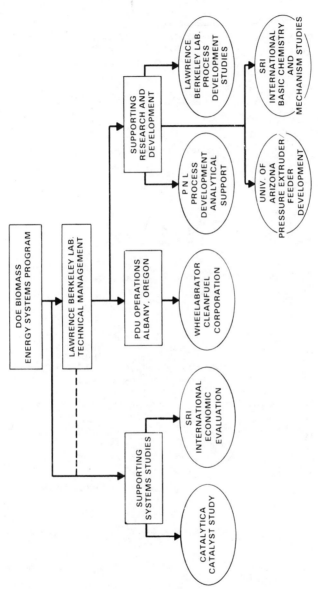

Figure 4.   Direct liquefaction systems (midterm)

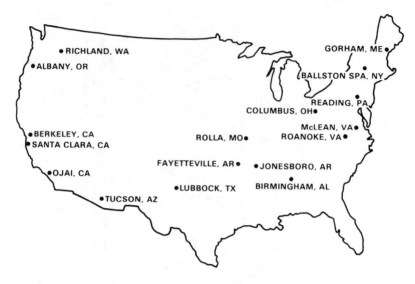

Figure 5.   Geographic distribution of thermochemical conversion projects

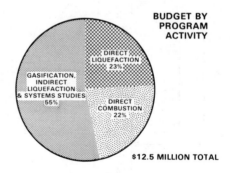

Figure 6.   Projected fiscal year 1979 budget—thermochemical conversion

National, Inc. The PDU is now being operated by the Rust Engineering Company division of Wheelabrator Cleanfuel Corporation.

The process development activities based on the original PERC flowsheet have been beset with many mechanical and operational difficulties and it has not yet been possible to obtain a pure wood derived oil which is free of the anthracene start up oil. Rust Engineering believes that the PDU needs modifications in order to develop the PERC process. However, a great deal of engineering design and materials compatability information has been accumulated which will be useful in designing future biomass liquefaction facilities.

Since October of 1978, LBL has been engaged in process research and development in support of PDU operations. LBL researchers now believe that the suitability of the oil produced from wood to serve as a vehicle oil is in question. Based on the results of bench scale experiments, they conceived a process in which wood chips (23 parts on dry basis) are prehydrolyzed in water (77 parts) containing 500 ppm sulfuric acid. The resultant slurry is pumpable and can be liquefied directly after the addition of sodium carbonate (1.15 parts) or other catalysts. The slurry, after being subjected to liquefaction, contains an aqueous phase and an oil phase containing small amounts of solid residue. The oil phase is separated as a product.

This optional process flowsheet was evaluated in a PDU trial run during the first week of May with moderate success. For the first time an oil entirely derived from wood was obtained. The oil has a heating value of 15,500 BTU/lb., contains about 82 percent carbon, 8 percent hydrogen, 10 percent oxygen, and less than 0.1 percent nitrogen. This test run is now being repeated to obtain a large batch of oil for characterization and upgrading purposes. Whether or not the Albany PDU is adequate to develop this process is not yet certain.

Supporting studies are being conducted by PNL, LBL, SRI, Catalytica, and the University of Arizona. The work area being investigated by each of these organizations is shown in Figure 4.

Program Objectives

A geographical distribution of current biomass thermochemical conversion projects sponsored by the Biomass Energy Systems Branch is shown in Figure 5.

The projected Fiscal Year 1979 budget for those thermochemical conversion activities managed by Lawrence Berkeley and Pacific Northwest Laboratory is $12.5 million. The distribution of funding by major activity subelement is shown in Figure 6. This does not represent the total thermochemical conversion budget for the Biomass Energy Systems Program. A few additional thermochemical conversion activities, which are not technically managed by PNL and LBL, are funded by the Biomass Energy Systems Program.

The projected Fiscal Year 1980 Biomass Energy Systems Program budget for thermochemical conversion activities (direct combustion, direct liquefaction, gasification and indirect liquefaction) is approximately $17 million.

Sources of Additional Information

The purpose of this paper is to provide a brief overview of the biomass thermochemical conversion activities of the Biomass Energy Systems Branch. Detailed papers on each of the projects mentioned in this overview were presented at the 3rd Annual Biomass Energy Systems Conference held in Golden, Colorado, June 4-7, 1979. Copies of the proceedings of this conference will be available from The National Technical Information Service (NTIS) as SERI/ TP-33-285, beginning approximately November 1, 1979

RECEIVED November 16, 1979.

AN OVERVIEW OF TECHNOLOGIES
FOR THERMAL CONVERSION OF WOOD

Preliminary Economic Overview of Large-Scale Thermal Conversion Systems Using Wood Feedstocks

STEPHEN M. KOHAN

Electric Power Research Institute, 3412 Hillview Avenue, Palo Alto, CA
94303

This paper presents a brief overview of the economics of conceptual technical schemes to convert woody biomass feedstocks into
other, more useful energy and chemical forms. It is becoming
increasingly clear that no single solution exists to the recurring
domestic energy crisis. Rather, regional energy supplies (some
renewable, some non-renewable) together with end-use energy
management practices, are likely to play important roles in
achieving stated national goals of energy self-sufficiency. Wood
is a regional, renewable resource. DOE's Fuels from Biomass
Branch is keenly interested in maximizing the cost-effective contribution of wood energy (as well as other forms of biomass) to
the nation's energy supply. DOE awarded SRI International a
contract (EY-76-C-03-0115 PA 131) in late 1977 to assist DOE in
determining promising biomass feedstock/conversion technology/
product options for purposes of R&D program planning. This paper
briefly reviews the data base for biomass thermal conversion
technologies developed as a necessary part of this program planning effort.
    This analysis employs uniform design and economic bases.
These are briefly reviewed, followed by summary economics for
the production from wood of electricity, steam, and cogenerated
products; intermediate-Btu gas (IBG) and substitute natural gas
(SNG); methanol; ammonia; fuel oil; and pyrolytic oil and char.
Several processing steps in these conversion schemes are conceptual or are at early stages of development by DOE, EPRI, GRI, and
others (an exception would be wood steam/electric power plants,
which are commercially used by the electric utility and wood products industries). Consequently, the economics presented here
may generally tend to be optimistic. Additional details of the
analyses can be found in Kohan and Barkhordar([1]); Jones, Kohan
and Semrau([2]); and Kohan and Dickenson([3]).

Design Bases

The base wood feedstock rate used in the study is 1.8×10^6 kg/d

(2000 ST/d) of green (50 wt% moisture) wood.

This value was selected as representing a near-term production goal. Product cost sensitivities over feedstock rates of
$0.9x10^6$kg/d to $5.4x10^6$kg/d of green wood are presented by Kohan
and Barkhordar([1]). The dry wood energy content is 22,200 KJ/kg.
Plant designs are generally self-sufficient in terms of purchased
energy (e.g., electricity) needs. Thirty days' feedstock storage
is provided on-site.

Economic Bases

Reference([1]) details the economic bases for the analysis.
These are summarized below.

A late 1977 cost basis is used. The plant facilities investment represents the capital cost of the plant as erected and
awaiting start-up. Constant-dollar economics are used. Plant
construction periods range from 2 to 4 years, depending on plant
complexity. Site-specific investments (e.g., townsites) and
across-the-board contingency factors are excluded from the analysis.

Constant-dollar economics are likewise used for the calculation of product revenue requirements. Two illustrative financing
extremes are considered: regulated utility, in which capital recovery factors are low and product markets would be non-competitive;
and non-regulated industrial, in which capital recovery factors
are high and product markets would be competive. For regulated
utility financing, a declining rate base analysis is used, with
the following financial parameters: 65% debt capital; 9% interest
rate on a 20-year loan; 15% return on equity capital; 10% interest
rate during construction; straight line depreciation; and 52%
federal and state corporate tax rate. For non-regulated industrial
financing, a discounted cash flow (DCF) analysis is used, with the
following financial parameters; 100% equity capital; 15 year plant
tax life; 15% DCF return; accelerated depreciation; and 52% federal
and state corporate tax rate.

The cost of wood delivered to the conversion facility is
assumed to be $0.95/GJ ($21/dry metric ton).

Electricity, Steam, and Cogenerated Products

Wood combustion to produce heat, steam, and electricity is an
old art. Rising oil fuel costs and other regional factors are
causing utilities such as Burlington Electric Co., Consumers Power,
Portland General Electric, and others to seriously consider instal-
ling 40-50 MW wood-fired steam/electric plants. Pacific Gas and
Electric, the Eugene Water and Electric Board, and others are
considering regional industrial joint ventures involving cogeneration (the simultaneous generation of electricity and steam).

Table I presents the estimated investment in new, grass-roots
facilities for the production of electricity, steam, or both pro-

Table I

WOOD COMBUSTION FACILITIES - ESTIMATED INVESTMENT

Principal Product:	Electricity	Steam	Cogeneration
Product Rate:	49.6 MWh/h	217,060kg/h	15.3×10^{12} J/d [1]
Plant Financing:	Regulated Utility	Non-Regulated Industrial	Regulated Utility
Investment, Millions of Dollars (1977)			
Plant Facilities Investment	51.1	27.9	35.1
Land	0.5	0.5	0.5
Organization, Startup Expenses	1.5	0.8	1.1
Interest During Construction	3.5	---	2.4
Working Capital	1.6	1.2	1.2
Total Capital Investment	58.2	30.4	40.3

[1] Total Product Basis

Basis: 1.8×10^6 kg/d green wood (50 wt.% moisture) feed rate

ducts. The investment in the steam-producing facilities (non-regulated industrial financing) is about half of that in the electric power plant because turbine-generator, steam condensing and
cooling tower facilities are omitted. The investment in cogeneration facilities is about midway between the other two investments
because turbine-generator facilities are included but steam condensing and cooling tower facilities are omitted. The cogeneration plant produces about 7 MW of electric power. For the electric power case, the plant facilities investment of $51.1 million
is about $1030/kW which is in reasonable agreeement with other
published estimates.
     Table II shows the estimated revenue requirements. Electri-
city and cogenerated products costs are based on regulated utility
financing, while steam costs are based on non-regulated industrial
financing. The cogeneration case costs are shown on a total product basis, since two products are involved. Figure 1 shows the
selling price of electricity as a function of the selling price
of steam for the cogeneration case.
     The prices of products from wood combustion appear to be
higher than their counterparts from coal combustion. A contributing reason may be the smaller sizes of the wood facilities contrasted with "typical" coal facilities (e.g., 50 MW for wood power
plants; 500-1000 MW for conventional coal power plants).
     Figure 2 shows the effect of wood cost and plant capacity on
electricity cost. As plant sizes increase from 25 to 150 MW,
electricity costs fall by 10 to 20 percent, reflecting some economies of scale. Kohan and Barkhordar[1] present analogous sensitivity curves for other products evaluated in this study.

Wood Gasification

     Wood gasification in small, fixed-bed producers was widely
practiced in the U.S. and Europe in the early 1900's. Today, the
Fuels from Biomass Branch of the Department of Energy is sponsoring the development of several generically different gasification
systems designed to exploit the higher reactivity of (high-alka-
line-content ash) wood as contrasted with coal. This field is
rapidly evloving. For the present study, a conceptual, high-
pressure, non catalytic, oxygen-blower fluidized bed gasifier was
selected for analysis. References [1] and [3] present additional
gasifier details.
     Table III presents the estimated investments for the produc-
tion of high pressure (2100 kPa) intermediate Btu gas (IBG) and
SNG from wood. Because of the low sulfur content of wood, no
sulfur removal facilities are included in the IBG design. Since
the methanation catalyst is sulfur-sensitive, sulfur in the synthesis gas in the SNG case is removed down to very low levels.
The estimated product rates for IBG and SNG are significantly
lower than corresponding production rates from "typical" coal
facilties (e.g., 260×10^{12} J/d SNG from coal).

Table II

WOOD COMBUSTION FACILITIES - PRODUCT REVENUE REQUIREMENTS

Principal Product:	Electricity	Steam	Cogeneration
	mills/kWh	$/GJ	$/GJ
Feedstock @ $0.95/GJ	16.0	1.23	1.25
Labor-Related	8.7	0.36	0.52
Purchased Materials	6.6	0.28	0.25
Fixed Costs	6.6	0.25	0.35
Plant Depreciation (20-yr)	8.1	---	0.44
Return on Rate Base & Income Tax[2]	14.5	---	0.79
Capital Charges for a 15% DCF Return	---	1.66	---
Total	60.5[2]	3.78(2.99)[3]	3.60[2]

[1]Total Product Basis

[2]20-Year Average Values

[3]Regulated Utility Financing

Basis: 1.8×10^6 kg/d green wood (50 wt.% moisture) feed rate

Table III

WOOD GASIFICATION FACILITIES - ESTIMATED INVESTMENT

Principal Product:	High-Pressure Intermediate-Btu Gas(IBG)	Substitute Natural Gas(SNG)
Product Rate:	14.3×10^{12} J/d	12.7×10^{12} J/d
Plant Financing	Regulated Utility	Regulated Utility
Investment, Millions of Dollars(1977)		
Plant Facilities Investment	48.9[1]	80.3
Land	0.3	0.3
Organization, Startup Expenses	2.4	4.0
Interest During Construction	3.7	6.1
Working Capital	1.5	2.0
Total Capital Investment	56.8	92.7

[1]Decreases to $39.5 million for low-pressure IBG production

Basis: 1.8×10^6 kg/d green wood (50 wt.% moisture) feed rate

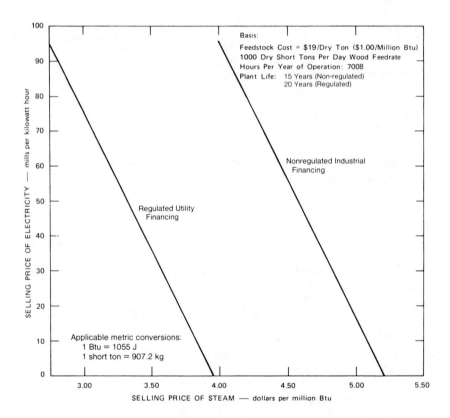

*Figure 1.   Production of electricity and steam (cogeneration) by wood combustion
—selling price of electricity as a function of selling price of steam*

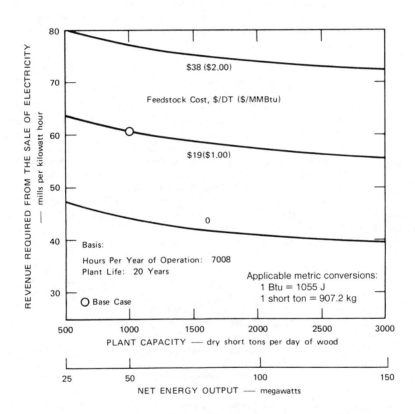

Figure 2. Production of electricity by wood combustion—effect of plant size on revenue required from the sale of electricity by a regulated utility

Table IV presents estimated revenue requirements for IBG and SNG using regulated utility financing. The low pressure alternative case shown for IBG produces 200 KPa (30 psig) pressure gas, which is suited only for across-the-fence sales to a utility or industrial customer. The wood-derived gas product prices are higher than their coal-derived gas counterparts partially for reasons of gas production rates. The costs of producing IBG from wood may be 15 to 50 percent greater than that of producing IBG from advanced coal gasification technologies; and 80 to 100% greater when considering the production of SNG. Catalytic wood gasification concepts being developed by Battelle Columbus Laboratories, Battelle Pacific Northwest Laboratories, and others may offer the potential to significantly reduce the costs of biomass or wood gasification.

Liquid Fuels Production

Liquid fuels producible by thermal conversion of biomass include fuel oil, pyrolysis oils, methanol (discussed under chemicals production), and others. This section discusses the production of fuel oils and pyrolysis oils from wood.

The concept for fuel oil production from wood involves the catalyzed removal of oxygen from the wood molecules by reducing gases (H_2, CO) at elevated temperatures and pressures. DOE is developing this concept at a pilot facility (2700 kg/d wood feed rate) at Albany, Oregon. Recent experimental work by Lawrence Berkeley Laboratory personnel suggests that the wood feedstock should be pretreated by mild acid hydrolysis before conversion. The economics presented below do not include costs for the pretreating step since this information was not available when these analyses were conducted.

Pyrolysis involves the decomposition of organic matter in the absence of oxygen. Pyrolysis is widely practiced in various industries (delayed coking in petroleum refining; batch coking of coal in the manufacture of iron and steel). The U.S. Environmental Protection Agency is sponsoring the development of pyrolysis technologies aimed at disposal of wastes and residues (both cellulosic and non-cellulosic). Large amounts of co-product char is produced during pyrolysis, and process economics are heavily dependent on the value assignable to the char. The conversion facility design in the present analysis was patterned after the Tech-Air technology because the technology had been developed to commercial prototype size (45,000 dry kg/day) and because of the availability of plant data. (Recent information suggests that the commercial prototype plant has been decommissioned).

Pyrolytic oils are acidic and may contain carcinogens. By analogy with heavy petroleum-derived oil fractions, the fuel oils produced by catalytic liquefaction may contain carcinogens. This suggests the need for special handling and storage precautions for the liquid fuels.

Table V presents the estimated investments for producing

Table IV

WOOD GASIFICATION FACILITIES - PRODUCT REVENUE REQUIREMENTS

| Principal Product: | Intermediate-Btu Gas | | SNG |
	High-Pressure $/GJ	Low Pressure $/GJ	$/GJ
Feedstock @ $0.95/GJ	1.33	1.19	1.51
Labor-Related	0.60	0.78	1.08
Purchased Materials	0.26	0.18	0.62
Fixed Costs	0.46	0.34	0.86
Plant Depreciation (20-yr)	0.59	0.42	1.07
Return on Rate Base & Income Tax[1]	1.03	0.77	1.91
Total-Regulated Utility Financing[1]	4.27	3.68	7.05
(Total-Nonregulated Industrial Financing)	(5.97)	(4.93)	(10.18)

[1]20-Year Average Values

Basis: 1.8×10^6 kg/d green wood (50 wt.% moisture) feed rate

Table V

LIQUID FUELS PRODUCTION FROM WOOD - ESTIMATED INVESTMENT

Principal Product:	Fuel Oil (Catalytic Liquef.)	Pyrolytic Oil,Char
Product Rate:	11.0×10^{12} J/d ($278m^3$/d)	14.9×10^{12} J/d[1]
Plant Financing:	Regulated Utility	Regulated Utility
Investment, Millions of Dollars (1977)		
Plant Facilities Investment	48.5	18.1
Land	0.3	0.3
Organization, Startup Expense	2.4	0.9
Interest During Construction	3.7	1.4
Working Capital	1.8	1.5
Total Capital Investment	56.7	22.2

[1]Total Product Basis (45% oil, 55% char)

Basis: 1.8×10^6 kg/d green wood (50 wt.% moisture) feed rate

liquid fuels from wood. The plant design for fuel oil production
involves high-pressure and-temperature slurry recycle concepts
and wood gasification to produce the reactant gases. The pyrolysis concepts involve low-pressure fixed bed pyrolysis, with all of
the pyrolytic gases burned on-site for wood drying and plant power
needs. Pyrolysis investment costs are in agreement with those
presented by Jones(4). Normalized plant investment costs of
$175,000 per daily $\overline{m}^3$ for fuel oil production exceed the $95,000
to $125,000 per daily m^3 values estimated for the production of
synthetic fuel oil from coal. A contributing reason is the significant difference in oil production rates (278 m^3/d for wood-
derived oil; 400 to 800 m^3/d for coal-derived oil).

Table VI shows the estimated revenue requirements using regulated utility financing. Costs for the pyrolysis case are shown
on a total product basis, since two products are involved.
Figure 3 shows the selling price of the pyrolytic oil as a function of the selling price of the char. The char is low in sulfur,
nitrogen, and ash, and may find applications as a compliance
boiler fuel, in water purification, and so forth. Char market
values were not established for this study.

The prices of wood-derived liquids appear to be higher than
prices of similar coal-derived liquids. A contributing reason
may be the lower production rates from the wood facilities than
from the coal facilities.

Chemicals Production

This section discusses the production of methanol and ammonia
from wood. Methanol is a clean-burning material that may find
widespread future use as an automotive fuel (directly or for
conversion to gasoline by the Mobil process); as a fuel for industrial or utility boilers, gas turbines, or fuel cells; as a chemical intermediate; or as a biological feedstock for protein.
Ammonia is an essential building block for synthetic nitrogen
fertilizers and finds widespread use in the production of synthetic fibers, explosives, and plastics.

The production of methanol and ammonia from wood involves
similar concepts: gasification followed by product synthesis.
The chemistry of the final product synthesis steps are different
for these two cases, resulting in different gas conditioning steps
between gasification and product synthesis.

Table VII presents the estimated investments. The methanol
case investment reflects the same gasifier type as used for the
IBG and SNG cases. A conceptual Chem Systems methanol synthesis
step is used. EPRI is sponsoring the development of the Chem
Systems technology(5). The ammonia case investment reflects the
same wood gasification concepts, employs pressure swing adsorption
for hydrogen gas purification (based on information provided by
the Linde Division, Union Carbide Corporation), and uses a conventional high-pressure ammonia synthesis loop.

Table VI

LIQUID FUELS PRODUCTION FROM WOOD - PRODUCT REVENUE REQUIREMENTS

Principal Product:	Fuel Oil (Catalytic Liquef.)		Pyrolytic Oil,Char
	$/GJ	$/m^3	$/GJ[1]
Feedstock @ $0.95/GJ	1.73	68.6	1.29
Labor-Related	1.14	45.0	0.82
Purchased Materials	0.42	16.6	0.08
Fixed Costs	0.60	23.9	0.17
Plant Depreciation (20-yr)	0.76	29.8	0.21
Return on Rate Base & Income Tax[2]	1.37	53.8	0.41
Total-Regulated Utility Financing[2]	6.02	237.7	2.98
(Total-Nonregulated Industrial Financing)	(8.25)	(325.6)	(3.66)

[1] Total Product Basis

[2] 20-Year Average Values

Basis: 1.8 x 10^6 kg/d green wood (50 wt.% moisture) feed rate

Table VII

CHEMICALS PRODUCTION FROM WOOD - ESTIMATED INVESTMENT

Principal Product:	Methanol	Ammonia
Product Rate:	11.6 x 10^{12} J/d	4.5 x 10^5 kg/d
Plant Financing:	Regulated Utility	Non-Regulated Industrial
Investment, Millions of Dollars (1977)		
Plant Facilities Investment	88.0	95.5
Land	0.3	0.3
Organization, Startup Expenses	4.4	4.8
Interest During Construction	6.7	---
Working Capital	2.1	2.2
Total Capital Investment	101.5	102.8

Basis: 1.8 x 10^6 kg/d green wood (50 wt.% moisture) feed rate

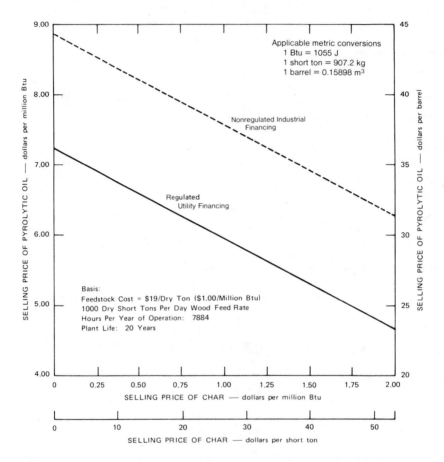

Figure 3.   Production of oil and char by pyrolysis of wood—selling price of pyrolytic oil as a function of selling price of char (regulated utility financing)

Table VIII presents the estimated revenue requirements. The methanol case uses regulated utility financing, and the ammonia case, non-regulated industrial financing, based on considerations of the likely markets to be served by these products. The estimated costs of producing methanol or ammonia from wood are higher than the costs of producing methanol or ammonia from coal. This is explained in part by production rates (the rates from wood-derived methanol plants are about one-tenth of those from "typical" coal-derived methanol plants; ammonia production rates are similar from the two resources since the ammonia market is typically demand-constrained); and partially by feedstock differences (e.g., green wood is half moisture).

Table VIII

CHEMICALS PRODUCTION FROM WOOD - PRODUCT REVENUE REQUIREMENTS

Principal Product:	Methanol	Ammonia
	$/GJ	$/mt
Feedstock @ $0.95/GJ	1.64	42.2
Labor-Related	1.28	35.2
Purchased Materials	0.85	20.5
Fixed Costs	1.03	28.9
Plant Depreciation (20-yr)	1.30	---
Return on Rate Base & Income Tax[1]	2.29	---
Capital Charges for a 15% DCF Return	---	203.9
Total	8.39 (12.13)[2]	330.7 (226.5)[3]

[1] 20-Year Average Values
[2] Non-Regulated Industrial Financing
[3] Regulated Utility Financing

Basis: 1.8×10^6 kg/d green wood (50 wt.% moisture) feed rate

Conclusions

The prices of wood-derived energy and chemicals products developed in this analysis tend to be higher than corresponding prices of coal-derived products. This is explained in part by smaller production rates from wood facilities than from coal facilities, which is in turn attributed to perceptions of the amount of feedstocks available to individual energy conversion projects; in part to feedstock differences (e.g., green wood is half moisture); and other factors. Wood is a regional resource;

coal is not. Wood fuel for proposed utility steam/electric plants may be obtained from forest cull wood and thinning operations. Thus forests are made more conducive to the production of commercial timber, and a residue disposal problem is solved by generating electricity. As energy prices continue to escalate, increasing uses may be found for regional biomass resources.

Acknowledgments

This work was performed at SRI International for the Department of Energy. The assistance of the DOE Technical Managers, Mr. Nello del Gobbo and Dr. Roscoe F. Ward, is gratefully acknowledged.

Literature Cited

1. Kohan, S.M., and Barkhordar, P.M., "Mission Analysis for the Federal Fuels from Biomass Program. Volume IV: Thermochemical Conversion of Biomass to Fuels and Chemicals," report prepared by SRI International for DOE under Contract EY-76-C-03-0115 PA131 (January, 1979) NTIS SAN-0115-T3
2. Jones, J.L., Kohan, S.M., and Semrau, K.J., "Mission Analysis for the Federal Fuels from Biomass Program. Volume VI: Mission Addendum," report prepared by SRI International for DOE under Contract EY-76-C-03-0115 PA131 (January, 1979).
3. Kohan, S.M., and Dickenson, R.L., "Production of Liquid Fuels and Chemicals by Thermal Conversion of Biomass Feedstocks," paper 78f to be presented at the 72nd AIChE Annual Meeting, San Francisco, CA, November 25-29, 1979.

4. Jones, J.L., "The Costs for Processing Municipal Refuse and Sludge," paper presented at the Fifth National Conference on Acceptable Sludge Disposal Techniques, Orlando, FL, January 31, 1978.
5. Sherwin, M.B., and Blum, D., "Liquid Phase Methanol," report prepared by Chem Systems, Inc., for the Electric Power Research Institute, EPRI AF-202 (August, 1976).

RECEIVED November 16, 1979.

Wood Gasification: An Old Technology with New Potential for the Eighties

JACK J. FRITZ[1], ABU TALIB, and JUDITH J. GORDON

The MITRE Corporation/Metrek Division, McLean, VA 22102

Gasification of wood and wood residues using air and steam as the gasifying medium to produce a combustible gas is a technology which has been commercially available since the turn of the century. However, the advent of inexpensive fossil fuels phased this technology out of the market. Today, there is a renewed interest in revitalizing this old technology. Approximately two dozen private organizations are attempting to enter the commercial marketplace. MITRE conducted a study to assess the commercial readiness of wood gasification in the near-term. Several organizations involved in the manufacture and development of gasification technology were queried. Current status and systems costs were found to vary considerably.

This paper summarizes the MITRE findings, giving both a state-of-the-art review and outlining policy related issues. Economic potential is assessed by considering a close-coupled wood gasifier as a source of gaseous fuel for an oil-and gas-fired boiler. The cost of wood-based fuel in three regions (Federal planning regions), is compared with the cost of conventional fossil fuels for 1985 and 1990 under different economic scenarios. Results indicate that wood gasification is competitive in the New England and Mid-west regions by 1985. In the Northwest region, high wood prices prevent wood-based gas from being cost competitive even in a long-term (1990) comparison. Some government participation in the form of further R&D and financial incentives will be required to accelerate commercialization.

Gasification Process Description

The gasification process converts a solid carbonaceous feed to a gaseous fuel that may be burned directly, upgraded to higher quality fuels or used as a feedstock for manufacturing chemicals, e.g., ammonia or methanol. However, the gas produced from an air and steam blown gasifier contains large amounts of nitrogen (nearly 50 percent by volume), and it is primarily suitable for on-site fuel application.

[1] Current address: Office of Energy, AID, Department of State, Washington, DC 20523

Reactors used for the gasification process are generally characterized by the method of contacting solids and gases. The principal reactor configurations are:

- Fixed Bed

- Stirred Moving Bed

- Fluidized Bed

- Entrained Flow

Of these configurations, only fixed-bed gasifiers have been used commercially for wood gasification. The most common design is an updraught fixed-bed reactor in which three distinct reaction zones, drying, pyrolysis and combustion, can be identified. Representative generic reactions are outlined below:

Drying Zone

100-200°C

Moist Wood + Heat → Dry Wood + Water Vapor

Pyrolysis Zone

200-500°C

Dry Wood + Heat → Char + CO + CO_2 + H_2O + CH_4 + C_2H_4 + other HC + pyroligneous acids + tars

Gasification and Oxidation Zone

Char + O_2 + H_2O → CO + H_2 + CO_2 + (Steam)   Heat

These reactions follow in sequential fashion as the wood descends by gravity through the gasifier.

A commercial gasifier using the above design is presented in Figure 1. This single stage fixed-bed gasifier is available from Davy Powergas Company, Houston, Texas. The gasifier is a steel cylinder; the upper part is refractory lined, and the bottom part is surrounded by an annular boiler which generates steam. The by-product steam is normally introduced with the air or oxygen to aid in the gasification reactions and to control the fire zone temperatures.

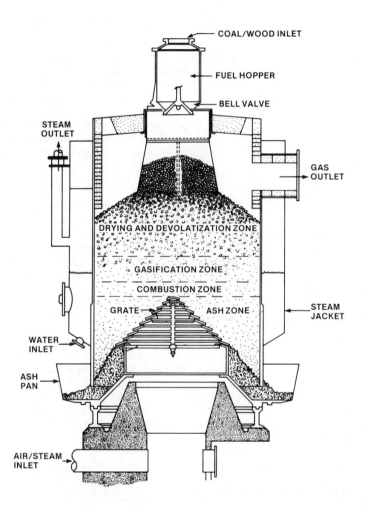

Figure 1. Single-stage fixed-bed wood gasifier

Wood chips (typically 4cm x 1cm) or other feed material (coal) is delivered to a bunker designed to contain an 8 to 12-hour supply at maximum rates. From the bunker the wood is automatically delivered by gravity through a fuel hopper into the gasifier. The gasifier is operated near atmospheric pressure. The gasifier contains an internal chute to distribute the wood chips to minimize gas channeling up through the fixed bed.

As the wood descends down through the gasifier it passes through a drying-devolatization zone, a gasification zone, combustion zone, and ash zone. Steam and air are used as the gasifying media, introduced through a rotating eccentric grate at the bottom of the gasifier to effect good distribution through the fixed bed.

The fuel gas produced from the devolatization and gasification of wood exits the gasifier at 120°C to 650°C depending on the moisture content of the feedstock. The exit gas contains tars and oils as well as the combustible CO, H_2 and CH_4 gaseous components. Typical gas analyses for air blown gasification of wood **are given** in Table I.

The product gas has a heating value of 4.4 to 6.5MJ/m^3 depending upon the feedstock, moisture content and operational variables. The gas is most likely to be used in a close coupled boiler, kiln or dryer after particulate removal in a simple cyclone. It is desirable to keep the product gas hot, usually 200°F to 400°F, to minimize condensation of the tars and to maximize sensible heat transfer.

Wood Gasification Status

Numerous commercial processes are available, with no single one currently possessing an obvious competitive advantage. Several firms have developed systems for specialized application such as agricultural residue conversion, wood residue conversion and municipal solid waste disposal. Most of these systems have capacities of less than 50 GJ/h of product gas, i.e., 80-90 oven dried tons per day (ODT/D) of feedstock. Larger systems gasifying more than 100 ODT/D of wood are similar in design to gasifiers developed prior to 1950.

Of systems currently in operation, few have demonstrated long-term successful performance. Technical problems related to the disposal of organic liquids formed during gasification and ash removal are being resolved. Advanced process development efforts also are underway at various research centers and universities. These processes are principally in the development stage and focus on the different aspects of fluidized-bed and catalytic gasification. Some of this work appears ready to be transferred to the private sector for commercialization. Increased cooperation between private developers and R&D institutions would aid in reducing remaining technical problems.

TABLE I
FIXED-BED WOOD GASIFIER
GAS ANALYSES

FEEDSTOCK ULTIMATE ANALYSIS (WT%):	VENTEAK WOOD
C	30.90
H	3.60
O	25.35
N	0.70
S	0.05
ASH	1.00
MOISTURE	38.40
TOTAL	100.00
HHV, KJ/KG	12,585

GAS COMPOSITION (VOL. %, DRY BASIS, TAR FREE)

H_2	13.0
CO	29.0
CO_2	6.6
CH_4	4.0
N_2 + A	47.4
H_2S + COS	----
TOTAL	100.0
HHV, KJ/M^3	6550
M^3 OF DRY GAS/KG FEED	1.07
TAR-OIL PRODUCT/KG FEED	0.081[a]

(a)   INCLUDES METHYL ALCOHOL, DOES NOT INCLUDE ACETIC ACID.

A summary of several gasification processes along with their current status is presented in Table II.

Commercial Application

Existing oil and gas-fired package boilers represent the most viable potential application for hot wood-based gas. The retrofit of these boilers to burn this gas does not present any unusual technical problems. It will generally include:

- Replacement of the existing burners

- Expanding the duct work for increased flue gas volume

- Adjusting the capacity of the induced draft fan or replacing it with a larger unit. (1, 2)

In most cases the retrofit of oil- and gas-fired package boilers to burn wood gas will involve minimum boiler derating. This is particularly true if the original boiler has a relatively large furnace. However, the present day oil- and gas-fired package boilers are designed to have very close tube spacing which results in a compact furnace. This will limit the extent of retrofit operation and will introduce some boiler derating when wood gas is used as fuel. The magnitude of boiler derating, depending on the original furnace design (relatively spacious or very compact) may range between 5 and 25 percent of the initial design capacity.

Economic Evaluation

Based on information supplied by manufacturers, capital costs of wood gasification systems vary from approximately $.30-$1.10/ MJ/D ($5,000 to $20,000/ODT/D) of capacity. (3) This wide variation is the result of differences in system capacity, inclusion of specific components, varying cost assumptions, and the requirement for land and facilities. The average costs are $.55-$.65/ MJ/D ($10,000 to $12,000/ODT/D) for a system which includes materials handling equipment, gasification equipment and ancillary systems such as controls but excludes a boiler, land, building, contingencies and fees.

Wood gas production costs are highly sensitive to feedstock cost and to a lesser extent to the system capacity. Some examples of published gas costs are presented in Table III. It is important to note that costing procedures differed for each case and that results are not directly comparable.

In summary wood gas can be produced at a cost of $2.00 to $3.00/GJ with a $1.00/GJ feedstock cost, or $3.00 to $5.00/GJ at a feedstock cost of $1.50/GJ.

TABLE II

STATUS SUMMARY OF DIFFERENT GASIFICATION PROCESSES

PROCESS DEVELOPER	HOME OFFICE	DEVELOPMENT STATUS	CAPACITY (ODT/D)	ENERGY PRODUCT(S)	REACTOR TYPE	CURRENT APPLICATION
American Fyr-Feeder	DesPlaines, IL	C	1-15*	LBG	SMB	Steam
Andco Inc.	Buffalo, NY	C	75-200*	LBG	MPB	Steam
Applied Engineering Co.	Orangeburg, SC	C	5	LBG	MPB	Boiler Retrofit
Battelle Columbus Lab	Columbus, OH	D	N.A.	MBG	EF	-------
Battelle Northwest Lab	Richland, WA	D	2	LBG	SMB	-------
Biomass Corp.	Yuba City, CA	D	3-4	LBG	FXB	-------
Bio-solar R&D Corp.	Eugene, OR	D	10	LBG	FLB	Drying
British Columbia Research	Vancouver, B.C., Canada	D	10	LBG	FLB	Boiler Retrofit
California, Univ., Davis	Davis, CA	D	18	LBG	MPB	-------
Canadian Ind. Ltd.	Kingston, Ont., Canada	D	24	LBG	N.A.	-------
Century Research Ind.	Gardenia, CA	D	N.A.	MBG/LBG	N.A.	Various
Davy Power Gas Inc.	Houston, TX	C	84	LBG	SMB	Drying
DeKalb Ag. Research	DeKalb, IL	D	N.A.	LBG	N.A.	Boiler Retrofit
Eugene Water & Electric Board	Eugene, OR	D	N.A.	LBG	N.A.	Boiler Retrofit
Forest Fuels Inc.	Antrim, NH	C	2.3-9	LBG	SMB	Boiler Retrofit
Garrett E.R.E. Inc.	Ojai, CA	D	3	MBG	SMB	-------
Halcyon Inc.	East Andover, NH	C	15-80	LBG	N.A.	Farm Utilization
Jamex Inc.	St. Peter, MN	D	8	LBG	SMB	Dual Fuel Engine
Nova Scotia Tech. College	Halifax, N.S., Canada	D	5	LBG	MPB	Dual Fuel Engine
Pioneer Hi Bred Int. Inc.	Johnston, IO	C	20	LBG	MPB	Dual Fuel Engine
Quebec Electro Marine Diesel Inc.	Beaconsfield, Que., Canada	C	15	LBG	MPB	Dual Fuel Engine
Texas Tech University	Lubbock, TX	B	0.25*	MGB	FLB	Synthesis Gas
Union Carbide Corp.	New York, NY	D	200*	MBG	MPB	Synthesis Gas
West Virginia University	Morgantown, WVA	B	N.A.	MBG	FLB	-------
Westwood Polygas Ltd.	Vancouver, B.C., Canada	D	60	LBG	MPB	Drying, Steam
Weyerhaeuser Co.	Tacoma, WA	B	N.A.	LBG	N.A.	-------
Wright Malta Corp.	Ballston Spa, NY	D	3	MBG	RK	Synthesis Gas

GLOSSARY

B - BENCH SCALE UNIT
C - COMMERCIAL SCALE SYSTEM, with units sold
D - DEMONSTRATION UNIT
LBG - WOOD GAS AT 4.4 TO 7.4 MJ/M^3
MBG - WOOD GAS AT 11.0 MJ/M^3

EF - ENTRAINED FLOW
SMB - STIRRED MOVING BED
FXB - FIXED BED
FLB - FLUIDIZED BED
MPB - MOVING PACKED BED

TABLE III

LBG AND MBG GAS COST AS REPORTED IN LITERATURE

PROCESS AND SYSTEM CAPACITY (ODT/D)	FEEDSTOCK COST ($/GJ)		
	1.00	1.50	2.00
LBG Air Gasification (850) [a]	2.56	3.31	4.04
MBG Oxygen Gasification (850) [b]	2.42	3.15	3.79
MBG Oxygen Gasification (850) [c]	4.45	5.08	5.72
MBG Catalytic Gasification (850) [c]	1.73	2.23	2.72
MBG Entrained Flow Gasification (850) [c]	4.62	5.16	5.70
LBG Air Gasification (84) [d]	3.31	--	--
LBG Air Gasification (65) [e]	2.23	--	--

(a) Salo, D., Gsellman, L., Medville, D., Price, G.,
     "Near Term Potential of Wood as a Fuel," MTR-7860, MITRE
     Corporation, Metrek Division, McLean, Va., July 1978.

(b) Bliss, C., Blake, D.O., "Silvicultural Biomass Farms,
     Vol. V, Conversion Process and Costs," MTR-7347, MITRE
     Corporation, Metrek Division, McLean, Va., May 1977.

(c) Wang, E., Cheng, M., "A Comparison of Thermochemical
     Gasification Technologies for Biomass," Proc. Energy from
     Biomass and Wastes Conf., IGT, Washington, D.C., August
     14-18, 1978.

(d) Preliminary Technical Information, "Low Btu Gas Production
     by the Power Gas Producer Unit," Davy Powergas Inc.,
     Houston, Texas, July 1978.

(e) Simons, H.A., Engineering Feasibility Study of the British
     Columbia Research Hog Fuel Gasification System, B.C.
     Research, Vancouver, B.C., May 1978.

In order to assess the relative economic competitiveness of wood-based gas for industrial application, an oil/gas-fired boiler retrofitted with two gasifiers was considered. The capital investment includes the installed cost of wood handling/feed preparation, the wood gasification units, and all the auxilliary and support facilities. An estimate is made of the cost required to retrofit the existing boiler in such a way as to avoid derating. (It is assumed that the existing boiler has a relatively large furnace so that a retrofit operation will not significantly derate the boiler.) This incremental investment is treated as part of the total capital investment. Plant costs are extracted from similar coal gasification systems.(4)

Total investment and annual operating costs were developed for a wood gasification fuel supply system capable of gasifying 220 ODT/D (Table IV). Two gasifiers supply fuel to an existing 45,350 kg/h steam producing plant. Total investment is approximately $5 million, and the annual operating cost, excluding the wood feedstock cost, is about $730,000.

The cost of gas from the industrial boiler retrofit is compared to the delivered prices to the industrial sector of distillate and residual oil and natural gas. Comparisons are presented for the Midwest, New England and Northwest regions for 1985 and 1990 under three economic scenarios. The scenarios are based on assumptions regarding escalations in the price of oil and the resultant impacts on the prices of other conventional fuels. Scenarios are classified as low, medium, and high, corresponding to oil prices of $14.50, $15.15, and $19.72 per barrel, respectively, in 1990. The scenarios are part of DOE's current policy and evaluation program.(5)

The major assumptions used in the gas cost estimates include:

- a mature technology plant

- inflation rate 6 percent

- effective income tax rate 50 percent

- investment tax credit 10 percent

- depreciable life 80 percent of project life

- double declining balance depreciation method for tax purposes

- debt/equity ratio for private competitive firm 30 percent/70 percent

- nominal after tax rate of return on equity of 16.1 percent

- nominal before tax cost of debt of 9.2 percent

TABLE IV

CAPITAL AND OPERATING COSTS FOR
A WOOD GASIFICATION/BOILER RERTOFIT SYSTEM
(In Millions of Dollars, 1978)[4]

BASIS: Two gasifiers linked to a single boiler

Feed Capacity (ODT/D)	220
Product Gas Output (MJ/hr)	132
Steam Production (kg/hr)	45,350
System Stream Factor	0.90
Boiler Efficiency (%)	80

Plant Investment

Installed Plant[a] Cost	4.14
Interest During Construction	.21
Total Installed Plant Construction	4.35

Non-Plant Investment

Land Cost	--
Start-Up Costs	.51
Working Capital	.16
Non-Plant Total	.67
Total Plant Investment Costs	5.02
Labor	.23
Material	--
Utilities	.05
Supplies	.13
Administrative and Overhead	.15
Insurance and Property Taxes	.17
Wood Feedstock Cost[b]	1.80
Total Plant Operating Costs	2.53

(a)   Includes equiment and boiler retrofit work.

(b)   An average cost of $25/dry ton is assumed.

Costs for the wood biomass feedstock were also estimated by applying the above assumptions to the harvesting equipment. In addition, different assumptions regarding operation productivity--variations caused principally by differences in types of forest stand--yielded a range of costs. This range is the basis of the differences between the low, medium, and high price scenarios for the wood feedstock. The wood biomass cost for each region represents a weighted average of the proportion of the various types of wood residues available in 1985 and 1990 (6) and their respective collection cost using present day wood collection technology. (7,8) The costs are escalated by average timber price increases projected by the Forest Service for each region considered. (9) Additional assumptions used in calculating wood feedstock costs are as follows:

- harvesting strategies: clearcutting to produce chips and residue collection

- technical and cost data for equipment from references (5,6)

- stumpage cost: $1.00/dry ton

- transportation cost: $.10/ton-km Northwest and $.08/ ton-km for Midwest and New England

- transportation distance: 75km

- operation schedule: 1600 hours/year

- annual cost escalation factors:

    .8% Midwest
    .5% New England
    4.3% Northwest

Results of the energy cost comparison are presented in Table V. In the Midwest, wood gas is estimated to cost $3.11 and $3.25 per GJ in 1985 and 1990, respectively, under the medium price scenario. In 1985, wood gas is cheaper than distillate oil and somewhat more expensive than residual oil and natural gas, but essentially competitive with all three fuels. In 1990, it is competitive with residual oil and natural gas and possesses a substantial economic advantage over distillate oil under the medium cost scenario. Results from the high price scenario in both years show wood gas at a disadvantage compared to natural gas, competitive with residual oil, and again, substantially cheaper than distillate oil.

TABLE V

COMPARISON OF PRICE OF WOOD GAS FROM ON-SITE GASIFICATION
PLANT AND DELIVERED PRICE OF CONVENTIONAL FUELS
FOR YEARS 1985 AND 1990 ($/GJ)

MIDWEST REGION	1985			1990		
	LOW	MEDIUM	HIGH	LOW	MEDIUM	HIGH
COST OF WOOD ($/ODT)[a]	18.00	22.00	30.00	20.00	24.00	33.00
REFERENCE TECHNOLOGY						
• Distillate Oil	3.33	3.41	4.50	3.46	3.93	4.81
• Residual Oil	2.85	2.86	3.73	2.84	3.21	4.03
• Natural Gas	2.95	2.93	3.21	3.13	3.01	3.14
LOW-BTU GAS BOILER RETROFIT						
• Wood-Fired	2.83	3.11	3.67	2.98	3.29	3.88
NEW ENGLAND REGION						
COST OF WOOD ($/ODT)[a]	18.00	22.00	29.00	29.00	24.00	32.00
REFERENCE TECHNOLOGY						
• Distillate Oil	3.47	3.51	4.64	3.55	4.03	4.95
• Residual Oil	2.89	3.24	4.85	2.82	3.22	4.08
• Natural Gas	3.28	3.28	3.58	3.50	3.37	3.51
LOW-BTU GAS BOILER RETROFIT						
• Wood-Fired	2.83	3.11	3.60	2.98	3.29	3.81
NORTHWEST REGION						
COST OF WOOD ($/ODT)[a]	33.00	40.00	53.00	41.00	49.00	66.00
REFERENCE TECHNOLOGY						
• Distillate Oil	3.69	3.51	4.62	3.55	4.00	4.65
• Residual Oil	2.90	2.87	3.81	2.85	3.18	3.79
• Natural Gas	3.17	3.01	3.64	3.31	3.19	3.32
PROPOSED TECHNOLOGY						
LOW-BTU GAS BOILER RETROFIT						
• Wood-Fired	3.87	4.36	5.26	4.43	4.99	6.17

(a) An oven dry ton (ODT) is approximately 18 GJ

In New England, the economics of wood gas are extremely attractive. Under the medium price scenario for 1985 and 1990, respectively, wood gas is projected to cost $3.11 and $3.25 per GJ. Wood gas is substantially cheaper than distillate oil and competitive with residual oil in both years under all scenarios. The same result holds in comparison to natural gas, with the exception of the high escalation scenario in 1990. Here the cost of gas exceeds that of natural gas by about $.20 per GJ.

Wood gas does not fare as well in the Northwest, remaining an expensive option through the year 1990. Wood gas at $4.36 and $4.99 per GJ in the medium price scenario in 1985 and 1990, respectively, is not competitive with any of the conventional reference fuels; this result does not change for any of the other scenarios considered.

Conclusions

The economic competitiveness of wood gas is highly sensitive to the availability of low-cost wood feedstock and therefore regionally dependent. The New England region where wood feedstock costs are relatively low, the economics of wood-based gas are extremely attractive. In contrast, in the Northwest, the higher wood feedstock cost makes it much less competitive. The Midwest enjoys relatively low wood feedstock and conventional fuel costs. On balance, wood-based gas appears to be a potentially attractive option in this region. Except in the Northwest region, the comparatively high escalation in conventional fuel prices helps enhance the competitiveness of wood-based gas in the near-term.

The results of our preliminary study are based on an evaluation of a wood gasification retrofit system for an industrial oil- and gas-fired boiler application. Some simplifying assumptions were made regarding the ease of retrofit and amount of boiler derating involved when switching to wood gas. These assumptions are justified for taking a first cut at the relative competitiveness of wood-based gas. However, an indepth technical and economic feasibility study must be conducted to confirm the viability of wood gasification for supplying industrial boiler fuel.

Literature Cited

1. CNG Energy Company and Dravo Corporation, "Equipment Convertibility and Interchangeability", paper presented at Industrial Fuel Gas (IFG) Training meeting, Pittsburgh, Pa., June 27-28, 1977.

2. Lewis, R. P., and Bress, D. F., "Installation of a
    FW-STOIC Gasifier for the Generation of a Clean Boiler
    Fuel", paper presented at the 85th National Meeting of
    the American Institute of Chemical Engineers,
    Philadelphia, Pa., June 4-8, 1978.

3. Fritz, J. J., Gordon, J. J., Henry, J. F., Nguyen, V. T.,
    "Status Review of Wood Biomass Gasification, Pyrolysis
    and Densification Technologies", MTR-8031, The MITRE
    Corporation, McLean, Virginia, July 1979.

4. Ashworth, R., Davy Powergas Company, private communications,
    March 1979.

5. Gouraud, J. S., "Phase II Commercialization Task Force
    Instructions", DOE Memorandum, 1979.

6. U. S. Department of Agriculture, Forest Service,
    Forest Statistics of the U. S. 1977 (Draft),
    Washington, D. C., 1978.

7. North Central Forest Experiment Station, Forest Service,
    Forest Residues Energy Program, St. Paul, Minnesota,
    March 1978, ERDA Contract No. E-(49-26)-1045.

8. Bergvall, J. A., Bullington, D. C., and Gee, L., Wood
    Waste for Energy Study, State of Washington, Department
    of Natural Resources, Olympia, Washington, 1979.

9. U. S. Department of Agriculture, Forest Service, An
    Assessment of the Forest and Range Land Situation in
    the United States (Draft), Washington, D. C., 1979.

RECEIVED November 16, 1979.

COMBUSTION TECHNOLOGY

Mass Burning of Municipal Solid Waste — Worldwide

PHILIP R. BELTZ

Battelle Columbus Laboratories, 505 King Avenue, Columbus, OH 43201

Municipal, commercial and light industrial waste is converted to energy daily in about 450 mass burning facilities worldwide. Of these, less than 20 operate in the U.S.A. Realizing this fact, Steve Levy and David Sussman of the U.S. EPA Office of Solid Waste developed a study program titled, "Evaluation of European Refuse Fired Energy Systems Design Practices." This paper presents research conclusions.

World Wide Inventory of Waste-to-Energy Systems

The "Battelle Worldwide Inventory of Waste-to-Energy Systems" shows (as of March 1979) 522 locations where daily operating plants have, are or will process waste into energy . Also included are several large pilot and demonstration plants. Of these about 450 are currently operating mass burning units. Over 99 percent of the world's municipal solid waste that is converted to energy is processed in a "Mass Burning Refuse Fired Energy System". Very few of the plants opened since World War II have closed. Plants typically operate for 25 to 40 years.

Hamburg has been producing electricity from household refuse since 1896. Other early refuse fired energy systems were Chicago (1877), Paris (1903), Zurich (1904) New York (1903), etc. Many of these systems have been replaced by a succession of refuse to energy systems.

Japan has more systems and processes more waste than any other country. However, due to the high moisture and plastics content (causing high temperature chloride corrosion), very little useful energy is produced for sale. The Central European countries of West Germany, France and Switzerland have concentrated on high temperature steam systems for electrical production and district heating. Many Scandinavians produce high pressure hot water for district heating. Sewage sludge is dried and usually is destroyed by the energy in refuse in 34 co-disposal systems worldwide.

0-8412-0565-5/80/47-130-059$05.00/0

   In the U.S., we have tried to develop new advanced chemical
systems while the remainder of the world has continued to build
mass burning energy recovery systems.

   At one plant, odors that are collected in a new rendering
plant are piped 100 feet to the refuse fired steam generator for
destruction. At other plants, flue gases from incinerating
industrial and hazardous waste are blown into the furnace/boiler
thus providing sufficient residence time at high temperatures.
In other words, a refuse fired steam generator can be an after-
burner--saving natural gas or fuel oil.

Composition of Refuse

   While there will be excursions below and above this range,
the moisture content normally varies from a low average of 22
percent to a high average of 32 percent. The average among six
facilities was 27 percent. European refuse has been rapidly
approaching the composition of American waste as Europe has continued to "modernize" its way of life.

   The heating values, which have risen dramatically since 1945,
are expected to begin stabilizing as citizens become more conservation conscious, as petroleum becomes more precious and as
material needs become satisfied. The dramatic rise in heating
values has adversely affected facility performance. Boiler tube
corrosion has increased. Downtime is up as well as maintenance
costs. Many units have eventually been derated. Some 10 to 15
year old plants have discontinued receiving "hotter" industrial
waste in an attempt to reduce the overall average heating value
to original design values.

Refuse Generation and Burning Rates Per Person

   Most facilities viewed accept household, commercial and
light industrial waste for burning: at a rate of 2.2 pounds per
person per day.

Development of Visited Systems

   The primary motivation for constructing refuse to energy
plants in Europe has been to replace an existing landfill, compost
plant or incinerator or to add additional incineration with heat
recovery capacity. However, dramatic changes in the world energy
supply will affect future attitudes about waste-to-energy. At
none of the 15 major and 15 minor visited plants did anyone
indicate to these authors that the primary motivation was connected with energy, i.e. cost savings from free fuel, energy
conservation or unavailability or other energy forms.

   Citizen and elected local official's perception of harmful
effects from landfills is greater in Europe than America. This
U.S. perception may be changing due to revelations about

"Love Canal", "The Valley of the Drums," etc. Revelations that 60 to 80 percent of the American municipal waste landfills have accepted hazardous waste are also bound to increase the American citizen's perception of the hazards of uncontrolled landfilling.

For many years, European federal governments have energetically supported refuse-fired energy plants at the same time that the U.S. Public Health Service and USEPA were innovatively developing and encouraging the sanitary landfill concept. Regarding potential leachate, the U.S. approach has been to "correct" the problem by operating a better landfill. Europe's approach has been to "avoid" the problem by burning the refuse and recycling the ash into useful products. Many European equipment vendors have not joined the American thrust towards co-firing and hence such technical options have not been readily available in Europe.

On both continents, net operating costs per ton, after sale of resources recovered, have always been two to four times greater in a refuse to energy plant than in a landfill of prevailing practice at the time.

Composting, very fashionable in Europe 10 to 20 years ago, has fallen as a chosen alternative because large volume, consistent product markets could not be maintained.

Ownership Organization and Personnel

Each of the 30 systems visited in Europe was owned by either the local municipality or a not for profit authority. Of the 15 plants studied in detail, cities own 8, public authorities own 6, and a nationalized electric company owns one. Private enterprise owns few daily operating resource recovery systems in Europe consuming municipal waste.

Potential economics of scale for large plants were often negated by establishing too many job categories and placing too many people into them.

The program of education, training and experience varies among countries. The Germans have a very rigorous program of schooling, shipboard boiler room experience, landbased powerplants experience, etc. Other countries, however, emphasize on-the-job training.

Capital Investment

Capital investment costs per daily ton of system capacity have increased 5 fold during the past 10 years. The 1960-1968 "capital cost per daily ton" ranged from $13,000 to $15,000 at three surveyed plants. The average for all 15 plants was about $35,000. Three later plants built in 1975 and 1976 averaged about $70,000. Plants in the early 1980's could initially cost over $100,000 per daily ton capacity.

Among the reasons for this extreme cost growth is that the American vendors of mass burning systems do not have as many "bell and whistle" design features as the European systems. The American purchaser has not previously feared corrosion enough to demand protective features. The American buyer often concentrates more on the lowest bid while the European buyer prefers a reliable system of which he and the community can be proud.

The essential difference, however, is momentum. With 275 European systems to visit and observe features, the European buyer knows and appreciates his options. To some extent, there may be peer pressures to have an excellent system. We Americans have not been exposed to enough facilities to have developed the same Continental appetite.

Expenses, Revenues and Net Disposal Costs

Communities that have insisted on extra "chute-to-stack" design and operating features to increase reliability have benefited by having low net disposal costs. The plants averaged $27 per ton for total gross expenses. With the exception of one very small plant, there seems to be little effect of plant capacity on total expenses per ton of refuse processed, i.e., there seems to be little if any economy of scale. Buyers, vendors, designers, citizen groups, etc. seem unable to resist the temptation to "spend" potential economies of scale on personnel, elevators, big offices, conference rooms, architecture, landscaping, multiple energy configurations, etc.

Revenues from sale of energy averaged about $7.50 per ton throughput. All 5 of the 15 systems receiving the highest revenue per refuse input ton are providing energy to district heating systems. Their average revenue is $9.76 per ton. All 5 of the 15 systems receiving the lowest revenue per refuse input ton are adversely affected by very competitive fossil fuel or nuclear electrical power stations. Their revenue averaged $5.30 per ton.

Net disposal costs per ton ranged from $6.27 up to $48.25 with the average being $16.38 per ton.

Refuse Handling

For furnace feeding, all plants observed have the system of "pit crane mixing and loading into the feed hopper". No plant observed has the American innovation of dumping refuse onto the floor, moving it by front-end loader onto conveyors, etc. Crane operators are situated in the stationary control rooms along the front, side or back of the pit. This is in contrast to many American systems where the cab is located on the crane. European designers have concern for operator safety should there be a pit fire and disease prevention from dusty pathologic atmospheres.

Source separation programs for newspaper, cardboard, bottles, cans, etc. will not substantially detract from energy production.

Most systems have doors between the truck tipping floor and the pit (1) to keep odors inside when the system is closed, (2) to permit a higher negative pressure and thus better odor control when operating and (3) to increase the effective volume of the pit for refuse storage against a few of the closed doors.

The only commercially operating municipal refuse systems in Europe known to these researchers which are not mass burners are in England. One is a suspension fired boiler at the IMI plant in Birmingham, England. Another is a Portland Cement kiln in England.

Most European grate vendors are skeptical regarding the long-term commercial viability of suspension fired systems instead of mass burning grate systems. One concern is the additional costs of preparing the refuse derived fuel (RDF). Another concern expressed by many vendors is the high temperatures usually experienced when co-firing with a conventional fuel such as coal, oil or gas. The flue gas temperature causes sticky deposits to form on boiler tubes. This can reduce heat transfer efficiency and occasionally can block sections of boilers. When these deposits are finally blown off, the unprotected surfaces suffer increased "high-temperature corrosion".

Observed grate capacities ranged from 3.33 tonnes (3.7 tons) to 24.6 tonnes (27 tons) per hour. Other furnaces, not viewed, can process as much as 50 tonnes per hour. Concentrated amounts of high energy-containing material may (1) melt the cast iron grates or (2) cause fire underneath the grates. Examples are magnesium chips, aluminum, plastic film, butter, etc. Progressive plant managers have suggested industrial waste sources separation programs.

Furnace Wall

Prior to 1957, all European refuse furnaces were lined with refractory and not with water tube walls. Beginning in 1957 with the Berne, Switzerland plant, the use of water tube walls as the furnace enclosure has increased substantially.

Refractory wall furnaces with waste heat boilers, being less expensive, are often appropriate for hot water or low temperature steam applications. Water tube wall furnace/boilers producing high temperature steam are usually chosen for high temperature steam systems producing electricity efficiently. The remarkable increase in refuse heating value since 1945 has exerted substantial influence on furnace wall design. Even controlled air injection or flue gas recirculation alone was not enough to lower flue gas temperature so that older equipment would survive.

Secondary (Overfire) Air

To obtain complete combustion and to abate smoke and corrosive carbon monoxide formation, the unburned volatile (or pyrolysis) gases rising from the fuel bed must be mixed rapidly with ample oxygen. This mixing must be done relatively near the fuel bed with the assistance of high velocity secondary air jets.

Boilers

At this time, 1977-1978, it appears that the current "best" boiler-furnace design in use for large, high pressure units is the completely water-tube-walled furnace and radiant section, studded and coated with thin plastic refractory in the intense burning zone, followed by one or more long, open, vertical radiation passes preceding a convention-type superheater and boiler-convection passes and economizer.

An emerging newer boiler design is to follow the tail water-tube-walled combustion chamber by a long superheater, boiler convection section and economizer. This is called the "dacha" boiler (after the female dashund dog) because of its extended horizontal configuration with several flyash hoppers. This permits tube cleaning by mechanical rapping and eases the labor of tube replacement when needed.

Above some ash deposit temperature, perhaps 744 C (1300 F), the protective deposit becomes chemically active and corrosion begins. There are several theories of corrosion causes. Three causes stand out: high temperature, HCl and CO. Investigators do not agree among themselves as to the "real" phenomena. Interestingly, the successful systems (those producing high temperature steam with little or no corrosion of boiler tubes) have design features and operating practices consistent with all the theories.

Recent work for EPA by the Battelle Columbus Laboratories postulate that corrosion can be lessened due to an interesting chemical phenomena. Sulfur (in coal, oil, sewage sludge or contaminated methane gases from landfills) has the effect of forming relatively harmless deposits that prevent chlorine from being so corrosive.

The report identifies over 33 design features and operating practices that reduce corrosion. A prudent design/operational will selectively use more than 10 features and practices but not waste money by utilizing all of them.

Excessive us of high-pressure steam soot blowers is a common source of tube erosion-corrosion. Other boiler cleaning methods less threating to boiler tubes are available such as mechanical rapping, shot cleaning, and compressed air soot blowing.

Supplementary Firing and Co-Firing of Fuel Oil, Waste Oil, Solvents and Coal

Supplementary firing of fuel oil, waste oil, or solvents is preferred when there is a need for emergency backup, routine weekend uses, preheating upon startup, prevention of dew point corrosion, legally destroying pathogens and other hydrocarbons, and for routine energy uses when the refuse fired energy plant is down.

However, supplementary firing may not be necessary when the energy user has his own alternative energy supply, when electricity is fed to a large electrical network, when treatment of refuse or sewage sludge can be postponed several days or when a regional plan mandates waste oil treatment and burning at another facility.

Many European vendors suggest that if refuse and a fossil fuel are to be fired in the same system they be fired in separate combustion chambers. Flue gases can later be united before entering the boiler convection section.

Air Pollution Control

The development of the modern water-tube wall furnace/boiler was in part due to the need for proper air pollution control of incinerators. Flue gases can be cooled with massive air dilution, water spray or boiler. Energy recovery in boilers is the preferred cooling method if a reasonable market can be assured or anticipated. The almost universally accepted method for particulate removal is the electrostatic precipitator. Scrubbers alone have usually failed to meet the particulate standards.

Reliability of electrostatic precipitators (ESP) has been excellent except where the inlet gases have been too cold or too hot. Entering flue gas temperatures must be kept above 177 C (350 F) to prevent dew point corrosion of electrostatic precipitators (ESP). But entering flue gas temperatures also must be kept below 260 (500 F) to prevent high temperature chloride corrosion.

The most stringent air pollution control standards have been set for Japan and West Germany. Their standards are much more stringent than in the U.S. Many persons questioned the health justification for the standards. Other European Federal environmental agencies have carefully viewed the standards and have either accepted the particulate but rejected the tight gaseous (HCl, HF) control or have adopted a wait and see attitude.

Technically, the U.S. standard for particulates of 0.08 grains per standard cubic foot adjusted to 12 percent CO_2

(180 mg/nM^3) is bettered by many plants achieving 0.03 to 0.05
g/SCF. Some Janapese plants and the Nashville Thermal Transfer
Corporation achieve even o.01 g/SCF.

The requirement of scrubbers for HCl removal for new plants
greatly increases original capital investment and has greatly
slowed commercialization of resources recovery in Germany. Of
the German plants visited having scrubbers, none was yet working
adequately and without corrosion. The Europeans seem to be more
concerned with heavy metals and organics in landfill leachate and
groundwater than they are with traces of heavy metal oxides from
the refuse fired energy plant stack.

Water pollution control is often not an issue except at
plants with scrubbers. Dirty process water can normally be disposed of in the ash chute quenching system.

Conclusion

The major conclusion is that the mass burning of unprepared
municipal solid waste in heat recovery boilers is well established, and can be a technically reliable, environmentally
acceptable and economic solution to the problem of disposal of
solid wastes. It is not as cheap as is currently available landfilling. However, when the cost is considered of upgrading
current landfills and establishing new landfills in accordance
with the expected Resource Conservation and Recovery Act (RCRA)
provisions, these mass burning waste-to-energy systems are
expected to compare more economically with true sanitary landfills.

Many conditions in the U.S. have been different, hence
waste-to-energy has not advanced as rapidly as in Europe. Some
of these differences will continue, but we are moving rapidly
to similar conditions in most of our metropolitan areas. Hence
the lessons that have been learned in 80 years of refuse fired
energy plant (RFEP) experience in Europe can be effectively
utilized by many U.S. communities.

RECEIVED November 20, 1979.

An Evaluation of Modular Incinerators for Energy Recovery from Solid Wastes

DAVID C. BOMBERGER and JERRY L. JONES

SRI International, 333 Ravenswood Avenue, Menlo Park, CA 94025

This paper is concerned with an analysis of modular or shop fabricated incinerators for mass burning of unprocessed (or as-received) solid wastes. It is based principally on work performed during 1978 by SRI International for the Navy Civil Engineering Laboratory at Port Hueneme, California (1). The burning of municipal solid waste (MSW) will be the main focus of the paper although other waste-burning applications will be described.

Modular incineration units are not generally available with single unit capacities greater than 25 to 30 tons/day. This capacity limitation is a result of the requirement that the units be small enough to be shipped by trucks. There are numerous types of modular incinerators but most are batch fed (approximately every 5 to 15 minutes) horizontal flow, cylindrical refactory-lined furnaces, with both a primary and a secondary combustion chamber. These two-stage combustors are often referred to as "controlled-air" or "starved-air" incinerators. The first stage operates with substoichiometric air, and the second stage serves as an afterburner to combust the gases (high in CO content and with other products of incomplete combustion) and organic particulate matter from the first stage. The two-stage combustion is supposed to eliminate the need for air pollution control devices. Other types of units include rotary kilns, basket grate units, and an auger bed unit (2, 3).

In this paper, information will be presented on the extent of current use of modular incinerators, the suppliers of such units, the performance of several units, and the estimated investment and operating requirements for these units. Plant capacities of up to 100 tons/day will be discussed. Multiple 25-ton/day units are specified for the larger plant sizes considered.

Background

Classes of Users

The market for modular incinerators has historically been comprised of the following users:

0-8412-0565-5/80/47-130-067$05.00/0

- Commercial users
  - Airports
  - Shopping centers
  - Large office complexes

- Institutional users (or government installations)
  - Hospitals
  - Military bases (wastes may be comparable to industrial or municipal solid wastes depending on the base functions)

- Industrial users
  - Warehouses (packaging trash and pallets)
  - Processors (cotton gin trash)

- Small municipal users (or government installations)

Following the completion of the work by SRI, an EPA-funded report was published that contains an extensive bibliography and analysis of the use of modular incinerators for various applications (4).

Extent of Use of Modular Incinerators

Based on the numbers of installations reported by the modular incinerator suppliers, probably more than 5000 units have been installed in the United States during the last decade. The exact number of the units that are still operational is unknown. It appears that less than 100 of the modular incinerator systems are designed for heat recovery.

Modular Unit Design

The design requirements for the modular units vary depending on the type of solid waste being burned. For units burning mainly cardboard, polymer, and wood-packaging wastes at commercial or industrial sites, it has not been necessary to install automatic ash removal equipment. Units at industrial sites may operate three shifts/day, six days/week, with cleanout of ash on the seventh day during a scheduled shutdown. Descriptions of the design, operation, and costs for several modular incinerators at industrial sites may be found in References 5 through 10. A modular incinerator that burns municipal refuse one day a week and plant trash five days a week is described in Reference 10.

Burning of municipal refuse will present numerous problems not commonly encountered with the burning of most low ash content (<5 wt%) industrial plant trash or commercial solid wastes. With an ash content that averages about 25%, municipal solid waste incinerators must have some mechanism for automatic ash removal if they are to operate on a 24-hour/day basis. Alternatively, the units may operate one or two shifts a day, cool down, and have the accumulated ash removed in a single operation before reloading for another burning cycle. Certain constituents of the waste, such as glass, may cause slagging and clinker formation, which can damage refractory or jam ash removal mechanisms. The variability of the waste is also likely to be more extreme than in the cases where commercial or industrial wastes are being burned.

Modular Units Burning MSW

Descriptions of selected modular incinerator installations burning municipal solid waste may be found in References 11 through 14. Table I shows a summary of most of the modular units installed between 1974 and 1978 with a number of planned install- ations also indicated. A survey of 14 manufacturers of small in- cinerators made in 1975 by Hofmann disclosed that before 1975, there were 37 municipal plants with a total of 95 individual modu- lar incinerators (11).

Based on the data we have collected and the data from previous studies, it appears that during the last ten years, the total in- stalled capacity of modular incinerators has been less than 2000 tons/day and that much of the capacity is no longer operational. Also, only about 25% of the installed capacity has included heat recovery, and most of these units have been installed during the last five years.

Considering that in 1970 there were 232 municipalities in the United States with populations ranging from 50,000 to 100,000, and 455 municipalities with populations ranging from 25,000 to 50,000, it appears that there are many potential users for small-scale re- source recovery systems (15). The above two groups of municipal- ities represented a total population in 1970 of over 30 million people and probably generated close to 40,000 tons/day of MSW. Thus, the current installed capacity for MSW-burning modular incin- erators with heat recovery (~500 tons/day) probably represents about 1% of the solid waste collected from municipalities with populations ranging from 25,000 to 100,000. (For comparison pur- poses, note that the total capacity for energy recovery from munic- ipal solid waste in the United States should be roughly equiva- lent to about 5% of the total waste collected by 1980.)

Potential Use of Modular Incinerators on Military Bases

Engineering groups within the Department of Defense (DOD) have been evaluating modular incineration technology for several years as a part of DOD energy conservation and environmental control programs. DOD engineers have been interested in the modular units for the same reason that small municipalities have an interest in such units--"modular incinerators are advantageous in that they are less capital intensive than their custom-designed, field- erected counterparts" (2).

To determine if serious problems are associated with modular incinerators, DOD groups have funded or actually conducted tests of operating modular incinerators (16, 17). As a result of the various field tests and engineering evaluations, the DOD has decided to install a number of modular incinerators to recover energy from solid wastes. One of these installations (with three incinerators) is located near Jacksonville, Florida, at a Navy base and will serve as a testing and demonstration facility. If the units prove to be a satisfactory answer to the solid waste disposal and energy recovery needs, then more units may be justified for use by the Navy.

Table I

SELECTED INSTALLATIONS OF SHOP-FABRICATED
INCINERATORS BURNING MUNICIPAL REFUSE (SINCE 1974)

Equipment Supplier	Site of Installation	Ash Removal	Designed with Heat Recovery	Air Pollution Control	Number of Units Installed	Total Facility Design Capacity[+] (ton/day)	Operating Schedule (shift/day)[+]	Plant Startup
Comtro Division Sunbeam Equip. Corp. Lansdale, PA	Pelham, NH	Manual	No	CAI**	2	16	1	1977
	Jacksonville, FL (Navy)	Automatic	Yes	CAI	3	75	?	1978
Consumat Systems, Inc. Richmond, VA	Blytheville, AR	Manual	Yes	CAI	4	48	1	1975
	N. Little Rock, AR	Automatic	Yes	CAI	8	100	3	1977
	Orlando, FL	Manual	No	CAI	2	100	1	1974
	Pahokee, FL	Manual	Yes	CAI	2	17	1	1974
	Siloam Springs, AR	Manual	Yes	CAI	1	21	1	1975
	Pittsfield, NH	Manual	No	CAI		5-6	1	Planned for 1978 or 1979
	Auburn, ME	--	Yes	--	--	150	--	1980
	Dyersburg, TN	--	Yes	--	--	100	--	1979
	Genesee Township, MI	--	Yes	--	--	100	--	1979
	Salem, VA		Yes					
Environmental Control Products, Charlotte, NC	Diamond International Groveton, NH	Automatic	Yes	CAI	1	~20 ton/day	3 (Unit burns municipal refuse only 1 shift/week)	1975
Kelley Company, Inc. Milwaukee, WI	Auburn, NH	Manual	No	CAI	1	5-6	1	1978
	Bridgewater, NH	Manual	No	CAI	1	5-6	1	1977
	Candia, NH	Manual	No	CAI	1	3-4	1	1976
	Canterbury, NH	Manual	No	CAI	1	5-6	1	1978
	Harpswell, ME	Manual	No	CAI	1	5-6	1	1978
	Kittery, ME	Manual	No	CAI	2	20-25	1	1977
	Meredith, NH	Manual	No	CAI	2	10-12	1	1975
	Nottingham, NH	Manual	No	CAI	1	3-4	1	1977
	Pittsfield, NH	Manual	No	CAI	1	5-6	1	1975
	Wolfeboro, NH	Manual	No	CAI	1	5-6	1	1977
	Suttan, NH	Manual	No	CAI	1	3-4	1	Planned 1979
	Lincoln/Woodstock, NH	Manual	No	CAI	1	5-6	1	Planned 1979
O'Connor Combustor Corporation Costa Mesa, CA	Gallatin, TN	Automatic	Yes	ESP	2	150	3	Planned 1980
	Dubuque, IA	Automatic	Yes	ESP	2	200	3	Planned 1980
Combustion Engineering* Windsor, CN	Numerous	Manual	No	CAI	--	--	--	Prior to 1975
Smokatrol	Crossville, TN		Yes			60		1979
CICO	Lewisburg, TN		Yes			60		1979

* Not currently active in the sale of this line of equipment.

+ Based on operating schedule indicated in the adjacent column.

** Controlled air incinerator with secondary combustion chamber

One shift may range from 6 to 10 hours of operation.

Note: For most equipment suppliers the foreign installations have not been listed.

Technical Analysis

At the time of the study, most of the installations in the
United States burning municipal refuse in modular incinerators had
been operating for only a few years or less. Most installations
did not have heat recovery. It is difficult to analyze a technology with so little operating experience. It is impossible to
determine unit lifetimes and maintenance requirements because no
unit has operated for much of its expected lifetime.

However, based on site visits, it is possible to make some
observations:

• The technology is still developing. The units supplied
  by the major suppliers are being improved with each installation and some field modifications are often required.

• Some units are experiencing problems with slag formation.

• Slag removal sometimes requires a jack hammer, which
  erodes the refractory.

• Heat recovery units (water tube boilers) have experienced
  problems with corrosion and buildup of soot on the heat
  transfer surfaces.

• A high level of operator skill is desirable. Temperature
  and burning control is achieved in large measure by the
  loading strategy. Units are not designed to provide the
  loading operator any direct information regarding the
  flame or stack condition. This information is provided
  to the operator by the shift supervisor, who has to make
  periodic inspections that involve opening an access door
  to the combustion chamber.

• Units with continuous ash removal are not producing a
  completely inorganic, non-putrescible ash.

• Even on relatively new units, refractory damage and
  evidence of overheating are visible.

• Based on a limited amount of test data, it appears that
  the units may not be able to meet specific local air pollution codes. Particulate stack plumes are visible, especially when operation requires an accelerated loading
  schedule to lower primary combustion temperatures.

As a result of these observations, several assumptions were
made for use in the economic analysis:

• Units operate at 90% of their rated capacity.

- Air pollution control devices may be required. Economics should be developed with and without air pollution control.

- Portions of the incinerator facility will have to be replaced after 12-13 years of operation.

Economic Analysis

The economics of modular incinerators was developed for a hypothetical location burning refuse typical of Naval installations. The modular units were assumed to be typical two-stage combustion devices. It was assumed that some source separation would be practiced to remove most of the glass and bulky items. The resulting refuse that was actually burned had a heating value of 11.72 x 10^9 joule/metric ton (10.1 million Btu/ton). This refuse is somewhat different from a typical municipal refuse, but the economics are not sensitive to the differences.

An important design assumption was that refuse would be collected 5 days a week for 52 weeks a year (260 days a year) but because of maintenance and downtime, refuse could be burned only 230 days a year. It was also assumed that units could only burn refuse at 90% of their rated design capacity. Three basic cases were considered: plants with rated capacities of 18.2, 45.5, and 90.9 metric tons/day, respectively. Table II details the actual daily and annual capacities of the installations. The small unit was evaluated for one- and two-shift per day operation with periodic ash removal whereas the larger plants were assumed to operate three shifts per day for five days a week with continuous ash removal.

In cases where heat recovery was practiced, the overall thermal efficiency was assumed to be 50%. The major heat loss was the hot flue gases, but other losses included sensible heat plus the unburned fixed carbon in the ash, and radiation losses from the incinerator unit. Figure 1 shows a summary of the mass and energy balances for a metric ton refuse input to the incinerator. Some auxiliary fuel consumption was assumed (based on discussions with system designers and the actual operating experience of users) for startup, temperature control, and pilot burners in the secondary combustion chambers.

Table III summarizes the basic plant operating requirements in terms of utilities, manpower, and ash disposal. Electric power consumption figures are provided for three different system designs: simple incineration without heat recovery or an air pollution control device, incineration with heat recovery but no air pollution control device, and incineration with both heat recovery and air pollution control (stack gases filtered through a baghouse). Table IV summarizes the economic bases.

18.2-Metric-Ton Incinerator

For the small installation, four basic cases were considered. These are detailed in Table V along with the capital expenditures required over the 25-year lifetime assumed for the project. Air

Table II

INCINERATION PLANT CAPACITIES
CONSIDERED FOR ECONOMIC ANALYSIS

| Plant Design Capacity | | Quantities of Refuse Burned [*] | | Operating Shifts |
Metric Tons/Day	Tons/Day	Metric Tons/Day	Metric Tons/Year	Per Day
18.2	(20)	16.4	3,763	1,2
45.5	(50)	40.9	9,408	3
90.9	(100)	81.8	18,816	3

[*]Based on burning 230 days per year.

Overall Heat Efficiency Is 50%

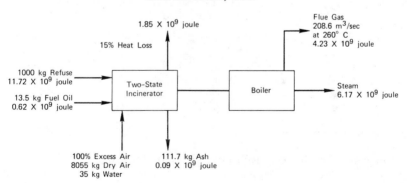

Figure 1.   *Mass and energy balance for an incinerator with heat recovery*

Table III

SUMMARY OF PLANT OPERATING REQUIREMENTS

	Amount Required
Auxiliary fuel	0.58×10^9 joule/metric ton of refuse burned (5% of refuse heating value)
Electric power	11 kWh/metric ton (10 kWh/ton) 22 kWh/metric ton (20 kWh/ton) HRB[*] 33 kWh/metric ton (30 kWh/ton) HRB, PCD[**]
Operating labor	2 men/shift
Ash disposal	0.1117 dry metric ton/metric ton refuse burned

[*]Heat recovery boiler included in installation.

[**]Particulate collection device included in installation.

Table IV

ECONOMIC BASES

<u>Mid-1978 Cost</u>

<u>Discount Rate = 10%</u>

<u>Economic Life</u>

Permanent buildings	25 years
Incinerator system	25 years (major maintenance and some replacement required halfway through life.)
Heat recovery boilers	25 years
Air pollution control devices	25 years

<u>Ash Disposal</u> $11.82/dry metric ton (in landfill)

<u>Maintenance Materials</u>   2.5% of the total plant investment[*]

<u>Purchased Utilities</u>

Water	$0.13/m^3 ($0.60/1000 gal)
Electric power	3.0¢/kWh
Fuel oil	$2.37/10^9 joule ($2.50/million Btu)

<u>Labor</u>

Operating labor	$6.0/hr
Supervision	20% of operating labor
Maintenance labor	2.5% of the total plant investment[*]
Administrative & support labor	20% of labor charges for operation, maintenance, and supervision
Payroll burden	30% of total direct labor costs

---

[*]For the particulate collection system, 10% was assumed to cover bag replacement.

Table V

ESTIMATED INVESTMENT COSTS FOR 18.2 METRIC TON
(20 TONS/DAY) MODULAR INCINERATION SYSTEM

Case	Estimated Investment Costs	
	Year 0	Year 13[*]
1 shift/day-- no heat recovery	$385,000	$190,000
2 shifts/day-- no heat recovery	$230,000	$120,000
1 shift/day-- heat recovery	$550,000	$190,000
2 shifts/day-- heat recovery	$325,000	$120,000

[*]Replacement of a portion of the incinerator facility.

pollution control (scrubber, fabric filter, or precipitator) is
not included. Table VI summarizes the operating costs for the
systems.

For the heat recovery cases, it was assumed that when the
unit was started every day, an hour would be required to remove
ash and to heat the unit up to working temperature. Steam would
be generated for 7 hours in the one-shift case and for 15 hours in
the two-shift case. Four hours of burndown on automatic control
would complete the daily cycle. It was assumed that this meant
that 7/11 of the refuse heat content was available for conversion
to steam in the one-shift case, and 15/19 was available in the
two-shift case.

Note that labor charges represent from approximately one-half
to two-thirds of the total operating costs (including capital
charges) for the four cases considered.

Without heat recovery included in a 18.2-metric-ton/day plant,
it appears from the data presented in Table VI that a one-shift/
day operation would be preferable to a two-shift/day operation.
Figure 2 shows that for a plant with heat recovery, the decision
between choosing a one- or two-shift operation is influenced by
the value of the steam produced. If the steam has a value of more
than $2/10^9$ joule, the two-shift operation is more attractive.
Almost 25% more steam can be produced from a two-shift operation
than from a one-shift operation. The addition of a particulate
control system increases the capital cost by approximately $80,000
and increases the operating cost by $7/metric ton.

Large Modular Installation

The 45 and 90 metric ton/day facilities were designed to include two and four individual modules, respectively. The modules
were operated three shifts/day, five days a week and included
automatic ash removal. Heat recovery was included in all of the
cases considered. Table VII details the capital costs for these
units, both without a particulate control and with fabric filters
for particulate control. Table VIII summarizes the operating
costs for the installations. For both sizes of installation, particulate control adds $4/metric ton to the operating cost. Figure
3 illustrates the effect of steam value on the operating costs.
The 90-metric-ton installation is significantly cheaper to operate
on a per ton basis than the 45-metric-ton installation because the
labor costs are almost equal for the two sizes.

The economic analysis indicates that operating costs are sensitive to the value of the steam generated. For example, if steam
were worth $4-5/10^9$ joule, the 90-metric-ton installation could
operate at the break even point with only a small tipping fee. At
steam values near $2/10^9$ joule, the 90-metric-ton installation
would have to charge a tipping fee of almost $20/metric ton to
break even.

Table VI

COSTS OF SMALL MODULAR REFUSE INCINERATION SYSTEM

Design capacity is 18.2 metric ton/day (20 ton/day)

	No Heat Recovery No Particulate Collection		Heat Recovery No Particulate Collection	
	1 Shift 5 days/week	2 Shifts 5 days/week	1 Shift 5 days/week	2 Shifts 5 days/week
Initial Capital Cost	$385,000	$230,000	$550,000	$325,000
Annual Operating Cost				
Materials and supplies and ash disposal	14,600 (10%)	12,000 (7%)	18,700 (11%)	13,100 (8%)
Labor	69,700 (50%)	111,800 (68%)	77,400 (46%)	113,700 (65%)
Utilities	6,400 (5%)	6,400 (4%)	7,700 (4%)	7,700 (4%)
Capital charges	48,600 (35%)	29,200 (21%)	66,400 (39%)	39,200 (23%)
Total annual cost ($/metric ton)	139,300 ($37)	159,400 ($42.3)	170,200 ($45.2)	173,700 ($46.2)

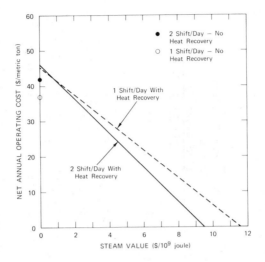

Figure 2.   Economics of operating an 18.2-metric ton/day (20 tons/day) refuse incinerator

Table VII

ESTIMATED INVESTMENT COSTS FOR
LARGE MODULAR REFUSE INCINERATION SYSTEMS
WITH HEAT RECOVERY

System Capacity (metric tons/day)	Year			
	0	6	13	18
45.5 (no particulate control)	965,000	4,000[*]	280,000[†]	4,000[*]
45.5 (with fabric filters)	1,064,000	4,000[*]	280,000[†]	4,000[*]
90.9 (no particulate control)	1,450,000	7,000[*]	420,000[†]	7,000[*]
90.9 (with fabric filters)	1,650,000	7,000[*]	420,000[†]	7,000[*]

[*]Replacement of refractory

[†]Replacement of portions of the incinerator facility.

Table VIII

COST OF LARGE MODULAR REFUSE INCINERATOR SYSTEMS

	45.5 Metric Tons/Day (50 tons/day)		90.9 Metric Tons/Day (100 tons/day)	
	Installed Capacity	Particulate Control	Installed Capacity	Particulate Control
Initial Capital Cost	$965,000	$1,064,000	$1,450,000	$1,650,000
Annual Operating Cost				
Materials supplies and ash disposal	36,600 (10%)	46,500 (11%)	61,100 (13%)	81,100 (14%)
Labor (3 shifts/day)	194,600 (53%)	207,300 (52%)	231,500 (44%)	239,500 (43%)
Utilities	19,100 (5%)	22,200 (6%)	38,300 (7%)	44,500 (8%)
Capital charges	115,500 (32%)	126,400 (31%)	173,700 (36%)	195,700 (35%)
Total	$365,800	$402,400	$486,600	$560,800
($/metric ton)	($39)	($43)	($26)	($30)

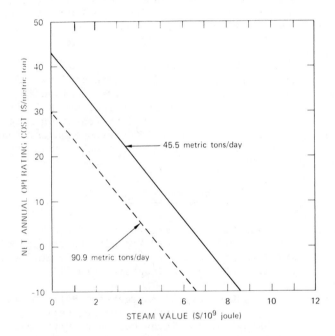

Figure 3.   Economics of operating large modular refuse incinerators with heat recovery and particulate control

Summary

The economic analysis indicates that the modular incinerators are expensive to operate. Even with steam values of $5/10^9$ joule, the very small units are likely to need tipping charges of $20/ metric ton to break even. Larger installations can do much better and, at steam values of $5/10^9$ joule, require tipping fees of only $0-10/metric ton. The principal cause of the high cost found by this analysis is the labor component. Labor represents over 50% of the operating costs for the very small units and only falls to 40% of the total operating cost at the 90-metric-ton/day size range. Total operating costs could be reduced by using less expensive labor and may actually be lower in some parts of the country where wage rates are low.

Acknowledgments

This paper is based principally on work done by SRI International and sponsored by the U.S. Navy, Civil Engineering Laboratory, Port Hueneme, California under Contract No. N00123-78-C-0868. The paper has been adapted from an oral presentation made by Dr. Bomberger of SRI to the National Conference on Environmental Engineering, American Society of Civil Engineers, San Francisco, California, July 1979.

LITERATURE CITED

1. Jones, J. L., and D. C. Bomberger, Jr., "Mass Burning of Refuse in Shop Fabricated Incinerators," draft report appendix prepared by SRI International for the Navy Civil Engineering Laboratory, Port Hueneme, CA (September 1978).

2. Hathaway, S. A., "Recovery of Energy from Solid Waste at Army Installations," U.S. Army Construction Engineering Research Laboratory, Champaign, IL, Technical Manuscript E-118 (August 1977).

3. Hathaway, S. A., and R. J. Dealy, "Technology Evaluation of Army-Scale Waste-to-Energy Systems," U.S. Army Construction Engineering Research Laboratory, Champaign, IL. Interim Report E-110 (July 1977), available through NTIS, AD/A-42 578.

4. Mitchell, G. L., et al., "Small Scale and Low Technology Resource Recovery Study," report prepared by SCS Engineers for the U.S. Environmental Protection Agency under Contract No. 68-01-2653 (January 1979).

5. Anguil, G. H., "Using Solid Waste as a Fuel," Plant Engineering, (November 13, 1975).

6. Miller, B. U., "Using Solid Waste as a Fuel," Plant Engineering (December 11, 1975).

7. Erlandsson, K. I., "Using Solid Waste as a Fuel," Plant Engineering (December 11, 1975).

8. "Companies Mine Energy from Their Trash," Business Week (August 2, 1976).

9. "Burning Waste for Heat in Horicon, Wisconsin," Solid Wastes Management (June 1976).

10. "Paper Mill Incinerator Solves Town Refuse Problem," Combustion (February 1977).

11. "Evaluation of Small Modular Incinerators in Municipal Plants," report prepared by Ross Hofmann Associates for the U.S. Environmental Protection Agency, EPA Report SW-1130 (1976).

12. Hofmann, R. E., "Controlled-Air Incineration--Key to Practical Production of Energy from Wastes," Public Works, 107(9): 72-79, 136, 138 (September 1976).

13. "City Finds Disposal Solution, N. Little Rock Picks Modular Incinerator for Steam and Income," Solid Wastes Management (March 1978).

14. Hart, C. H., "Modular Incineration Units: Hot New Disposal Equipment for Municipalities," Solid Wastes Management (July 1978).

15. Statistical Abstract of the United States, 1973, U.S. Department of Commerce (July 1973).

16. Kleinhenz, N., and H. G. Rigo, "Operational Testing of a Controlled Air Incinerator with Automatic Ash Handling," report prepared by Systems Technology Corp. for the Navy Civil Engineering Laboratory, Port Hueneme, CA (November 1976).

17. Hathaway, S. A., J. S. Lin, and A. N. Collishaw, "Field Evaluation of the Modular Augered-Bed Heat-Recovery Solid Waste Incinerator," U.S. Army Construction Engineering Research Laboratory, Champaign, IL, Technical Report E-128 (May 1928).

RECEIVED December 3, 1979.

Energy Recovery from Wood Residues by Fluidized-Bed Combustion

J. W. STALLINGS[1]

Energy Incorporated, P. O. Box 736, Idaho Falls, ID 83401

MIKE OSWALD

Energy Products of Idaho, P. O. Box 153, Coeur d'Alene, ID 83814

Dramatic increases in costs of energy during the decade of the seventies have provided the economic incentive for the forest products industry to recover energy from their wood residues rather than purchasing fossil fuels. The Fluid Flame® System was designed by Energy Incorporated and its subsidiary, Energy Products of Idaho, to accomplish this task utilizing fluidized-bed technology. Thirty-one of these units are now in commercial operation burning various types of wood residues and recovering energy in the form of steam, hot gases, or a combination of the two. The total installed output is equivalent to 1.62×10^{12} J/hr (1,540 MM Btu/hr) of thermal energy or 559,000 kg/hr (1,232,000 lb/hr) of steam.

Applications have varied according to the energy needs of the individual customer. In steam production, fire-tube boilers are generally employed for systems providing less than 13,600 kg/hr (30,000 lb/hr) of steam, while water-tube boilers are used for larger installations. Either steam or hot gases from the Fluid Flame® systems have been used in plywood dryers, tube dryers, and both direct and indirect hot-gas dry kilns. In one installation, a steam turbine generates up to 2 megawatts of electricity on demand. This amount is equivalent to fifty percent of the total energy available from the facility.

Fluidized-Bed Technology

The term fluidized bed refers to a layer of sand-like particles suspended by the upward flow of a gas stream. A point of equilibrium is reached at which the upward drag force of the gas is equal to downward force of the weight of the particles. The fluidized gas-solid mixture exhibits a high degree of turbulence similar to boiling water. Additional fluidizing air increases the circulation of the particles and further expands the height of the bed.

Fluidized-bed technology is ideally suited to energy recovery from wood wastes for a number of reasons. The high heat

[1] Current address: SRI International, 333 Ravenswood Ave., Menlo Park, CA 94025

0-8412-0565-5/80/47-130-085$05.00/0

transfer rates and high surface areas per unit volume ensure
virtually complete combustion within the burner. The large
amount of energy stored in the hot bed material enables combustion of fuels with relatively high moisture contents. The abrasive action of the turbulent bed material continuously exposes
unburned surface area of the fuel to the intense heat and the
oxygen supply. Thus, a wide size distribution of fuel can be
fed into the burner without loss in combustion efficiency.

Wood residues currently burned in commercial units include
plywood trim and sander dust, wood shavings, whole log chips,
bark, and log yard wastes. Coal or agricultural residues can
also be burned in the system either separately or in combination
with other fuels.

System Description

A schematic diagram of a typical system for steam production, including approximate mass and energy balances as well
as capital and operating costs, appears in Figure 1.
After screening to remove particles larger than a three-
inch nominal size, the wood residues are transferred into a
live-bottom storage hopper by means of a pneumatic conveyance
system. Hopper sizes vary with the amount of storage desired.
Those manufactured by Energy Products of Idaho employ hydraulic
drives to provide adequate torque and power to a sweep auger.
From the bottom of the storage bin hopper, a screw conveyor transfers the wood residues to a metering bin. The rate
of discharge from the storage bin hopper to the metering
bin is controlled by level indicators on the sides of the
metering bin. The rate of fuel fed to the burners from the
metering bin is controlled by burner demand. The feed pipe
enters the combustion cell from the top and protrudes at an
angle from four to eight feet into the vapor space to insure
even distribution of the feed on the bed.
The combustion cell is fabricated out of carbon steel and
is lined with high-temperature block insulation covered by a
layer of castable refractory.
The original combustion systems employed sand as a bed
material. However, excessive attrition necessitated the search
for materials which were more thermally stable and chemically
inert and thus would not elutriate from the bed. Currently, two
proprietary bed materials are employed. One is a naturally
occurring stable mineral, and the second is a raw material used in
the production of fire bricks. With these improved bed materials,
attrition is minimal. In fact, particulate from the wood residues

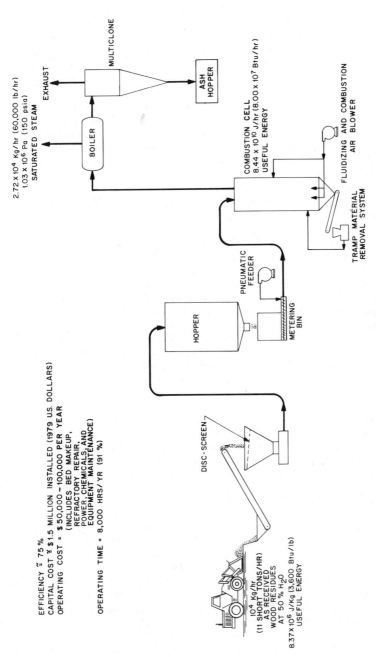

Figure 1. Flow diagram

often builds up in the bed and necessitates removal of portions of bed material.

Grid plates were used in the initial units to distribute fluidizing air to the system. However, the growing need for a tramp removal system and the excessive thermal stress on the plates led to the design of a manifold for air distribution instead of plates. An inverted vibrating cone below the manifold was included to gradually remove bed material from the system in order to screen off large pieces of inorganics, such as rocks and stones. The screened bed material is then reinjected back into the burner. In a further improvement of this design, a double static cone assembly utilizing gravity flow has been installed to cut costs and maintenance. This removal system was designed to allow tramp material up to 10.2 cm (4 in.) in cross section to be removed from the combustion cell without shutdown. Larger pieces cannot fit in the spaces between the manifold sections.

The system described here uses hot combustion gases from a fluidized-bed combustion cell in conjunction with a package water-tube boiler for steam production. The combustion gases are then released to the atmosphere. More recent designs have employed venturi scrubbers for gas cleanup when necessary. However, the final decision on a gas cleanup method depends on the specific regulations for each location.

Some advantages of the radiant effect on heat transfer can be gained by placing the boiler directly on top of the combustion cell. Burning takes place both in the combustion cell and in the boiler. This approach is effective with both fire-tube and water-tube boilers.

The Fluid Flame® energy systems are designed to run themselves without the need for full-time operators. The system is fully automated and is equipped with annunciators to alert operators when problems arise. A number of electrical process controllers are used to maintain constant temperatures in the bed and in the vapor space of the combustion cell in order to provide steam at constant pressure. Signals from these controllers dictate the feed rate from the metering bin and the amount of excess air fed into the system. The temperature is controlled in order to maintain constant parameters throughout the system and to keep temperatures below the slagging temperatures of the ash. Normal temperature variation is less than $+5C^o$ ($9F^o$), and the response time to process variations is rapid.

The maximum turn-down ratio for an operating system is three-to-one. Operation in an on-off mode can also be employed to effectively increase this ratio. Another approach is the use of modular units. However, this latter option will increase capital costs considerably.

The advantages of a fluidized-bed system over conventional systems are many. Wood residues with moisture contents as large as 63 percent can be burned without the need of a supplemental fuel, other than during startup. The absence of grates removes a major maintenance problem, as does the use of an air distribution system within the fluidized bed. The large heat sink in the bed allows for automatic startup after shutdowns of up to sixteen hours.

The oldest Fluid Flame® system has been in commercial operation for approximately six years. The original refractory is still in place but a new coating was added after five years. With improved techniques in refractory installation, systems which have not been in operation for five years have a projected refractory lifetime of approximately ten years, assuming proper care and maintenance.

System Efficiency

The calculated efficiency of the steam system as described here is approximately 75 percent. This value is derived by using the total useful energy of the fuel as the energy input. The heat content of the wood is approximately 1.9×10^9 J/kg (8300 Btu/lb) on a dry basis, which is equivalent to 9.6×10^6 J/kg (4100 Btu/lb) on a wet basis. If the assumption is made that 1.2×10^6 J (500 Btu) is needed to boil off the water in each pound of fuel, the the energy left in the fuel which can be used for steam production is 8.4×10^6 J/kg (3600 Btu/lb).

Variations in efficiency are caused by differences in feed-water temperature and moisture content of the wood residues. The efficiency could be increased by employing the waste heat from the combustion gases to preheat either the fluidizing air or the boiler feedwater. Waste heat could also be employed to partially dry the fuel.

System Economics

Data for the payback period in years for the system described herein are presented in Figure 2. These results are for the 8.44×10^{10} J/hr (8.00×10^7 Btu/hr) system and are based on replacement of natural gas with wood residues. The cost of the natural gas is assumed to be $2.09/10^9$ J ($2.20/10^6$ Btu). These calculations were made to show the payout period in terms of replacement of natural gas. Labor and depreciation were not included, and an operating cost of $100,000/year was assumed. For systems where the wood residues are owned by the operator, the payout period has generally been less than two years for a twelve-month operation.

Commercial Facilities

Thirty-one facilities are now in commercial operation. Figure 3 includes a map showing the locations of these burners. The capacities vary from 1.27×10^{10} J/hr (1.2×10^7 Btu/hr to

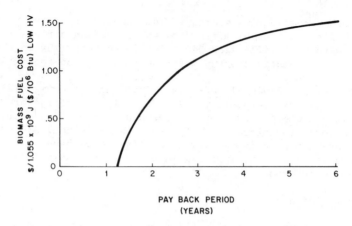

Figure 2.   Payback from biomass replacement of natural gas at $2.09/10⁹ J
($2.20/10⁶ Btu) for a system with a capacity of 8.44 × 10¹⁰ J/hr (8.00 × 10⁷
Btu/hr)

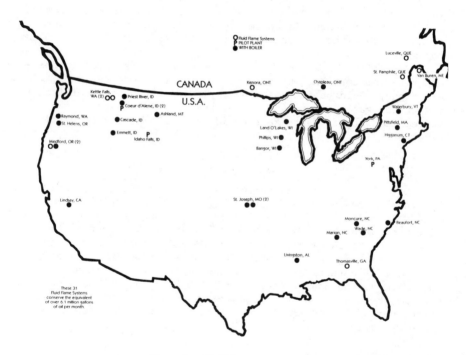

Figure 3.   Fluid flame energy systems

1.27×10^{11} J/hr (1.2×10^8 Btu/hr).

As the perceived market value of organic residues increases, changes will be made to the basic system to increase efficiency. Further developments are also planned to improve the overall economics of the system. If operated in a starved-air mode with a fire box downstream from the combustion cell, capital costs for the burner and boiler per unit throughput can be decreased, and more advantage can be taken of the radiant effects on heat transfer. As the costs of energy continue to increase relative to the basic price indices, the economics of fluidized-bed combustion of biomass to recover energy will continue to improve.

RECEIVED November 16, 1979.

Fluid-Bed Combustion of Solid Wastes

ROBERT H. VANDER MOLEN

Combustion Power Company, Inc., 1346 Willow Road, Menlo Park CA 94025

Solid waste disposal has traditionally been an economic cost to industry, institutions, and municipalities. Numerous disposal methods have been used with combustion, for purposes of volume reduction, i.e., incineration, an acceptable method in limited applications. The current energy picture in the United States has resulted in many industries, institutions, and municipalities taking a new view of solid wastes; in many instances they are now seen as energy sources.

One combustion approach which has received wide interest as a means of converting solid wastes to energy is the use of the fluidized bed combustor (FBC).

The fluidized bed reactor is a relatively new approach to the design of high heat release combustors. The primary functions of the air-fluidized inert bed material are to promote dispersion of incoming solid fuel particles, heat them rapidly to ignition temperature, and to promote sufficient residence time for their complete combustion within the reactor. Secondary functions include the uniform heating of excess air, the generation of favorable conditions for residue removal, and the ability to reduce gaseous emissions by control of temperature or by use as a gas-solids reactor simultaneously with thermal conversion.

The fluidized bed reactor greatly increases the burning rate of the refuse for three basic reasons:

1. The rate of pyrolysis of the solid waste material is increased by direct contact with the hot inert bed material.

2. The charred surface of the burning solid material is continuously abraded by the bed material, enhancing the rate of new char formation and the rate of char oxidation.

3. Gases in the bed are continuously mixed by the bed
    material, thus enhancing the flow of gases to and
    from the burning solid surface and enhancing the
    completeness and rate of gas phase combustion
    reaction.

A significant advantage of the fluidized bed reactor over
conventional incinerators is its ability to reduce noxious gas
emission and, finally, the fluidized bed is unique in its
ability to efficiently consume low quality fuels. The relatively high inerts and moisture content of solid wastes pose
no serious problem and require no associated additional devices
for their removal.

For over ten years Combustion Power Company has been conducting experimental programs and developing fluid bed systems
for agencies of the Federal Government and for private industry
and institutions. Many of these activities have involved
systems for the combustion of solid waste materials. Discussed
here will be three categories of programs, development of
Municipal Solid Waste (MSW) fired fluid beds, development of
wood waste fired fluid beds, and industrial installations.

MSW development work began in 1967 with a program to convert waste directly into electrical energy by combustion in a
pressurized fluid bed combustor (PFBC) and expansion of the
hot gases through an industrial gas turbine. This work was
funded entirely by the U.S. Environmental Protection Agency
(EPA) until 1974 when the U.S. Department of Energy (DOE),
then the Office of Coal Research, began funding tests on PFBC
coal combustion. Currently, Combustion Power Company has a
subcontract from Stanford University to conduct development
tests on a CPC owned fluid bed boiler (FBB) using MSW as the
sole fuel. Stanford has an EPA Research Grant to support this
work and much of the Grant money came from the DOE's Office of
Solar Energy and Conservation through an interagency transfer
of funds. All of the wood waste fluid bed development work
has been funded by the Weyerhaeuser Company or by CPC. This
work began in 1973 when Weyerhaeuser paid for a feasibility
study on the use of the PFBC to convert wood waste to electrical energy. In early 1974 a 100 hour wood burning test was
conducted using the EPA's PFBC equipment. In 1975 CPC became
a wholly owned subsidiary of Weyerhaeuser Co. and various
research programs have been conducted to date using the FBC as
a gas-solids reactor to suppress corrosion causing alkalai
vapors and to develop proprietary fluid bed hardware. A
current test program involves the use of wood wastes as fuel
for the thermal degradation of sodium based pulp mill sludge.

This research and development work on wood wastes has led
to the design and construction of two large industrial fluid
bed combustors. In one of these, a fluid bed is used for the
generation of steam with a fuel that was previously suited only
for landfill. Rocks and inerts are continuously removed from

this combustor using a patented system. The second FBC is designed to use a variety of fuels as the source of energy to dry hog fuel for use in a high performance power boiler. Here the FBC burns green hog fuel, log yard debris, fly ash (char) from the boiler, and dried wood fines to produce a hot gas system for the wood dryer.

The following sections discuss the above programs and applications in more detail.

Municipal Solid Waste Programs

CPU-400. In 1967 CPC began the development of an overall municipal solid waste management concept for the U.S. EPA known as the CPU-400 program. In 1970, construction began on an 82,000 kg/day pilot plant at CPC's development facilities in Menlo Park, California. The pilot plant consisted of four major subsystems; Figure 1 illustrates the interconnection of the solid waste processing and handling subsystem, the hot gas subsystem, and the turbo-electric subsystem, while the control system is both a part of each of the other three subsystems and it is also a separate subsystem in itself, causing the other subsystems to interact properly with one another and respond correctly to external commands.

In preparation for the pilot plant design and testing, numerous tests were conducted in subscale fluid bed combustors to develop feed and gas cleanup hardware and to evaluate process conditions. The majority of these tests were accomplished using a 0.20 m² FBC. Early testing of the pilot plant was conducted at atmospheric pressure conditons. Operation of the CPU-400 pilot plant on MSW at its design point of 4.5 atmospheres was short lived.

After performing a series of initial checkout runs at high pressure and determining nominal performance on solid waste, an attempt was made to satisfy a two-hour test requirement before proceeding to longer tests. In this test, operating conditions were sufficiently stable that a slow, steady increase in compressor discharge pressure was apparent. A review of data logged in this and preceding tests indicated that the effective flow area in the compressor turbine had apparently been reduced. Pressure drops across other hot gas system components, including the power turbine just downstream of the compressor turbine, had not changed. Since surge margin had been reduced to levels which precluded further successful operation, the compressor turbine stator sections were removed for inspection and cleaning.

The inspection revealed significant reduction of the nozzle area produced by hard deposits. The deposits were much thicker on the concave portions of first stage blades than on the second stage blades. The intervening first stage rotor

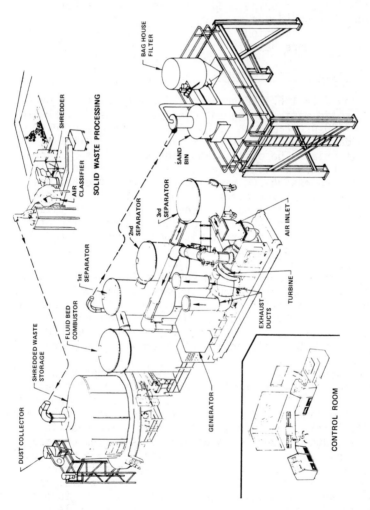

Figure 1. CPU-400 pilot plant

blades were also inspected and showed somewhat lighter deposits
uniformly distributed on concave surfaces. The lightest deposition occurred on the second stage rotor blades.

In order to identify the problem more precisely, testing
was initiated to determine the characteristics of the deposits.
Pilot plant tests were conducted on municipal solid waste fuel
at inlet temperatures of 1010 K, 867 K, and 839 K and using
aluminum-free fuel such as wood waste. Deposit analysis indicates that the aluminum and glass are the primary elements
involved in the deposit growth.

While the MSW fired CPU-400 system has not evolved to a
state of industrial application, much of the work accomplished
in that program has been the foundation for other work both at
CPC and elsewhere.

During the CPU-400 program a laboratory was established to
perform many of the required experiments on material samples
drawn from the process operation.

Laboratory samples and adjustments based on long term
pilot plant mass balance measurements have yielded the following
average weight distributions for the processed MSW during a 24-
hour and a 48-hour test period.

Ash-free combustibles	0.489
Moisture	0.351
Inerts	0.160

During a 48-hour low pressure test, a series of 24 bomb
calorimeter experiments in the laboratory on dried parallel
samples produced an average higher heating value of 16,480 kJ/kg.
Using an approximate ultimate analysis of $C_{30}H_{48}O_{19}$ for
the combustible fraction and converting to a lower heating
value based on the combustibles only, a corresponding average
value of 18,480 kJ/kg of combustibles is found. This result is
in very good agreement with values determined by applying heat
balance relations to observed pilot plant temperature and flow
measurements. It also correlates well with expected values for
cellulose-like material such as is approximated by the $C_{30}H_{48}O_{19}$
formulation.

The FBC in the CPU-400 was designed to operate under adia-
batic conditions. That is, no energy was extracted from the
FBC and thus the combustor ran at high excess air levels in order
to maintain bed temperature control. Typical steady state operating characteristics of the FBC, using MSW having the described
fuel properties, are shown on Table I.

A set of instruments was installed for on-line concentration
measurements of six specific constituents of the exhaust gas.
Concentrations of the seventh constituent, HCl, were determined
by laboratory titration of individual samples using the Volhard

method. The gas sampling, conditioning, and distribution systems
are integrated with the instruments and analog recorders in a
mobile, rack-mounted complex.

TABLE I
PILOT PLANT OPERATING CHARACTERISTICS

Operating Characteristic	Low-Pressure 48-hr Test	Typical High-Pressure Test
Solid waste feed rate	19 kg	45 kg
Moisture	38%	27%
Inerts	15%	18%
Combustor bed temperature	889 - 1000 K	967 - 989 K
Combustor freeboard temperature	1006 - 1089 K	1000 - 1028 K
Exhaust/turbine inlet temperature	1033 ± 6 K	1006 ± 6 K
Superficial velocity	1.6 m/s	1.8 m/s
Combustion efficiency	99% +	99% +
Power generated	0	1000 ± 100 kW

Measurements from the recent pilot plant test are presented
in Table II. All instrument records were relatively steady and
free from apparent anomalies considering the potential heterogeneous composition of the fuel form. The twenty-four discrete
laboratory samples (two-hour intervals) and subsequent HCl
analyses showed more variance from a low value of 84.5 ppm to a
high of 256.9 ppm.

TABLE II
PILOT PLANT EMISSION DATA

Constituent	Low Pressure	High Pressure
O_2	13.4%	16.1%
CO_2	5.8%	5.2%
CO	30 ppm	30 ppm
CH_x	0 ppm	2 ppm
NO_x	139 ppm	100 ppm
HCl	161 ppm	63 ppm

A 230 sq m solid waste receiving building was constructed
on a 510 sq m pad to provide for the 82,000 kg/day throughput
required during continuous operation of the pilot plant. The
processing facility utilizes two vertical axis shredders of
56 kW and 75 kW in size. Facility tests have been conducted to
insure the required throughput and a nominal rate of 4500 kg/hr
has resulted. The air classifier typically drops out about 16
percent of the processed material while another 6 percent
reduction results due to the moisture loss during the shredding
operating. A total of 934,000 kg of municipal waste has been
shredded, 481,000 kg in this processing facility.

The shredded and air classified solid waste fuel form is a
mixture whose visual appearance is homogeneous and dominated by
identifiable paper products. A typical kilogram of the material
consists of 320 g of water, 440 g of ash-free combustibles,
and 240 g of inerts (including ash). The latter two fractions
are typically subdivided as seen in Figure 2. As would be
expected, all of these values are subject to considerable
variation. As an example, moisture content in the final fuel
form ranges from 10 to 40 percent by weight depending upon the
origin of raw material (e.g., residential or commercial sources),
time of year, weather conditions, and municipal collection
policies.

SSWEP. In October of 1978 a program designated as the
Stanford Solid Waste Energy Program (SSWEP) was initiated.
CPC, under subcontract to Leland Stanford University, will
conduct combustion tests, at low excess air to simulate boiler
conditons.

The main objective of the test program is to burn processed
MSW in a fluidized-bed boiler environment to investigate combustion characteristics, heat transfer, corrosion and erosion, and
control parameters. The tests, are being conducted in a CPC-
owned 0.7 sq m fluidized-bed combustor which has been reworked
to incorporate water tubes for heat extraction in both the bed
region and exhaust. Additional tubes will be air cooled to
simulate boiler temperatures and to observe erosion and corrosion
phenomena. Metallographic examination will be made of these
materials.

The program plan is to accomplish 400 hours of testing
consisting of two (2) 50-hour tests and one (1) 300-hour test.
Material elutriated from the fluid bed will be recycled back to
the bed in two (2) tests to investigate its effect on fouling.
Offgases will be sampled for total particulate, particle distribution O_2, CO_2 CO, HCl, CH_x, NO_x, and SO_x, for characterization of
potential environmental impact.

The two 50-hour tests were completed and a test report
submitted to Stanford in early August 1978. These tests were
conducted at various operating conditions to probe the expected
boundaries of the operating envelope. Stable conditions were

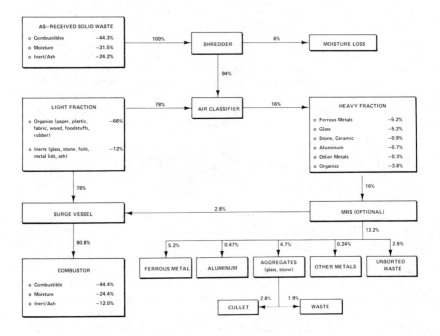

Figure 2.   MSW processing facility materials flow diagram

TABLE III
SSWEP PARAMETRIC TEST RESULTS
DATA SUMMARY

TEST POINT	SUPERFICIAL VELOCITY m/s	EXCESS AIR %	FREEBOARD PRESSURE kPa	TEMPERATURES		COMBUSTION EFFICIENCY
				BED K	FREEBOARD K	
SSWEP 1/1	2.2	26.6	2.5	992	1124	96.3
SSWEP 1/2	1.8	11.0	3.0	948	1123	95.6
SSWEP 1/3	1.9	34.1	2.7	980	1147	90.6
SSWEP 1/4	1.6	53.3	12.9	976	1136	96.2
SSWEP 1/5	1.6	53.1	23.6	965	1150	97.4
SSWEP 1/6	1.6	(1)	22.9	918	1062	97.0
SSWEP 2/1	2.1	40.5	3.5	982	1026	96.9
SSWEP 2/2	2.0	74 (2)	2.5	1029	1097	97.5
SSWEP 2/3	1.5	51 (2)	2.0	1026	1143	97.9
SSWEP 2/4	1.2	29 (2)	1.2	1016	1149	87.0
SSWEP 2/5	1.3	65.9 (3)	1.5	1039	1089	97.5

Notes: (1) Not recorded
(2) Calculated (Gas analysis sense line leaking)
(3) Orsat analysis

attained above an excess air level of approximately 40% and with
a settled bed depth of over 1 m. Table III contains a summary of the eleven operating points attained during these
two tests.

The MSW processing facility built for the CPU-400 program
was used to prepare the fuel for this test program also. Since
the single feed line to the 0.7 sq m combustor is only 7.6 cm
in diameter, the shredder choke bars were set to produce a
finer material size. Also, during SSWEP #1, numerous feed line
plugs resulted due to plastic lids of 7.6 cm diameter and larger.
Subsequent modification of the air classifier air damper settings
allowed the virtual elimination of these lids from the light-
fraction carry over. Properties of the derived fuel are shown
on Table IV.

TABLE IV
SSWEP PARAMETRIC TEST RESULTS
MSW PROPERTIES (AS FIRED)

	AVERAGE	RANGE
Moisture Fraction (%)	28.2	15.9 - 40.0
Ash Fraction(%)	11.9	8.2 - 16.0
Volatile Fraction (%)	52.2	44.1 - 59.8
Fixed Carbon Fraction (%)	7.6	4.7 - 10.0
Higher Heating Value (kJ/kg)	11,900	11,900 - 14,200
Sulfur (%)	0.15	0.12- 0.19
Chlorine (%)	0.16	0.12- 0.25

Wood Waste Programs

CPU-400 Test. In August 1973 a short checkout test was
conducted operating the CPU-400 pilot plant on waste wood. The
results were very encouraging and no turbine blade deposits or
erosion was detected. In January and February 1974, a 100-hour
demonstration test was run but in contrast to the previous
test, turbine blade deposition occurred. For purposes of
completion of the planned test, the problem was solved by
incorporation of a turbine cleaning system based on periodic
injection of milled walnut shells. Post test inspection revealed,
however, that erosion was significant and other measures would
be required for depend-able long duration operation. Studies
indicated that the problem depends strongly on the alkalai/silica
ratio of the wood content. Investigation showed that this
ratio is species dependent and that a favorable ratio exist in
Ponderosa pine bark (used in August 1973) but not in white fir
bark such as predominated in the 100-hour test. Table V shows
pertinent data from this test. Work has continued, over the
years, to obtain more information on the problem of alkalai
vapor formation and on a means to suppress this formation. A
new 0.5 sq m fluid bed was constructed and has been used to

conduct well over 1000 hours of tests using various additives to encourage the suppression of condensible alkali
vapors. The results have been encouraging but the details
are proprietary

TABLE V
CPU-400 WOOD TEST DATA

Total wood burn time	101 hours
Average load	905 kW (91,000 kWh)
Total wood consumes	254,000 kg
Average feed rate	42.2 kg/min
Average moisture	42.6% (40.3 to 45.9)
Average heating value	19,360 kJ/kg (18,840 to 19,900)
Average bed temperature	1004 K
Average freeboard temperature	1074 K
Average turbine inlet temperature	991 K

SO_2 ⎤
NO_x ⎥ Test averages
CO ⎥
CH_x ⎦

0.0
0.113 g/MJ
45 ppm
5 ppm

In a test program currently underway, dried wood fines
will be used as fuel for the thermal degradation of a sodium
based pulp mill secondary sludge. Key questions to be
answered are: What is the minimum temperature required to
destroy odors? What is the maximum bed temperature allowable
to avoid bed agglomeration and fusion? and, What are equilibrium levels of sodium and calcium in the bed and their
effects on process operation? These tests will be conducted
using the 0.20 sq m FBC.

Industrial Applications

Boiler Application. At the Weyerhaeuser facility in
Longview, Washington, a survey of potential fuels indicated
that a major quantity of energy was available in the form of
waste wood from their log sort and storage operation. This
material consists of the bark, limbs, and log butts that are
broken off during log yard operation. At the time of this
survey, this material was being cleaned off the log decks
and hauled off to landfill areas near Longview. Therefore,
by using this materials as fuel, landfill costs could be
reduced along with the reduction in power plant fuel costs.
Weyerhaeuser developed the specification shown on Table
VI for a combustor to utilize this energy resource.

TABLE VI
LONGVIEW FUEL SPECIFICATION SUMMARY

Debris Flow

Moisture (% wood)	55
Combustible (kg/hr)	5,390
Moisture (kg/hr)	11,000
Inerts (kg/hr)	3,600

Fly Carbon (Char)

Combustible (kg/hr)	540
Inerts (kg/hr)	540

Total Fuel Flow (kg/hr)   21,000

Further, the specification required the handling of limbs
and log butts up to 7.6 cm in diameter and to 0.91 m
long. Also, the inerts could come in the form of
rocks up to 20 cm maximum cross-section. CPC designed,
built, and installed an FBC with special features to meet
these specifications.
The installation and schematic are pictured on Figure 3.
The FBC specifications are shown on Table VII.

TABLE VII
LONGVIEW FBC SPECIFICATIONS

Internal Gas Flow Area	35.4 m^2
External Diameter	907 cm
Overall Height	768 cm
Bed Height	61 cm
Operating Exit Temperature	1260 K
Excess Air	65%

Debris fuel is fed to the combustor through a single above
bed feed point by use of a conveyor. Fly carbon is pneumatically injected. To avoid defluidization caused by an
accumulation of the large inerts, a patented rock removal
system continuously clears the bed of high density, over-
sized solids. The Longview FBC encountered numerous
difficulties during startup. During its first year of
operation major changes were made to improve system reliability. This combustor is now fully operational.
Drying Application. A second industrial FBC system
(Figure 4) has been designed and installed at Cosmopolis,
Washington to provide hot gas to a rotary dryer; the dryer
in turn provides dried wood fuel to a boiler that supplies
steam for inplant power generation. Drying the fuel

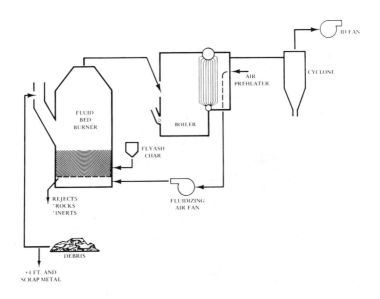

Figure 3.   View and schematic of FBC boiler installation

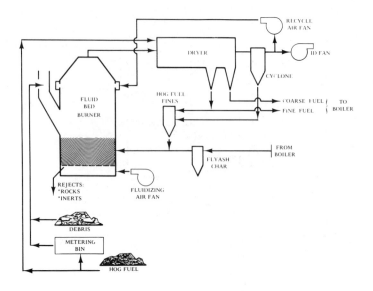

Figure 4.   View and schematic of FBC hog fuel drying installation

allows operation of the existing boiler at rated conditions without auxiliary fuel such as oil. This fluid bed combustor can be operated on mill wastes, on a portion of the fuel being supplied to the dryer, or on secondary sludge. This unit is rated at 70 GJ/hr. Fluid bed burners are typically limited to a turndown ratio of three to one; this dryer system, however, has been designed for a turndown ratio of four to one, accomplished by zone control of the fluidizing combustion air.

In this application, it is necessary to blend the FBC exhaust gas with recycled gas in order to reduce the gas temperature to the dryer. This blending is accomplished in the upper zone of the fluid bed unit in order to eliminate a separate blend chamber and to reduce the refractory requirements in the combustor. Figure 4 shows a view of this installation and its schematic.

These industrial systems are demonstrating the versatility of the fluid bed in using waste materials for fuel. Through such systems, it is possible to provide alternate sources of energy for mill operations.

RECEIVED November 16, 1979.

Energy Recovery from Municipal Solid Waste and Sewage Sludge Using Multisolid Fluidized-Bed Combustion Technology

WAYNE E. BALLANTYNE, WILLIAM J. HUFFMAN, LINDA M. CURRAN, and DANIEL H. STEWART

Battelle's Columbus Laboratories, 505 King Avenue, Columbus OH 43205

This study was initiated to investigate the potential for energy recovery from municipal solid waste and sewage sludge using Battelle's multi-solid fluidized bed combustion (MS-FBC) technology. The MS-FBC technology was chosen because it represents an advanced fluidized-bed combustion technology that is designed to operate with heterogeneous solids while achieving higher throughputs and larger turn-down ratios than conventional fluidization. The technology, originally developed for coal, was thought to be highly adaptable to the diverse solids that are inherently difficult to process when using municipal solid waste (MSW) and domestic sewage sludge (DSS).

Steam generation from MSW and DSS has been studied by many investigators, but fouling of heat transfer surfaces can be a major problem. Thus, this research study also sought to investigate direct steam generation using hot sand to vaporize the water in DSS (4 percent solids) in a conventional fluidized bed that is an integral part of the MS-FBC technology.

The research program incorporated several items and the two tasks discussed in this paper were:

Task 1. Obtain experimental data using a MS-FBC pilot plant to validate calculated energy recovery.

Task 2. Demonstrate the technical feasibility of direct steam generation using DSS as 4 percent solids.

Background on MS-FBC Technology

Multi-solid fluidized bed combustion (MS-FBC) was initially developed by Battelle(1) as an advanced fluidization technology for achieving high rates of combustion of high sulfur coal while eliminating the need for flue gas desulfurization or extensive coal cleaning. This coal technology was shown to be feasible during numerous runs in the 15 cm pilot plant system described in this paper and it was then further developed over a 1500

cumulative hour operating period using a 4.4 metric ton/day pilot
facility. Other feedstocks (e.g., wood) have also been shown to
be compatible with MS-FBC and the basic processing capability has
recently been adapted to the gasification of wood.(2).

The basic MS-FBC concept incorporates an entrained fluidized
bed superimposed on an inert, dense-phase fluidized bed as shown
in Figures 1 and 2 for the MSW/DSS application. As currently
operated in the MSW/DSS experiments, the dense phase is African
iron ore. This dense bed remains in the comb stor and its height
essentially defines the combustion zone (760-870 C); its high
density permits dense-phase turbulent fluidization to be achieved
at gas velocities exceeding 9.0 m/sec.

The entrained phase is commonly silica sand and its addition
to the dense-phase bed is generally credited with improving the
quality of the dense-phase fluidization. The entrained phase also
moderates the combustion by absorbing the heat of reaction. After
passing through the dense-phase region, the entrained phase along
with flue gas and ash pass through a free-board region and are
separated. The heated sand is then transferred a conventional
fluidized bed shown in Figures 1 and 2. This bed is operated at
gas velocities less than 0.39 m/sec. Steam tubes can be incorporated into this fluidized bed for the production of high pressure
steam (Figure 2). In one MSW/DSS application, no tubes are employed and direct-contact heat transfer between hot sand and a 3-4
percent DSS mixture is accomplished (Figure 1). The cooled, entrained material is then returned to the combustor where the cycle
starts over again. The valve shown in the sand return-line is
used to control the rate of the entrained-phase recirculation
which, in turn, controls the amount of combustion that can be
permitted in the combustor.

Some of the more significant features of the MS-FBC process
that have been experimentally proven at the 4.4 metric ton/day
scale using coal as a feedstock include the following items:

(a) High specific rates of combustion (grams feed/hr·cm^2
grate area) due to the high amount of heat removal achieved at
velocities greater than 9 m/sec.

(b) "Turn-down" of 3:1 can be achieved at 20 percent excess
air by adjusting the rate of recirculation of the entrained phase;
this recirculation control eliminates the need for bed-slumping
that is common to regular fluidized bed combustors.

(c) Different types and sizes of coal feed can be handled
with minimum feed preparation; lump-coal approximately 2.5 cm in
diameter is preferred.

Description of Equipment

A process flow schematic of the combustor, sand disengager,
and the EHE used in this investigation is given in Figure 3.

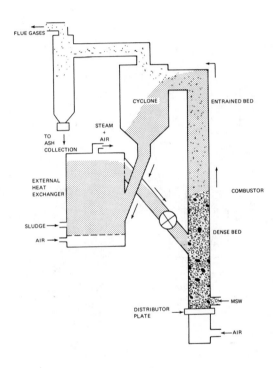

Figure 1. Battelle's MS–FBC process with contact evaporation of domestic sewage sludge

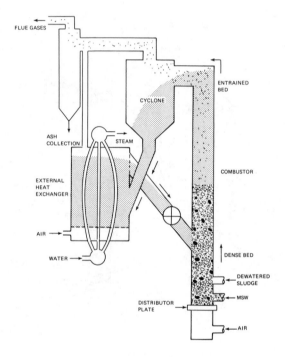

Figure 2. Battelle's MS–FBC process with steam generation in external heat exchanger

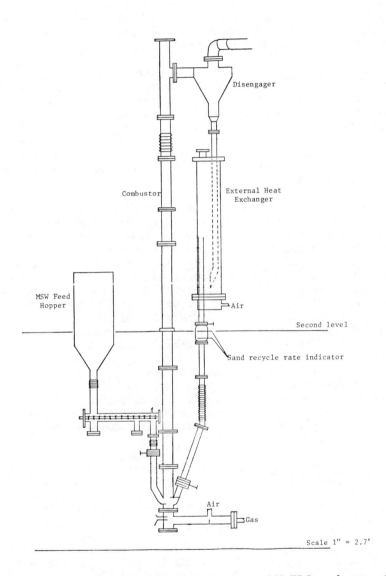

Figure 3. Thermocouple and pressure tap locations on MS–FBC combustor and EHE

MSW Feed System. The MSW feed system consisted of a sealed
weigh hopper supported by load cells, a 10.2 cm diameter by 102
cm long screw feeder powered by a variable speed DC motor, and
a 7.6 cm diameter feed chute fitted with an expansion bellows and
a knife gate valve. An air cooler was installed to keep the feed
chute near the combustor from overheating.

DSS Feed System. The DSS feed system consisted of an agi-
tated, 55-gallon polyethylene supply tank for the sewage sludge
and standard piping and pumps. The feed system supplied two air-
sparged tubes located 25 cm and 46 cm above the distributor plate
in the EHE. Each tube contained six, 2.36 mm diameter holes
drilled horizontally (3 on each side) through which DSS and air
were fed. Because the clay and lime additives used in the study
had a tendency to settle out in the supply tank, a recirculation
line was also installed.

Combustor. The combustor assembly of the MS-FBC system consisted of the combustion chamber and the gas-fired burner. The
combustion chamber consisted of a vertical, 15 cm I.D. by 6.7 m
high pipe constructed of Type 304 stainless steel. The distributor plate was a 0.63 cm thick disc located on the bottom flange
perforated with 96 uniformly-spaced holes of 0.238 cm diameter.
The recycled silica sand entered the combustor through a
7.62 cm diameter pipe located approximately 10 cm above the distributor plate. This pipe was installed at an angle of approximately 20 degrees to the combustor and was equipped with a slide
gate valve to regulate the sand flow from the EHE into the combustor.
All materials in the combustor and gas burner were of Type
304 stainless steel. Entry ports into the combustor were also
constructed from stainless steel. The combustor, sand disengager,
EHE, and sand recycle leg were wrapped with two layers of Fiber-
frax (registered trade name of Carborundum Corporation) insulation to reduce heat losses.

Sand Disengager. The sand disengager was a modified 61 cm
cyclone equipped with a spoiler into which air could be introduced to classify the fly ash from the heavier entrained sand.
Type 304 stainless steel was used in the construction of the
disengager.

EHE. The EHE consisted of a 35.6 cm diameter by 2.13 m high
fluidized bed. A sketch of the EHE is included in Figure 1. The
distributor plate was 0.63 cm thick and had 128 holes, each 0.238
cm in diameter uniformly spaced across the plate. An adjustable
overflow tube was located inside the EHE to regulate the height of
the bed. The EHE was constructed from Schedule 10 carbon steel
pipe.

Control System. The operating control system contained standard rotameters, manometers, gauges, thermocouples, recorders, etc., for controlling/monitoring the MSW feed air, EHE nozzle sparge air, disengager spoiler air, pressure tap purge air, natural gas flow, combustor air flow, EHE fluidizing air, bed pressures, DSS feed, and temperatures.

The flue gas monitoring system consisted of analyzers and recorders for continuous monitoring of SO_2, CO_2, CO, NO, and O_2 from either the combustor or EHE. A digital readout indicator provided the weight of MSW in the hopper. SCR controls were provided for the MSW screw feeder and the DSS feed pump.

Materials

The feedstock selected as the most applicable for the pilot facility was pelletized MSW. The pellets, obtained from the Teledyne National Company, had been preprocessed to remove metal and most of the moisture before pelletizing. A representative analysis based on an average of several analyses is presented in Tables I and II. The pellets were preferred over shredded MSW due to the relative ease of handling in the 10 cm diameter screw feeder used in the pilot facility. A larger combustor system would not be limited to the use of a pelletized feed.

TABLE I.   TYPICAL MSW ANALYSIS

Item	Weight %
C	39.90
O (by diff.)	33.98
H_2O	13.50
Ash	6.80
H	5.40
N	0.30
S	0.12
	100.00
Heating value (as received)	4252 cal/g
Bulk density	0.48 g/cm^3

The DSS injected into the EHE was obtained from the concentrator underflow at Columbus' Jackson Pike wastewater treatment plant. Within a few hours prior to each run, the sludge was loaded into 55-gallon drums which were trucked to Battelle and emptied into the DSS feed tank. The sludge was nominally 3 to 4 percent solids. A typical analysis of the metal content is provided in Table II. The heating value of the dried DSS solid was

TABLE II. TYPICAL CONCENTRATION OF METALS AND OTHER CONSTITUENTS
IN MSW, SEWAGE SLUDGE, AND SCALE

Compound or Element	Dried MSW Weight Percent	Sludge(a) Weight Percent	Disengager Scale (Run 213) Weight Percent
SO_4	0.16	0.05	
Cl	0.08	0.02	
Al	0.8	0.04	1.0
Fe	0.2		3.0
Si	2.0		90.0
Na	0.5	0.02	1.0
Ca	0.8	0.05	3.0
K	0.5	0.01	0.5
Mg	0.1	0.01	0.2
Mn, Pb	0.02-0.03		0.02
Cr	<0.02	0.01	<0.02
Ba,Ni,Cu,Ag, Zr,Sr	≤0.01 (each)		0.02
Ti	0.1		0.01

(a) Weight percent as-received basis from the sewage treatment
plant; 3.75% solids; 55% ash.

2220 cal/g and the ash content was 55 percent. During runs where
limestone or clay was used as an additive, it was mixed with the
DSS in the feed tank at 1 or 2 percent by weight. The clay and
limestone were of commercial quality.

African iron ore (-6+12 mesh) was used as the dense bed
material in the combustor and silica sand (-20+50 mesh) was used
as the entrained bed material.

Operating Procedure

Prior to each run, the MSW hopper was loaded with MSW pellets;
the DSS feed tank was filled and the lime or clay (when needed)
was mixed with the DSS; the flue gas analyzers, pressure, and
weight indicators were calibrated; and the combustor and EHE were
loaded with their respective dense and entrained bed material. A
650-675 C fluidizing gas stream was generated using a natural gas
burner and was used to heat the bed.

When the temperature of the dense bed reached 675 C, the MSW
feed was started. After the bed temperature reached 870 C, the
EHE sand was permitted to flow into the combustor to maintain 870
C within the dense bed. The natural gas to the burner was turned
off.

After a steady state had been reached, temperature and pressure data were collected as well as the MSW feed rate to the combustor, DSS feed rate to the EHE, fluidizing air to both the

combustor and EHE, ash collection rates, and flue gas analyses.
Flue gas EHE gas and ash samples were also collected.

Results

   Task 1:   Experiments with MSW and Water Feedstocks.   The first
part of the experimental effort focused on shakedown of the system
and demonstration of the feasibility of operating the MS–FBC with
the pelletized MSW and generating steam from water injected directly into the external fluidized bed.
   The initial runs, Runs 200–202, were startup and shakedown
runs in which operation of the process and analytical equipment
was tested.   Solids handling and recycle flow rates, temperature
constraints, and operational characteristics of the system were
also determined.
   After this startup phase, Runs 203 through 208 were conducted
while feeding pelletized MSW to the combustor and injecting water
into the EHE to generate steam.   In these runs, the goal was to
achieve the most stable, steady-state operating conditions and
maximize the energy recovery via steam generation.   Operating conditions were varied to determine the qualitative effect of such
items as fluidizing velocities feedrates, EHE temperature, and
EHE liquid distribution.
   By the end of Run 208, a total of 36.5 hours of system operating time had been logged during which both MSW and water were
simultaneously fed.   Superficial combustor air velocities were
varied from 7.6 to 11 m/sec and MSW feed rates were varied from
32.2 kg/hr to 39 kg/hr.   EHE fluidization characteristics and temperatures and superficial air velocities of 0.24 m/sec to 0.61 m/
sec.
   As a result of these preliminary process definition runs, the
following technical accomplishments and conclusions could be cited:
   1.   The process concept was shown to be feasible using pelletized MSW and water as feedstocks.
   (a)   Four of the runs lasted in excess of 7 hours, with
the longest run lasting almost 10 hours after which the system was
voluntarily shut down.   Steady state operation of the EHE at 150–
162 C totaled over 9 hours.
   (b)   Combustor temperatures could be satisfactorily controlled by varying the sand recycle rate.   Temperatures within the
combustor were relatively stable and constant within 100 C.
   (c)   The thermal inertia of the 400 pounds of recirculating sand dampened temperature excursions caused by feed rate
variations and helped insure reasonably uniform operation of the
system.   This thermal inertia was thought to be relatively important because of the non-uniformity of the energy content of the
MSW feed.
   2.   Direct steam generation in the EHE appears feasible.
   (a)   The seal legs through which the sand flowed into and
out of the EHE sufficiently isolated the EHE from the sand disen-

gager and combustor. This was evidenced by flue gas analysis of
the EHE combustor effluents as well as pressure in the EHE.

(b)   EHE bed temperatures remained reasonably constant
throughout the bed during water injection. A typical temperature
profile within the bed was constant within 15 C. This stability
attested to good liquid distribution within the bed. Steady-state
water injection rates into the bed reached 56.8/hr or approximately 0.6/hr of bed volume. This exceeded loadings of 0.5/hr
achieved in other liquid injection fluidized bed systems(3).

3. Key operating conditions for the system were established.

(a)   The maximum MSW feed rate for the combustor was
approximately 38.6 kg/hr or 0.21 kg/hr·cm^2. Above this value,
fluidization in the bed became non-uniform and combustion efficiency, as evidenced by flue gas analysis, dropped. The recommended
feed rate for the MSW appeared to be approximately 34 kg/hr or
0.18 kg/hr·cm^2.

(b)   The recommended operating superficial air velocity
in the combustor was between 9.1 and 10.7 m/sec. The recommended
superficial air velocity for the EHE was approximately 0.33 to
0.36 m/sec, based upon current data.

(c)   Stable operation of the EHE bed could be achieved at
temperatures between 135 C and 165 C. Below 121 C, the operation
became too difficult as the EHE suffered from localized quenching.

4. Experimental steam production Run 208 was shown to be
approximately 1.91 g/g dry, ash-free MSW as compared to an original
estimate of 5.5 g/g. When adjusted for calculated heat recovery
from flue gas (there is no economizer in the experimental system),
excessive heat losses from an uninsulated lower portion of the
disengager, and the difference between experimental steam temperature and the original assumption of 121 C steam, the adjusted experimental steam rate was 5.61 g/g which is close to the original
estimated rate(4).

Task 2:   Experiments with MSW and DSS Feedstocks. The second
experimental part focused on the operation of the MS-FBC system
while feeding DSS to the EHE. Unlike the previous runs made with
water and beyond nonprocess failures, operational problems with
plugging of the DSS injection ports and excessive plugging in and
around the disengager were encountered.

Plugging of the DSS injection ports occurred in Runs 210, 211,
and 213. In Runs 210 and 211, the problem was concluded to be
insufficiently sized orifices through which the DSS was injected.
These ports were enlarged. The problem in Run 213 was attributed
to using 24-hour old DSS which had increased in viscosity and
plugged the 1/2-inch injection lines. In all other runs, fresh
DSS was used daily.

Subsequently, scale formation in the disengager caused an
interruption in the flow of recycle sand and forced shutdown of
Run 215 after 5 hours of operation. There was no evidence of
plugging in the EHE. Slight scale formation, in fact was initially

observed after Run 213 and an analysis of the scale indicated
relatively high levels of sodium and magnesium (see Table II for
analysis). The scale formation was believed to be attributable to
the high alkali concentrations (particularly sodium ion) in the
DSS which formed low melting eutectics with the carbonates, sulfates, and chlorides present in the DSS and MSW.

For example, one eutectic of $Na_2Cl_2-Na_2SO_4-Na_2CO_3$ exists at
a temperature as low as 612 C. The eutectics are tacky and can
form scale on vessels or can cause agglomeration which results in
eventual defluidization of the bed. With a silica sand bed or
with SiO_2 in the sludge or MSW (about 2 percent by weight), the
Na_2SO_4 or NaCl could react as follows:

$$Na_2SO_4 + 3SiO_2 \rightarrow Na_2O \cdot 3SiO_2 + SO_2 + 1/2\ O_2$$

$$2NaCl + 3SiO_2 + H_2O \rightarrow Na_2O \cdot 3SiO_2 + 2HCl$$

to form a very viscous, sticky sodium silicate glass that could
also cause scaling and defluidization.

This scaling problem is not unique, however, to fluidized–bed
combustion systems that burn salty sludges. It is possible to reduce the combustion temperature below 612 C, but this temperature
limit affects combustor efficiency. One currently demonstrated
method of mitigating the problem without altering combustion conditions is to add a fine suspension of clay, lime compounds (5).

To evaluate the feasibility of forming these higher melting
silicates and minimizing the formation of scale in the MS–FBC
system, limestone ($CaCO_3$)was added to the DSS feed in Run 216 to
form devritite (MP=1030 C); the reaction is:

$$Na_2O \cdot 3SiO_2 + 3SiO_2 + 3CaO \rightarrow Na_2O \cdot 3CaO \cdot 6SiO_2\ .$$

The results of adding limestone were encouraging as the system was
operated with no ostensible plugging for 7 hours. The system was
then dismantled for a thorough visual inspection of the disengager
and connecting lines. Only slight traces of scale formation were
found. It is particularly important to note that 2/3 of the sand
used in this run was from Run 215. The use of this "old" sand was
thought to be a rather severe test of the assumed chemistry and
use of limestone. Thus, the absence of scale in the disengager
after Run 216 was thought to be quite significant.

In both of these runs, a stoichiometric amount (no excess) of
limestone was added as indicated in the above reaction. The amount
of sodium in both the MSW and DSS (Table II) was used for calculating the stoichiometry. The ratio of limestone to as–received
MSW and dry DSS was 0.047 g/g and 0.5 g/g, respectively.

Run 217 was intended to be an extension of Run 216 to demonstrate prolonged operation of the system. The system was operated
using sand from Run 216 for about 13 hours. Then, combustor pressures built-up, the MSW became difficult to feed, and the system
had to be shutdown. On dismantling the system, a plug was found
about half-way up the combustor. The plug contained high amounts
of silica and salt and may have existed prior to the addition of

limestone, gradually growing in size with each run. The belief
that the plug existed prior to Runs 216 and 217 was supported by
the fact that there was minimal scale formation elsewhere in the
combustor and disengager.

After cleaning the combustor, Run 218 was made. The major
goal of this experimental run was to show that agglomeration/
scaling could be minimized over an extended period by adding commercial clay to the DSS feed since clay would be a more economical
inhibitor. The clay is believed to react with the alkali salts
present in the DSS to form albite or nephaline(5), double alkali
salts (Mp-1108 and 1280 C, respectively) which melt above the combustion temperature. The reactions for the clay were assumed to
be as follows:

$$Na_2O \cdot 3SiO_2 + Al_2O_3 + 3SiO_2 \rightarrow Na_2O \cdot Al_2O_3 \cdot 6SiO_2$$
$$Na_2O \cdot 3SiO_2 + Al_2O_3 \rightarrow Na_2O \cdot Al_2O_3 \cdot 2SiO_2 + SiO_2 \quad .$$

Clay was added at the stoichiometric quantity needed for the
sodium reactions and the ratios to MSW and DSS were 0.0235 g/g as-
received MSW and 0.25 g/g dry DSS. During the run, considerable
difficulty with the MSW feed system was encountered. Upon shutdown and inspection of the MSW, it was found that large metallic
objects were present in the hopper and screw and were probably
hindering the MSW feed to the combustor.

During uninterrupted feed periods, the MS-FBC system operated
smoothly as evidenced by the temperature profile in Figure 4. Furthermore, as can be seen by the stable EHE bed temperature profile,
the stability of MS-FBC permitted quick recovery after the interruptions. Total operating time with feed was 31 hours and the DSS
feedrate was 38 l/hr or 0.4 l/hr·l of bed volume.

The objective of minimizing scale formation was met because
no significant build-up of sand or other materials could be detected in the combustor, external boiler, piping, and cyclones
after shutdown and inspection. The problem of agglomeration and
build-up was apparently minimized by the addition of clay.

Other Results

Relatively high CO levels prevailed throughout most of the
runs. This was not regarded as a serious problem and did not
receive much attention. That is, high CO levels are common to the
15 cm diameter pilot plant because it does not have sufficient
combustor freeboard height to ensure combustion of CO. This high
CO emission was also a deficiency in burning coal, but, in runs
made with coal in a larger MS-FBC pilot plant (4.4 metric ton/day)
with additional freeboard, acceptable emissions of CO were achieved
and the same result is fully expected to be true when burning MSW.
Emissions of NO_x and SO_2 were within Federal limits.

The combustor efficiency of the MSW and DSS in the experimental system was high and approached 99+ percent in some runs.

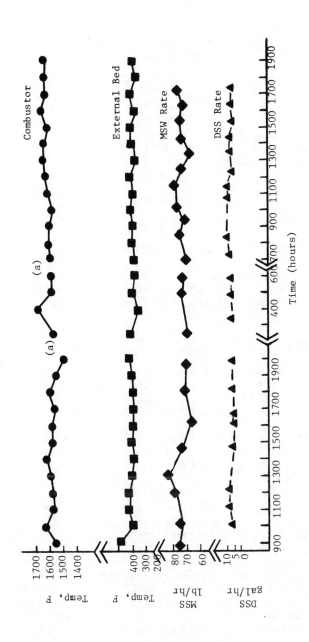

(a) Data interrupted due to difficulties with MSW feed
     caused by foreign objects in the MSW.

Figure 4. Six-inch MS–FBC operating conditions for run 218

This high efficiency is indicated by the low amount of unburned carbon that was determined to exist in various streams during Run 212 (Table III).

TABLE III.   ANALYSIS FOR ASH AND TOTAL CARBON
IN SOLID STREAM FROM RUN 212

Stream	Present Ash	Total Carbon
Recycled sand	99.7	0.2
Sand in EHE	99.7	0.2
Sand in EHE cyclone	99.1	1.9
Sand in combustor cyclone	98.6	1.3
Combustor	100.0	<0.01
Fresh sand	99.9	<0.01
Sand taken from disengager	99.9	<0.01

The amount of excess air required, with acceptable SO_2 and NO_x emissions, was approximately 20 percent. Conventional fluidized combustors have been operated with 50-100 percent excess air (e.g., Reference 6). Thus, the MS-FBC system may be an efficient means of combusting MSW, provided that toxic/hazardous organics are also destroyed.

Overview of Experimental Results

In general, the experiments with DSS feed to the direct-contact, external heat exchanger of the MS-FBC process were operated at conditions comparable to those established with the water feed. Thus, these two sets of experiments indicated that the use of MS-FBC technology for energy recovery from MSW and DSS applications remains to be established. It is noted that MS-FBC technology is being actively pursued by a private company using another feedstock. This process is competitive with other conventional systems for the intended application. In general, MS-FBC technology is considered to be a technically feasible option for MSW and DSS, but one that would require further investigation at a larger scale.

At this time, it is thought further investigations of MS-FBC for energy recovery from MSW and DSS should be concentrated on a larger scale and/or long-term operation (>150 hrs) to fully establish that (1) DSS with high salt concentrations can be processed, (2) environmental emissions requirements can be met, (3) shredded, not pelletized, MSW can be used as a feedstock, and (4) efficient energy recovery can be achieved even with diurnal changes. In addition, a complete and detailed economic evaluation should be undertaken.

Acknowledgements

This work was completed through U.S. DOE contract Number EM-78-D-01-5136. The overall management and direction of the project is under the Urban Waste Technology Program (D. Walter, Chief), Office of Conservation and Solar Applications. The technical monitor of the Project is Cynthia M. Powers, Argonne National Laboratories.

Literature Cited

1. Nack, Herman and Liu, Ke Tien, U.S. Patent 4,084,545, April 18, 1978, assigned to Battelle Development Corporation.
2. Feldmann, Herman F., "Conversion of Forest Residues to a Methane Rich Gas", Energy From Biomass Wastes Symposium, Washington, D.C., August 14-18, 1978.
3. Zenz F. A., Personal communication to W. E. Ballantyne, November 1978.
4. Huffman, William J. and Ballantyne, Wayne E., "Energy Sludge Using Multi-Solid Fluidized Bed Technology", Proposal, August, 1978.
5. Wall, C. J., Graves, J. T., and Roberts, E. J., "How to Burn Salty Sludges", Chemical Engineering, April 14, 1975.

RECEIVED November 16, 1979.

DENSIFICATION TECHNOLOGY

The Preparation and Properties of Densified Refuse-Derived Fuel

HARVEY ALTER and JAY A. CAMPBELL

National Center for Resource Recovery, Inc., Washington, DC 20036

Refuse-derived fuel (RDF) generally refers to the product of the mechanical (or chemical plus mechanical) processing of municipal solid waste (MSW) to produce a specification output. Densified refuse-derived fuel (d-RDF) is the product of the mechanical compaction of some form of RDF to agglomerated pieces which are sufficiently cohesive to sustain storage and handling. The term "densified" is used in the generic sense to include all forms of compaction, such as extrusion or rolling to produce briquettes, pellets, cubettes, etc. Generally, d-RDF is a fuel for stoker boilers analogous to lump coal. The concept of RDF and d-RDF can be extended to other waste-derived or biomass fuels (1).

The pilot-scale preparation of d-RDF has been described (2). This paper is to update and expand the information and to describe the objectives and operation of some of the unit processes used and the properties of the d-RDF. It should be an objective of processing to maximize the quality of the fuel, even at the expense of quantity, so as to maintain the fuel specification. Thus, a mass of waste, m, is converted to a mass of fuel, m', so that m > m', but the heats of combustion are $\Delta H_m < \Delta H_{m'}$, which is the objective of processing. The yield is limited by the composition of m and the limits of the law of conservation of mass-energy; also, allowance must be made for the energy input for processing (3). The concept of processing to maximize yield is opposite to the traditional objective of waste management of maximizing disposal.

The d-RDF Process Flow

The process flow used has been described (2). MSW (principally from household sources) was shredded, screened, air classified, shredded a second time, and passed through an animal feed densifier modified for the purpose. The primary shredding has been described (4); brief descriptions of the other unit processes follow. The major pieces of equipment are described in Table I.

0-8412-0565-5/80/47-130-127$05.00/0

Table I. Description of ETEF d-RDF Preparation Equipment

ITEM	MANUFACTURER - MODEL	SOURCE	HP - DESCRIPTION	FUNCTION
1. Primary Shredder	Williams Patent Crusher Co. Model 780	D.C. Gov't.	1000 hp horizontal hammer-mill, 24-230 lb hammers, grates 9 x 12 in. 720 rpm (no load)	primary shredder
2. Screen Feeder	Triple/S Dynamics Texas Shaker (with ball tray)	Mnfr. loan	7 x 20 ft, 10 hp, 1/4 in. punched plate on 7/16 in. staggered centers	screen shredded MSW to remove fines
3. Air Classifier	NCRR - Research Unit	NCRR built	20' high x 84" wide x 16" deep w/ baffles; 125 hp fan for 18,000 ACFM nominal flow	light/heavy separation to remove metals & glass & concentrate organics
4. Cyclone	Fisher-Klosterman Model XQ52 plus Carter-Day Rotary Valve Model AN-24	Purchased	27' high, 8 1/2' long x 6' diam. barrel. De-entrainment box 5' x 4' x 4' w/24" diam. outlet. Rotary valve is 22" diam., 2 hp	de-entrainment of light fraction
5. Pneumatic Transport System	Nadustco Fan & Cyclone	Purchased/ NCRR built	10,000 ACFM fan - 60 hp; positive pressure - 14" sq. duct	transport light fraction to secondary shredder
6. Secondary Shredder	Heil Model 42 D	Mnfr. loan	verticle shaft - modified w/ stators, 150 hp	size densifier feedstock
7. Livebottom Feeder	California Pellet Mill	Lease	52" long x 52" wide x 36" high, 4 screws, 7 1/2 hp, vari-speed	surge capacity and feed rate control
8. Densifier	California Pellet Mill Model 8162-B1	Lease	30" I.D. die, 1/2" holes, 170 rpm. - 150 hp; 10" & 13" O.D. corrugated rolls; 7 1/2 hp Even Flo feeder and 15 hp mixer screws	

Screening. A vibratory screen was used as a feeder for the shredded MSW to the air classifier. It was also intended to remove material <6.4 mm as a means of reducing the ash content of the RDF. Small particulate inorganic material in shredded MSW will be conveyed by the air stream along with the fuel fraction because the elutriation velocities are generally in excess of the settling velocities of the particles. Fan (5) showed that for a simple model of an air classifier, operating at 20 m/s, the light fraction will contain ~5 wt% inorganic particles, ~5 mm in size. The shredded MSW used to prepare the d-RDF contained 30.4 wt% material <6.4 mm for the average of 10 determinations with a CV = 28.4% in agreement with prediction (4), of which 82.6% was inorganic (average of 14 determinations, CV = 8.92%) (6), so there is an obvious advantage to removing this material. (CV = coefficient of variation = (standard deviation/mean) x 100.)

The flat vibrating screen used had to be partially blocked to avoid air leaking into the air classifier. The effective screening area was 2.1 x 2.4 m and apparently easily blocked by the feed of flake-like particles, at least judging from the relationship found between feedrate, F (short tons/h) and recovery of material <6.4 mm, $0 \leq R \leq 1$, which was $R = 0.64/F^{0.42}$. The results indicate that a screen at some point can remove a significant quantity of ash-forming material but that the type of screen used in the preparation of the fuel described here was improper. A rotary screen may be an improvement.

Air Classification. An adjustable configuration, vertical zig zag air classifier was used; the dimensions are given in Figure 1. Detailed results and methods of evaluation have been described (7). By way of illustration, Figure 2 describes the yield of light fraction as a function of elutriating air velocity. Configurations 2 and 4 appear to be useful compromises in terms of producing a high yield of fuel fraction at low air velocities, hence with a lower probability of elutriating large quantities of inorganic particles.

Secondary Shredding. The sizing of the feedstock and its proper metering significantly affect pellet mill performance, wear, and product quality. The primary shredded air classifier light fraction must be further decreased in size prior to densification to reduce milling action on the material by the die and roller assemblies. Otherwise, there are increases in wear and power consumption and a decrease of pellet mill capacity. Also, smaller size feed leads to higher bulk densities and improved flow.

During early operations of the secondary shredder, several configurations of swing hammers and fixed stators were tried to reduce the particle size of the product, particularly random textile pieces, to a level that minimized jamming and stalling of the densifier.

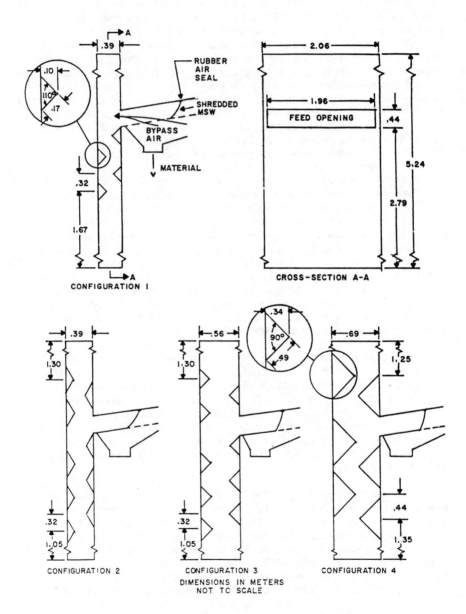

Figure 1. *Vertical air classifier test configurations (not to scale; all dimensions in meters)*

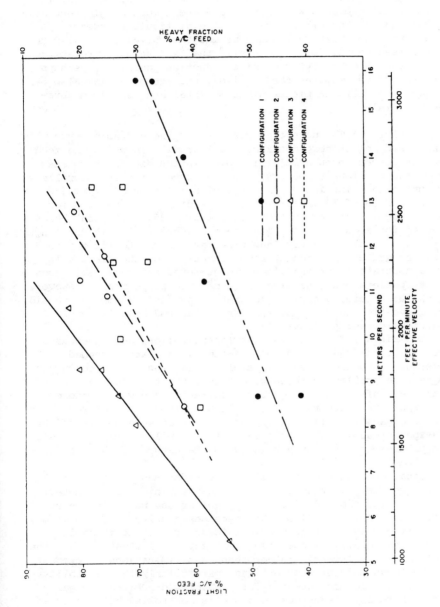

Figure 2. Research air classifier light fraction yield as a function of air velocity

The particle size distribution of the secondary shred product is shown in Figure 3 in Rosin-Rammler form, similar to primary shredded material (4). Although 88 wt% of the material is
<32 mm, textile pieces were larger and are almost singly responsi-
ble for jamming the mill. Textiles are difficult to shear
through the die apertures, wedge behind the rollers, and slip until the feed cavity fills and stalls the feed screw or rotating
die. Although other types of size reduction equipment, perhaps
knife shredders, may cut the textiles, the equipment may also be
damaged by occasional pieces of tramp metal and wear from abrasive materials.

Densifier. The densifier used was a ring extrusion type
animal feed pelletizer, described earlier (2), in which the feedstock is forced through multiple holes in the die as it rotates
around two stationary rollers. Several types of dies and rollers
were utilized. The inside diameter of all dies was 762 mm (30
inches). Dies of 13 mm (1/2 inch) hole diameter were used for
most of the 1270 Mg production. A die with 25 mm (1 inch) holes
was used for about 100 Mg. The die thickness (which is the length
of the die land) and entry land taper are chosen to fit the feedstock properties and product requirements. For the small hole
die, a moderate taper and 114 mm length produced reasonably good
quality pellets over the wide range of feed properties. However,
for the 25 mm die, a similar moderate taper and 127 mm length did
not. Investigation of a more severely tapered hole is required
to achieve denser pellets.

Two sizes and two surface styles of rollers have been used.
Of two rollers (254 and 330 mm diameter), the larger seemed
preferable in being less susceptible to jamming. Both corrugated
and textured roller surfaces were tried; the textured was polished
by the feed after only a few hours operation and was abandoned.

Experience with the pellet mill highlighted concerns with regard to capacity, reliability, and wear, that were unexpected or
not experienced in other applications of similar equipment. All
such problems are believed related to the nature of the feedstock
(8).

Usual operation of the mill (except for MSW) is for the
largest particle size to be smaller than the die hole diameter.
For the application described here, 40% of the particles were
larger than the die. Also, the feedstock density, 16 - 48 kg/m^3,
is significantly lower than animal feeds, resulting in an increased sensitivity of operation to feedrate fluctuations and the
likelihood of volumetric rather than mass limitations on capacity.

The mean moisture content, and particularly the variability
in moisture, are much greater for producing d-RDF than animal
feeds. Normally, feedstock moisture is controlled between 12 and
18 wt%.

The content of inorganic particles (hence abrasiveness) when
producing d-RDF is high, unless there is effective screening. If

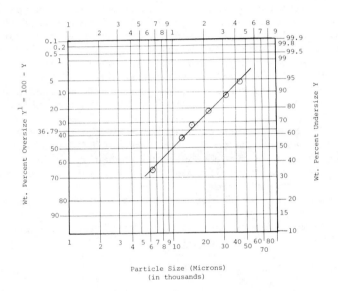

Figure 3.   Particle size distribution of secondary shred light fraction—Rosin–Rammler

not, the quality of the pellet can be degraded and the machine
wear increased.

Owing to a combination of these factors, particularly particle size and feedstock density, the capacity of the pellet mill
used is believed to be 4 - 5 Mg/h. At higher throughputs, the
frequency of jams due to feed surges or textiles is too high.

Power Consumption. A previous report (2) showed the power
consumption of the pellet mill as a function of throughput. Subsequent work has shown there were errors in these measurements and
the result is hereby withdrawn. Later work, using processed
office waste with added moisture (75% of the samples contained
10 - 30 wt%, 22% of the samples contained 30 - 50 wt%, and the remainder 10 wt%) showed that the relationship between power consumption P(kW), and feedrate to the densifier F (short tons/h, dry
wt basis), could be represented by: $P = 50.5F + 21.7$, for 63 determinations, $0.3 \leq F \leq 1.7$ Mg/h, with a correlation coefficient
$r = 0.964$ and a standard error of the regression line of 5.7. The
intercept should be equal to the idling power of the machine which
was separately measured as 24 kW. There was no discernible relationship between power consumption and moisture content.

Wear. Measurements of wear rates on the rollers and dies
were attempted but due to frequent changes of dies and rolls, related to changes in production plans or equipment failures, there
are no statistically verifiable data. One set of rollers, which
was used with several dies, lasted 500 Mg. One die, which had
been ground twice during use and used with several sets of new
rolls (which might promote the wearing of both), had a life of
about 1000 Mg. Removal of inorganic particles would decrease
wear.

Physical Properties of the d-RDF

During the 2-1/2 year investigation and production period,
the d-RDF was characterized by measuring moisture, ash, pellet
and bulk densities, pellet length, content of fines and integrity,
properties chosen because of their probable relationship to the
use of d-RDF as a stoker fuel. The values of these properties,
and the relation to corresponding values for coal, are necessary
to predict and understand results when d-RDF is transported,
mixed, and fired.

The measurement techniques were chosen to parallel, as much
as possible, similar ones used for coal. Most test methods con-
formed to ASTM procedures where available or adaptable to d-RDF.
The sampling and analysis procedures for each property measurement are reported elsewhere (9).

The average properties for the d-RDF produced are summarized
in Table II for the two sizes (diameter) of pellets. A previous
report displayed such data as histograms, giving the distribution
of properties (2).

Table II: Physical Properties of Densified Refuse
Derived Fuel as Produced

Property	1/2" Diam. Pellets			1" Diam. Pellets		
	$\bar{x}$	(n)	CV,%	$\bar{x}$	(n)	CV,%
Moisture, %, D.W.	22.9	(40)	38.9	22.8	(13)	21.9
Ash, % D.W.	23.4	(55)	17.5	22.8	(13)	17.1
Pellet Density, g/cm^3	1.01	(16)	44.5	0.76	(7)	19.7
Bulk Density, kg/m^3	575	(49)	12.5	432	(10)	6.7
-3/8" Fines, % A.R.	14.9	(15)	47.6	9.6	(10)	52.1
Mean Length, mm	15.9	(15)	15.7	-	-	-

Notes: D.W. = dry weight
A.R. = as received
$\bar{x}$ = mean of property
n = number of determinations
CV = coefficient of variation

The data of Table II indicate that pellet production is not
sensitive to contained moisture over a wide range. Some moisture
is needed for die lubrication and pellet cohesion; high moisture
content results in a scaly, loose-bound pellet which breaks more
easily in handling than others. Considering both equipment operations and pellet properties, the preferred moisture content is
judged to be in the range 15 - 20%.

The ash content is a combination of the intensive (inherent
ash in combustible material) and extensive (free inorganic materials). Intensive ash is approximately 8 wt%, illustrating the
potential for screening the feedstock someplace in the process
flow.

Pellet density is important to assure proper flow through
the storage and volumetric feed systems of stoker boilers. The
density will also affect burning rate in the boiler. The particle density of coal is of the order of 1.3 g/cm^3. Table II indicates a mean pellet density 10 - 20% below that for coal.

Lower moisture tends to produce higher pellet densities,
probably as a result of increased friction through the die and
thus greater compaction of the feed. Higher moisture reduces
friction, absorbs some of the heat of compaction, and results in
poorly bound pellets. Die hole inlet taper appears to be the
major control of pellet density; a more severe taper preferred
for denser pellets.

High bulk density is necessary because fuel is fed by volume but heat input to the boiler is by mass. For stoker coal, the density ranges 720 - 880 kg/m^3 (45 - 55 lbs/ft^3). Table II shows the 13 mm pellet bulk density to be 20 - 30% less but with the narrowest range of any property (lowest CV), a beneficial consistency for boiler feed control. The bulk density of the 25 mm pellets was much lower, as discussed above in relation to the die taper. However, there was a short production period with another die that produced denser 25 mm pellets.

The fines content is arbitrarily defined as <10 mm material, instead of <6 mm as with coal. This conservative definition is more inclusive of flaky particles in d-RDF. A low fines content is desired for increased bulk density, improved material flow, and reduced dusting. However, as with coal, some fines are necessary to achieve a mix of suspension and on-the-grate combustion. Table II lists an average fines content somewhat lower than the 20 - 40 wt% range typical of stoker coal. Note that these measurements were taken on the as-produced pellets, and any abuse during transport, storage, and feeding will increase the fines in the fuel as fed to the boiler. However, the pellet mill output can be screened, and the screen underflow returned to the mill, as is done with animal feeds to reduce fines.

The pellet lengths were limited by the position of the breaker plate installed on the interior surface of the cover of the die. However, the centrifugal force from the rotating die and impacts from loose pellets within the shell cover of the die, cause the pellets to break off at random lengths. Also, included pieces of plastics and textiles form planes of weakness and the pellets break shorter.

The integrity of d-RDF can be defined as its ability to maintain its dimensions (hence its bulk density and fines content) during transport, storage, feeding, and other handling, not forming dust, and meeting specifications. To evaluate the relative amount of abuse pellets can withstand without loss of integrity, a drop-shatter test was adapted from a test for coke (ASTM D3038-75). In this procedure, pellets are dropped on a steel plate from a 6 foot (1.8 m) height. Drop-shatter tests were run on 13 and 25 mm pellets from production and from storage. The change in distribution of pellet lengths within a sample, both before and after the drop-shatter, is shown as Figure 4. Some results are in Table III.

Table III shows that for the as-produced 13 mm and 25 mm pellets there is no significant change in mean length (Student t test and F variance ratios). Curiously, the distribution shifted significantly (χ^2) indicating skewing. Judging from Figure 4 and other data (9), both symmetric and assymetric breaking of larger pellets seems to be occurring. Although only two results are shown for as-produced pellets, the same result was observed on a much larger number of trials. This led to increasing the number of drops from 2 to 4 to 10, still with no significant changes in mean length, but a steady increase in fines. Apparently, after

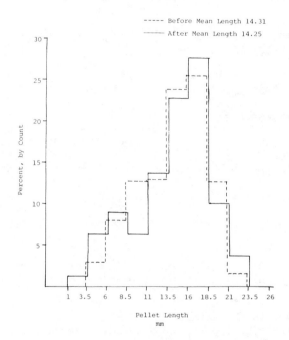

Figure 4. Pellet length distribution sample A (½ in. diameter—2 drops)

a certain level of abuse, most weak pellets have broken and fur-
thur abuse simply chips pieces from the ends. Similar data for
pellets of high moisture content show them to be more susceptible
to breakage, as expected.

Table III: Drop-Shatter Test Results

Tests of Distribution: Variance ratio (F test); Means (t test);
    Distribution (χ^2 test)

Sample	Description	Moisture	Statistical Test			Mean Length	Fines
			F	t	χ^2	$\overline{x}$	f%
A	13 mm, as produced	11	–	–	–	14.31 mm	9.8
	2 drops		*	*	**	14.25	10.0
B	25 mm, as produced	21	–	–	–	11.50	11.5
	2 drops		*	*	**	10.36	12.0
C	13 mm, storage yard	15%	–	–	–	10.28	17.9
	2 drops		*	*	**	9.32	22.1
	4 drops		*	*	**	9.38	24.2
	10 drops		*	*	**	9.59	25.6
D	13 mm, storage yard	34%	–	–	–	6.90	N/A
	2 drops		*	*	**	5.98	N/A
	10 drops		*	*	**	5.46	N/A

 * Hypothesis of no change accepted at 95% level of certainty
** Hypothesis of no change rejected at 95% level of certainty

A practice in the animal feed industry to control integrity
is to immediately cool pellets as-produced, resulting in harder,
tougher pellets. Immediately cooled d-RDF has been observed to
harden, as well, and this practice should be considered for full-
scale facilities.

Fuel Properties

Samples of the feed to the densifier were supplied to ASTM
E-38.01 as the material for round-robin analysis of RDF proximate
and ultimate fuel properties. The samples were generated on
three separate occasions over a period of 17 months by diverting
the output from the secondary shredder. A pile of approximately
100 kg was repeatedly coned and quartered and aliquots of 1 to 2
kg were packaged in double polyethylene bags and tied. Several
bags of each sample were sent to participating laboratories.

The analytical results are summarized in Table IV, listing ultimate and proximate properties, and in Table V, listing ash fusion temperatures. Both tables combine the results for inter- and intra-laboratory replicates. E-38.01 will report the results of statistical analysis and precision statements (10). Table IV includes values for the CV of just Sample 3 as an indication of the precision of these measurements.

Table IV: Properties of Refuse-Derived Fuel

Property	Sample 1 6 labs $\bar{x}$	(n)	Sample 2 6 labs $\bar{x}$	(n)	Sample 3 12 labs $\bar{x}$	(n)	CV,%
Calorific Value* MJ/kg	17.7	(24)	15.9	(32)	17.4	(48)	3.54
(Btu/lb)	(7624)		(6848)		(7491)		
Moisture, wt %	20.53	(24)	19.81	(32)	30.09	(44)	5.66
Dry Weight Basis, wt %							
Ash	15.70	(28)	25.28	(32)	22.17	(48)	11.7
S	0.34	(28)	0.24	(32)	0.48	(48)	17.4
C	46.33	(32)	40.02	(32)	42.60	(44)	3.14
H	6.17	(32)	5.37	(32)	5.84	(44)	4.98
N	0.76	(32)	0.64	(28)	0.77	(44)	9.21
Cl_t	0.58	(20)	0.38	(28)	0.57	(44)	40.0
Cl_s	0.35	(20)	0.24	(24)	0.27	(42)	25.1

Notes: Cl_t = total chlorine; Cl_s = determined as Cl^- after water extraction

n = number of determinations   * Dry basis
CV = coefficient of variation
$\bar{x}$ = mean of property

Table V: Average Ash Fusion Temperatures, d-RDF (ASTM D-1857) Average of 20 determinations by 5 laboratories

	Reducing Atmosphere	Oxidizing Atmosphere
Initial Deformation	1024°C	1065°C
Softening	1063	1092
Hemispherical	1097	1131
Fluidity	1182	1193

Storage of d-RDF

A large quantity (ca. 1000 Mg) of d-RDF had to be stored and stockpiled for a test burn for periods of ca. 1 year. Storage was outdoors, under cover of plastic sheeting, on a 100 m^2 asphalt pad. The pile was originally 2 m high and later increased to 4 m. Moisture evaporated continuously and some condensed on the under surface of the plastic sheeting, forming a "cap" 15 - 30 cm thick of wet pellets on the surface of the pile.

The temperatures at several locations inside the pile were generally 40° - 70°C, with perhaps a trend of lower temperatures in older parts of the pile, but no clear relationship between depth or age of the pile and temperature.

After 10 months of storage, several small smoldering seams (about 100 x 100 cm in cross-section) of pellets became obvious on the sides of the pile and caused ignition of the plastic sheeting and of the tires used as anchors. The pellets themselves did not ignite, perhaps because of their close packing and limited oxygen access. The seams ran deep into the pile; the affected pellets were black and brittle, presumably partially oxidized or pyrolyzed. The seams were in both the old and new parts and adjacent to both wet (>130 wt% moisture) and dry (<4 wt%) d-RDF.

The reasons for the formation of the "hot" (>150°C) seams is not clear. Conjecture is that the biological exotherm from decomposition (composting) caused a chemical exotherm, perhaps from an oxidant in the pellets. The d-RDF was made from household waste and some few portions may have contained an oxidant, such as nitrates in garden fertilizer.

Samples of pellets (~2 kg) in open mesh bags were so placed in the pile that they could later be retrieved and the properties of the d-RDF measured. Typical results are shown in Table VI illustrating what was generally observed; both moisture and fines content may increase or decrease with storage.

Table VI: d-RDF Storage Effects

Storage Duration	Moisture wt%	-3/8 Fines % A.R.	Mean Length mm
		Sample 1	
0	35	10	13.5
5 Weeks	26.2	12.5	12.3
	25.4	11.2	12.9
	11.5	9.6	12.2
	50.0	14.3	11.9
	20.0	9.0	1.3
	10.2	17.0	9.5
		Sample 2	
0	15.5	6.7	16.8
21 Weeks	4	8	10.8
	32	7	11.3
	57	27	8.8

Importantly, d-RDF from the storage pile was used for a large test burn in an industrial stoker plant with only minor materials handling problems (e.g., dusting from open conveyors) and no fuel problems. Nonetheless, it is not recommended that d-RDF be stored for 12-14 months or stored in deep piles.

Acknowledgments

The work reported here was supported by the U.S. Environmental Protection Agency, Office of Research and Development, through the Municipal Environmental Research Laboratory, Grant R80415001, Mr. Carlton Wiles, Project Officer, and Grant R803901, Mr. Donald Oberacker, Project Officer.

Thanks are due our colleagues, particularly Mr. John Arnold, Mr. Ralph Reinoldi, and Mr. William Schlag.

The pellet mill was supplied through the courtesy of the California Pellet Mill Co. The secondary shredder was loaned by the Heil Co. and Newell-Dunford, Ltd. We thank the District of Columbia, Department of Environmental Protection, for their cooperation.

Literature Cited

1. Reed, T. and B. Bryant, Densified Biomass: A New Form of Solid Fuel, SERI-35, Solar Energy Research Institute, Golden, CO, 1978, 30 pp.
2. Alter, H. and J. Arnold, "Preparation of Densified Refuse-Derived Fuel on a Pilot Scale," in Proceedings, Sixth Mineral Waste Utilization Symp., E. Aleshin, Ed., U.S. Bu. Mines and IIT Res. Inst., Chicago, 1978, pp. 171-7.
3. Alter, H., Science, 189, 175 (1975).
4. Stratton, F.E. and H. Alter, J. Environ. Engr. Div., ASCE, 104, 93-107 (1978).
5. Fan, D.N., Resource Recovery & Conservation, 1, 141-50 (1975).
6. Alter, H., Materials Recovery from Municipal Solid Waste: Air Classification, Upgrading RDF, Aluminum and Glass Recovery. Final report to U.S. Environmental Protection Agency, Office of Research and Development, Grant R803901, Cincinnati, Ohio, 1979; ibid., paper presented at the Fifth Annual Research Symp., Land Disposal and Resource Recovery, U.S. Environmental Protection Agency, Office of Research and Development, Orlando, FL, March 1979, proceedings in press.
7. Campbell, J.A., in ibid., Chp. 3.
8. Wiles, C., Paper presented at the Fifth Annual Research Symp., Land Disposal and Resource Recovery of Wastes, U.S. Environmental Protection Agency, Office of Research and Development, held in Orlando, FL, March 1979, proceedings in press.

9.	Campbell, J.A., *Preparation and Properties of Densified Refuse-Derived Fuel*, Final report to U.S. Environmental Protection Agency, Office of Research and Development, Grant R80415001, Cincinnati, Ohio, 1979, in preparation.
10.	Love, J. and C. Wiles, Paper presented at the Fifth Annual Research Symp., Land Disposal and Resource Recovery of Wastes, U.S. Environmental Protection Agency, Office of Research and Development, held in Orlando, FL, March 1979, proceedings in press. Also, the ASTM E-38.01 development of RDF test methods is supported by EPA Contract 68-03-2528, Office of Research and Development, Municipal Environmental Research Laboratory, C. Wiles, Project Officer.

RECEIVED November 16, 1979.

The Eco-Fuel II Process: Producing a Storable, Transportable Fuel for Co-Combustion with Oil

FLOYD HASSELREIIS

Combustion Equipment Associates, Inc., 555 Madison Avenue, New York, NY 10022

Background

The oil embargo of 1972 woke up a nation which was complacent about its energy resources but at the same time rapidly becoming aware of its environment. The sudden increase in fuel prices directed our attention to the values lying untapped in our solid wastes, and the synergism of converting a waste disposal problem into an energy resource.

At that time, CEA had just completed a state-of-the-art facility to incinerate fumes from six plastics-treating ovens using fossil fuels in a thermal oxidizer, and recovering the heat to generate the steam needs of the Spaulding Fibre plant in Tonawanda, New York, (designed, built and operated by CEA because industrial firms are reluctant to get into development of new technology foreign to their normal business). During the contract negotiations, the year before, it was discovered that disposal of both liquid and solid wastes from this plant were costly to dispose of and valuable as a fuel: consequently the plant was designed to make use of these fossil fuel substitutes, and has done so since 1972.

Combustion of liquid wastes in a Thermal Oxidizer turned out to be relatively simple compared with the problems encountered in processing and burning solid industrial wastes, including phenolic resins, impregnated with paper and canvas, and various forms of paper. Once these problems were overcome, requiring among other things development of the Double-Vortex burner to burn shredded solids in suspension, other fuels were accepted, including waste hardwood, and tires and battery cases in limited quantities. These activities led CEA to enter into a contract with East Bridgewater, Ma, to build and operate a Municipal Waste Disposal Facility.

0-8412-0565-5/80/47-130-143$05.00/0
© 1980 American Chemical Society

The East Bridgewater facility (also serving Brockton), was built with an incinerator, (conventional except that it has bag filters for emission control), and as an alternate disposal means, a Resource Recovery facility. In 1972 CEA entered into an agreement with Arthur D. Little (ADL) to develop processes for emission controls from stationary sources and also for treatment of Municipal Solid Wastes.

The East Bridgewater facility was originally built to produce a Refuse-derived-fuel (RDF) using a process along the lines of the Bureau of Mines College Park Pilot Plant, using a Flail Mill, horizontal air classifier, secondary shredder, storage bin, and finally a Rader pneumatic conveying system to feed the minus-1-inch shredded fuel directly to a Double-Vortex burner for suspension firing. There was no economic incentive to the operation of this process, since it was devoted to disposal without recovery of the heat value of the fuel. In addition, the amount of MSW delivered to the plant turned out to be half the projected amount. Coupled with the many difficulties encountered in processing MSW, the ECO-FUEL®I process was essentially abandoned and attention devoted to developing a better process to make a marketable fuel product.

The RDF production using the two-stage shredding process and the product RDF had many disadvantages:

- Shredders have high operating costs
- High moisture fuel is hard to burn efficiently
- High ash fuel causes excessive slagging in boilers
- Storage retrieval and transportation is difficult and expensive due to high moisture, low density and poor flow characteristics
- Fuel has low heatig value due to high moisture and ash
- Abrasion is excessive due to high glass content
- It is difficult to control combustion due to variable moisture and particle size.
- Low Energy recovery efficiency is low

Research and development work started in 1972 led to the conclusion that while many of the above disadvantages could be reduced or eliminated by drying and cleaning the RDF, ECO-FUEL®I still had a large size particle, low bulk density (4 - 7 lb/cu ft), and was not suited to storage, retrieval and transportation.

Economic studies showed that unless the fuel could be stored and transported economically its value was greatly diminished. Also, the value of the product as fuel was greatly reduced by the difficulties in combustion and the resultant low overall energy recovery.

These problems precipitated a decision to concentrate on development of a dry, dense fuel which would be storable, have a high density, and be readily burned in existing boilers, burning fossil fuels including not only coal but oil.

The result was the discovery that a high density powder, dubbed ECO-FUEL II, could be produced by treating the partially comminuted MSW with small amounts (around 0.5%) of readily available commodity chemicals which at temperatures under 150°C would embrittle the cellulosic materials in MSW and, with low levels of grinding energy, produce a powdered fuel of high density. The process is protected by patents.

In January 1974 a pilot plant was built at ADL to develop the design parameters for a full-scale plant to produce this Fuel. The feed-stock for the pilot plant was ECO-FUEL®I from the East Bridgewater facility which served as the 'front-end' to receive and process MSW for the embrittling process.

The pilot plant proved that a powdered fuel could be produced by treatment with chemical agents followed by grinding the treated material in a ball mill using circulating hot balls to dry and grind the material. On this basis the full-scale plant was built as an add-on to the East Bridgewater facility, and started producing the new fuel in 1975.

Table I - Properties of ECO-FUEL® II

Percent By Weight (as fired)

Carbon	41.6	-	47.3
Hydrogen	5.5	-	6.3
Oxygen	33.9	-	38.6
Nitrogen	0.6	-	1.5
Ash	7.0	-	15.0
Sulfur	0.3	-	0.9
Chloride	0.1	-	0.7
Water	1.0	-	5.0

HHV   13,430 Joules/gram

Note: the properties of MSW vary, consequently small variations can be expected in the fuel product.

The properties of the ECO-FUEL®II product are listed in Table 1. The bulk density was increased about seven times to 35 lb/cu ft, the moisture reduced to about 3%, and the ash to below 15%. The majority of the product is retained on a 200-mesh screen, and is similar to pulverized coal. The heating value of ECO-FUEL® II is 7500 BTU/lb (13,430 J/g).

The combustion characteristics of ECO-FUEL®II were first briefly demonstrated in a 40,000 lb/hr industrial boiler and subsequently East Bridgewater facility product was burned on a regular basis in an oil-fired Utility boiler (designed for coal but lacking up-to date emission controls) in Waterbury, Connecticut.

In 1975 CEA joined with Occidental Petroleum (OXY) to build the Bridgeport RR Facility under contract with the Connecticut Resource Recovery Authority (CRRA), entirely with private funds. OXY had planned to produce a product essentially like ECO-FUEL® I, but the new technology prevailed: Bridgeport was designed and built by CEA using a second-generation ECO-FUEL®II process.

Description of the Bridgeport ECO-FUEL®II Process

Figure 1 shows a simplified process flowsheet. Actually the plant, designed for 1800 Tons per day (TPH) has two lines, one having a flail mill, the other a hammermill. The metals and potential glass recovery systems serve both lines.

The MSW is discharged from 75-yard transfer trailers of 7-yard packer trucks from an elevated Trucking Module to the Tipping Floor. Front-end-loaders (FEL) move the MSW to storage or to picking areas. After picking to remove oversize bulky waste, including tires and other objectionable materials, the FEL pushes the MSW onto pan-conveyors which feed the plant.

The raw MSW then enters the Primary Trommel which removes the small materials (minus 12 cm) which do not require size-reduction, and the glass, passing the large materials over the screen to the flail mill or hammermill. The magnetic materials are removed from the stream before it enters the air classifier, also called the Air Density Separator (ADS). Here the heavy materials are dropped, including the glass, while the light materials, including most of the aluminum, are carried up to the cyclone, and dropped into the Secondary Trommel, which removes the fine glass and sand.

Drying of the MSW is carried out partly in the ADS system, the remainder in the Ball Mill, after the chemical agents have been added in the discharge end of the Secondary Trommel.

A process heater is used to generate hot gases for drying in the ADS and for heating the balls circulated through the Ball Mill. Fossil fuel and low-grade RDF may be used to fire the Process Heater.

By using the products of combustion generated by the Process Heater, in combination with the water vapor generated by drying, the atmospheres used to dry the MSW are kept at low oxygen levels which do not permit exothermic heat generation while processing the MSW.

The MSW, ground to powder in the Mill, is separated from the circulating balls at the mill outlet, and conveyed by low-oxygen gases (below 8% by volume), to the Product Cyclone which drops the product to the Product Screen. The ECO-FUEL®II product is then cooled while conveying it to the Product Filter, which in turn discharges it to Storage Silos.

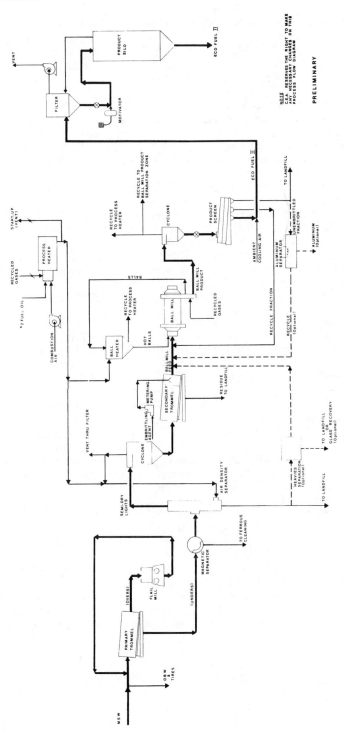

Figure 1. Process flowsheet

The product no longer contains the low-temperature volatiles and organic materials which cause degradation during storage, nor the moisture necessary for these reactions. ECO-FUEL II can be stored indefinately at temperatures below 60 C. The plastics in MSW have also been treated by the balls which enter the Ball Mill at about 200 C, reducing them to powder.

The Product Screen overs are either recycled for further combustibles recovery or rejected. The aluminum fraction in the overs can be removed readily by OXY's Recyc-AL Aluminum magnet.

The six storage silos each hold 300 Tons of Product. This is then discharged from the silos to Dry-solids Transport Trailers, like those used to transport cement and dry chemicals, which carry the fuel to the fuel users. At the user site, the fuel is delivered by pneumatic conveying equipment to storage silos.

The Bridgeport Facility has as its major Fuel customer the United Illuminating Company, which has storage and transport equipment to feed the ECO-FUEL®II to Cyclone Furnaces on their Babcock and Wilcox boilers. The boilers burn oil, and are expected to be able to burn up to 40% ECO-FUEL®II in one 60 MW boiler now, and in a second larger boiler when the MSW received by the Facility justifies it.

The energy recovery efficiency of the ECO-FUEL®II process is high by comparison with other alternate routes to combustion of Refuse-derived-fuels, for reasons which may not be so apparent.

The ECO-FUEL® II process produces a dry material which burns efficiently in inherently efficient Utility boilers. Due to the investment of fuel energy in drying the fuel prior to burning in the boilers, additional fuel does not have to be burned in the furnace to provide the heat necessary for evaporating the moisture in the RDF. The dry fuel assures rapid and complete combustion, high flame temperatures, low excess-air requirements, and minimum boiler stack losses. Thus the boiler efficiency can be about 80%, the same as achieved by firing oil alone, as compared with 60 to 70% for boilers burning high moisture RDF. In addition, Utility power plants have high heat rates, using 5.5 kcal/KW hr, as compared with 8.9 kcal/KW hr, or more for the smaller power plants associated with 'dedicated' boilers burning refuse as fuel. In the final analysis the losses determine the energy efficiency. The dry-fuel process has fundamentally smaller losses than wet-fuel processes.

The material losses of alternate RDF processing plants depend mainly on the sophistication and plant. This should depend upon economics: if it is economically justified to recover more combustibles, it should be done. There is a limit to the investment and operating costs which can be justified to recover the last Calorie. The same applies to recovering aluminum and glass. At this date fuel energy is the valuable product: ferrous metals have low economic value, aluminum recovery has dubious economic value, and glass recovery does not seem to be justified at all, in the Bridgeport locale.

The Bridgeport system could produce over 30 MW of power at its design rating of 1800 TPD of MSW, after deducting the 90 KWH/Ton gross power used by the plant.

Summary: The Bridgeport Facility has produced ECO-FUEL® II since July, 1979. The fuel product is burned with oil in a Utility boiler and at minor industrial sites. The Connecticut Resource Recovery Authority has made a major step toward its forward-looking and ambitious objective: refuse disposal by environmentally and economically sound methods.

RECEIVED November 20, 1979.

Small-Scale Source Densification of Navy Solid Waste

DONALD BRUNNER

Civil Engineering Laboratory, Naval Construction Battalion Center, Port Hueneme, CA 93043

In the Navy, as everywhere else, the disposal of solid waste is becoming more and more costly as environmental regulations become more restrictive. This paper will examine some of the economic and technical aspects of a new concept for reducing the overall cost for solid waste management.

The problem of increasing cost of solid waste management within the Navy is not the same as in the private sector. When the Navy is directed to meet specific requirements of new regulations or executive orders, Congress often does not appropriate additional funds or allocate new manpower to accomplish the new requirements. Increased costs cannot be passed on to the consumer since the Navy is the consumer. Expenses must be absorbed into the existing Navy budget thereby diverting funds from the primary mission.

The Navy generates approximately 1.75 million tons of solid waste annually. Current costs are approximately $30 per ton for collection, transportation, and disposal. This represents an annual expenditure of approximately $50 million. The Navy could staff up, operate, and maintain a squadron of aircraft at this price. With these costs expected to double or even triple in the next few years due to more restrictive environmental regulations, it becomes prudent for the Navy to develop techniques for managing their waste that are both environmentally and economically acceptable.

The collection process consumes about 70% of all the dollars expended to manage Navy solid waste as illustrated in Figure 1. The high costs of collection derives from the characteristic fluctuation in the generation rate, the nature of the assortment of materials involved, the variable and generally low density and the requirement to remove it from the premises at frequent regular intervals. Solid waste, for the most part, is an organic material having a high surface area to mass ratio. In the presence of moisture, it becomes biologically active fostering the development of unacceptable and noxious decomposition by-products. It can be a fire hazard as well as a breeding ground for flies and rodents.

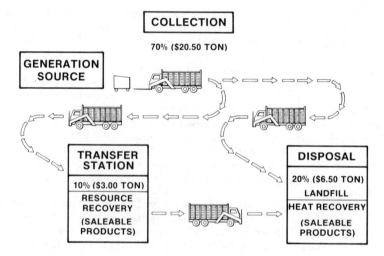

Figure 1.   Cost of solid waste management unit operations

It can attract birds, animals and human scavengers and is subject to becoming windblown and unsightly. Hospital wastes add a further dimension to waste management in that they are subject to becoming a significant vector for pathological bacteria and viruses and therefore often require special handling.

Historically the only available solution to most of these problems has been frequent, regular removal of the waste from the generating source. Cost effective mating of equipment and manpower to the highly varying waste generation rates has, of necessity, been subordinated to health and safety considerations. Hence the disproportionate amount of dollars devoted to the collection activity.

A study of the problem indicates that processing solid waste to a high density, self encapsulated, biologically stable slug at the generating source has the potential for reducing collection frequency up to 18:1. This will allow major reductions in collection equipment, manpower, and containers by increasing the number of loads per crew day (1, 2). Additional savings can also accrue in disposal by increasing land fill capacity, reducing pollution problems and most significantly by increasing available disposal options. New options will include disposal as a refuse derived fuel and environmentally sound disposal of shipboard generated wastes at sea.

Feasibility

The Civil Engineering Laboratory began investigating the feasibility of densifying waste into self-encapsulated biologically stable slugs in 1974. At that time, two extrusion concepts were investigated: ram extrusion and screw extrusion (3).

Ram extrusion was examined using a modified commercially available Mil Pac System shown in Figure 2. A schematic showing the operation of the system is presented in Figure 3. Unprocessed waste was first fed into a vertical shaft hammermill where it was reduced to a nominal 2-inch particle size. The shredded refuse was then pneumatically conveyed to the charge hopper of the extruder where hydraulic rams forced the material through 2-inch square extrusion dies. Continuous processing with this system was never realized. The extrusion dies jammed often as the characteristics of the refuse changed. The densified pellets or slugs produced by this process varied considerably. The experiment did demonstrate that high density slugs could be produced. Densities as high as $92\#/ft^3$ (1.47×10^3 kg/m^3) were achieved. The high density slugs were characterized by a glazed surface which tended to bind them together. Also because of the batch type processing resulting from the reciprocating action of the ram, the slugs contained transverse shear planes or laminations. The slugs could easily be broken across these planes into square chips. The

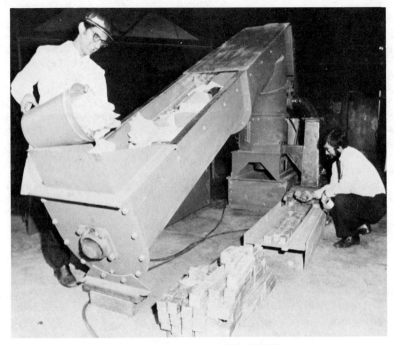

Figure 2.   Ram extrusion modified Mil Pac system

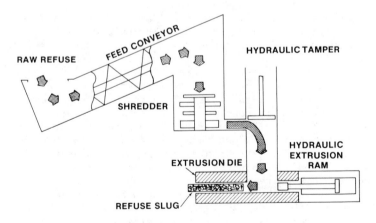

Figure 3.   Schematic of ram extrusion process

thickness of these chips depended on the quantity of material
processed during each cycle of the ram. For these experiments,
the nominal chip thickness was 0.25 inch (6.35×10^{-3} m).

The effects of the second process, screw extrusion, was
investigated using a 3-inch (7.62×10^{-2}-m) diameter single screw
extruder similar to those used by the plastic and rubber industries. In this concept, shown in Figure 4, the rotation of the
screw provided the pressure necessary to force the shredded solid
waste through the cylindrical extrusion die. The waste slugs
produced by this process were more continuous and not characterized by transverse shear planes exhibited by the ram extruded
slugs. Also, because the action of the screw tended to "work" the
material and compact it before entering the die, high temperatures
were generated in the barrel of the screw and the resulting slugs
tended to have a harder and more cohesive surface finish. Figure
5 shows comparative samples of ram and screw extrusion products.

Additional findings of these early investigations showed
that:

(a) The density, structural cohesiveness, and surface
    finish of the product from either process can be
    controlled by regulating the processing tempera-
    tures and controlling the resistance offered by
    the extrusion die.

(b) Both densification concepts effectively reduced
    the biologically active surface area and impart a
    degree of biological stability to the refuse.

(c) The moisture content of the refuse is a significant
    processing parameter. It affects the resistance
    developed within the extrusion die, and therefore,
    the density and cohesiveness of the resulting pro-
    duct. Variations in moisture content can cause the
    die to jam and thus inhibit continuous processing.

These preliminary tests were designed to identify the processing parameters and product characteristics, but were not of
sufficient depth to quantify the relationship between processing
characteristics and product characteristics. They did show the
need and direction for further research.

At the time CEL was performing its feasibility investigations
other experiments were also in progress (4, 5). For the most
part, these experiments were directed at centralized processing of
solid waste into small pellets for co-firing with coal. Commercial utilization of pelletizers for solid waste processing was
being explored by a number of companies including Papakube Corporation in San Diego, California. In the summer of 1978, CEL processed Navy wastes from the San Diego area at Papakube. The cubes
were made using a modified John Deere extruder (shown in Figure 6)
and measured approximately 1 inch square (2.54×10^{-2} m) and 1.5

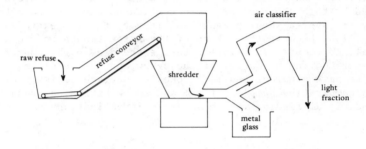

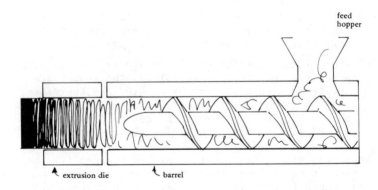

Figure 4.   Screw extrusion process

Figure 5. Samples of ram- and screw-extruded solid waste

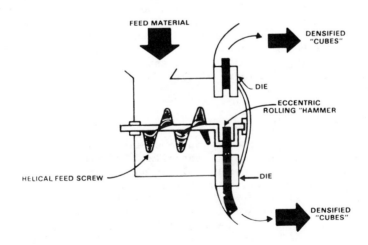

Figure 6. Schematic of John Deere/Papakube cuber

inches (3.8x10^{-2} m) in length. Some minor measurements were made
during the processing of the wastes but the primary objective of
the tests was to obtain densified material for performing physical
and chemical tests and analyses of the product. Preliminary
results indicate the moisture of the cubes varied between 20% and
36% and the energy content between 4,400 (10.2x10^6 J/kg) and 6,000
(14x10^6 J/kg) Btu/lb. A report of the processing and subsequent
physical and chemical tests of the cubes is nearly completed (6).
        A portion of the cubes were brought to CEL and stored adja-
cent to shredded solid waste to obtain an indication of the relative biological stability of the densified material. The cubes
were stored in 40-yard (30.6-m^{-3}) drop boxes and covered with
black plastic. Probes were inserted at various points in the
containers to allow extraction of gas samples and to measure
fluctuations in temperature. Results of temperature fluctuation
over a 75 day storage period are shown in Figure 7. Note that the
temperatures of the cubes track very closely with those of the
ambient air temperatures indicating very little in the way of
composting of the refuse was taking place. On the other hand, the
temperature of the shredded refuse was consistently higher than
the ambient, attaining an average temperature close to 50°C,
indicating a much higher level of biological activity. These
tests demonstrated the biologically stabilizing effect of densification and the potential for long term storage without adverse
environmental effects. A report on these storage tests is currently being prepared and should also be available in the near
future (7). More in-depth analysis of the biological aspects of
densified waste storage are planned at our Solid Waste R&D test
site currently being constructed at NAS Jacksonville, Florida. We
will also be using the test site to establish processing criteria
and criteria for solid forms of waste-derived fuel for use by the
Navy.

Application/Benefits

        Densification processing can be applied basically either at
the generating source or at a centralized location. We have
emphasized the generation source location in our studies.
        The operation benefits of source densification are derived
from the technical objectives of high density self-encapsulation
and biological stability.
        High density has the obvious effect of decreasing the volume
of the waste. This in turn allows a decrease in the size of the
storage area whether it is a dumpster, a truck or a cell in a
landfill. Densification to 65#/ft^3 (104 kg/m^3) represents an
approximate volume reduction of 18:1 over that of unprocessed
refuse. The higher density also implies a reduced surface area
and therefore a reduced fire hazard and lower biological activity.

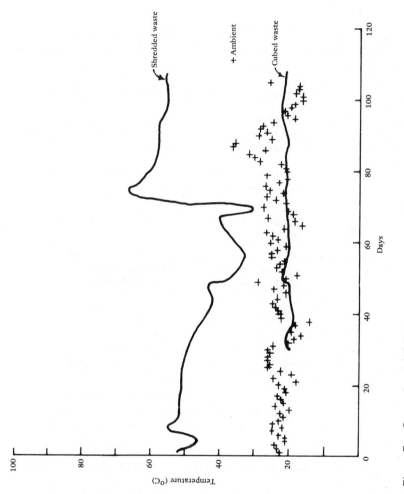

Figure 7. Comparison of daily temperatures for cubed and shredded refuse during storage tests

Self-encapsulation means the refuse becomes encased in a hard skin or shell which is developed by the refuse itself. Self encapsulation offers the benefit of reduced water absorption and therefore lower biological activity which impedes the development of noxious odors and leachate. Self-encapsulation eliminates the problem of blowing litter and makes the waste less attractive to scavengers. Encapsulated refuse also lends itself well to mechanized handling.

Biological stability allows the refuse to be stored for long periods by delaying the breeding of vectors, the formation of noxious odors or hazardous gases, and the fire hazard of spontaneous combustion. If the processing temperature is sufficiently high densification may reduce the public health hazard associated with solid waste, especially those wastes emanating from hospitals.

The operational benefits to be gained by placing the densifier at the generating source are realized during collection and land disposal operations. Collection benefits are as follows:

(a) Increased time interval between collections without increasing the number and size of containers.

(b) Better utilization of container capacity and a reduction in the number of containers required at a given location.

(c) Reduced number of collection stops to fill a truck. This increases the number of collection trips per day per truck thereby reducing the number of collection crews and trucks required.

(d) Increased size of each load by eliminating the packer mechanism on trucks.

(e) Increased load density and more uniformity of load size thus providing a higher level of equipment utilization.

(f) Complement a source segregation effort to recover saleable products such as high grade paper, cardboard, and metals while processing the residue for disposal.

Disposal of densified waste offers the following benefits:

(a) Reduced operating costs. The high density, uniform size and easily manageable slugs can be spread evenly and require little compacting and therefore can be accomplished in considerably less time.

(b) Less cover material. The biological stability and unobtrusive nature of the slugs would enable the use of less cover dirt and reduce the requirement for daily covering.

(c) Extension of landfill life. The higher densities and reduced cover requirements enable more refuse to be stored in the same land volume.

(d) Improved pollution control. Elimination of blowing litter, noxious odors, fires, and vector infestation plus reduced leachate. Slowed biological activity would reduce production of and danger from explosive gases.

Economics

Based on the potential benefits for source densification an economic analysis was performed to quantify the dollar savings (1). A model of the collection process was developed that allowed comparison of the cost of conventional waste collection with proposed densified waste collection.

This model was based on the following considerations:

(a) The amount of solid waste found in a container at the time it is serviced and the space it occupies are characteristics that control the basic investment in containers and vehicles and establish the nominal manpower level for solid waste management.

(b) Maintaining the quality of service requires an additional investment in equipment and manpower as a contingency against the high variability of the waste density and generation rate. The best available estimate for this contingency is the reported container utilization factor at a particular activity.

The model consists of mathematical expression developed for calculating: (1) truck utilization and investment in containers and vehicles based on container utilization, (2) trips per collection day, and (3) collection costs.

Parameters pertinent to solid waste collection costs used in the mathematical expressions are as follows:

1. Length of collection day and week (hr/day)

2. Container collection frequency (times per week)

3. Payload of collection vehicle (tons)

4. Vehicle availability (%)

5. Waste density in the container (lb/yd) (kg/m^3)

6. Container size (yd^3) (m^3)

7. Equipment capital, operation, and maintenance costs ($)

8. Economic life of equipment (years)

9. Labor rates ($/hr)

10. Quality of service (percentage of time containers are overfilled)

11. Travel time to motor pool from route (min)

12. Time required to empty one container (min)

13. Travel time to and from disposal site (min)

14. Time to offload vehicle (min)

15. Waste generation rate (ton/week)

In addition to these parameters, two constraints were placed on the system: (1) the vehicle must be returned empty to the motor pool at the end of each day, and (2) bulky wastes are separately collected.

The model was exercised for a conceptual system and a family of curves plotted showing the relationship of truck utilization to collection costs as a function of weekly waste generation rate. Results are shown graphically in Figure 8 for conventional and densified waste collection. The lowest costs for either system occur along the optimum benefit lines. These lines correspond with maximum truck utilization represented by the points of discontinuity in the curves. These discontinuities result from the integer number of trucks required in any collection system. Given one truck, the cost per ton goes down until maximum truck utilization is reached. At this point, another truck must be purchased to collect more tons per week with a resultant step increase in cost per ton and a lowering of truck utilization.

The magnitude and frequency of the discontinuities are valid only for systems operating within the constraints and parameters specified for the conceptual system. The points of discontinuity can be shifted laterally on the graph by changing truck and container size. Thus the same minimum cost can apply to any size operation. The optimum benefit available through source densification is therefore represented by the minimum cost per ton between conventional and densified collection. For conventional collection under this conceptual system, the minimum cost is $23.90/ton and for densified collection $6.30/ton - a difference of $17.60/ton.

The savings realized depends upon how much of the $17.60/ton is consumed by the capital, operation, and maintenance cost of the densification equipment. Using a capital cost goal of $4,000 for

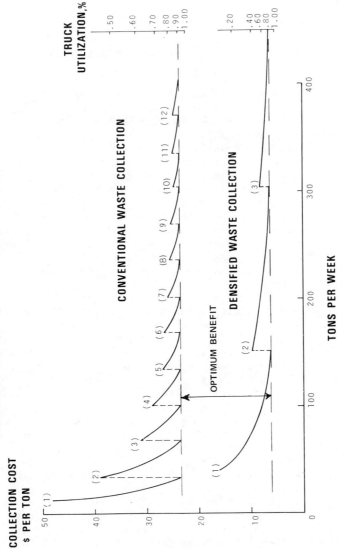

Figure 8.   Comparison of costs for conventional and densified waste collection

the densification unit and an average daily processing rate of 400 pounds, the total cost for densification was calculated to be $11.78/ton. This leaves a net savings of $4.42/ton.

Follow-On Technical Effort

The development of specifications for the design and con- struction of a system for source densification of solid waste that will meet the established size and cost objective requires exten- sive knowledge of the parameters governing the densification process.

Jamming of material within the extrusion die has been the single most limiting factor to continuous low cost densification processing. We have experimented in a number of ways in attempts to quantify the relationships among critical parameters associated with the jamming phenomena (8, 9).

To aid in our investigation an electrically actuated hydraul- ically operated ram/die densification device was designed and constructed to allow observation of the process under controlled conditions. The device, shown in Figure 9, functioned by forcing refuse into the die tube with the hydraulic ram and compacting it against material already in the die. The die was a constant 2.25-in. (5.7x10^{-2}-m) diameter and designed to enable changes to its length. Also, the hydraulic ram was designed to allow changes in its hydraulic pressure as well as stroke length and penetration depth within the die.

Efforts to overcome the jamming first centered on increases in the ram penetration within the die. The theory being that as the force necessary to extrude the waste increased, it could be controlled by shortening the length of the slug of refuse remain- ing within the die. The resistance force offered by the die is a function of the normal pressure on the die wall and the inside surface area. Reducing the slug area in contact with the die wall would reduce the total normal force and therefore the resistance offered by the die. Different length die tubes were also used in the test mechanism. Shorter die tubes results in poorly formed slugs. Longer die tubes eventually resulted in die jamming that often could not be overcome by simply increasing power to the ram.

Results have indicated that the densification of the waste takes place almost entirely at the die entrance. The impact action of the ram produces highly localized stress and distortion in the die walls near the entrance. Friction causes the stress to remain even after the ram load has been removed. Progressive cycles of the ram transport the distortion down the longitudinal axis of the die thus forming the densified material into a slight cone shape. (This situation is illustrated in Figure 10.) Even- tually the force necessary to transport the cone shaped material through the die exceeds the force available and the jamming phe- nomenon occurs. Or if the available force is high enough, the die fails.

Figure 9.   Experimental ram densification device

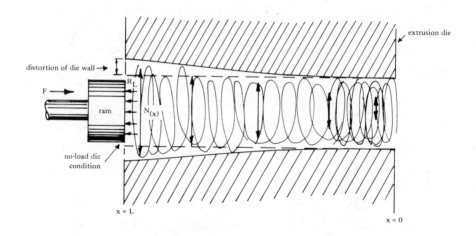

$$R_{(x)} = F P C_f \times N_{(x)}$$

where: $R_{(x)}$ = die resistance

P = die perimeter

C_f = coefficient of friction

$N_{(x)}$ = normal force on die wall (a function of x)

F = force available for densification

R_{max} occurs at R_L

If $R_{max} \geqslant$ F, jamming will occur

Figure 10.   Distortion of die wall during extrusion processing

The refuse was also varied in an attempt to understand the pre-processing requirements to reduce the propensity for jamming. Variables included degree of size reduction, composition and moisture content. Experience has shown that the most difficult to control is refuse moisture content. Moisture content requires the development of a detection and control devices capable of operating under highly variable conditions. Laboratory tests have indicated a moisture content of 12%-16% produces a cohesive high density slug. The other parameters can be controlled for the most part using presently available technology, although these techniques are costly to implement on a small scale.

We've experienced mixed results and do not have all our data analyzed so are not in a position to make many definite statements at this time concerning the parameters governing the extrusion process. However, we have concluded that to enable continuous production of densified solid waste with reasonably consistency in the cohesiveness of the slugs the densification system must be designed to accommodate the wide variance in refuse parameters, rather than processing the waste to a homogeneous feedstock that meets exacting requirements of existing machines. Economics demand that densification at the generating source avoid the costly pre-processing necessary to achieve a homogeneous feedstock.

Conclusion

The feasibility of producing a high density, self-encapsulated biologically stable slug of refuse has been demonstrated.

The economic and many pollution control benefits of densifying the solid waste at the generating source have been examined and a design goal of approximately $4,000 for a 400-lb/day system established to achieve a savings of $4.42/ton.

The extremely variable and complex nature of the material makes it difficult to process. For successful economical source densification, the system must be designed to accommodate the material variations.

CEL is working at developing the basic data needed to quantify the relationship governing densification and to develop design criteria for a system that will accommodate the wide variations in material properties.

Literature Cited

1. Brunner, Donald E., Braswell, J., and Saam, R. D. Technical Memorandum M-54-76-28, Application and Cost Analysis of Refuse Densification Processing; Civil Engineering Laboratory: Port Hueneme, Calif., Dec 1976.

2. Lory, Ernest E. Technical Memorandum M-52-76-05, A Review of the Biostatic Effectiveness of Refuse Densification; Civil Engineering Laboratory: Port Hueneme, Calif., Apr 1976.

3. Brunner, Donald E. Report No. 65-75-14, Refuse Densification - Phase I, Experimentation; Civil Engineering Laboratory: Port Hueneme, Calif., Jul 1975.

4. Hathaway, Stephen A. "Mechanics of Densified Refuse-Derived Fuel," Third International Conference on Environmental Problems in the Extractive Industries, Dayton, Ohio, Nov 1977.

5. Alter, H., and Arnold, J. "Preparation of Densified Refuse Derived Fuel on a Pilot Scale," Proceedings, Sixth Mineral Waste Utilization Symposium, Illinois Institute of Technology Research Institute and U.S. Bureau of Mines, Chicago, Ill., 1978.

6. Cal Recovery Systems, Inc. Contract Report (in preparation), Densification of Navy Wastes.

7. Lingua, M. Technical Memorandum (in preparation), Storage Tests of Densified Refuse Derived Fuel; Civil Engineering Laboratory: Port Hueneme, Calif.

8. Boogay, M. A. Technical Memorandum M-54-76-25, Extrusion Parameters for Refuse Densification; Civil Engineering Laboratory: Port Hueneme, Calif., Nov 1976.

RECEIVED November 16, 1979.

Biomass Densification Energy Requirements

T. B. REED — Solar Energy Research Institute, Golden, CO 80401

G. TREZEK — University of California, Berkeley, CA 94720

L. DIAZ — Cal Recovery Systems, Inc., Richmond, CA 94804

Biomass materials, especially residues, are in many ways an attractive renewable fuel. However, they suffer from low volume energy content and occur in a wide variety of forms unsuitable for fuel use. Densification of biomass to pellets, briquettes, logs, or dense powders yields clean, renewable fuels with energy densities comparable to coal.(1–4) Yet the cost and energy required for densification must be considered in deciding whether densification is practical in a given situation. A knowledge of the dependence of this energy on various operating parameters also will make possible design of better densification processes.

The purpose of this study was to determine the work required for densification under various laboratory conditions and to compare this to the energy consumed by practical operating equipment.

Experimental Procedure

The apparatus used to study densification by compression is shown in Figure 1 (a). It consists of a steel die and piston 2.5 cm in diameter capable of being heated to various temperatures during pressing. The travel of the piston was measured as a function of applied pressure. Ten-gram samples of minus 10 mesh pine sawdust dried at 110 C were held at temperature for 15 min after initial cold compaction at 200 psi, and the pressure was then increased in 300-psi increments to 10,000 psi. The resulting densities are shown as a function of pressure in Figure 2 for temperatures from 100–225 C. Runs were made also at 250 and 300 C, but wood was heavily pyrolyzed in the process and the results were discarded.

Elemental and proximate analyses were run on the initial sawdust and a 225 C pellet and are shown in Table I. The energy contents of the sawdust and the resulting pellets are shown in Table II.

The experiment just described was designed to simulate the densification process by direct compression, a batch process. However, many commercial densification machines provide continuous extrusion of pellets. Sufficient pressure is built up to cause the material first to densify and then to flow through a constricting nozzle. The apparatus of Figure 1 (b) was designed to simulate this extrusion process. Twenty-gram samples

0-8412-0565-5/80/47-130-169$05.00/0

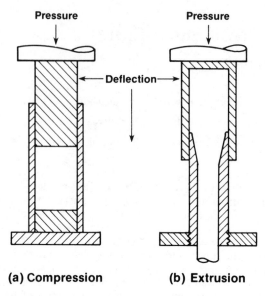

(a) Compression   **(b) Extrusion**

Figure 1.   Densification test configurations

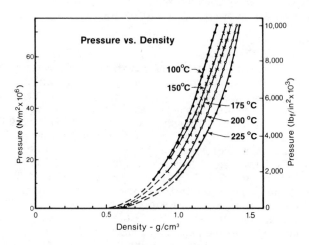

Figure 2.   Pressure vs. density

Table I. Analysis of Pine Sawdust and 225 C
Pellet From Compression Experiments

Experiments[1]	Dry Sawdust[2]	225 C Pellet
Proximate Analysis[3]-%		
Ash	0.83	0.99
Volatile Matter	83.19	85.05
Fixed Carbon	15.98	13.96
Sulfur	0.06	0.06
Ultimate Analysis-%		
Carbon	49.89	52.09
Hydrogen	6.40	6.28
Oxygen	42.48	40.36
Nitrogen	0.36	0.22
Chlorine	—	—
Sulfur	0.05	0.06
Ash	0.86	0.99
Heating Value - kJ/g	19.3	21.4
Btu/lb	8290.0	9200.0

[1] Analysis and heats of combustion by Hazen Assoc., Golden, Colo.

[2] Average of 3 runs.

[3] The C, H, O correspond to an elemental formula $CH_{1.57} O_{0.65}$ for sawdust and $CH_{1.32} O_{0.58}$ for the 225 C pellet.

Table II. Heats of Combustion of Pine Sawdust and Pellets
Made at Various Temperatures at 10,000 psi Pressure

Form	Temp. °C	Density g/cm³	Weight Remaining %	Energy Content kJ/g	Energy Content Btu/lb
Pine sawdust	—	—	100.0	19.3	8290
Pellets	100	1.278	99.3	—	—
Pellets	150	1.340	97.6	19.6	8420
Pellets	175	1.374	96.5	19.6	8440
Pellets	200	1.422	95.6	19.6	8450
Pellets	225	1.435	91.2	21.4	9200
Pellets	250	1.316	71.2	23.0	9890

of screened solid municipal waste (SMW) were loaded into the container and heated to the desired temperature for 15 min. Pellets were then extruded at a rate of 5, 10, or 20 cm/min and the pressure was recorded as a function of deflection. The deflection versus pressure is shown in Figure 4 for four temperatures and in Figure 5 for three rates of extrusion at room temperature. The resulting pellets were well formed, especially those made at higher temperatures, and had a density of about 1.0 g/cm³.

The work done during densification is given for both processes by:

$$W = A \int_0^X P \ dx = A \int_0^X P \ dx \qquad (1)$$

where P is the applied pressure, x is the sample thickness, and A is the cross sectional area of the die and piston. In the compression apparatus of Figure 1 (a) the density at each point was calculated from:

$$P = m/xA \qquad (2)$$

where m is the sample mass. The work of compression is shown as a function of density, calculated from these equations, in Figure 3. The total work of extrusion was obtained by integrating the curves of Figures 4 and 5 and is shown in Table III.

Table III. Work of Extrusion of RDF[1]

	Energy	
	kWh/tonne	kWh/ton
Temperature[2] °C		
25	7.76	7.06
93	6.09	5.54
149	6.23	5.67
204	4.45	4.05
Extrusion Rate[3] cm/min		
5	7.76	7.06
10	10.93	9.95
20	10.90	9.92

[1] From integration of Figure 5.

[2] Extrusion rate 5 cm (2 in.)/min.

[3] At 25 C.

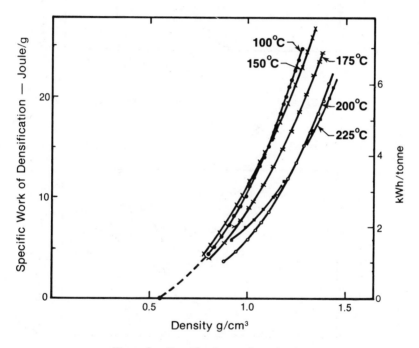

Figure 3.   Densification work vs. density

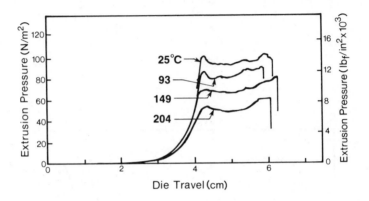

Figure 4.   Extrusion temperature vs. pressure

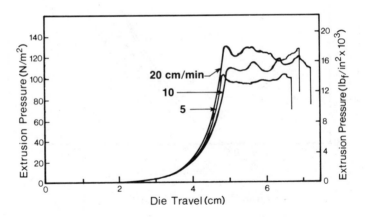

Figure 5.   Extrusion rate vs. pressure

Discussion of Results

Three basic types of pressure application are used in commercial densification processes: (1) Straight compression in a die; (2) Extrusion through a constriction; and (3) Shear of precompacted material to produce heat and flow under pressure. Approximate energy consumptions supplied by the manufacturers are compared to the laboratory tests reported here in Table IV but it must be stressed that these figures are only approximate, depending critically on type of material, size, temperature, etc.

Table IV. Comparison of Reported Energy Requirements
For Commercial Densification Apparatus with
Laboratory Results

			Work	
	Material	Density	kWh/tonne	kWh/ton
Compression				
In Laboratory[1]	Sawdust	1.0	4.0	3.6
	Sawdust	1.2	6.6	6.0
Commercial[2]	Sawdust	‒1.2	37.4	34.0
Extrusion				
In Laboratory[3]	MSW	1.0	7.76	7.06
Commercial[4]	MSW	1.0	16.4	14.9
	Sawdust	1.0	36.8	33.5

[1] This study, 2.5 cm pellet, Figure 1 (a) at 100 C and Figure 3.

[2] From specifications of 150 hp Hausmann briquettor No. FH 2/90/200 for 8 cm diameter log.

[3] This study, 1.2 cm pellet made at 25 C, 5 cm/min, Figure 1 (b) and Table III.

[4] From Ref. 1, data supplied by California Pellet Mill Corp.

The work of compression measured in this study is seen to be lower by a factor of two to ten than that consumed in operating compression machines. This is to be expected because the work measured here does not include motor and bearing losses associated with commercial equipment, and the measurements were made under idealized conditions. Given these differences, the agreement is satisfactory and the laboratory results probably represent a lower limit to the work required.

An important result of this study is the finding that the work and pressure of compression or extrusion can be reduced by a factor of about two by preheating the feedstock to 200-225 C before densification. This requires extra thermal energy for complete drying and to heat the biomass (heat capacity about 1.8 J/g-C) to the higher temperature; however, these are offset by lower electrical power costs, lower equipment costs because of the lower pressure requirements, possibly reduced die wear due to improved lubricity of the biomass at increased temperatures, and increased fuel value due to complete water removal and prepyrolysis. These factors must be tested at the commercial scale before any conclusions can be drawn on the desirability of preheating feedstock.

The analyses reported in Tables I and II show that the pellet made at 225 and 250 C had considerably higher energy contents than those made at lower temperatures. The energy content of the pellet made at 225 C is 20.2 J, essentially all of the energy in the sawdust from which it was made (91.2% of 21.4 J/g). This suggests that there is a "prepyrolysis" reaction for biomass, similar to that which occurs for lignite, in which CO_2 and H_2O are driven off with little or no energy loss.

Many commercial densification machines use extrusion rather than straight compression, because extrusion can be adapted to continuous rather than batch processing. An examination of Figures 4 and 5 suggests that there is an initial stage in which the feedstock is compressed to a pressure sufficient to overcome the static friction at the throat. At this point the pressure drops slightly to a value necessary to overcome the sliding friction encountered as the biomass passes through the constriction. The data of Figure 4 and Table I clearly show the same trends observed in Figures 2 and 3—that the work of densification drops by a factor of about two as the feedstock is preheated to 200 C.

Although the data are not directly comparable because they were taken on different feedstocks, it seems clear that the work required for compression is less than the work required for extrusion. In commercial equipment, the friction involved in extrusion performs a useful function—the heating and drying of the pellets.

Although the data in Figure 5 and Table I suggest that the work of extrusion increases with increasing rates, the effect is small here. Other data taken in pilot plant and commercial operation generally show a dramatic decrease of extrusion work with flowrate.

Conclusions

• The pressure required for densification is reduced a factor of two by heating to 225 C.

• The energy required for both compression and extrusion is decreased by a factor of about two as temperature increases from room temperature to 225 C.

• The energy content of the pellets rises with temperature of densification.

- Energies measured for densification in the laboratory are comparable to but smaller than those required in commercial equipment.

Acknowledgements

We would like to thank the Coors Ceramic Division for performing the compression experiments and Mr. John Lawson of Hazen Research for the proximate and ultimate analyses.

References

1.   T. Reed and B. Bryant, "Densified Biomass: A New Form of Solid Fuel." SERI report 35. The Solar Energy Research Institute, Golden, CO 80401. July 1978.

2.   T. Reed and B. Bryant, "Energetics and Economics of Densified Biomass Fuel (DBF) Production." In "Biochemical Engineering No. 181," Vol. 74, p. 26, J. M. Nystrom and S. M. Barnett, eds. (AIChE Symposium Series) 1978.

3.   R. A. Currier, "Manufacturing Densified Wood and Bark Fuels." Report 490, Oregon State University Extension Service, July 1977.

4.   Cohen and Parrish, "Densified Refuse Derived Fuels." Bulletin 6, No. 1, National Center for Resource Recovery, Washington, D. C., Winter 1976.

RECEIVED November 16, 1979.

Densification Systems for Agricultural Residues

THOMAS R. MILES and THOMAS R. MILES, JR.

T. R. Miles, Consulting Engineers, 5475 S. W. ArrowWood Lane, Portland,
OR 97225

The marketing and utilization of many agricultural residues can be significantly expanded by densification into bales, cubes or pellets. End uses and supply logistics usually determine their particular requirements of physical form and "packaging". Densification itself is only one of a sequence of operations which always includes collection, storage and transport – and can also involve treating, grinding and drying. Because residue is a waste, one obviously limits processing to the minimum. Since residue utilization is essentially a series of materials handling operations, not only these steps but also the shaft and thermal energy required must be kept to a minimum. If truck transport and storage are the only criteria, there is no need to densify to a greater compaction than 16 Lb/CF (256 Kg/m^3), since at this density a truck reaches both its volume and weight limits. Nearly all stalks and straws can be field baled in one step to this density with currently available equipment.

Densification is accomplished in several quite different ways: stalks, stems and leafy materials can be chopped or ground; stalks, stems and chips can be aligned or "laminated" to substantially increase bulk density; mechanical vibration and pressure can be used in combination with the above methods for higher densities. To maintain achieved density, the material must be bonded, or restrained (as in wire-bound bales) or not disturbed (as in the aligned or laminated states). The "Bulk Density" (B. Dens.) resulting from a process is more significant than the specific particle density.

Significant factors controlling densification are moisture, particle size, form, fibrous or nonfibrous nature, leafy legumes vs. stiff waxy resilient straws, and binders, starch or sugar content, type and quantity of binder.

0-8412-0565-5/80/47-130-179$05.00/0
© 1980 American Chemical Society.

In our experience, the use of <u>COMBINATIONS</u> of materials to facilitate densification - particularly for cubing and pelleting - cannot be over-emphasized.

The following must be borne in mind when considering densification: (a) Solutions are always specific to a particular material, site and end use. (b) The use of a given residue is always affected by the material it can replace or the availability of other residues and many special crops or products competing in the same market. (c) Selection of densification systems must take into account the entire process from production of the residue to its delivery and use in its final form. (d) Physically handling a residue will account for higher costs and more energy than the individual densification systems.

Most of the data and conclusions following are the result of our nationwide experimental and production engineering in the harvest, processing and transport systems for agricultural and wood residues over the past 31 years. Since these materials are all natural - and thus variable - values are presented in ranges.

Characteristics of Residues

Crop Residues:
Straws, Stalks, Vines, Shells, Pits, Fibers
7500 BTU/Lb, 10-20% M, 2T/A, B. Dens. 10-16 Lb/CF baled
Must bear collection costs aside from crop.
Multi-uses as feeds, fibers, fuels, fertilizers.
Seasonal supply, limited harvest period.
Most are perishable without covered storage.
Do not produce clean burning fuels without processing.
Straws do not densify without binders.
Best utilized at Area-Integrated-Conversion-Centers.

Manure:
Feedlots and Dairies, Poultry and Hog Farms
6500 BTU/Lb DM, 80-90% M, 1-8 Lb DS/Animal/Day
60 Lb/CF slurry very perishable, use-at-site specific.
Low solids (12-15%) make drying costly.
Twice BTU yield as Biogas (methane) vs dry/dir. combust.
Relatively few large concentrations of animals.
Has on-farm, small scale potential.

Forest Residues:
Mature, Diseased or Fire-damaged Trees
4500 BTU/Lb, 50% M, 20-50T/A, B Dens 18-22 Lb/CF chipped
Storage on stump, thinnings, undesirable specie.
High capital cost of roads, access, right-of-way.
Other characteristics similar to logging residues.

Logging Residues:
Limbs, Tops, Shatter, Chunks
4500 BTU/Lb, 50% M, 20-50T/A, B Dens 18-22 Lb/CF chipped
Collection costs vary widely with site.
Must be chipped at landing to allow bulk hauling, hand-
ling and storage, with added capital cost.
Chips ready for boiler, to gasify or convert.
Potential domestic fuel thru distributors.
Quantities per acre depend on specie and age.
Store outside, low degradation.
Competing fiber uses: pulp, board.
Whole-harvest chipping for fuel/fiber.

Sawmill Residues:
Shavings, Bark, Sawdust
4500 BTU/Lb, 50% M, 2T/MFBM, B Dens 16-22 Lb/CF "hogged"
Collection and transport subsidized by lumber.
Already reduced to usable form.
Ready for direct combustion or gasification.
Convenient to convert to other forms at millsite.
At or near other logical Energy Conversion sites.
Steady year-round supply, stores well outside.
Sawdust/bark mix good domestic fuel (in use up to 1950).
Burns well without drying (4500 BTU/Lb net).

MSW, RDF:
Collected Municipal Solid Wastes, Refuse Derived Fuel
5000 BTU/Lb, 30-40% M, 5 Lb/Cap/Day, B Dens 12-20 Lb/CF
Generated in cities - near energy need.
Subsidized collection and transport - low cost.
Separation and recovery of metals, etc.
Supply constant and proportional to population.
Store only under cover.
Logistics demand prompt separation and use in large
facilities.
Institutional and economic constringencies abound.

Residue Combustion

    Several furnaces (1 to 6 M BTU/H) to use bulk or
baled residues are being developed here and abroad for
on-farm or small commercial use for grain or seed drying - ours among them. Proven small sloping grate
designs for domestic and small commercial furnaces
(200-1000 K BTU/H - 211-1055 MJ/H) use sawdust pellets
and small cubes. Larger industrial dutch ovens,
sloping or pinhole grates or cells can be efficiently
fired with wet (>40% M) residue fuels including wood
hog fuel, cubed RDF or straw, shells and pits in various mixes - and in combination with coal - with

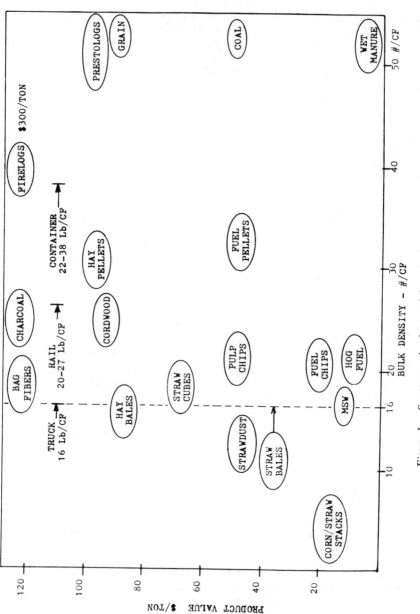

Figure 1. Spectrum of values and densities for various products

pneumatic or rotostokers. Suspension firing, either
tangential cyclonic or linear, is restricted to low
moisture (<20% M) finely ground residues and in combination with a "reliable" fuel such as coal, or with an
oil or gas pilot. Wood and crop residues in ground or
chip form can be co-fired with coal more economically
than using fuel pellets - $1.90 vs $2.61/K Lb steam/hr.

It is practical to suspension "over-fire" a dutch
oven or grate system with residues but characteristi-
cally higher emissions from suspension firing must be
considered. The higher percentage (4-6%) of low-melt-
ing ash in ag-residues poses special grate problems
vis-a-vis wood at 1%.

Gasifiers almost universally require higher
density (16 Lb/CF - 256 Kg/m^3), low moisture (20%)
fuels for efficiency, even feeding and uniform gasification.

Typical Densification Sequence

COLLECT STACKS/BALES	TRANS GRIND DRY	DENSIFY CUBES, COBS PELLETS, LOGS	TRANS STORE	SPECIFIC END USES

Some Densification Systems Available Today

Several densification systems are in use today.
Note that most existing systems are producing high value
products for feed or fiber from the best available
materials.

System	Principal Use
Orientation and Vibration:	Pulp Chips, Coal
Size Reduction:	All Materials
Field Stacks and Balers: John Deere, Moline IL Hesston, KS New Holland, PA Vermeer, Pella IA Foster, Madras OR J.A.Freeman, Portland OR	Forages and Residues
Cubers: John Deere, Moline IL Lundell, Cherokee IA Osborn, Moses Lake WA Calif. Pellet Mills, San Francisco CA	Hay, Cobs, Whole Ration Feeds

Pellet Mills: Concentrate, forage feeds, fuels
Ring Die -
  Sprout Waldron, Muncy PA
  California Pellet Mills, San Francisco CA
  Simon Barron, Gloucester, England
  Esbjerg Mølle, Esbjerg, Denmark
  Volkseigen Betrieb Muhlenbau, Dresden, Germany
Roller Head -
  Amandas Kahl, Hamburg, Germany

Extruders, Briquetters: Wood and composite, firelogs,
                         Hay briquettes
Piston Press -
  Fahr, Gottmadingen, Germany
  DeSmithske A/S, Alborg, Denmark
  Hausman Glomera (Agnew Products, Grants Pass OR)
  Combi-Press, Switzerland
Screw Compactor -
  Canadian Car (Pacific), Vancouver, B.C., Canada
  Taiga Industries, San Diego CA
  Presto-Log, Lewiston ID
  Spanex Sander, Uslar, W. Germany

Orientation and Vibration

     Residue geometry will sometimes permit a signifi-
cant increase in bulk density. In 1950 we found that
by pneumatically "snowflaking" flat wood chips into a
railcar, load density was increased from 18 Lb/CF up to
26 Lb/CF (291-421 Kg/m^3). Slightly overloading the cars
allowed further compaction from the running vibration of
the car. This 40% increase in compaction permitted
optimum load weights which greatly extended the shipping
radius. These techniques, now used universally in pulp
chip handling, are often ignored when considering "new"
residues.
     Cubed or extruded crop residues can be oriented and
packaged-baled to increase bulk density. As yet there
are few markets for bagged crop residues.
     The Molitorisz Hayroll winds grass or alfalfa hay
into a cigar-like roll which is cut into short lengths,
taking advantage of stem orientation to achieve density
(1). The roll is similar to the microstructure of a
cellulose molecule. But we found it impossible to make
in the field with stiff brittle straw.

Size Reduction

     Proper Material Form, Size and Composition are es-
sential for good densification. Baling can densify most
bulk materials, but agricultural residues in their bulk

form are resistant to the compression, heating and
material flow required for further densification.
Straws are structural tubes of highly crystalline
cellulose protected by a waxy outer surface. In order
to change this geometry for further densification the
tubes must be split by hammers or attrition milling.
During fine grinding ($\frac{1}{4}$") the denser wax and soluble
contents are redistributed into the fine fraction.
Leaves are shattered or shredded. If straw is coarsely
ground (1$\frac{1}{2}$") for cubing, a chemical or steam treatment
is required to dissolve waxes.
   Chopping or grinding is best done close to the use
point in preparation for direct combustion or processing.
   Coarse chopping crop residues for in-plant handling
and processing varies with the product requirements.
Reel chop from the field can come in 1" to 4" lengths
with bulk densities of 2 to 4 Lb/CF (32-64 Kg/m^3). For
straws we found Japanese feed markets required 2" to 3"
lengths (3 Lb/CF, 10 HP Hr/Ton) but domestic feed, fuel
and fiber products could be made using a 3/8" screen
(5 Lb/CF - 20 HP/Ton/Hr). Hammermills are better than
shredders. Tub grinders have low efficiency and are
best for mixing. Cornstalks, husks and leaves can be
cubed readily using a 3/4"-1$\frac{1}{2}$" screen.
   Leaves and vines must also be shredded for densification. Mint straw, hop vines, cotton gin trash and
municipal refuse can be coarsely cut and cubed with dry
materials when possible.

   Strawdust. We find that finely grinding straw into
"strawdust" which will pass a 1/8"-1/4" screen is a
useful form for suspension firing (2). Its bulk density
(8-12 Lb/CF - 129-194 Kg/m^3) allows a short haul in
self-unloading trailers, which can readily meter the
fuel into a suspension firing system. We demonstrated
this method in particleboard chip dryer and boiler applications. Some residues such as flax shives and dry
sawdust are close to this state without grinding. Tri-
Valley Growers in California is using this technique
with peach pits for seasonal process steam generation in
a converted natural gas boiler (3).
   Grinding long straw to 1/4"-1/8" consumes approximately 40 HP Hr/Ton (107 MJ). Careful selection of
the hammermill and knives can minimize horsepower consumption.

   Sawdust, chips and shavings are three dimensional
structures with good compression strength and few inherent binders. They are best fired as is, dried if

necessary. For cubing or pelleting they must be ground
or "fibrized" and combined with amorphous bark, lignin
or other binders.

Fibers can be cubed or extruded. Cotton motes,
flax fiber, dry refined wood or straw fibers can be
cubed to 16 Lb to 18 Lb/CF (256-288 Kg/m^3). With good
moisture distribution and properly designed infeed systems, hydrogen bonding and edge glazing can be employed
to cube fibers without losing fiber quality.

Wet Fibers can be used to densify ground materials.
Complete mixing of wet pulp or fiber with ground straw
in the proper proportions will not only bind the ground
material but increase cube density by 18 to 20%.

Stacks, Bales and Bale Compressors

The most expedient methods for rapid field removal,
on-farm use and short hauls (2 miles) are breadloaf
stacks of flailed straw (2 to 3 Lb/CF - 32-48 Kg/m^3)
and large round bales. Although large in size (8' x 10'
x 20' or 6' in diameter by 5' long) they store well outside in most climates. A large bale or stack burner is
now available in the midwest for on-farm grain drying
(4).

The Rectangular Bale is still the most versatile
package for collection, storage and delivery of field
crop residues to an industrial user. The three-wire
"restrained" straw bale (18"x22"x48") is about 10 Lb/CF
(160 Kg/m^3); weighs about 100 Lbs (45 Kg); can be made,
stacked and unstacked mechanically; can be handled manually if necessary; can be handled in units of 56 bales
(2.5 Tons) for storage, loading and unloading. Slightly reinforced balers can readily produce 15 Lb/CF bales
(240 Kg/m^3) for optimum truck weights.
The large Hesston baler produces a 1200 Lb (545 Kg)
straw bale 4'x4'x8', 10 Lb/CF (160 Kg/m^3) requiring a
stack or large bale grinder at the use end and somewhat
special handling equipment.
Balers use a ram pressure of 30 to 50 psi (0.2-
0.4 MPa) on the end of the bale at 50 cycles per minute
to produce 10 Lb/CF (160 Kg/m^3) straw bales. The
higher density baler (15 Lb/CF - 240 Kg/m^3) uses a ram
pressure of 50 to 100 psi (0.4 - 0.7 MPa) with a higher
strength wire or plastic twine. Some field rate is
sacrificed for higher density.
Bale compactors have been used for increasing bale
density. In Oregon a large press was built by Gender

Machine Works to compact 16 standard 2-wire bales (14"x
18"x48", 6 to 8 Lb/CF - 96-128 Kg/m^3) into one large
1200 Lb bale (545 Kg) for export. The press reached a
suitable export or container density of 27 Lb/CF
(433 Kg/m^3) after applying pressures up to 1200 psi
(8.3 MPa). The bales were bound with metal straps and
the product containerized for export. After 2 seasons
the Japanese buyers conceded they could not handle the
bales; and the company producing them went broke.
     In England, the Bradbury Bale Compactor was devel-
oped to make a small package from regular two-tie straw
bales (5). The press compresses an 18"x14"x36" bale to
approximately 12" in length. Density is increased from
6 Lb/CF (96 Kg/m^3) to 18.5 Lb/CF (296 Kg/m^3). The resultant cube retains the maximum weight the English
farmer is willing to handle (40 Lbs - 18 Kg) but presents some problems in loading and storing. Other
research in the UK is directed toward achieving a
15 Lb/CF (240 Kg/m^3) field bale from straw (6).
     Fibers are readily baled with wire or into paper or
plastic bags. Refined wood and straw fiber for hydro-
mulch, shredded recycled paper for insulation or mulch,
and RDF are commercially extruded with piston balers at
50 to 100 psi (0.4-0.7 MPa) producing 16 to 18 Lb/CF
(256-288 Kg/m^3). Cotton motes, linters and flax fiber
are baled at similar pressures into wire- or twine-
bound bales. Under certain conditions fibers can be
cubed.

Pelleted Agricultural Residues

     Pelleted Byproduct dusts or meals such as seed and
grain cleanings are currently the principal agricultural
residues which are pelleted. They contain higher protein and more readily digestible cellulose compared with
other residues. Since minimum preparation is required
they exemplify the most economic form of residue for
densification. Purchased at $15 to $20/Ton, cleanings
are made into $45 to $50/Ton feed products for horses,
dairy and small animals and exported throughout the
world (7). Our 15 such producers in the Willamette
Valley are usually associated with seed cleaning plants
or feed mills. Fuel pellets would be a good use for
damaged or contaminated seed cleanings. As yet we have
no local markets.
     The Dakota Pellet Company makes fuel pellets from
flax shives (8), which are sold to the city of Water-
town, South Dakota and several local industries for
$40 to $50/Ton. As a low cost pulp mill residue, the
shives and dust from decorticated flax are a preprocessed residue.

Cubes, Pellets and Extruded Logs.

   Compressed and extruded products from biomass residues represent a high level of material preparation and
processing. Wet residues must be mixed with dry residues or excess moisture evaporated. Materials must be
ground to a minimum size. Mixing, conveying and feeding
materials to a press or extruder must be carefully organized.
   The specific density of the final product is increased to as much as 1.3 (80 Lb/CF). Bulk densities
are increased to a range of 16-40 Lb/CF (256-641 Kg/m^3)
depending on the material.
   In order to relieve the stresses of compression at
higher densities by thermoplastic processes instead of
with wire or twine, several conditions must occur. A
balance between the pressure, and retention time must be
found so that the material is exposed to temperatures of
200° to 300°F (93-150°C), and pressures of 1000 psi to
1500 psi (6.9-10.3 MPa) at moisture levels of 12% to 25%
(9, 10). Moisture and agglutinant substances such as
soluble sugars, starches, extractives, phenolic acids
and lignins which will plasticize at these conditions
must be evenly distributed throughout the material. The
geometry of the material must allow a uniform fiber
matrix and intimate contact between adhesive surfaces
during compression.
   Natural binders occur to varying degrees in agricultural and wood residues. Low in lignin, straws
contain no good binders in sufficient quantity for the
lower pressure systems. But the addition of 20% cornstalks, alfalfa, hay, bark or other soluble-rich materials produces good cubes and pellets. Acids or bases
(4-5%) will partially hydrolyze or saponify cellulosic
materials for self-bonding. Lye treated straws cube as
readily as alfalfa hay but chemical additions are dangerous, corrosive, costly and contribute to slagging in
furnaces.
   Lignin sulfonate byproducts from sulfite pulping
make excellent binders for densified products (11, 12).
Like other adhesives they require complete blending and
adequate heating to make durable products. In excess
they increase pelleting horsepower. Although available
in quantity in the Northwest ligninsulfonates are
presently burned for lack of markets.
   Although sawdust contains sufficient lignin for adhesion at extrusion temperatures, the addition of bark
with its diverse fibers, phenolics, lignin and waxes is
often necessary for good pelleting or cubing (13).
   Extruded logs reach higher temperatures and pres-

sures which causes more complete heat transfer and apparently permits compaction with fewer binders.

Composition of Residues

	Grass Straw	Hay	Softwood	Bark
Cell Soluble Matter	22%	56%	15%	17%
Lignin	8	7	30	41
Hemicellulose	25	8	20	20
Cellulose	41	25	34	22 *
Ash	4	4	1	(.1 to 10)
% Dry Matter	100	100	100	100

*Ash picked up in logging and storage can amount to 10% to 18% of mill residue bark.

While in intimate contact with binding surfaces, adhesives need time to cool to properly anneal the compressed material and evaporate excess moisture. It is difficult to make durable products with cubers and pellet mills that have tumbling outfeeds. With ram briquetters, piston cubers and roller head pellet mills it is possible to allow more residence time in the dies and undisturbed cooling time at the outfeed before "testing" the product's durability by dropping, rolling and tumbling it in materials-handling equipment.

Field pelleting has proven impractical for these and for logistic reasons.

Pressure and Resistance

The optimum conditions of densification are obtained by a balance between the pressure applied on a material and the resistance opposing it. Residual compression stresses must be opposed by wire or plastic ties, or relieved by adhesion.

Compression tests are a useful first step in designing densification equipment. Maximum pressures required for the desired springback or bound densities can be obtained in a ram compression chamber if heat and moisture levels are also simulated. A test on flax tow demonstrated that 100 psi (0.7 MPa) ram pressure would be required to make 16 Lb/CF (256 Kg/m^3) wire bales. In practice most field balers which exert 50 psi (0.4 MPa) on dry fibers or straw produce 8 to 10 Lb/CF bales (128-160 Kg/m^3); 100 psi (0.7 MPa) balers make 13 to 18 Lb/CF (208-288 Kg/m^3) bales on dry fibers and straws. Even at these low pressures too much moisture in residues like peavines can overheat the bale in storage to cause spontaneous combustion.

Cubes, pellets and extruded logs or briquettes require much higher pressures to take advantage of

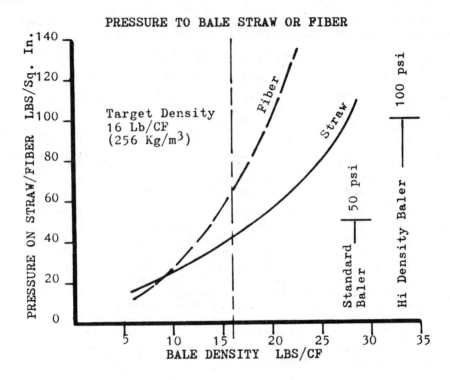

thermoplastic binding. Cubers and pellet mills exert 1000-3000 psi (7-20 MPa) on the material, ram or screw extruders may be higher. System pressures extend up from 6000 psi (42 MPa).

Pressure alone does not densify. The high momentary pressure of pocket wheel briquetters is useful for easily compressible materials with high levels (30%) of glue. Charcoal briquettes are about one-third starch. But there is not sufficient residence time at adequate temperatures and pressures to bind straw.

Precompression is also useful in densification. Field and stationary balers use methods of precompression.

We used gear cubers for precompression and processing export feed cubes. The Japanese customer required long fiber in this product at high densities. NaOH was used as a binder and lubricant. Dry straw was chopped and treated with H_2O and NaOH. This mix was cubed with the alkaline hydrolysis taking place in the cuber. The warm product (16 Lb/CF - 256 Kg/m^3) was packed in 88 Lb (40 Kg) charges into a ram extruder. Each charge was extruded into a burlap bag. The bagged product (42 Lb/CF - 673 Kg/m^3) was allowed to cool and evaporate excess moisture (heating continued for 2 to 3 days) before loading. Extrusion pressures on the straw reached 800 psi (5.5 MPa).

Power Requirements for Densification

We have assembled the following information from experience and from manufacturers. Equipment is usually sized by experience on forage feeds. Willson explored thermodynamic limits for wood pellets (9) and arrived in the 50-100 HP/Ton/Hr range we find in the field.

Power economies can be made in densification as in other processes by proper attention to material size and composition, moisture application, blending, material orientation and feeding, and sizing electric motors. Power requirements for the entire production system are usually two to three times that of densification alone.

Demand ranges from 37 to 100 KW/Ton (133-360 MJ/T) for densifiers and 55 to 240 KW/Ton (198-864 MJ/T) for a densified product.

Power Required for Densification

Method	Densifier Only Connected HP HP/Ton	KW/T	Total System* KWH/Ton
Field Collection			
Stacks and Bales	5-15	4-11	0.5-0.9gal#
Hammermill Straw			
to 1/2"	30	22	40
to 1/8"	50-60	37-45	55-75
Cubers			
Straw + NaOH	50-75	37-56	55-80
Hay or Corn	50	37	110-120
Straw + Hay	60	45	120-130
Cotton Gin Trash	80	60	135-140
Straw + 10% Binder	100	75	180
Pellet Mill - Ring Die			
Straw + Binder	50-85	37-64	90-120
Preheated Bark + Wood	40-60	30-45	70-140
Ambient Bark + Wood	100	75	80-110
Extruders	80-130	60-95	80-240

*Includes material preparation and auxiliary motors, less mobile fuels in plant.
#Gallons diesel @ 140,000 BTU/gal; 20-37 KW thermal (74-133 MJ). 1 KWH = 3.6 MJ.

The Densification Balance Sheet

The overall energy efficiency of densification
systems may be misleading. Only 10 to 25% of input
energy is lost in making a biomass fuel pellet, even
if electrical energy is thermally generated from the
same residue when used as feedstock.

In processing sawmill residue, wet bark has to be
dried and ground. Some agricultural residues such as
straw must be ground and treated or combined with another material as a binder in order to be cubed or
pelleted for use. Drying and grinding bark, or grinding
and treating straw are processing costs that result in
similar overall process plant costs for these two very
dissimilar materials. Both are abrasive, requiring
extra equipment maintenance expense.

Capital costs and overhead vary among producers,
and include expected return on investment. Fuel and
power vary geographically. Labor and maintenance are
significant processing cost factors as shown in the
rough cost array that is a result of our experience with
cubing and pelleting systems for various residues.

Raw material costs for sawmill residues such as
bark, and agricultural residues such as straw, do not
include any value for the materials. Straw cost does
include baling, however. Production is based on 6 TPH.

	$/Ton	%
Capital (including 20% ROI)	$8.50	35%
Fuel and Power (incl. 105-112 KWH/T)	5.50	23
Labor	4.50	19
Repair and Maintenance	4.00	17
Overhead and Supervision	1.50	6
Total Processing Costs	$24.00	100%

Per Ton of Dry Product

	Bark (Sawmill)	$/MBTU	Straw	$/MBTU
Raw Materials	$13.00	$0.77	$35.00	$2.33
Processing	$24.00	1.41	24.00	1.60
Total Cost FOB	$37.00	$2.18	$59.00	$3.93

The minimum cost of feed pellets of seed cleanings
is $40/T or $2.67/MBTU. Fuel pellets of bark are selling for $38/T delivered, $2.24/MBTU; the sale price is
close to our cost shown above. Direct cost of gather-
ing and bringing field and forest residues to some
point where they can be processed is between $1.50 and
$2.40/MBTU. It appears that we must concentrate on the
lower cost mill residues and wait for market changes to
stimulate collection and densification of field and
forest residues for industry.

Conclusions

The least densification and the most direct use of agricultural or wood residues makes the most sense from an energy and economic point of view. We recommend that residue producers or users look closely at their raw materials and design combustion processes around low density, low energy processes, which combine residues for their highest and best use.

Literature Cited

1. Molitorisz, J., P. O. Box 129, Medina WA 98039.
2. Miles, T. R. The Combustion of Straw - Mobile and Stationary; The Combustion Institute, 1977 Spring Meeting Western States Section (WSS/CI Paper 77-15)
3. Stobbe, Klaus. Tri/Valley Growers, P. O. Box 948, Modesto CA 95353.
4. Stormor, Inc., P. O. Box 198, Fremont NE 68025.
5. Catt, W. R. Moving Straw from Field to Factory; 1976 Report on Straw Utilization Conference, A. R. Staniforth, Ed.; Ministry of Agriculture, Fisheries and Food, Oxford, England.
6. Klinner, W. E. Some Practical Problems of Straw Densification and Recovery. 1978 Fourth Straw Utilization Conference, MAFF/ADAS, Oxford, England.
7. Broeder, R. K. Successful Marketing of Pelletized Field Residues. Proceedings Forest and Field Fuels Institute, 1977. Winnipeg, Manitoba, Canada. Bio-Mass Energy Institute, Inc., P. O. Box 129, Postal Station "C", Winnipeg, Man. R3M 3S7, Canada.
8. Dakota Pellet Company, personal communication.
9. Willson, G. Biomass Densification: A State of the Art Report on the History, Chemistry, Thermodynamics and Machinery of Densification. July 1979. To be published. P. O. Box 3091, Richmond, Nelson, New Zealand.
10. Wellons, J. D. and Krahmer, R. L. Self Bonding in Bark Composites. Wood Science, Oct. 1973 6 (2).
11. Currier, R. A. and Laver, M. L. "Utilization of Bark Waste", PB 221876, Oregon State University, Corvallis OR 97331, 1973. Prepared for the E.P.A.
12. Odell, G. "Straw Utilization by Pelleting: A Cost and Feasibility Test". Glen Odell, Consulting Engineer, Portland OR 97204. 1974. Prepared for The Oregon Field Burning Committee.
13. Page, G. E. "Straw Handling and Densification". Agricultural Experiment Station Circular 647. 1974. Oregon State University, Corvallis OR 97331.

RECEIVED November 16, 1979.

Replenishable Organic Energy for the 21st Century

JOHN L. STAFFORD and ANDREW D. LIVINGSTON

Guaranty Fuels, Inc., P.O. Box 748, Independence KS 67301

During the past several years, the "Economic Feasibility" and the "Technical Feasibility" of a biomass derived fuel source has been argued and generally concluded. Certainly the "Technical Feasibility", limited in definition to the answer to the question, "Can it be done with existing techniques?", is established. Many studies have been exercised to address the question of the economic justification of a commercial operation for the purpose of producing a specification fuel derived from biomass raw material. The answers assembled by these studies have differed because the assumptions used to define the question have been different. A recently encountered expression, used in description of an entirely different human effort, is nevertheless appropriate in the context of human effort in general; "...while not perfect...is probably as near to perfection as possible." There must come a time in the development of an enterprise when an end to study is called and the proposal subjected to a test. The feasibility of the densification/refinement of fuels from residual biomass sources is currently being tested.

The Guaranty companies are engaged in this "Feasibility" test. The quotation marks used are for the emphasis that the technical feasibility and economic feasibility are for this discussion separated, when in truth they cannot be. There are natural laws which govern the behavior of materials and processes of science. These laws are not invented nor created but are discovered. There are within our economic system a set of analogous "natural laws". Efficient production and delivery of goods is provided by one of these natural laws of the economic system. A statement of this law is: "Nothing happens until something is sold." Sold means that the two or more parties involved in the proposal are satisfied and agreed that it is to their mutual benefit to consumate the proposal. The "Feasibility Test" is not to examine the conclusion drawn from specific hypotheses, but to test the existence of those pretexts. Guaranty Fuels, Inc., was formed to test the feasibility of a commercial enterprise producing densified biomass fuels. The companion company, Guaranty Performance Co., Inc., having determined to the limits of its own definition of the diminishing return of investigative study, that

0-8412-0565-5/80/47-130-195$05.00/0

the process is "Technically Feasible".

Ultimate technological development of the refining of biomass fuels is a long way from conclusion. It can and is being done. The first commercial embodiment by necessity has adapted many existing process technologies. The drying and suspension burning of biomass are two examples of proven technology that have been immediately incorporated. Size reduction by hammermill and densification by pellet extrusion are current techniques. Improvement and optimization of these consecutive steps are under development. Our entry into the business of refined biomass fuels is in hindsight viewed ingenuous. How else could we attempt to produce a specification, refined fuel product with no receiving specification for raw materials? The refining process removes the more troublesome variables of moisture, density, and particle size, but energy will not be contained in the product unless it was first in the raw material. The raw material source has a pronounced effect on the overall system production capacity and the product quality.

Benefits of the Use of Densified Biomass Fuels

Burning refined, woodbased fuels in existing boilers allows immediate benefit to the operator by reduced emission of particulate that for stoker fired systems has demonstrated to be within state air quality limits. No costly conversion is required since existing stoker systems are used unchanged and pulverized coal systems require minimal change. The inconvenience of "hog fuel" or green wood chips mainly derives from material handling problems that result from a minimally specified fuel. Refined fuel pellets provide a convenient, uniform fuel which meets property limits specified by the fuel supply contract. Thus the operator may anticipate the fuel properties and burning characteristics.

It has been estimated that 50% of the energy needs of large industrial users, typified by currently coal-fired 250,000 pound per hour steam generators, could be satisfied by wood sources that are now wastes, an environmental burden. The use of these wood fuel sources in substitution for the coal now burned would reduce sulfur dioxide emission in proportion to the degree of substitution without increasing smog forming nitrous-oxide or particulate emission. These waste wood sources may be collected, processed by refining, and conveniently supplied.

Atmospheric pollutants are reduced by the large scale employment of the biomass refined fuel program through first- and second-order effects. The reduction of fuel burning source emissions has been cited and is the primary effect. The waste materials that without the program would have remained an environmental burden, would also have been allowed to be destroyed by decomposition, rot, or combustion. The only controls imposed of these processes are so as to contain them within an assigned area. The contribution of these processes to atmospheric pollution are being studied by others and thus far have exposed at least one interesting datum: Carbon monoxide contribution by the minimally controlled combustion of biomass has been much greater than would be

predicted by data recently published by the EPA. This CO in the
atmosphere reacts rapidly with the hydroxyl radical, OH. Since OH
reacts with nearly all pollutants, and thereby "cleanses" the
atmosphere, the role of CO is then to "use up" the available OH
and allow the accumulation of other pollutants. The controlled
combustion of biomass refined fuels, derived from these otherwise
waste sources, completes the oxidation of carbon, and this is the
second-order effect. The conversion of wastes to refined biomass
fuels benefits the atmosphere's natural function of cleansing
itself.

Densification Plant Processes

Basic process steps in the production of ROEMMC® fuel pellets
are DRYING of the wet raw material, SIZE REDUCTION of the dry material, and DENSIFICATION of the dry, fine material. The thermal
energy for the drying process is derived from the combustion of
fines separated from the dry, fine material. The only external
energy input to the production of pellets is electrical. The
hardware embodiment of these processes is described in these para-
graphs, as employed in the Stillwater, Minnesota, plant. This
plant was designed to have an output production rate of 150 tons
per day, continuous operation.

Densification is accomplished by a 300 HP California Pellet
Mill Model 7162-3. The pellets are formed by extrusion through
nominally 3/8 inch diameter radial holes in a cylindrical die.
Pressure is exerted on a "pad" of the material to be densified by
rolls which have fixed shafts. The pellet die rotates about the
pressure rolls. Pellet length is controlled by a "cut-off" knife
positioned to clip the pellets as they exit the rotating die. The
exiting material stream from the pellet mill is mechanically lift-
ed to a pellet cooler, which discharges to a screen to remove
under-size pellet pieces and fines. The fines and pieces are
recycled to the pellet mill infeed system. Finished pellets are
conveyed by high pressure air system to tank storage.

Size Reduction is by grinding through a screen using a 300 HP
Champion Model 18304 "Magnum" hammermill. Screen opening sizes
ranging from 5/64 inch diameter to 1/4 inch diameter have been
used. Ground material from the hammermill is conveyed by air to a
cyclone receiver on the pellet mill infeed system. Fines are
separated from this material stream by screening to be used as dry
fuel.

Drying of the wet raw material is performed in a Guaranty
Performance Co. 10 feet diameter by 32 feet long, three-pass rotary drum dryer. Dry material is conveyed in the dryer exhaust
gases to a cyclone receiver which discharges into the hammermill
infeed system. Drying is accomplished primarily by convection
heat transfer between the hot dryer gas and the drying medium
(solids). The hot dryer gas is produced by the combustion of the
fine material screened from the hammermill output, in a cyclonic
suspension ROEMMC burner system. The burner in this application
is a wood-fired air heater.

Heterogenous Biomass Raw Materials

The nature of the raw materials for biomass residue derived
fuels is certainly heterogenous. This is somewhat demonstrated by
the size distributed fuel properties of Peanut Hulls. Table I
lists the fuel analyses of samples prepared by sieve separation.
This material leaves the peanut processing plant with a relatively
low moisture content, average value for the reference sample is
7.83%. Weather protection during transportation and storage would
preserve this low moisture and may eventually, but cannot currently be assured. Occasionally a simple process step can be employed
to significantly upgrade the product. In this instance, by screen
separation, a major portion of the non-combustible fraction (ash)
can be removed while sacrificing a small portion of the fuel heating value. A three-way screening operation is invisioned that
would separate the raw material stream into: 1) Large size for
densification processing, 2) Intermediate size for process
(dryer) fuel, and 3) Undersize to be discarded. The components,
from an ideal screening process would result in the material
streams listed in Table II. Discarding half the raw material ash
content may be accomplished for 5% of the heating value. Although
higher moisture, the intermediate "cut" is still an acceptable
dryer fuel.

TABLE I
PEANUT HULLS, FUEL ANALYSIS DISTRIBUTION BY SIZE

Passing US No.	Retained US No.	Fraction Total Weight	Moisture	Ash	Heat Value BTU/LB
	8	.710	.0676	.0277	8015
8	16	.102	.0689	.0311	7920
16	30	.061	.0669	.0498	8110
30	50	.068	.1585	.1928	6971
50	100	.051	.0367	.4521	5026
100		.008	.0313	.6375	3385
Average			.0783	.0672	7750

TABLE II
THREE-WAY SPLIT, PROCESS MATERIAL STREAMS

Stream Final Use	Component Stream Analysis			Fraction Total Analysis		
	Moisture	Ash	Heat Value	Mass	Ash	Heat Value
Product	.07	.03	7970	.778	.325	.800
Dryer Fuel	.11	.08	8070	.143	.175	.150
Discard	.07	.43	4970	.079	.500	.050

Densified biomass fuel in the form of extruded pellets have
been produced from a variety of raw materials and by systems that
vary the sequence of process steps from that described herein.
These variations in the processes should have little effect on the
composition of the combustible fraction of the product pellets, A
survey of fuel properties of a number of these fuel pellets is
illustrated in Figure 1. When the moisture and ash are removed

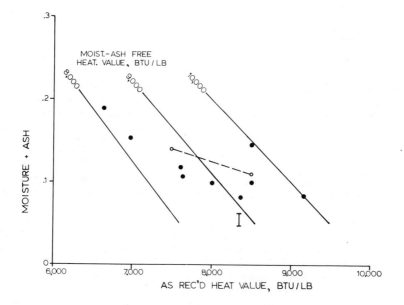

Figure 1.   Pellet fuel analyses

from the fuel, the balance is combustible material. As the
moisture + ash fraction of the fuel increases, the "as-fired"
heating value on a "moisture-free and ash-free basis" might be
expected to be fairly constant on the assumption of a simple mixing theory. The data of Figure 1 would suggest, however, that the
addition of moisture and ash content is accompanied by a loss in
the moisture- and ash-free heating value. This effect would
result for the degradation of the fuel raw material by partial
oxidation of the combustibles. Some of the samples analyzed for
the Figure 1 survey were of pellets produced from sawdust raw
material taken from the bottom of the sawdust discard pile of a
saw mill. These had obviously suffered some decomposition. Fuel
analyses of these samples support the partial oxidation
hypothesis.

Process Equipment Capacities

The dryer air-heater system was somewhat over-sized for the
evaporation load imposed by the average 45% moisture wet raw
material. The significant adjustments made were those required so
that the dryer would be able to induce all of the "low-fire" products of combustion. A fundamental principle of the burner design
is the control of the combustion chamber gas temperature in order
to prevent the melting or fusion of the solid non-combustible
(ash). The burner system includes a refractory-lined cyclone
furnace from which the dry ash is removed. The control of the gas
temperature to avoid slagging is accomplished by the use of large
amounts of "excess" combustion air. Although there is no thermal
efficiency loss for this air-heater application, there results a
large quantity of gaseous combustion products that the dryer
system must be able to totally induce for there to be no thermal
loss. The dryer exhaust fan was sped-up to further aid the system
in using all the product gases.
A necessary compromise has been suggested for the optimum
boundary conditions to impose on the hammermill and pellet mill
systems. Hammermill horsepower should be reduced by drying the
material to a lower moisture and by the use of a larger hammermill
screen opening. Both of these adjustments are expected to reduce
the quality and production throughput capacity of the pellet mill.
The original arrangement of the plant was to dry the infeed
biomass, grind it, and accumulate the ground material in a surge
bin. Material was metered from this bin to the pellet mill and as
fuel to the dryer air-heater. It is generally accepted that the
ground material for pelleting should be produced by grinding
through a screen with openings less than the pellet diameter, and
that extreme fines are not sufficiently compacted in the pellet
mill. Smaller particle size production from the hammermill may by
virtue of the greater specific energy input produce a greater
moisture reduction associated with the size reduction and this may
contribute to the failure to compact the extreme fines.
Production experiments were performed to demonstrate the
effect of ground material size on the pelleting operation. Pellet

quality was determined by judgement of pellet density, hardness, durability, and flowability. Pellet approximate length and visual appearance indicate combinations of desirable properties. "Good" pellets appear shiny, slick, or smooth, while "Bad" pellets are dull finished, rough, or scaly. Near optimum pellet length is 3 to 4 times the pellet diameter. Ground material particle size was adjusted in steps by replacing the grinding screen in the hammer-mill. Screen opening diameters of 5/64, 1/8, 3/16, and 1/4 inch were used. Pellet diameter was 3/8 inch. In general an optimum hammermill screen opening size was not demonstrated. Although an optimum of 3/16 inch diameter was indicated for one species of bark material, most materials demonstrated continuously improved pellet quality and horsepower economy for larger screen opening size.

The particle size experiments demonstrated pellet quality was improved by larger sized particles. The dryer energy source, cyclonic suspension burner, prefers small size particles with rapid ignition and short residence requirement for complete com-bustion. Therefore, the best material for each process, pelleting and suspension burning, is produced by screening the ground mater-ial. Currently the dry material is ground through a 1/4 inch diameter hammermill screen opening and separated by a 22 mesh screen into dryer fuel and pellet mill feed streams.

Some further relief of the hammermill would be achieved by performing this separation before the hammermill. A second unit or a replacement unit of greater capacity may have to be the solution of what now appears to be the plant "bottle-neck". This arrangement of pellet mill infeed properties permits the pellet mill to operate slightly in excess of the plant design throughput rate for the short period allowed by the surge bin feeding the pellet mill. It is believed that the pellet mill would be capable of maintaining very near the design production rate on a daily average basis. An average rate of 120 tons per day is used in the cost references to follow.

User Experience with Densified Biomass Fuels

Densified fuel pellets may be used as boiler fuel in place of coal in stoker-fired furnaces. Direct substitution of fuel pel-lets was demonstrated in two informally reported instances with no furnace adjustments. The particulate emission from each of these installations were monitored with the emission rate results shown in Figure 2. As a comparison, the North Carolina Administrative Code allowable particulate emission rate schedule (a function of heating rate) is also shown. The environmentally acceptable part-iculate emission contributes to the economic justification of the conversion of small-to-medium size steam generators to the use of pelleted biomass fuels. Two such installations are now using pelleted fuels on a routine basis, and a third has concluded the economic benefit by entering into a contract agreement to purchase pelleted fuel.

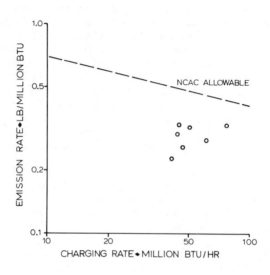

Figure 2.   Stoker fired boiler emissions grate burning pelleted fuels

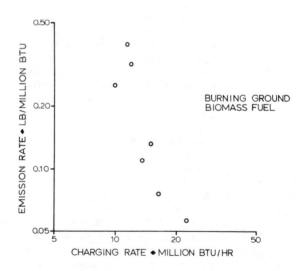

Figure 3.   Demonstration unit roemmc burner emission rate

Pulverized coal-fired steam generators should experience similar emission rates as the ROEMMC burner which is a cyclonic suspension burning system. Several emission tests have been performed on the demonstration burner unit in Independence, Kansas. The results of these tests are shown in Figure 3. For these tests the burner system was operated without the recovery of energy from the gaseous products of combustion. Emission sampling was done in the relatively high temperature gas stream exiting the cyclone furnace.

Production Costs Considerations

The "Feasibility Test" being conducted must demonstrate for success that a market may be established for the raw material that competes with the alternative uses or disposition of the material. This market will define a schedule of raw material prices such as is indicated by Table III. At the same time, a successful test result will demonstrate that the product densified fuel competes with alternative fuels, establishing a product sales price on the order of the Table IV invitation. Between these schedules, there must be the money to accomplish the physical processing and attract the financing. The "test" is not to presume either of these schedules, but to demonstrate the three-fold mutual benefit: 1) A market for the raw material, 2) An attractive investment, and 3) An acceptable price for the product energy.

TABLE III

RAW MATERIAL

Guaranty Fuels, Inc. shall purchase sized wood and other biomass materials on the following conditions:
1. Must be delivered in self-unloading trucks (5 ton or over).
2. NO metal, glass, rocks, or dirt shall be in materials.
3. All material must be 3 inch all sides or less.
4. Moisture in material must not be over 55%
5. A scale ticket must come with each load.
6. Price per ton: Sawdust (wet)..........................$5.00
                   Whole tree chips.......................$6.00
                   Bark (hogged)..........................$3.00
                   Dry hogged wood........................$6.00
                   Sawdust (dry)..........................$6.00
7. Terms: Invoiced, paid weekly.

The production cost break-down itemization includes, in addition to the above indicated cost of raw material and cost of financing, the cost of energy for the process, labor, maintenance and repairs, insurance, taxes and royalities. These costs have been variously listed in the study reports. Estimates range from approximately 15 dollars per ton to over 20 dollars per ton of finished product. It is not the purpose here to present a detailed balance sheet or income statement. There has been far too little operating time of the Stillwater plant to establish the long-term costs yet. The replacement of machine parts has been a

result of speed changes to evaluate capacity, system rearrange-
ments, and so forth in addition to the replacement of worn or
broken parts. The production status of the plant has been dependent on the status of the negotiations required to establish the
balance suggested by the price schedules of Table III and Table
IV. This aspect of the operation has settled down a bit and a
longer run of experience operation is now under way.

TABLE IV

SALE OF ROEMMC FUEL PELLETS

Specifications:
      BTU per pound............................8,200 Average
      Moisture.................................10% Maximum
      Ash......................................5% Maximum
      Density (LB/CF)..........................38 Average
      Fines (by weight)........................5%
Price:
      Are F.O.B. Bayport, Minnesota
      On Contract, 10 ton or more per day........$31.00 per ton
      NO Contract, 1 to 10 ton...................$35.00 per ton

The energy costs, for the plant as defined, are more easily
established. A long run of operation is not required to yield a
fairly good idea of the production capacity of the given equipment
arrangement. It was stated before that the production capacity of
the plant, dependent to a degree on the specific raw material, is
120 tons per day sustained operation. Thermal energy requirement
for the process is supplied by the combustion of fines separated
from the process. The externally supplied energy is electrical
power. A total of 1114 connected horsepower is distributed among
the production subsystems (Table V).

TABLE V

HORSEPOWER DISTRIBUTION

	Connected Horsepower	Running Amperes
Dryer System	140.5	161.5
Burner System	88.5	103.
Hammermill System	378.	423.5
Pellet Mill System	378.2	413.5
Storage Transfer	87.5	47.5
Air Quality System	16.	15.2
Utility Systems	25.3	32.5
TOTALS	1114.0	1196.7

Air quality influence is actually somewhat greater since the
collection equipment imposes additional pressure loss on the air
systems. The listed running amperes are for a normal load on each
of the affected motors. As such they represent a typical motor
current total, adding to about 90% of the nameplate amperes.
Electrical energy demand represents an energy account of 2.5 million BTU per hour. On a specific unit of production basis using

the 120 tons per day capacity, this is:

222.8 HP hours/finished ton, or
0.5 million BTU/finished ton.

Two significant differences exist between densification processes available today. Pelleting and Cubing are extrusion processes relying on the forces of friction at the extrusion die surface to compact the material. Another process of Briquetting uses the force between opposed roll-dies to compact the material. Comparison production tests were performed using the roll-die, briquetting machine. Two series of briquetting machine runs were performed, one on a small sample shipped to the machine manufacturer, and a more extensive test with the machine on-loan at the Stillwater plant. Energy costs were no greater per ton of finished product. Further tests may demonstrate less energy cost than for the pellet mill operation. The finished product was acceptable for air system transfer and tank storage. Maintenance and wear costs require considerable conjecture for comparison at this stage.
The urgency of conversion from fossil-based energy sources is in order to prepare for the inevitable depletion of these non-renewable sources. Preparation and orderly conversion to renewable energy sources may forestall the panic-stricken scramble for depleted sources within the next decade. The United States Central Intelligency Agency projects world demand for oil to reach production capacity in the next few years. Sharply rising price will effectively ration available supplies without regard for Middle-East politics or balance-of-payment. The supply of biomass to support the process densification/refinement described herein will not support the energy needs of all the steam production of this country. The use of these fuel pellets may not prove to be the best alternative source for everybody. The direct combustion of biomass fuel pellets may eventually be opposed as a poor use of the feedstock, supported by second-law arguments. The course we take as individuals must be according to our visionary aptitudes. For the present, for the economic conditions and technology existing as the depletion of fossil sources is felt, Densified Biomass fuels have been and are gaining in acceptance as a means of limiting the ever escalating cost of energy with environmental acceptance.

References

"Burn Trees Not Coal to Reduce SO_2", SCIENCE NEWS, March 17, 1979.
"Plant Burning is Major CO Source", SCIENCE NEWS, March 17, 1979.
"The International Energy Situation: Outlook to 1985", U. S. Government Printing Office, S/N 041-015-00084-5, 1977.

PYROLYSIS, GASIFICATION, AND LIQUEFACTION PROGRAMS IN THE UNITED STATES

Gasoline From Solid Wastes by a Noncatalytic, Thermal Process

JAMES P. DIEBOLD

Naval Weapons Center, China Lake, CA 93555

The need to use organic wastes as an alternate feedstock for the production of petrochemicals and/or gasoline is becoming increasingly apparent as existing oil fields become depleted and new oil is found primarily at greater depths and/or further out at sea. The relative value of crude oil has risen very sharply in the 1970s and is widely predicted to continue to do so. Solid organic wastes as boiler fuels will compete only with coal and will not be as economically attractive as if they were an oil substitute. Consequently, the conversion of organic wastes and biomass into synthetic liquid hydrocarbons could serve to greatly reduce the dependency on foreign oil of the United States as well as that of a great majority of the Third World nations which are very capable of growing biomass. The economics of the process to be described are relatively attractive now and will improve as the price of oil escalates.

Several different processes have been proposed for the conversion of solid organic wastes to a more usable liquid or gaseous form to be utilized as fuel or petrochemical feedstocks. Principally the commonly discussed processes involve the biological conversion to alcohols, the catalytic chemical conversion to methanol or Fischer-Tropsch liquids via a carbon monoxide and hydrogen synthesis gas, or the thermochemical formation of gases or oxygenated liquids by pyrolysis. Pyrolysis is defined as the decomposition of organic material at elevated temperatures. The process to be described uses a very special case of pyrolysis.

The China Lake process employs an adaptation of the petrochemical pyrolysis process used to crack hydrocarbon oils to make primarily ethylene with some propylene, butylene, and aromatic by-products. However, rather than using crude oil or naphtha as the feedstock, the process uses oxygenated feedstocks such as municipal solid waste which is primarily cellulosic in nature. This results in pyrolysis gases containing similar product distribution of hydrogen, ethylene, methane, propylene, butylenes, and aromatic compounds, but with carbon oxides also present. After purification of the pyrolysis gases, the olefinic products (ethylene, propylene, butylene, etc.) may be used in the same manner as if they had been derived from petroleum feedstocks. Of course, the yield of olefins per unit weight of oxygenated cellulosic feedstocks is lower than that obtained from the hydrocarbons found in petroleum feedstocks. The advantage of the process is that organic wastes and biomass materials are renewable resources that are found or can be raised within many countries which now must import the hydrocarbons they need for transportation and petrochemical feedstocks. The widespread implementation of this process to form olefins from cellulosic materials could serve to slow the dramatic rise in crude oil prices. With the product distribution available by very rapid and selective pyrolysis, a wide variety of petrochemical products can be made with existing technology such as the catalytic formation of alcohols, polyethylene, polypropylene, or gasoline by the polymerization of the olefin mixture.

Most of the effort at China Lake was directed toward demonstrating, at the bench scale, that polymer gasoline could indeed be made noncatalytically from the olefins formed by the selective pyrolysis of municipal solid waste (MSW). Funding for the bench-scale demonstration was provided by the Industrial Environmental Research Laboratory (IERL) of the Environmental Protection Agency, beginning in 1975 (EPA-IAG-D6-0781).

The China Lake process to make polymer gasoline from trash involves: (a) the selective pyrolysis of finely ground organic wastes to form gases containing the desirable olefins; (b) the compression and purification of those gases to concentrate the olefins; and (c) the noncatalytic polymerization of the olefins to make a synthetic crude oil containing about 90% high octane gasoline, with some fuel oils and lubricating oils as the by-products. All char and by-product fuel gases formed such as hydrogen, carbon monoxide, and methane would be consumed for process energy.

Selective Pyrolysis to Olefins

Equipment. The pyrolysis reactor that evolved in this program is one in which the residence time is minimized by exposing the feedstock to the pyrolysis temperature in such a way as to have extremely high heating rates; i.e., an entrained flow reactor using a powered feedstock having a particle size on the order of 250 μm (about half the size of table salt). The evolved bench-scale reactor consists of a 2.5-cm screw feeder to meter the feed. As the feed is metered to the reactor, a carrier gas (normally carbon dioxide) fluidizes it and pneumatically transports it to the steam ejector. The steam ejector pumps the entrained particles into the pyrolysis reactor which consists of a long stainless steel tube 1.9 cm in diameter. Reactor lengths of 2 and 6 meters have been evaluated. The pyrolysis reactor tube was formed into a helix having a diameter of about 38 cm. This compact reactor is located inside of a gas fired kiln formed by loosely stacking foamed alumina insulating blocks. Rather than using electricity as the energy source for such a small system, a gas-fired kiln was used to demonstrate that the needed heat transfer rates into the tubular reactor were attainable in this fashion. The kiln temperature in the pyrolysis reactor zone is normally 100 to 200°C higher than the temperature of the pyrolysis stream at the reactor exit. After the pyrolysis stream exits the reactor it passes to an 8-cm-diameter cyclone separator which removes most of the char. The gaseous stream then passes to the water quench drum to precipitate the tar vapors and steam. After an additional scrubbing step, the gases pass either to the flare for disposal or to the compressor after filtering. The pyrolysis reactor is operated at atmospheric pressures. The cleaned pyrolysis gases are analyzed with an automated, multicolumn gas chromatograph capable of analyzing for the hydrogen, carbon oxides, and hydrocarbons up through pentane. The pyrolysis system schematic is shown in Figure 1.

Experimental Results. Selected experimental results from pyrolyzing material derived from municipal solid waste (ECO II by Combustion Equipment Associates) are shown in Table I. The ECO II used contained about 10% ash by weight. As can be seen, the pyrolysis experiments were characterized by reactor exit temperatures of 700 to 840°C, relatively dilute conditions of 55 to 172 grams of dry feed per cubic meter of pyrolysis gas stream at the reactor exit conditions, and short residence times of 42 to 195 msec. The residence time was estimated by assuming that: the entrained solids had the same velocity as the turbulent gases; the pyrolysis gases were formed at the entrance of the reactor; and the gases attained the exit temperature at the entrance of the reactor. Due to the relative dilution of the pyrolysis gas stream, the last two assumptions would not be expected to result in large errors in the residence time. Although the residence time was not a rigorously derived parameter, it is indicative of the relative times involved.

Attempts to correlate the data in Table I with pyrolysis temperature, moisture content of feed, feed rate, and ratio of carrier gas to feed did not produce major trends. It was noted, however, that the dilution of the pyrolysis gas seemed to have a very significant effect on the amount of desirable hydrocarbons formed as shown in Figure 2 for the 6-m coil and in Figure 3 for the 2-m coil. The 6-m coil produced slightly better results at less dilute conditions than the 2-m coil, possibly due to more complete pyrolysis of the larger feed particles. There did not appear to be an advantage to residence times less than about 150 msec, although previous experimentation in which the pyrolysis gases were held for about 5 seconds at the pyrolysis temperature indicated an increasingly rapid disappearance of the propylene and butylenes at temperatures above about 725°C.[1] In fact, the small propylene and butylene contents decreased at the higher temperatures even with the shorter reactor.

The correlation of olefin production and dilution shown in Figure 2 for the 6-m coil is better than that shown in Figure 3 for the 2-m coil. This is probably because the 2-m reactor data are more extensive and, therefore, differences in the feedstock composition and particle size due to stratification of the heterogeneous MSW in the barrel would become more important.

Carbon monoxide was the predominant gas specie by weight in the pyrolysis gases. For this reason, it was chosen for a basis of comparison to determine if the other gaseous products varied substantially relative to each other. Table II shows the molar quantities of the gaseous pyrolysis products relative to carbon monoxide. Interestingly enough, these ratios of products are considerably more constant than the overall yields per unit weight of feedstock. The most consistent ratios were those of carbon dioxide, methane, and ethylene; although it should be noted that the carbon dioxide consistency was based on a sample size of only three experiments whereas the rest of the data was based on 53. (Since substantial amounts of carbon dioxide carrier gas were often lost in the feeding system, the vast majority of the experiments were unsuitable for the generation of quantitative carbon dioxide data.)

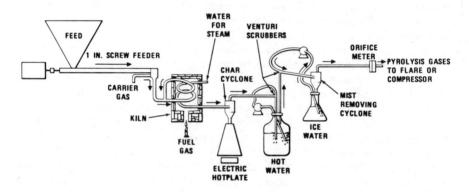

Figure 1.   Pyrolysis schematic (12)

TABLE I. Experimental Pyrolysis Data.

Run	Length of pyrolysis reactor, m	Wt-% H₂O in feed	Pyrolysis temp, °C	Wet feed, kg/hr	Carrier gas, m³/kg dry feed	Steam, kg/kg dry feed	Dry feed, gm/m³ gases	Residence time, sec	CH₄	C₂H₂	C₂H₄	C₂H₆	C₃H₆	C₃H₈	C₄H₈	C₅⁼/C₆⁺	H₂	CO	Char	Comments
					Conditions							Results with MSW (ECO II) Product/100 gm dry ECO II, gm								
21	6.1	1	700	2.84	0.40	1.0	126	0.195	5.6	0.5	7.6	0.7	2.8	0.6	2.2	3.5	0.7	29.6	11.7	
22	6.1	1	740	4.22	0.25	0.7	172	0.178	3.2	0.5	6.1	0.5	2.1	0.1	1.7	3.2	0.5	22.0	...	
23	6.1	1	810	3.09	0.31	0.9	120	0.171	4.2	0.6	8.1	0.8	3.3	0.2	2.5	1.8	0.7	30.1	...	
24	6.1	1	750	3.13	0.54	1.2	95	0.133	5.1	0.8	9.4	1.0	3.3	0.2	2.2	3.2	0.8	36.8	23.0	
25	6.1	1	730	1.77	0.95	2.4	56	0.139	4.5	1.0	11.4	0.9	3.9	0.2	2.5	3.4	0.9	35.9	...	Smelly tars
26	2.1	1	~600	...	...	...	...	...	...	...	...	...	...	...	...	...	...	...	...	
27	2.1	1	730	3.43	0.49	1.1	111	0.050	2.5	0.5	6.7	0.6	2.0	0.1	1.8	3.7	0.6	22.2	16.1	
28	2.1	1	780	1.86	0.90	1.9	62	0.052	3.5	1.3	9.4	0.6	2.3	0.1	1.5	3.0	0.9	27.4	12.5	N₂ carrier
29	2.1	1	770	2.31	0.88	1.6	68	0.045	5.2	1.6	9.5	0.9	2.5	0.1	2.5	1.9	1.2	42.5	22.7	
30	2.1	1	760	1.91	0.98	2.3	56	0.049	4.1	1.2	11.3	0.7	2.3	0.2	2.1	5.0	0.9	34.8	17.9	
31	2.1	10	720	2.16	0.87	1.9	67	0.053	3.1	0.9	9.1	0.8	2.0	0.1	2.0	5.8	0.7	27.1	12.7	
32	2.1	10	700	2.14	0.88	1.9	68	0.054	3.0	1.0	9.1	0.7	2.4	0.1	2.4	3.9	0.7	27.0	14.3	
33	2.1	10	760	2.34	0.80	1.8	68	0.049	3.8	1.2	10.5	0.8	2.5	0.1	2.3	2.3	0.8	32.3	20.4	
34	2.1	10	830	2.72	0.69	1.6	70	0.044	4.3	1.4	10.7	0.7	2.0	0.1	1.8	2.9	0.9	34.7	10.3	No cyclone
35	2.1	10	810	2.12	0.89	2.3	55	0.043	3.9	1.4	10.7	0.8	1.7	0.1	1.5	5.2	1.0	32.3	17.2	
36	2.1	10	820	2.57	0.71	1.7	66	0.044	4.6	1.3	10.3	0.8	2.4	0.1	2.0	2.9	1.0	38.3	...	
37	2.1	10	820	3.47	0.54	1.2	86	0.042	3.9	1.3	9.1	0.7	1.6	0.1	1.6	3.4	0.9	30.2	25.0	
38	2.1	10	770	3.79	0.49	1.1	102	0.046	3.3	0.5	8.1	0.4	2.8	0.2	2.4	4.4	0.7	29.6	17.2	
39	2.1	10	840	2.20	0.84	1.9	60	0.046	4.5	1.4	10.0	0.6	2.1	0.1	1.4	2.8	1.1	34.3	18.2	(Char includes tars)
40	2.1	10	750	2.29	0.82	1.8	69	0.051	3.3	1.1	9.2	0.8	2.8	0.1	2.4	3.9	0.8	29.6	15.8	

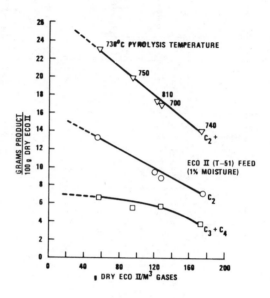

Figure 2.   Products vs. concentration, 6.1-m pyrolysis coil (12)

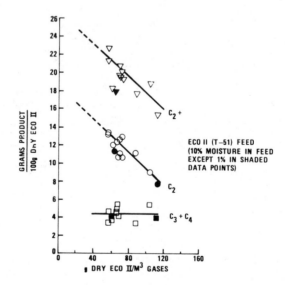

Figure 3.   Products vs. concentration, 2.1-m pyrolysis coil

TABLE II. Gaseous Pyrolysis Products; Molar Ratios and Weight - Percent.

Product	Product, moles/moles CO	σ_x	$\frac{\sigma_x}{\overline{x}}$	N	Volume, percent	Weight, percent
CO	1.00	. . .	. . .	. . .	45.6	50.8
H_2	0.35	0.08	0.23	53	16.0	1.3
C_2H_4	0.28	0.05	0.18	53	12.8	14.2
CH_4	0.24	0.04	0.17	53	11.0	6.9
CO_2	0.16	0.02	0.13	3	7.3	12.7
C_3H_6	0.045	0.015	0.33	53	2.1	3.4
C_2H_2	0.035	0.010	0.29	53	1.6	1.6
C_4H_8	0.028	0.011	0.39	53	1.3	2.9
C_2H_6	0.019	0.005	0.26	53	0.9	1.0
C_5+	0.031	0.022	0.71	53	1.4	5.2

σ_x = standard deviation.

$\frac{\sigma_x}{\overline{x}}$ = coefficient of variation

N = sample size

Compared to the ratio of ethylene to carbon monoxide, the ratios of acetylene, ethane, propylene, the butylenes, and the $C_5^=+ C_6^+$ fraction (presumed to be aromatics), and hydrogen to carbon monoxide were more variable due to their tendency either to form secondary products or to be secondary products.

The consistency of the pyrolysis gas composition suggests that those pyrolysis conditions which favored ethylene production were actually favoring the overall production of all gases. In Figure 4 it is seen that the same parameter that favored ethylene production did indeed also favor overall gasification. However, since the amount of char formed (including about half ash) was fairly constant at $20 \pm 5\%$ by weight, the unmeasured amount of tars, water, and water soluble organics presumably varied accordingly to close the mass balance.

The average overall empirical formula for the gaseous pyrolysis products is very nearly the same as that for the ECO II feedstock. In fact, if the feedstock were to be completely gasified to the observed gaseous products, the elemental balances would be 95% closed for hydrogen and 106% closed for oxygen. This implies that the other pyrolysis products in the form of char, water, tars, and water soluble oxygenated organics would also have the same overall chemical composition as the feedstock. If the dry, ash-free feedstock were to be completely pyrolyzed to the same average gaseous products, the ethylene yield would be 14% and the C_2+ yield would be 28% by weight. These approximate yields are predicted as maximum yields at infinite dilution as can be seen in Figures 2 and 3.

Discussion. Pyrolysis of cellulosic materials appears to proceed via several different concurrent competing reactions whose relative rates are very temperature dependent. At low temperatures the predominant reaction is one of dehydration to form char and water vapor. At intermediate temperatures, the predominant reaction is chain cleavage to form levoglucosan tars. However, above 650°C the gasification reactions to form combustible volatiles predominate.[2] These combustible volatiles include hydrocarbons, oxygenated organics, carbon oxides, and hydrogen. The change from the tar forming reaction to the gasification reactions is accompanied by a change in activation energy from 25 to 14 kcal/mole and the pre-exponential factor is reduced from $2.64 \cdot 10^5$ sec^{-1} to $3.0 \cdot 10^3$ sec^{-1}. Apparently the reactions are still first order.[3] Of primary interest to this program were the hydrocarbons. The data generated in this program revealed that when the pyrolysis system was dilute, significantly more gases were produced. These gases had a relatively constant composition, although secondary reactions could shift the composition of some components. The overall elemental composition of the gases was remarkably similar to the feedstock. This indicates that the gasification reaction itself has at least two major competing reactions: (a) to form gases; or (b) to form relatively stable oxygenated compounds (many of which would be water soluble) as shown in Figure 5.

Gasification of an organic molecule involves the extensive rupturing of the original bonds to result in smaller molecules and/or fragments of molecules called free radicals. As the original molecule is heated, low level energy surges are thought to be traveling up and down the molecule. When the surges collide at a random location, sufficient energy would then be present at that random location to rupture the bond. If the material is heated slowly to an intermediate temperature, the

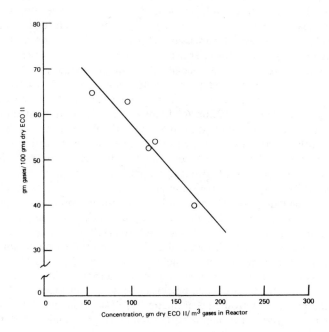

Figure 4. Gaseous products vs. concentration, 6.1-m pyrolysis coil (excludes H_2O and CO_2)

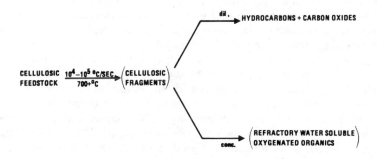

Figure 5. High temperature pyrolysis reaction overview

energy surges could perhaps gain slowly in intensity until the weaker bonds would be more selectively ruptured to form the levoglucosan tars widely reported. However, with rapid heating perhaps many larger energy surges are transferred to the molecule. The probability of random chain scission is thus increased, and the relative importance of levoglucosan as a pyrolysis intermediate is decreased. In any case, whether or not levoglucosan is a significant intermediate, it is thought that, at the thermal conditions employed in the process, the pyrolysis products very rapidly form a primordial cloud of small molecular fragments or free radicals, many of which are oxygenated hydrocarbons as shown in Figure 6. If the reaction conditions are at relatively low pressures and are dilute (low partial pressures), the free radicals are hypothesized to further pyrolyze with the formation of hydrogen and hydrocarbon free radicals as well as carbon oxides. The hydrogen and hydrocarbon free radicals would then randomly combine to form hydrogen, methane, ethylene, propylene, etc., as shown in Figure 7. At pyrolysis temperatures above 700°C, the primary pyrolysis products begin to immediately polymerize to form secondary products such as benzene, styrene, and toluene, as well as multiple ringed aromatic tars as shown in Figure 8. (Using a long residence time of a few seconds to crack hydrocarbons was reported to lead to the evolution of hydrogen and an aromatic product containing about 50% tars.[4]) If the pyrolysis products are held for an extended period of time, the aromatic products continue to grow in size to form macromolecules containing very little hydrogen, which are known as carbon black as shown in Figure 9. Part of the char could thus be derived from the aromatic pyrolysis products. The dilution of the pyrolysis reaction serves to minimize the bimolecular secondary tar and char forming reactions involving the olefins.

The dilution of the pyrolysis reaction also apparently has another very important effect. Returning to Figures 6 and 7, it is thought that the primary reaction competing with the gasification is a second order reaction (bimolecular) involving the oxygenated, primordial, free radicals. It is postulated that, if these oxygenated free radicals collide and react with each other, stable oxygenated compounds would be apt to evolve which could be of a refractory nature and survive the high temperature exposure. In order to maintain the elemental balance, these refractory organic compounds would need to be highly oxygenated.

If it is assumed that ignition of a compound involves the combustion of the products of decomposition, then the question of relative thermal stability can be qualitatively compared by consideration of the autoignition temperature of that compound. It is interesting to note that many oxygenated compounds have autoignition temperatures (thermal stabilities) equal to or higher than that of ethylene. These thermally stable compounds include phenols, triethylene glycol dimethyl ether, formic acid, ethyl formate, and acetic acid. Between the stability of ethylene and propylene are acetone, methyl ethyl ketone, and methyl acetate.[5] These oxygenated compounds tend to have a fairly high solubility in water. This solubility tends to allow them to be relatively unnoticed in the aqueous phase of the pyrolysis condensates and helps to explain the lack of closure of the mass balance for most of the pyrolysis experiments which had low gasification.

The qualitative agreement of the autoignition values with relative thermal stability can be illustrated with the compounds acetone and acetaldehyde. Acetone has an autoignition temperature of 538°C, whereas the value for acetaldehyde is

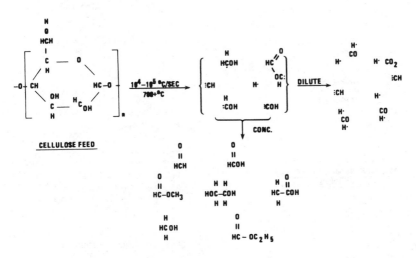

Figure 6. *Competing high temperature pyrolysis reaction details*

Figure 7. *Formation of primary pyrolysis products*

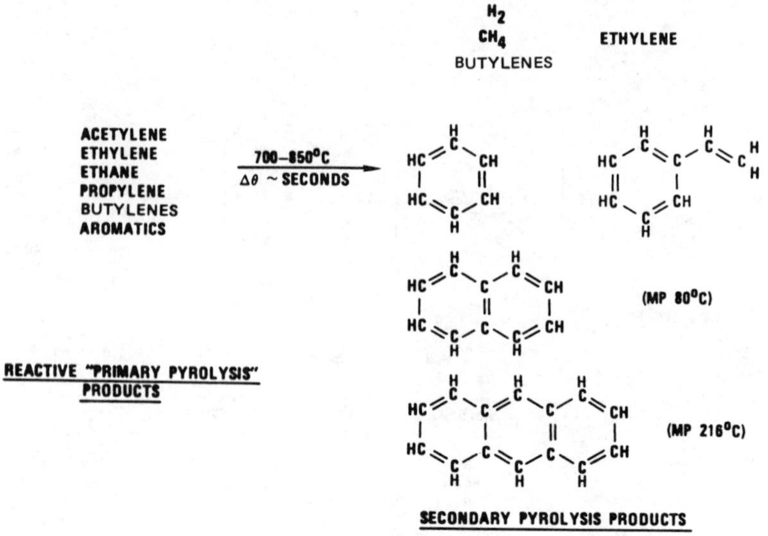

Figure 8. *Formation of secondary pyrolysis products*

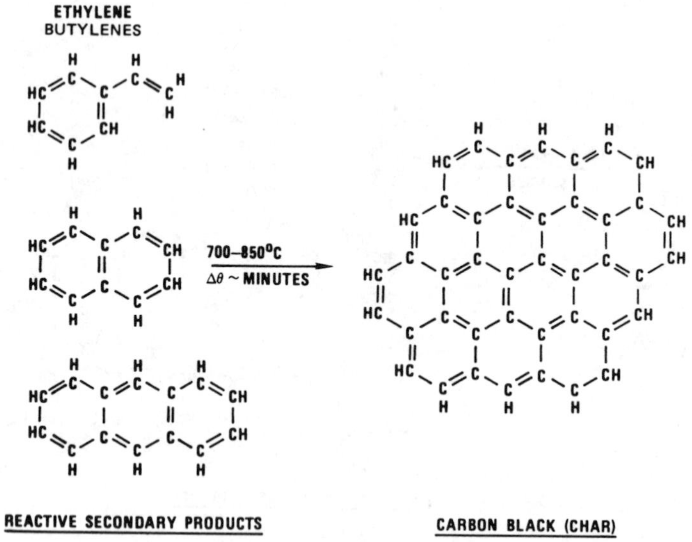

Figure 9. *Formation of char*

only 185°C. At 750°C the half-life of acetone is calculated to be 0.290 second,[6] whereas the half-life of acetaldehyde is only 0.006 second.[7] Interestingly enough, the reported acetone decomposition products for short pyrolysis times had nearly the same ratio of ethylene to carbon monoxide as noted herein for ECO II feedstock, although the acetaldehyde decomposed to give only methane and carbon monoxide.

It is doubtful that the oxygenated organic formation could be explained by a mechanism involving incomplete pyrolysis because the gasification reaction to olefins did not appear very dependent on cracking severity; e.g., residence time and temperature. If the oxygenated water soluble compounds were products of unimolecular degradation, the relative dilution should have had no effect on yields and the more severe pyrolysis reactions would have been expected to increase the overall gasification, but these trends were not observed. It should be noted that the hypothesis for pyrolysis in the entrained bed reactor is based only on experiments using ECO II as feedstock. Pure cellulosic feedstock or biomass feedstocks may behave differently when pyrolyzed in this manner.

For a discrete particle of organic feedstock to undergo the selective gasification, a very rapid heating rate appears to be necessary. Referring to Table I, many of the residence times were in the neighborhood of 0.050 second. The temperature rise was from 25°C to about 750°C during pyrolysis. This results in a heating rate greater than 15,000°C/sec. If the particle were to attain the pyrolysis temperature part way through the pyrolysis reactor, the heating rate would be much higher. High heating rates minimize the time during which the particle is acted upon by competing reactions favored at lower temperatures which do not form the desired olefin-containing products.

In order to achieve this high heating rate, the following are necessary: (1) the surface-to-volume ratio of the feed must be high; (2) the thickness of the particle must be relatively low to reduce the thermal gradient in the particle, and (3) the rate of heat transfer into the particle surface must be very high. The first two requirements can be met with a very small particle size, and/or a "two-dimensional" particle (paper), and/or a "one-dimensional" particle (fiber). The heat transfer requirement needed for good gasification is a function of the reactor design and apparently can be met with pure radiation[8] or with fluidized beds[9] and entrained flow reactors which combine radiation, conduction, and convection.[10]

The molecular fragments or free radicals formed by the shattering of the cellulosic molecules are small enough to lose their cellulosic identities in the process. This implies that the free radicals will react to form a similar product distribution; the ancestral origins or feedstock would be of secondary importance since the water and carbon oxides are relatively inert. Thus, it would be expected that the relative hydrocarbon product distribution would have similar trends whether the free radicals were made from cellulosic, hydrocarbon, or kerogen (shale oil) feedstocks. The weight percent conversions will, of course, have a direct correlation to feedstock characteristics. As shown in Figure 10, the pyrolysis product distribution obtained from naphtha,[11] oil shale,[10] and from MSW (ECO II) are indeed remarkably similar considering the very different nature of the feedstocks.[12]

In summary, when the cellulosic molecule is rapidly heated to the 700°C temperature regime, very reactive free radicals are formed. These free radicals are

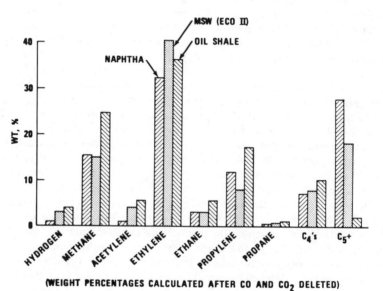

American Society of Mechanical Engineers

Figure 10.   Pyrolysis products comparison using various feedstocks (12)

very short lived and form the more stable primary pyrolysis products. If the system is relatively dilute (low free-radical partial pressure), the formation of olefins rather than aromatic tars and oxygenated organics is favored. If the pyrolysis reaction is not stopped or quenched, the primary reaction products themselves react to form aromatic tars. With more time, the molecular weight of the tars increases so that they are solid in nature to become carbon black. At this point thermodynamic equilibrium is being approached and the products are those of the coking ovens: hydrogen, carbon monoxide, char, water, carbon dioxide, and methane. To maximize gaseous hydrocarbon production (e.g., ethylene) it appears necessary to have rapid heating rates, low partial pressures, and short residence times.

High Pressure Gas Purification

The goal of the gas purification portion of the program was to reject the hydrogen sulfide, carbon dioxide, carbon monoxide, hydrogen, and methane from the pyrolysis gases. This would provide a product gas stream consisting primarily of low molecular weight olefins (ethylene, propylene, butylene, etc.). Since petrochemical reactors are operated at elevated pressures, it was assumed that the pyrolysis gases would be compressed to some intermediate pressure for purification. The effect of hydrogen sulfide in the purified gas stream fed to a reactor would be to react with the olefins to form odiferous mercaptans or to poison most catalytic processes. The presence of methane or hydrogen in the purified gases would reduce the effective olefinic pressure and slow bimolecular reaction rates. Gas purification can be accomplished at cryogenic temperatures by distillation techniques. However, it was reported to be more energy efficient to use solubility differences of the gases at ambient or elevated temperatures and pressures. The absorption of low molecular weight olefins into a light oil from gases formed by the pyrolysis of propane has been commercialized and ethylene recoveries of 99% reported.[13] These propane pyrolysis gases were very similar in general composition to those made from cellulosic pyrolysis except for the oxides of carbon found with the latter. Carbon dioxide has a solubility in light aromatic oils approaching that of ethylene, so it must be removed first.

Figure 11 shows the high pressure gas purification system used to concentrate the olefins. The carbon dioxide was removed very selectively with a hot carbonate solution at a system pressure of about 3100 kPa. This resulted in a relatively pure stream of carbon dioxide being rejected. The semipurified pyrolysis gases then passed to the hydrocarbon absorber where the hydrocarbons other than methane were absorbed. The low molecular weight olefins were then boiled off the absorbing oil to result in a concentrated product stream. At this point in the process, several different uses for the olefins can be imagined: (a) the olefins could be further purified and marketed as pure compounds; (b) the olefin mixture could be passed through a series of catalytic reactors to form alcohols or hydrocarbon liquids; or (c) the olefin mixture could be passed through a noncatalytic, polymerizing reactor to form polymer gasoline. This latter option is the one which was pursued due to its apparent simplicity and the extensiveness of the gasoline market.

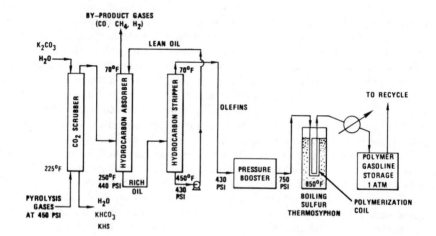

Figure 11. Pyrolysis gas purification and polymerization to gasoline (12)

Polymerization to Gasoline

The polymerization of olefins to form a liquid product primarily composed of gasoline was researched in the early 1930s when the cracking of crude oil to make gasoline resulted in a large amount of gaseous by-product olefins. Although phosphoric acid was found to polymerize propylene and butylene fairly readily, ethylene was fairly resistant to its catalytic effect. In contrast, at the relatively more severe conditions required to polymerize olefins without catalysis, ethylene was found to polymerize more readily than propylene or butylene.[14] Since the primary olefin formed by pyrolysis of organic wastes was ethylene, thermal polymerization to make gasoline was selected.

Early experiments in the program were made with pure ethylene which produced a synthetic crude oil containing 90% gasoline which was shown to have an unleaded ASTM motor octane of 90. A polymer gasoline sample held at ambient temperature for 3-1/2 years visually appeared not to change color, although some commercially produced gasoline turned to a dark orange during this time. The distillation curve for the liquid product appeared to be fairly continuous and reflected the relatively extensive rearrangement of the molecules at the 450 to 500°C reactor temperature until a stable molecular structure was attained, such as highly branched or cyclic. (Fortunately, these complex molecular structures result in high octane ratings.) Synthetic crude made from relatively impure pyrolysis gases had a very similar physical appearance and distillation curve. With efficient processing of the pyrolysis gases, a polymer gasoline yield is projected of about 0.18 liter/kg of feed (MSW containing 60% by weight organic material), which is about one petroleum barrel per ton of MSW.

Summary

The ability to convert organic wastes to pyrolysis gases containing relatively large amounts of olefins has been demonstrated by using a very selective pyrolysis process. Pyrolysis appears to involve a large number of competing reactions, most of which do not favor the formation of olefins. Rapid heating rates, short residence times, small feed particle size, and dilute pyrolysis conditions appear to favor the formation of olefins rather than char, tar, or water soluble organics. Although the olefins produced could be used for petrochemical feedstock, their potential for a straightforward conversion to high octane gasoline was demonstrated using 1930s technology. The economics[12] and the commercialization potential of the process[15] look very favorable and have been reported previously in detail.

Acknowledgements

The financial support of the Environmental Protection Agency (IERL) made the bench scale demonstration of the process possible, and much of this paper will be appearing in the forthcoming final report. Dr. Charles B. Benham (presently at SERI) was instrumental in the polymerization demonstration with pure ethylene and with the early phases of the program from which this process evolved.

References

1. Diebold, J.P., C.B. Benham, and G.D. Smith. "R&D in the Conversion of Solid Organic Wastes to High Octane Gasoline," *AIAA Monograph on Alternate Fuel Resources,* Volume 20, 1976. (Western Periodicals Co., No. Hollywood, CA.)

2. Shafizadeh, F. "Pyrolysis and Combustion of Cellulosic Materials", *Adv. in Carbohydrate Chem., 23,* 419-475. (1968)

3. Brink, D.L., M.S. Massoudi, and R.F. Sawyer. "A Flow Reactor Technique for the Study of Wood Pyrolysis," presented at the 1973 Fall Meeting of the Western States Section of the Combustion Institute in El Segundo, CA (October 1973).

4. Cadman, W.H. "II. Semi-Industrial Production of Aromatic Hydrocarbons from Natural Gas in Persia," *Ind Eng Chem, 26,* No. 3, 315-320. (1934)

5. Hilado, C.J. and S.W. Clark, "Autoignition Temperatures of Organic Chemicals," *Chem Eng,* 4 Sept 1972. pp. 75-80.

6. Hinshelwood, C.N. and W.K. Hutchinson. "A Homogeneous Unimolecular Reaction - the Thermal Decomposition of Acetone in the Gaseous State," *Proc. Roy. Soc.,* 111A, 245-257. (1926)

7. Hinshelwood, C.N. and W.K. Hutchinson. "A Comparison between Unimolecular and Bimolecular Gaseous Reactions. The Thermal Decomposition of Gaseous Acetaldehyde," *Proc. Roy. Soc.,* 111A, 380-385. (1926)

8. Lincoln, K.A., "Flash-Pyrolysis of Solid-Fuel Materials by Thermal Radiation," *Pyrodynamics,* 2, 133-143, Gordon and Breach, No. Ireland. (1965)

9. Finney, C.S. and D.E. Garrett. "The Flash Pyrolysis of Solid Waste," 66th Annual AIChE meeting, November 1973, paper 13-E.

10. Sohns, H.W., et. al. "Development and Operation of an Experimental, Entrained Solids Oil-Shale Retort," BUMINES, RI 5522 (1959)

11. Prescott, J.H. "Pyrolysis Furnace Boosts Ethylene Yield by 10-20%," *Chem Eng,* July 1975, 52-53.

12. Diebold, J.P. and G.D. Smith. "Noncatalytic Conversion of Biomass to Gasoline," presented to the ASME sponsored Solar Energy Conference in San Diego, March 1979. 79-Sol-29.

13. Kniel, L. and W.H. Slager. "Ethylene Purification by Absorption Process," *Chem. Eng. Prog., 43,* No. 7, 335-342. (1947)

14. Sullivan, F.W. Jr., R.F. Ruthruff and W.E. Kuentzel. "Pyrolysis and Polymerization of Gaseous Paraffins and Olefins," *Ind Eng Chem, 27,* No. 9, 1072-1077. (1935)

15. Diebold, J.P. and G.D. Smith. "Commercialization Potential of the China Lake Trash-to-Gasoline Process," presented at the Engineering Foundation Sponsored Conference "Municipal Solid Waste, the Problems and the Promise," held in Henniker, NH, July 1979.

RECEIVED November 16, 1979.

Fuel Production from Wastes Using Molten Salts

R. L. GAY, K. M. BARCLAY, L. F. GRANTHAM, and S. J. YOSIM

Rockwell International Corporation, Energy Systems Group,
8900 De Soto Avenue, Canoga Park, CA 91304

The disposal of municipal and industrial wastes has become
an important problem because the traditional means of disposal,
landfill, has become environmentally much less acceptable than
previously. In addition, special incinerator systems are required
to meet environmental standards for disposal by incineration.
Disposal of wastes by landfill or incineration also includes a
potential loss of energy sources and, in some cases, valuable
mineral resources. New, much stricter regulation of these disposal methods will make the economics of waste processing for
resource recovery much more favorable.

One method of processing waste streams is to convert the
energy value of the combustible waste into a fuel. One type of
fuel attainable from wastes is a low heating value gas, usually
3.7-5.6 MJ m^{-3} (100-150 Btu/scf), which can be used to generate
process steam or to generate electricity.

Described here are some results which show the feasibility
of processing wastes in molten salts into usable fuels. The
molten salt acts as a reaction medium for the conversion of the
waste into a low heating value gas (by reaction with air) and the
simultaneous retention of potential acidic pollutants in the
molten salt. The waste is converted to a fuel gas by reacting it
with deficient air, that is, insufficient air for complete conversion to CO_2 and H_2O. Results are presented for a high-sulfur
oil refinery waste, rubber, wood, leather scraps, and waste X-ray
film. These waste streams represent a small segment of the large
variety of wastes which may be processed using molten salts.

Process Description

A flow diagram of the Rockwell International process for
fuel production from wastes is shown in Figure 1. In this process, combustible waste and air are continuously introduced
beneath the surface of a sodium carbonate-containing melt at a

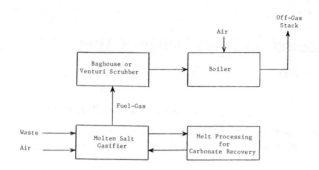

Figure 1.   Molten salt process for fuel production from wastes

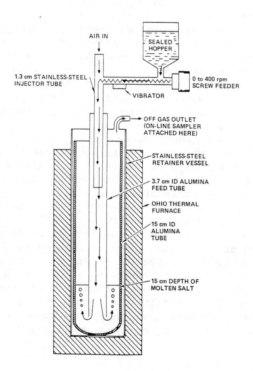

Figure 2.   Bench-scale molten salt gasifier

temperature of 900–1000°C. The waste reacts chemically with the
molten salt and air to produce a pollutant-free combustible gas
containing mainly CO, H_2, N_2, and a small amount of CH_4 and C_2H_6.
The product gas is cleaned of particulates using a baghouse or
venturi scrubber. It is then burned in a boiler to produce
steam or as an alternative, the product gas may be burned in a
gas turbine as part of a combined-cycle process.

Sodium carbonate is a stable, nonvolatile, inexpensive, and
nontoxic material. It is the intimate contact of the waste
material with the molten salt and air that provides for complete
and rapid reaction of the waste.

Any gas that is formed during gasification is forced to pass
through the basic carbonate melt. Halogens in the waste (for
example, chlorine from chlorinated organic compounds) form the
corresponding sodium halide salts. Any sulfur in the waste is
retained as sodium sulfide, and any ash in the waste will also be
retained in the melt. The temperatures of operation are too low
to permit a significant amount of nitrogen oxides to be formed by
fixation of the nitrogen in the air. Also, at these operating
temperatures, odors and infectious material are completely
destroyed.

A portion of the sodium carbonate melt is withdrawn from the
molten salt reactor, quenched, and processed in an aqueous recovery system. The recovery system removes the ash and inorganic
combustion products (mainly sodium salts such as NaCl and Na_2S)
retained in the melt. Unreacted sodium carbonate is returned to
the molten salt furnace. The ash must be removed when the ash
concentration in the melt approaches 20 to 25 wt % in order to
preserve the melt fluidity. The inorganic combustion products
must be removed before all of the sodium carbonate is completely
converted to noncarbonate salts. For the case of a waste containing a valuable mineral resource, the valuable mineral resource is
retained in the melt during the gasification process and may be
recovered as a by-product of the regeneration process.

This molten salt gasification process is the basis for the
Rockwell International molten salt coal gasification process. A
900-kg-h^{-1} (1 ton/h) process development unit pilot plant has
been built and is being tested under contract from the Department
of Energy. This plant includes the gasifier and a complete
sodium carbonate recovery and regeneration system.

Experimental

Bench-Scale Test Apparatus. A schematic diagram of the
bench-scale test apparatus is shown in Figure 2. This apparatus
is used to test the feasibility of gasifying wastes before pilot-
scale tests are performed. The waste throughput of the bench-
scale apparatus is 1 to 3 kg h^{-1} (2 to 6 lb/h). Approximately
5.5 kg (12 lb) of molten salt are contained in a 15-cm ID by
76-cm high alumina tube placed inside a type 321 stainless steel

retainer vessel. This stainless steel vessel, in turn, is contained in a 20-cm ID, four heating zone Ohio Thermal furnace.
Each heating zone is 20 cm in height, and the power to each zone
is controlled by a silicon-controlled rectifier circuit. An
additional flat plate heater is located at the bottom of the
vessel. Furnace and reactor temperatures are recorded on a
12-point Barber-Colman chart recorder.

Solids are pulverized in a Wiley laboratory knife-blade mill
to less than 1 mm in size. These solids are then metered into a
1.3-cm OD stainless steel injection tube by a screw feeder. The
speed of the screw feeder is variable between 0-400 rpm. The
solids are mixed with process air in the injection tube. The
solids-air mixture comes out of the end of the 1.3-cm OD tube and
into the annulus of a 3.7-cm ID alumina tube. This alumina tube
extends to within 1.3 cm of the bottom of the 15-cm ID alumina
containment vessel. During normal operation of the reactor, the
depth of the molten salt expands from a quiescent level of 15 cm
to an expanded depth of about 30 cm.

In the case of liquids, a laboratory peristaltic pump is
used to pump the liquid into the 1.3-cm OD stainless tube. The
feed rate of the liquid is controlled by the pump speed.

Heat balance of the bench-scale reactor is maintained in one
of two ways. If the feed material is of relatively high heating
value (greater than 14.0 MJ kg^{-1}, 6000 Btu/lb), a small air
cooler at the bottom of the furnace is used to maintain the
vessel at the operating temperature. If the waste material is
low in heating value, the melt temperature is maintained by an
electrically heated furnace, or by adding auxiliary fuel to the
waste.

Off-Gas Analysis. Gas samples are initially cleaned of
particulates and dried to 2% moisture before analysis. Carbon
monoxide and carbon dioxide are measured continuously using a
Horiba Mexa-300 CO analyzer and a Horiba Mexa-200 CO$_2$ analyzer.
Syringe samples are taken downstream of the CO$_2$ analyzer for gas
chromatographic analysis. A room temperature molecular sieve 13X
column is used to analyze for carbon monoxide, oxygen, and nitrogen. A Poropak Q column at 130°C is used to analyze for carbon
dioxide, methane, ethane, ethylene, sulfur dioxide, and hydrogen
sulfide.

Molten Salt Test Facility. A 90-kg-h^{-1} (200 lb/h) pilot
plant at the Rockwell field laboratory at Santa Susana is used
for larger scale testing. The molten salt gasification vessel is
made of Type 304 stainless steel and is 4.6 m high with a 0.9 m
inside diameter. The inside of the vessel is lined with 15-cm-
thick refractory blocks. The gasifier contains 900 kg of sodium
carbonate, which has an unexpanded bed height of 0.9 m. During
operation, the bed height approximately doubles. A natural gas
preheater is used to heat the vessel on start-up.

A flow schematic for the Santa Susana molten salt test facility is shown in Figure 3. Salt is loaded into the vessel from a carbonate hopper and a weigh belt feeder. Waste materials are shredded and crushed using a rotary knife shredder and a hammermill and are then metered into a pneumatic feed line with a variable-speed auger.

Product gases produced in the molten salt vessel unit flow through refractory-lined ducting to a spray cooler and into a secondary combustor. The gases are burned in the secondary combustor and the off-gas is cleaned of any particulate matter with a baghouse or a venturi scrubber. The off-gas is then released to the atmosphere.

A photograph of the molten salt test facility is shown in Figure 4.

Results

Gasification of Acid Pit Sludge. Crude or partially refined crude is treated with sulfuric acid during refining. This produces a waste product which contains substantial amounts of water (12-40 wt %), sulfur (2-14 wt %), ash (1-55 wt %), and significant combustible material (11.6 to 20.9 MJ kg^{-1}) (5000-9000 Btu/lb). Normal incineration of this waste is not economic due to the high water, ash, and sulfur content. Therefore, the waste is generally ponded and not processed. The nickel and vanadium content and energy value of this waste product, as well as the value of the large area of land dedicated to current disposal techniques, offer significant economic potential for improved disposal processes.

Data on the gasification of acid pit sludge are given in Table I. In this test, the water-sludge mixture was heated to the consistency of light oil (about 90 to 95°C) and fed to the reactor without removing any water from the as-received material. A gas containing a higher heating value of 8.57 MJ m^{-3} (230 Btu/scf) was produced. During this test, some auxiliary heat was furnished by the outer furnace to maintain the operating temperature. It is estimated that a steady-state test without auxiliary heat would produce a gas with a higher heating value of 6.33 MJ m^{-3} (170 Btu/scf). Emissions of H_2S were below the limit of detectability (40 ppm).

Gasification of Rubber Tires. Conventional incineration of rubber tires produces a large amount of particulates which contain unburned hydrocarbons. As an alternative to incineration, molten salt gasification was tested. Two gasification tests were made using buffings from a rubber tire. The results of these tests were averaged and are presented in Table I. Since the rubber contained organic sulfur that would form Na_2S in the melt, the Na_2CO_3 melt originally contained 6 wt % Na_2S to approximate steady-state conditions. Using a stoichiometry of 33% theoretical air

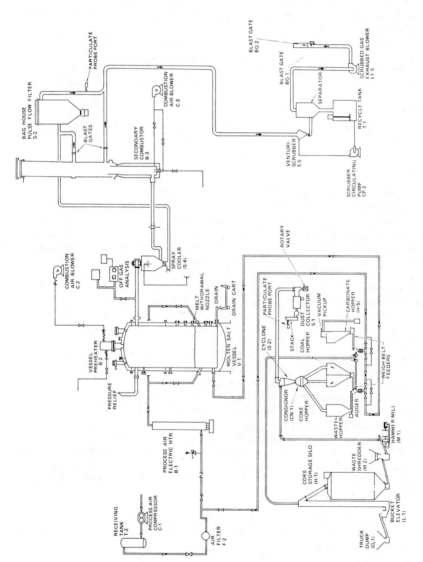

Figure 3.   Schematic of the molten salt test facility

Figure 4. Santa Susana molten salt test facility

TABLE I

GASIFICATION OF WASTES

Waste Material	Temperature (°C)	Waste Feed Rate (kg h⁻¹)	Air Feed Rate (m³ h⁻¹)[a]	Composition of Off-Gas (Vol %)[b]					Higher Heating Value (MJ m⁻³)[a]
				CO_2	CO	H_2	CH_4	C_2	
Acid Pit Sludge	935	3.4	4.0	16.0	7.0	15.8	8.4	2.5	8.57[d]
Rubber Tires	920	0.8	2.6	4.0	18.4	16.0	2.4	1.1	5.77
Wood	951	1.0	3.3	14.5	20.3	21.1	3.0	0.9	6.74
Leather Offal	935	2.0	3.6	16.5	13.7	12.9	2.2	ND[c]	4.03
Film	1015	2.4	7.2	16.5	12.0	11.7	2.6	0.2	3.99
	958	3.0	4.0	16.0	18.3	14.1	5.2	1.2	6.67

[a] Gas volumes are defined at conditions of 101.3 kPa (1 atm) pressure and 15.6°C (60°F).

[b] Remainder is N_2.

[c] Not determined.

[d] Auxiliary heat provided by the electric furnace during testing. Estimated 6.33 MJ m⁻³ gas would be produced without auxiliary heat.

(100% theoretical air is required to oxidize the material completely to CO_2 and H_2O), a product gas with a higher heating value of 5.77 MJ m^{-3} (155 Btu/scf) was made. No H_2S or sulfur-containing gases (less than 30 ppm) were detected in the off-gas.

Gasification of Wood. Pine wood was chosen as representative of typical biomass. A typical composition of pine wood on a dry basis is 51.8 wt % carbon, 6.3 wt % hydrogen, 0.5 wt % ash, and 41.3 wt % oxygen. The pine wood tested here was sawdust with a moisture content of 2.8 wt %. The heating value is typically 21.2 MJ kg^{-1} (9100 Btu/lb). A pure sodium carbonate melt was used with 30% theoretical air. Gasification of wood was rapid and complete with production of a gas with a higher heating value of 6.74 MJ m^{-3} (181 Btu/scf), as given in Table I.

Gasification of Leather Offal. Chrome-leather tanning scraps or offal are a waste product of the tanning and leather industries. These wastes are usually the result of trimming operations and contain 3.5 wt % chromium and about 50 wt % water. The wastes cannot be incinerated because malodorous substances and carcinogenic chromium-containing particulates are formed. Landfill disposal is presently used.

The tanning scraps used for these tests were air-dried and contained 37.2 wt % carbon, 5.7 wt % hydrogen, 12.8 wt % nitrogen, 2.5 wt % chromium, and approximately 5 wt % nonchrome ash. The heating value of the chrome-waste was approximately 18.8 MJ kg^{-1} (7800 Btu/lb).

The analysis of the off-gas from leather gasification is given in Table I. Approximately 50% of theoretical air was used. The product gas had a heating value of 4.03 MJ m^{-3} (108 Btu/scf). The NO_x emissions were only 40 ppm even through the leather contained 12.8 wt % organic nitrogen. The chromium emission in the form of particulates was 0.3 mg m^{-3} (0.00013 grain/ft^3), which corresponds to a chromium retention of 99.998%. This chromium could be recovered and recycled to the tanning process by processing the salt bed.

Gasification of Waste Film. Two series of tests were conducted with waste X-ray film. The first series was performed in the bench-scale gasifier; the second series was run in the molten salt test facility. Bench-scale gasification of film at 51% and 22% theoretical air produced product gases of 3.99 MJ m^{-3} (107 Btu/scf) and 6.67 MJ m^{-3} (179 Btu/scf), respectively. Small pellets of pure elemental silver (greater than 99.9% pure Ag) were recovered from these tests.

In the pilot test, 6,800 kg (15,000 lb) of waste X-ray film were burned at the rate of 90-105 kg h^{-1} (200-230 lb/h). The average air feed rate was 290 m^3 h^{-1} (180 scfm). The air/film ratio corresponded to 50% theoretical air. Since the main purpose of this test was to recover silver, no attempt was made to analyze the off-gas. After the film had been gasified, the silver and salt were drained from the vessel and 160.5 kg (354 lb) of 99.97% pure metallic silver were recovered.

Conclusion

The Rockwell International molten salt process for gasification of wastes with resource recovery has been shown here to be well-suited for the processing of a variety of wastes. A variety of waste forms may be processed, that is, solids, liquids, and solid-liquid mixtures. The process is suitable for applictions which involve either small or large throughputs. The gasification medium, sodium carbonate, is stable, non-volatile, inexpensive, and nontoxic. Sulfur-containing pollutants are retained in the melt when sulfur-containing wastes are gasified. In the same manner, halogen-containing pollutants are retained during gasification of halogen-containing wastes. The gasification of a high-nitrogen-content waste (leather scraps) produces very little NO_x in the off-gas. Valuable minerals may be recovered by processing of the salt after gasification of mineral-laden wastes. In general, the molten salt process is best applied to waste materials involving potential pollutants (such as sulfur or chromium) or to wastes where gasification and resource recovery are important (such as the recovery of silver with simultaneous gasification of X-ray film).

RECEIVED November 16, 1979.

Biomass Gasification at the Focus of the Odeillo (France) 1-Mw (Thermal) Solar Furnace

M. J. ANTAL, JR.

Princeton University, Princeton, NJ 08544

C. ROYERE and A. VIALARON

C.N.R.S., Odeillo, France

Recently completed research at Princeton University (1-3) has shown biomass gasification to be a three step process:

1. Pyrolysis. At modest temperatures (300°C or more, depending upon heating rate) biomass materials lose between 70% and 90% of their weight by pyrolysis, forming gaseous volatile matter and solid char. As discussed in this paper, very high heating rates enhance volatile matter production at the expense of char formation. Recent Princeton publications review mechanistic and kinetic research on cellulose (4), lignin (5), and wood (6) pyrolysis in more detail.

2. Cracking/Reforming of the Volatile Matter. At somewhat higher temperatures (600°C or more) the volatile matter evolved by the pyrolysis reactions (step 1) reacts in the absence of oxygen to form a hydrocarbon rich synthesis gas. These gas phase reactions happen very rapidly (seconds or less) and can be manipulated to favor the formation of various hydrocarbons (such as ethylene). Rates and products of the cracking reactions for volatile matter derived from cellulose, lignin, and wood are now available in the literature (1, 3, 5, 6).

3. Char Gasification. At even higher temperatures char gasification occurs by the water gas and Boudouard reactions, and simple oxidation:

$$C + H_2O \rightarrow CO + H_2 \quad \text{(water gas)}$$

$$C + CO_2 \rightarrow 2CO \quad \text{(Boudouard)}$$

$$\left.\begin{array}{l} C + \tfrac{1}{2}O_2 \rightarrow CO \\ C + O_2 \rightarrow CO_2 \end{array}\right\} \quad \text{(oxidation)}$$

Because pyrolysis (step 1) produces less than 30% by weight char for most biomass materials, the char gasification reactions (step 3) play a less important role in biomass gasification than steps 1 and 2.

0-8412-0565-5/80/47-130-237$05.00/0

Figure 1 illustrates the critical role played by step 2 in cellulose gasification. By increasing the temperature achieved by the gas phase volatile matter from 500°C to 750°C, the carbon conversion efficiency η_c is increased from $\eta_c = 0.1$ to $\eta_c = 0.76$. Here the carbon efficiency η_c is defined by η_c = carbon in permanent gases ÷ carbon in solid feed.

The carbon conversion efficiency is emphasized here because (for non-oxidative conversion processes) η_c provides an excellent measure of how well the energy and chemical content of the solid fuel feedstock is converted to gaseous fuels and chemicals. Because about 20% of the carbon initially in the cellulose is carried by the char product of step 1, the data presented in Figure 1 indicates that for gas phase temperatures above 750°C and residence times of 2 sec or more, permanent gases and char are essentially the only products of cellulose gasification. Less than 4% of the feedstock carbon is carried by the condensible fraction of the reactor effluent.

Because char commands a low market value, there is some incentive to increase gas production at the expense of char formation. The traditional approach is to use the water gas, Boudouard, and combustion reactions (step 3) to gasify the char produced by step 1. An alternative approach is to rapidly heat solid biomass feed, modifying the pyrolysis mechanism (step 1) and reducing the initial formation of char by the pyrolysis reactions. The latter approach has been emphasized in the research reported here.

The work of Broido (7), Shafizadeh (8), Lewellen (9) and others (10) has shown cellulose pyrolysis to be describable in terms of a competitive mechanism:

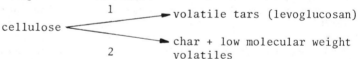

Two pyrolysis reactions compete to consume the cellulose; however only one reaction produces char. The first reaction is favored by high temperatures and rapid heating, producing combustible volatile matter at the expense of char formation. The second reaction is favored by low temperatures and slow heating of the cellulose. Thus chemical reactors designed to provide very rapid heating ("flash pyrolysis") of solid cellulosic feedstocks can minimize char formation with the potential of significantly increasing gas yields.

With but one exception, past studies of the flash pyrolysis of cellulose using laboratory equipment have relied on radiation to achieve the high heating rates required. Lincoln (11) used pulsed carbon arcs and xenon flash tubes to achieve the complete volatilization of cellulose by pyrolysis reactions. Berkowitz-Mattuck

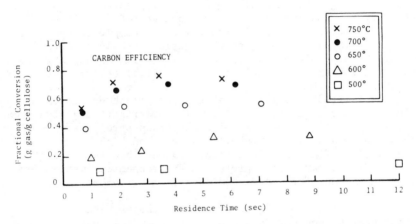

Figure 1. Cellulose carbon conversion efficiency η_c as a function of gas-phase residence time for various gas-phase temperatures

and Noguchi (12) used carbon arc and solar furnace radiation
sources to achieve flash pyrolysis of cotton cellulose. They
project no char formation for radiant fluxes in excess of 120 w/cm^2.
Martin (13) used a high current carbon arc radiation source for his
experiments, and noted that char formation was reduced to 4% of the
original cellulose weight with a radiant flux level of 49 w/cm^2.
Martin speculates that no char formation should occur for radiation
fluxes exceeding 400 to 4000 w/cm^2. Exploratory research at
Princeton has recently achieved 99% volatilization of cellulose
with a flux level of 30 w/cm^2. Other researchers (14-15) have also
employed radiation sources to achieve very rapid heating of cellulosic materials; however their results are not pertinent to this
research.

Lewellen et al. (9) at M.I.T. studied flash pyrolysis of
cellulose using a bench scale, electrically heated screen. They
found that for heating rates ranging between 400 and 10,000°C/sec
in an inert helium atmosphere the cellulose completely vaporized by
pyrolysis reactions, leaving no char residue. Only by extended
heating of the cellulose at 250°C was the M.I.T. group able to
produce some char (2% by weight of the initial cellulose).

Only limited data is available on the flash pyrolysis of lignocellulosic materials. Rensfelt et al. have reported a significant
reduction in char yields following the flash pyrolysis of various
biomass materials (17). Diebolt has obtained high gas yields from
Eco Fuel II by flash pyrolysis (18). These results suggest that
flash pyrolysis may be the preferred thermochemical method for
obtaining gaseous fuels and chemicals from all biomass materials.

The fact that most flash pyrolysis studies have used
radiant heating suggests solar heat as a natural means for effect-
ing flash pyrolysis of biomass feedstocks. Since solar radiation
has a characteristic temperature of almost 6000°K, it can be used
to achieve very rapid heating of opaque, solid particles. Earlier
studies (2) have shown that the quantity of heat required for
biomass gasification is small: less than 1 Gj per Mg of dry solid
feed. Consequently, small amounts of solar heat can be used to
process large quantities of biomass. Finally, a recent study for
the President's Council on Environmental Quality (19) concluded
that the use of solar heat for flash pyrolysis of biomass appears
to be more economical than conventional gasification processes.
References 19-24 discuss in greater detail these and other reasons
for using solar heat to gasify biomass. Experiments described in
this paper were undertaken to explore the use of concentrated solar
radiation for the flash pyrolysis of biomass.

APPARATUS AND PROCEDURE

The Odeillo 1000 kw$_{th}$ solar furnace has been described in
detail by Royere (25); consequently only a summary will be given

here. The furnace consists of sixty-three 6 x 7.5 m heliostats which track the sun and redirect the solar radiation onto a 2000 m^2 parabolic concentrator which focuses the radiation within a circle of about 20 cm radius. The heliostats are composed of eleven thousand three hundred forty 50 x 50 cm flat, back surfaced mirrors and the parabola consists of over 9000 mechanically warped mirrors. Radiant flux levels of 1600 w/cm^2 are available at the focal point, and temperatures in excess of 3800°C can be obtained.

The biomass flash pyrolysis reactors, designed and fabricated at Princeton, followed in part a coal gasification reactor design described by Badzioch and Hawksley (26), and Howard (27). A vibrating feeder with a capacity of about 1 g fed about 0.01 g/s of powdered biomass material into a flow straightener (see Figure 2). A small flow (40-100 ml/min) of He was used as a carrier gas, and partially fluidized the bed of biomass material in the feeder. Exiting the flow straightener, the biomass material was entrained by flowing superheated steam (about 3 g/min) and carried into the solar flux passing through the wall of the reactor. Char residue was collected in a bucket and weighed. Exiting the quartz reactor, the steam was condensed in a tar trap/condenser and permanent gases were collected in 1.2 l teflon bags for analysis by gas chromatography. Following an experiment, the volume of the inflated bags was measured by water displacement and the volume of gas produced during the experiment was calculated.

Two of the flash pyrolysis reactors used for the experiments were fabricated from Amersil TO8 commercial grade 50 mm OD quartz tube, and one from 25 mm OD vycor tube (Corning Glass Co.). The flow straightener/female joint was made of quartz for one of the 50 mm OD reactors, and pyrex for the other two reactors. The feeder was made primarily from plexiglas. Teflon tube (6 mm ID) was used to connect the reactor to the pyrex tar trap.

Biomass samples taken to France for use with the reactor included Avicel ® PH 101 and 102 powdered microcrystalline cellulose, ground corn cob material, hardwood and softwood sawdust, Kraft lignin, and Eco Fuel II. The corn cob and softwood material was sieved and samples with particle sizes <75 µm, 75-150 µm, 150-212 µm, 212-425 µm, and 425-710 µm were available for the experiments. The Avicel ® PH101 cellulose has an average particle size of 50 µm, and the PH102 has an average size of 100 µm.(28). The Kraft lignin was a very fine powder (size <75 µm), while the Eco Fuel II and the hardwood sawdust consisted of particles with sizes ranging up to about 700 µm.

Additional experiments using a long coil of stainless steel tube immersed in a fluidized sand bath at 800°C were originally planned to investigate the gas phase cracking chemistry (step 2) of the volatile matter derived from the flash pyrolysis experiments. Although this apparatus was available in France, experimental

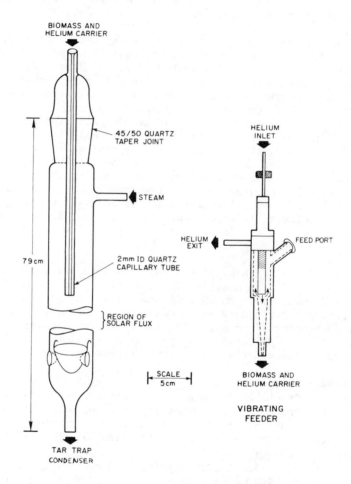

Figure 2.   Quartz reactor and vibrating feeder schematics

difficulties with the flash pyrolysis reactor prevented experimentation with the cracking reactor.

RESULTS

Low Flux Experiments

The first 50 mm OD quartz reactor with pyrex flow straightner was wrapped with electrical heater tape and installed behind a large water cooled shield. The shield had a 10 cm diameter circular hole located in it for admission of the solar flux into the reactor. Behind the reactor, a water cooled semi-circular (cross section) cylindrical backplate prevented radiation from entering the work area through the back of the reactor. Unfortunately, the first reactor cracked at the pyrex/quartz joint when the upper portion of the reactor was accidentaly heated to a temperature in excess of its 100°C design operating temperature. A second, all quartz 50 mm reactor was immediately fabricated at Princeton and flown to Odeillo to serve as backup for the 25 mm vycor reactor.

The vycor reactor was installed and greater care was taken not to exceed the design operating temperature. Upon exposure to a flux of about 200 w/cm^2 with flowing steam and helium (no biomass) a large bubble (approximately 50 mm diameter) formed in the vycor reactor and it bent slightly so that its axis was no longer perfectly linear. Reflected radiation carried down the conduit formed between the shield and the semi-circular cylindrical backplate melted the heater tape, fusing it into the vycor reactor tube. Remnants of the heater tape were removed and several successful experiments were conducted at low (200 w/cm^2) flux levels with the Avicel PH102 cellulose and the 75-150 μm ground corn cob material.

The first test with cellulose (Experiment 1.05) used no steam flow. Because the evolved volatile matter cracked on the vycor walls depositing an opaque, black material thereon, the wall temperature of the reactor rapidly climbed to temperatures above 1000°C. Gasification data is presented in Table 1 for the 0.54 g of cellulose fed into the reactor during the test. Subsequent, brief exposure of the reactor to the solar flux with steam (but no biomass) flow cleaned the reactor's walls nicely.

A second test (Experiment 1.08) with cellulose in steam flowing at 3 g/min was perfomed, and no clouding of the reactor's "window" was observed. Data for the gasification of 0.48 g of cellulose is presented in Table II. A third test (Experiment 2.08) using corn cob material was also performed, and the data obtained is listed in Table III. Following the third experiment a small crack developed where the heater tape had fused into the wall, and work with 25 mm OD vycor reactor terminated.

TABLE I. Results of Experiment 1.05 - Cellulose Pyrolysis

	Solar Gasification*	Prior Princeton Research
Carrier Gas	Helium	Steam
I	200 w/cm^2	---
β	27000°C/sec	2.0°C/sec
T_g	925°C	750°C
t_g	10 sec	4.0 sec
η_c	0.52	0.73
Product	Yield (mass fraction)	Yield (mass fraction)
H_2	0.013	0.013
CO_2	0.044	0.127
CO	0.380	0.420
CH_4	0.044	0.073
C_2H_4	0.026	0.056
C_2H_6	0.003	0.010
C_3H_6	0.0	0.0
Total gas	0.51	0.699
Char	0.01	0.1
Calorific Value of Gas	3.9 x 10^6 cal/m^3	4.4 x 10^6 cal/m^3

*See text for explanation of β, T_g, t_g, η_c

TABLE II. Results of Experiment 1.08 - Cellulose Pyrolysis

	Solar Gasification*	Prior Princeton Research
Carrier Gas	Steam	Steam
I	200 w/cm^2	---
β	27000°C/sec	2.0°C/sec
T_g	450°C	500°C
t_g	1 sec	1.3 sec
η_c	0.12	0.10
Product	Yield (mass Fraction)	Yield (mass fraction)
H_2	0.004	0.001
CO_2	0.015	0.066
CO	0.101	0.045
CH_4	0.004	0.003
C_2H_4	0.005	0.0
C_2H_6	0.0	0.001
C_3H_6	0.0	0.0
Total gas	0.129	0.116
Char	0.01	0.1
Calorific Value of Gas	3.3 x 10^6 cal/m^3	

*See text for explanation of β, T_g, t_g, η_c

TABLE III. Results of Experiment 2.08 - Powdered Corn Cob
Pyrolysis

	Solar Gasification*	Prior Princeton Research
Carrier Gas	Steam	Steam
I	200 w/cm^2	---
β	$27000°C$/sec	$2.0°C$/sec
T_g		$750°C$
t_g	1 sec	4 sec
η_c	0.6	0.6
Product	Yield (mass fraction)	Yield (mass fraction)
H_2	0.027	0.005
CO_2	0.177	0.200
CO	0.397	0.256
CH_4	0.056	0.063
C_2H_4	0.026	0.034
C_2H_6	0.0	0.007
C_3H_6	0.0	0.003
Char	0.007	0.22
Calorific Value of Gas	3.4×10^6 cal/m^3	4.2×10^6 cal/m^3

*See text for explanation of β, T_g, t_g, η_c

Contrary to expectations, small quantities of char were collected after each experiment. Soot formation was evident, and particularly prominent following the corn cob experiment. Coking of the volatile matter on the reactor wall beneath the window rendered the lower part of the reactor opaque. The water condensate was a light yellow color with a disagreeable (but not strong or particularly foul) odor. No tar was observed in the condenser, but the teflon tube connecting the reactor to the condenser was blackened.

High Flux Experiments

Before the second 50 mm OD quartz reactor arrived from Princeton, several exploratory experiments were performed using a 50 mm OD vycor tube with rubber corks and steel inlets for the biomass and steam, and a gas exit. The reactor functioned well when exposed to the solar flux (1000 w/cm^2) and some semi-quantitative data was obtained; however because the rubber corks partially melted under exposure to the radiation flux carried up and down the walls of the vycor tube by total internal reflection, little significance could be attached to the data.

The second 50 mm OD quartz reactor was first tested with only steam and helium flow in the full solar flux. The wall temperature, of the reactor stabilized at 320°C and no difficulties were encountered. For the high flux experiments the use of heater tapes was found to be unnecessary: the solar flux entering through the 10 cm hole in the shield was sufficient to maintain the entire reactor above 100°C!

During the first experiment with Avicel® PH102 cellulose the reactor wall temperature gradually rose to 850°C (after 150 seconds of biomass flow) in a solar flux of 1000 w/cm^2. Table IV summarizes the results of this experiment (Experiment 1.13). Char formation was still evident, but no soot was observed.

An experiment with corn cob material failed to yield results because the vibrating feeder plugged shortly after the experiment began. Water was added to the steam generator, and a second attempt was made with the corn cob material. However, immediately following admission of steam into the reactor and before the vibrating feeder was activated, the reactor wall temperature suddenly jumped to over 1200°C. Subsequent experiments failed because the high wall temperature (ultimately exceeding 1500°C for an experiment using Kraft lignin) caused a steam overpressure in the reactor forcing moisture into the vibrating feeder and plugging the flow straightener with biomass material. The reactor was removed and an examination of its walls revealed whitish deposits on the interior wall of the reactor. The bulk of

TABLE IV. Results of Experiment 1.13 - Cellulose Pyrolysis

	Solar Gasification* Steam	Prior Princeton Research Steam
Carrier Gas	Steam	Steam
I	1000 w/cm²	---
β	140000°C/sec	2.0°C/sec
T_g	↑850°C (after 150 sec)	600°C
t_g	2 sec	3 sec
η_c	0.29	0.23
Product	Yield (mass fraction)	Yield (mass fraction)
H_2	0.005	0.002
CO_2	0.029	0.083
CO	0.274	0.160
CH_4	0.004	0.012
C_2H_4	0.0	0.007
C_2H_6	0.0	0.003
C_3H_6	0.0	0.0
Char	0.04	0.1

*See text for explanation of β, T_g, t_g, η_c

these deposits was easily removed by scraping the wall with a
wooden dowel; hence they were not due to devitrification. However,
some devitrification may have occured. Because no biomass entered
the reactor, the entering steam was the only source of foreign
material available to form the whitish deposits. Subsequent
investigation revealed that yellow colored tap water had been mis-
takenly added to the steam generator immediately prior to the
experiment. Because no more reactors were available, the experimental aspect of the research ceased after the successful cellulose pyrolysis experiment.

DISCUSSION OF RESULTS

Before discussing the experimental results, it is helpful to
consider several characteristic times representative of chemical
and physical processes occuring within a pyrolyzing biomass
particle. These characteristic times shed some light on solar
flash pyrolysis phenomenon.

Characteristic Times

A "small" solid particle encountering an intense radiation
flux of intensity I is subject to rapid heating at a rate β which
may be estimated by

$$\beta \simeq \frac{3aI}{4\rho c_p} \frac{1}{r} \tag{1}$$

where a is the particle's absorptivity, ρ its density, c_p its
specific heat and r its radius. Values of these physical properties
for cellulose are listed in Table V. Equation (1) assumes that the
particle's internal temperature is uniform and that heat loss from
the particle is negligible. Calculated values of β based on Eq.(1)
for the various experiments are given in Tables I-IV, and summarized
in Table VI for the Avicel micro crystalline cellulose.

TABLE V.   Physical Properties for Cellulose (11)

ρ	=	0.47 g/cm^3
c_p	=	0.32 cal/g°C
α	=	0.002 cm^2/s
a	=	$0.93 - 0.00036$ $(T - T_a)$
ΔH (pyrolysis)	=	$+ 88$ cal/g

TABLE VI.   Characteristic Times for Solar Flash Pyrolysis of
            Avicel Ⓡ PH102 Cellulose

I (w/cm^2)	β (°C/sec)	T_p (°C)	t_p (ms)	t_d (ms)	t_c (ms)	t_{hr} (ms)
200	27000	840	41	3	13	20
1000	140000	960	9	0.6	13	2

Pyrolytic weight loss of cellulose in steam can be described by the following rate equation (4):

$$\frac{dV}{dt} = k(V^*-V)^n; \quad k = A \exp(-E/RT) \tag{2}$$

where $V^* = (w_i-w_f)/w_i$, $V = (w_i-w)/w_i$, w_i is the initial sample weight, w_f the final sample weight, w the time dependent sample weight, A the pre-exponential constant, E the apparent activation energy, n the apparent reaction order, R the universal gas constant, and T the time dependent sample temperature. Values for cellulose are (4): $A = 1.7 \times 10^9$ sec^{-1}, $E = 34$ kcal/gmol and $n = 0.5$.

Assuming the solid's heating rate β to be uniform and linear, the temperature T_p at which the maximal rate of weight loss occurs satisfies the following relationship:

$$T_p \approx \frac{-E}{R \log\left(\frac{\beta E}{ART_p^2}\right)} \tag{3}$$

No more than one iteration is usually necessary to obtain a good value for T_p. Finally, the characteristic time t_p for pyrolysis of a small solid sample can be estimated by the formula

$$t_p \approx T_p/\beta \tag{4}$$

Values of T_p and t_p for the solar experiments with Avicel cellulose are listed in Table VI.

Recognizing the values of t_p listed in Table VI, it is worthwhile to note that the residence time of the solid biomass particles in the intense solar flux was less than 0.3 sec for the low flux experiments, and less than 1.0 sec for the high flux experiment. Thus, more than ample time was available for completion of the pyrolysis reactions in the intense solar flux.

Another characteristic time of interest is the inverse reaction rate of pyrolysis at temperature T_p:

$$t_d = \frac{1}{A \exp(-E/RT_p)}$$

The value of t_d provides a measure of the length of time required for the pyrolytic disappearance of a particle heated instantaneously to temperature T_p. As indicated in Table VI, the very high reaction rate at temperature T_p leads to very short disappearance times t_d.

Because of the very rapid heating rates and short time scales evidenced in Table VI, heat and mass transfer within the reacting particle become a subject of great interest. The characteristic time t_c for conduction of heat into a spherical particle is given by

$$t_c = \frac{r^2}{\alpha} \qquad (5)$$

where r is the particle's radius and α its thermal diffusivity (29). Values for t_c are given in Table VI for the Avicel cellulose. These results indicate that conduction alone would not be sufficient to assure uniform temperatures within the 100 μm diameter Avicel cellulose particles for the high flux experiments. However, cellulose is not perfectly opaque, and it seems likely that entry of radiation into the body of the particle would tend to establish uniform temperatures. Clearly, the extinction coefficient of various biomass materials should be measured to provide for a more quantitative investigation of this phenomenon.

A measure of the characteristic time required for the solar flux to provide the pyrolytic heat of reaction is given by

$$t_{hr} = \frac{4\Delta H \rho r}{3aI}$$

As indicated in Table VI, $t_{hr} < t_p$ for the Odeillo experiments; however $t_{hr} \gg t_r$ indicating that even with the large solar flux it is very difficult to supply heat to the particle rapidly enough to meet the large heat demand caused by the high reaction rate at temperature T_p. It would appear that during pyrolysis the biomass particles experienced some self cooling due to the inability of the solar flux to meet the extraordinary demand for heat occuring within the particle.

Comparison with Prior Research

At first data presented in Tables I-IV seemed puzzling because, in spite of the very low char yields, gas yields from the solar flash pyrolysis experiments were lower than had been anticipated from the results of earlier work. The explanation of this dichotomy lies in the role played by the gas phase reaction chemistry in the production of permanent gases. In most of the Odeillo experiments the gaseous volatile matter was not subject to high temperatures; consequently gas yields were low because cracking of the volatile matter did not occur.

Tables I-IV provide ballpark estimates of the temperature T_g (based pyrometry data) reached by the gaseous volatile pyrolysis products, and the residence time t_g of the volatile products at temperature T_g. A comparison of the Odeillo data with earlier Princeton research (1-3) using similar gas phase conditions (but much lower solid's heating rates) is given in Tables I-IV. Since the Odeillo reactor did not provide for good control of the gas phase conditions, the relative agreement of the two sets of results given in Tables I-IV is as good as could be expected.

Generally the mass fraction yields of the permanent gases are relatively similar, except CO_2 which is low by a factor of three in the cellulose experiments. Thus the Odeillo results are in concord with the introduction of this paper, which discussed the critical role of the gas phase chemistry (largely unaffected by the high solar flux) in determining the final product distribution.

Table VII summarizes the results of earlier work on flash pyrolysis. The data of Martin (13) and the low flux data of Lincoln (11) agree well with the "comparable" Odeillo experiment (Table II), except that the CO_2 yield in the Odeillo work is very low. The high flux data of Lincoln (11) is in relatively good agreement with the "comparable" Odeillo experiment (Table IV).

TABLE VII. A Summary of the Results of Prior Research on the Flash Pyrolysis of Cellulose

Investigator	Martin	Lincoln	Lincoln
I	184 w/cm^2	46 w/cm^2	12600 w/cm^2
Carrier Gas	He	He	He
Product	Yield (mass fraction	Yield (mass fraction)	Yield (mass fraction)
H_2	>0.004		
CO_2	0.12	0.11	0.04
CO	0.13	0.13	0.37
CH_4	0.01		
C_2H_4	0.01		
C_2H_6	0.002		
Total Gas	>0.27		
Char		0.03	0.01

The very low char yields for the low flux experiments (Tables I-III), and the high yield for the high flux experiment (Table IV) are a subject of interest. The reactor used in the low flux experiments bent slightly under exposure to the intense solar flux; consequently some of the char may not have fallen into the bucket. However, very little char was observed within the reactor outside of the bucket following the experiments. On the other hand, copious amounts of soot were present. More char was obtained in the high flux experiment (using a larger bucket properly oriented below the flow straightener), and very little soot was formed. The increased formation of char may have been the result of premature initiation of the pyrolysis reactions by (lower intensity) scattered radiation contained between the shield and the backplate surrounding the reactor. The cavity containing the reactor acted as a conduit for the radiation, carrying it up and down the length of the reactor. This unfortunate artifact of the experimental design could not be avoided because of necessary safety precautions in using the large solar furnace.

CONCLUSIONS

Biomass materials (powdered, microcrystalline cellulose and ground corn cob material) have been successfully gasified in a windowed chemical reactor operating at the focus of the Odeillo 1 Mw$_{th}$ solar furnace. The quartz window survived radiant flux levels in excess of 1000 w/cm^2; however impurities carried by the steam flow into the reactor ultimately clouded the window. Some devitrification may also have occured. Future experiments should be designed to insure that no mineral matter is deposited on the chemical reactor's windowed area.

Pyrolytic char yields of the Odeillo experiments were quite low: ranging between one and four percent. Gas yields were also relatively low, but condensible yields were high. These results reflect the important role played by the gas phase chemistry (largely unaffected by the high solar flux) in the production of permanent gases from biomass.

The Odeillo results are in good agreement with earlier flash tube and arc image furnace work associated with the simulation of thermonuclear weapons effects. They are also in concord with earlier Princeton work on biomass gasification.

A consideration of the characteristic times for chemical kinetic and heat transfer phenomenon within a rapidly pyrolyzing particle indicate that heat transfer (not chemical kinetics) is the rate limiting step. However, the thermochemical and optical properties of biomass materials are poorly understood and much more experimental work must be completed before definitive conclusions in this important area can be made.

Because the use of concentrated solar radiation for direct gasification of biomass materials results in the formation of little or no char without reliance on the water gas or Boudouard reactions, solar flash pyrolysis of biomass holds unusual promise for the economical production of liquid and gaseous fuels from renewable resources.

ACKNOWLEDGMENTS

At Princeton, much of the preliminary experimental work was done by Mr. W. Edwards and Mr. F. Conner. The plexiglas feeder was fabricated by Mr. B. Reavis and the quartz reactors by Mr. H. Olson. Professor W. Russel offered helpful insights into the role of characteristic times discussed in this paper.

In France, the gas analysis was ably performed by Dr. Rivot. Dr. Tuhault, Mr. Ribeill and Mr. Peroy were of great assistance in operating the experiment.

Helpful information on the optical properties of quartz was

obtained from Mr. H. Hoover (Corning Glass), and Amersil Corporation.
The Avicel microcrystalline cellulose was obtained from Dr. L. Jones
with the FMC Corporation, and the ground corn cob material from
Mr. S. Bosdick with DeKalb AgResearch.

Support for this work was obtained jointly from the Solar
Thermal Test Facilities Users Association and Centre National De
La Recherche Scientifique. The authors also wish to thank F. Smith
and M. Gutstein for their interest in this work.

LITERATURE CITED

1.   Antal, M. J. "The Effects of Residence Time, Temperature and
     Pressure on the Steam Gasification of Biomass", American
     Chemical Society Division of Petroleum Chemistry, Honolulu,
     1979.

2.   Antal, M. J. "Synthesis Gas Production from Organic Wastes
     by Pyrolysis/Steam Reforming", IGT Conference on Energy from
     Biomass and Wastes, Washington, D. C., 1978.

3.   Antal, M. J.; Friedman, H. L.; Rogers, F. E.; "A Study of
     the Steam Gasification of Organic Wastes", (Final Progress
     Report, U.S.E.P.A.)", Princeton University, Princeton, N. J.,
     1979.

4.   Antal, M. J.; Friedman, H. L.; Rogers, F. E. "Kinetic Rates
     of Cellulose Pyrolysis in Nitrogen and Steam", to appear in
     Combustion Science and Technology, 1979.

5.   Kothari, V. S.; Antal, M. J.; Reed, T. B. "A Comparison of
     the Gasification Properties of Cellulose and Lignin in Steam",
     to appear.

6.   Mattocks, T. MSE Thesis, Princeton University, Princeton,
     N. J., 1979.

7.   Broido, A. "Kinetics of Solid Phase Cellulose Pyrolysis" in
     "Thermal Uses and Properties of Carbohydrates and Lignins",
     Shafizadeh, F. et al. (ed.), Academic Press, New York, 1976.

8.   Shafizadeh, F.; Groot, W. F. "Combustion Characteristics of
     Cellulosic Fuels" in "Thermal Uses and Properties of Carbo-
     hydrates and Lignins", Shafizadeh, F. et al. (ed.), Academic
     Press, New York, 1976.

9.   Lewellen, P. C.; Peters, W. A.; Howard, J. B. "Cellulose
     Pyrolysis Kinetics and Char Formation Mechanism", 16th
     International Symposium on Combustion, Cambridge, 1976.

10. Arseneau, D. F. Can. J. Chem., 1971, 49, 632.

11. Lincoln, K. A. Pyrodynamics, 1965, 2, 133.

12. Berkowitz-Mattuck, J. B.; Noguchi, J. B. Journal of Applied Polymer Science, 1963, 7, 709.

13. Martin, S. "Diffusion Controlled Ignition of Cellulosic Materials by Intense Radiation", 10th International Symposium on Combustion, Cambridge, Mass. 1964.

14. Shivadev, U.; Emmons, H. W. Combustion and Flame, 1974, 22, 223.

15. Hottel, H. C.; Williams, G. C. Industrial Engineering and Chemistry, 1955, 47, 1136.

16. Nelson, L. S.; Lundberg, J. L. J. Phys. Chem., 63, 434.

17. Rensfelt, E.; Blomkvist, G; Ekstrom, C.; Engstrom, S.: Espenas, B-G.; Liinanki, L. "Basic Gasification Studies for Development of Biomass Medium Btu Gasification Process", IGT Conference on Energy from Biomass and Wastes, Washington, D.C., 1978.

18. Diebolt, J.; Smith, G. "Conversion of Trash to Gasoline", NWC Tech. Pub. 6022, Naval Weapons Center, 1978.

19. Antal, M. J. "Biomass Energy Enhancement, A Report to the President's Council on Environmental Quality", Princeton University, Princeton, N. J. 1978.

20. Antal, M. J.; Rodot, M.; Royere, C.; Vialaron, A. "Solar Flash Pyrolysis of Biomass" in Proceedings of the ISES Silver Jubilee International Congress, Atlanta, Ga., 1979.

21. Antal, M. J. "Tower Power: Producing Fuels from Solar Energy" in Towards a Solar Civilization, Williams, R. (ed.), M.I.T. Press, Cambridge, Mass. 1978.

22. Antal, M. J. "The Conversion of Urban Wastes" in Energy Sources and Development, Barcelona, Spain, 1978.

23. Antal, M. J.; Feber, R. C.; Tinkle, M. C. "Synthetic Fuels From Solid Wastes and Solar Energy" in Proceedings of the World Hydrogen Energy Conference, Miami, Fla., 1976.

24. Feber, R. C.; Antal, M. J. "Synthetic Fuel Production from Solid Wastes", EPA-600/2-77-147, Cincinnati, 1977.

25. Royere, C. "CNRS Solar Furnace" in Proceedings of Annual Meeting Technical Sessions, Solar Thermal Test Facilities Users Association, Golden, Colorado, 1978.

26. Badzioch, S.; Hawksley, P. G. W. Ind. Eng. Chem. Process. Des. Develop., 1970, 9, 521.

27. Kobayashi, H.; Howard, J. B.; Sarofim, A. F. "Coal Devolatilization at High Temperatures", 16th International Symposium on Combustion, Cambridge, Mass., 1976.

28. "Avicel PH Microcrystalline Cellulose", Bulletin PH-6, FMC Corporation.

29. Carlslaw, H. S.; Jaeger, F. C. "Conduction of Heat in Solids", Clarendon Press, Oxford, 1959.

RECEIVED November 20, 1979.

Operation of a Downdraft Gasifier Fueled with Source-Separated Solid Waste

S. A. VIGIL, D. A. BARTLEY, R. HEALY[1], and G. TCHOBANOGLOUS

Department of Civil Engineering, University of California, Davis, CA 95616

THE SOLID WASTE PROBLEM

The disposal of solid waste is a critical problem faced by most communities. It is also costly in terms of capital and energy resources. In 1971, 109 million metric tons of solid waste were collected in the United States at a total cost of $2.64 billion for collection and disposal. By 1985, this quantity is expected to increase to between 150 and 200 metric tons per year with an annual cost between $4.02 to $5.06 billion.(1).

Although the composition of municipal solid waste varies greatly both geographically and by season, an analysis of data from many cities in the United States shows a remarkable similarity. Typically, municipal solid waste in the United States contains about 40 percent combustible material by weight.

Although resource recovery and recycling are receiving considerable attention today, most solid waste is still disposed of in a sanitary landfill. Even though it is widely recognized that placing solid wastes in a landfill is a misuse of potentially valuable resources, this practice will continue until economic and environmentally acceptable alternatives are found. One such alternative, the gasification of the paper fraction of solid waste is considered in this report.

Recently, interest has developed in pyrolysis, thermal gasification, and liquefaction (PTGL) processes for the recovery of energy. The PTGL processes offer the advantage of producing a usable fuel (gas or liquid) from the combustible portion of solid waste. Jones, (2), provides an overview of the many processes currently under consideration. Unfortunately, most of these processes have one characteristic in common. They are more complex than conventional incineration systems. They are also very capital intensive. For example, one such system, the PUROX Process, developed by Union Carbide Corporation, is economic only in capacities ranging from 363 to 1814 metric tons per day of solid waste (3). The corresponding size of cities vary from about 200,000 to 1,000,000 in population. Because of the cost and complexity of such systems, they are not cost-effective for smaller cities.

---

[1] Current Address: Los Angeles County Sanitation District, Los Angeles, California.

0-8412-0565-5/80/47-130-257$05.00/0

A PTGL SYSTEM FOR SMALL COMMUNITIES

Since small communities (< 100,000 population) cannot afford either conventional mass fired incinerators or the new generation of PTGL processes, they currently have few options for solid waste disposal other than landfills. Unfortunately, landfill sites near communities are getting scarce, and the cost of operating them is rapidly increasing.

One possible system which could be used by smaller communities is to combine source separation of solid waste with a simple PTGL process for energy conversion of the paper fraction of the waste. Since typical solid waste contains over 40 percent paper, energy recovery from the paper fraction could reduce landfill requirements by the same amount.

A low cost system for smaller communities is shown in Figure 1. It consists of a shredder to reduce the size of waste paper, a densification system to convert the shredded waste paper to a dense fuel "cube," the gasification reactor, a gas cleanup system, and an engine-generator to convert the gas to electrical power.

SOURCE SEPARATION OF SOLID WASTE

It has been assumed for the past few years that the recycling of waste paper into newsprint or low quality paper is both economically and ecologically sound. However, current energy prices, coupled with the fluctuating nature of newsprint prices, are making energy recovery a viable option. Also, since only prime, hand selected, clean newsprint is suitable for recycling purposes, far more waste paper is available for energy recovery since cleanliness is unimportant.

In a source separation system, residents are requested to place bundles of newspaper out with their weekly trash collection. The newspapers are picked up by the regular collectors and carried in special containers or racks on the trash trucks. Other cities use a smaller separate vehicle to collect paper and other recycable materials. Usually cities require that newspapers be tied into bundles. Magazines, paper bags, and food packaging are not accepted. Such a system is currently operated by the city of Davis, California as shown schematically in Figure 2.

A source separation system designed to recover waste paper as a fuel could be less restrictive and as a result, a higher proportion of a given community's waste paper could be recovered. Such a system is shown in Figure 3. In this system, only a combustible fuel fraction and aluminum cans are recovered. No attempt is made to recover steel cans and glass because marketing of these components is difficult and seldom economic for smaller communities. Shredding the aluminum cans reduces on site storage requirements and increases the market value of the aluminum from $0.51/kg to $0.73/kg .

DENSIFICATION OF SOLID WASTE

Although source separated solid waste could be directly fired in some types of furnaces or PTGL reactors, there are many advantages to densifying the waste before combustion. These advantages include:
1. Increased storability

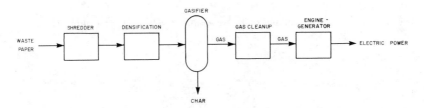

Figure 1. Solid waste gasification system for small communities

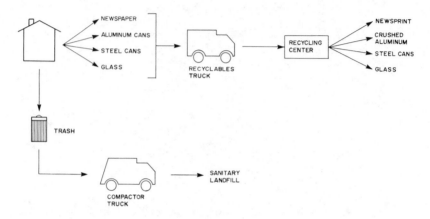

Figure 2.   Conventional source-separation system—Davis, California

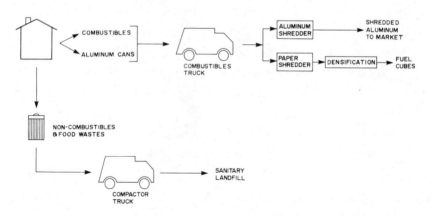

Figure 3.   Proposed combustible fuel source-separation system

2. Increased transportability
3. Densified waste can be burned in stoker-type boilers designed for coal or the simplest types of PTGL systems (i,e., packed bed)

The simplest type of densification system consists of a shredder followed by an agricultural type "cubing" machine. These machines were originally built to produce densified animal feeds, but can be modified easily to produce solid waste fuel cubes. One of the first systems of this type was operated during the early 1970's in Fort Wayne, Indiana, to produce a solid waste fuel for the city power plant (4). A more recent application of densification was demonstrated by the Papakube Corporation of San Diego, California, which manufactures solid waste cubes as a boiler fuel (5).

GASIFICATION

Gasification is an energy efficient technique for reducing the volume of solid waste and the recovery of energy. Essentially, the process involves partial combustion of a carbonaceous fuel to generate a combustible fuel gas rich in carbon monoxide and hydrogen. The reader is referred to reference (6) for an excellent review of gasification reactions.

The inventor of the process is unknown, but stationary gasifiers were used in England in the early 1800's (7). By the early 1900's, gasifier technology had advanced to the point where virtually any type of cellulosic residue such as rice hulls, olive pits, straw, and walnut shells could be gasified. These early gasifiers were used primarily to provide the fuel for stationary gasoline engines. Portable gasifiers emerged in the early 1900's. They were used for ships, automobiles, trucks, and tractors. The real impetus for the development of portable gasifier technology was World War II. During the war years, France had over 60,000 charcoal burning, gasifier equipped cars while Sweden had about 75,000 wood burning cars. With the return of relatively cheap and plentiful gasoline and diesel oil, after the end of World War II, gasifier technology was all but forgotten. However, in Sweden, research has continued into the use of wood fueled gasifiers for agriculture (8) and currently, downdraft gasification of peat is being pursued actively in Finland (9).

In the United States, gasification technology was, until recently, virtually ignored. In the early 1970's, work was started in the United States on "pyrolysis" systems for energy recovery from solid wastes. Many of these "pyrolysis" systems are actually complex adaptations of the simple gasification process. For example, the BSP/Envirotech multiple hearth pyrolysis system (10) and the PUROX process (3) are in reality gasification systems. The reader is referred to (2) and (11) for an in-depth review of current research into pyrolysis and gasification systems.

Reactor Types

Four basic reactor types are used in gasification. They are:
1. vertical packed bed
2. multiple hearth

3. rotary kiln
4. fluidized bed
Most of the early gasification work in Europe was with the packed bed type reactors. The other types are favored in current United States practice, with the exception of the PUROX oxygen blown gasifier. The simple vertical packed bed type reactor has a number of advantages over the other types including simplicity and relatively low capital cost. However, it is more sensitive to the mechanical characteristics of the fuel. Eggen and Kraatz, (12), discussed the merits and limitations of vertical bed gasifiers in detail. For this work, a co-current flow, packed bed vertical reactor (also called downdraft gasifier) was chosen. As shown in Figure 4, fuel flow is by gravity with air and fuel moving co-currently through the reactor. At steady state, four zones form in the reactor. In the hearth zone, where air is injected radially into the reactor, exothermic combustion and partial combustion reactions predominate. Heat transfers from this zone upward into the fuel mass, causing pyrolysis reactions in the distillation zone and partial drying of the fuel in the drying zone. Actual production of the fuel gas occurs in the reduction zone, where endothermic reactions predominate, forming CO and H_2. The end products of the process are a carbon rich char and the low-BTU gas.

Gasification of Solid Waste

As mentioned earlier, downdraft gasifiers are simpler to construct and operate than the other reactor types, but they have more exacting fuel requirements which include:
1. moisture content < 30 percent
2. ash content < 5 percent
3. uniform grain size
Since wastes can be dried prior to gasification, excessive moisture can be overcome. However, ash content and grain size are more difficult to handle. When the ash content is higher than 5 percent solidified particles known as clinkers tend to form which can cause severe maintenance problems. Excessive fine material in the fuel can cause mechanical bridging in the fuel hopper. One method of overcoming these problems is to use more complex reactors sulch as the Envirotech Multiple Hearth System or a high temperature slagging gasifier, such as the PUROX process, in which the ash is melted. Although these approaches work, they are costly and complex.

A lower cost approach is to utilize the simplest reactor type, the downdraft gasifier, and tailor the fuel accordingly. A suitable fuel can be made by densifying the paper fraction of source separated solid waste to produce a densified refuse derived fuel (d-RDF), that has low moisture content, low ash content, and uniform grain size.

SMALL SCALE ENERGY CONVERSION

The low-BTU gas from a downdraft gasifier can be utilized in several ways. The simplest technique is to burn the gas with stoichiometric amounts of air in a standard boiler designed for natural gas. This

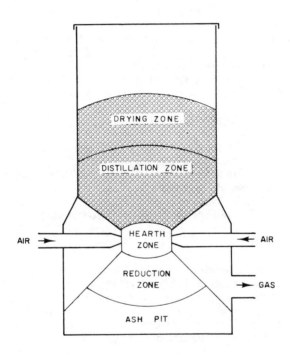

Figure 4. Downdraft gasification

requires minor modifications to the burner head to allow for more combustion air and enlargement of the gas feed pipes to account for the lower energy content of the gas ($\simeq$ 5.6 MJ/m) as compared to natural gas ($\simeq$ 37.3 MJ/m)

Another approach is to cool and filter the gas and utilize it as an alternative fuel for internal combustion engines. Skov and Papworth, (2), described the operation of trucks, buses, and agricultural equipment in Europe with gas produced using portable wood fueled gasifiers. Gasifiers can also be used to operate diesel engines as described in (6) and (9).

In a solid waste gasification system for a small community, the low-BTU gas could be burned in a dual-fueled engine-generator. Two modes of operation are possible:

1. Operation of the gasifier-engine-generator set at the fuel cubing site. Electricity produced in excess of local requirements would be fed into the local power grid.
2. Production of fuel cubes at a central location and transport of the cubes to satellite gasifier systems in other locations.

THE EXPERIMENTAL SYSTEM

To demonstrate the feasibility of operating a downdraft gasifier with densified solid waste, an experimental gasifier was designed and constructed. The complete system consists of three subsystems:

1. Batch fed downdraft gasifier
2. Data aquisition system
3. Solid waste shredding and densification system

Downdraft Gasifier

The gasifier is a batch fed unit with a 46 cm diameter firebox. A cross section of the gasifier is shown in Figure 5. The unit is based on agricultural waste gasifiers designed and built by the Agricultural Engineering Department at the University of California, Davis Campus (13 and 14). The gasifier was designed to operate at a rate of 23 kg/hr The gasifier is shown in operation in Figure 6.

Data Aquisition System

Five type K (chromel-alumel) thermocouples are installed in the gasifier as shown in Figure 5. Additional thermocouples are installed on the air inlet and gas outlet lines. Temperature readings are recorded on paper tape with a Digitec Model 1000 Datalogger.

Shredding and Densification Equipment

A 5 HP hammermill rated at 9 kg/hr was used to shred recycled newsprint. The shredded newspaper was densified with a John Deere Model 390 stationary Cuber rated at 4.5 metric tons per hour. Since the gasifier and shredder used in the experiments only have a capacity of 23kg/hr and 9kg/hr., respectively, an obvious mismatch exists in the

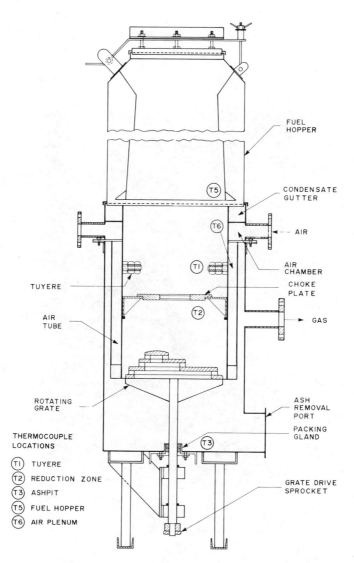

Figure 5. Cross section—laboratory-scale solid waste gasifier

Figure 6.   Solid waste gasifier in operation

equipment. However, in full scale systems, the components would be matched in capacity.

LABORATORY TESTS

Proximate analyses of the fuel and char were run using a standard laboratory drying oven, muffle furnace, and analytical balance according to ASTM Standard Methods. Ultimate Analysis for percent C, H, N, S, and O in the fuel char, and condensate was conducted by the Chemistry Department, University of California, Berkeley. The energy content of the fuel and char was determined with a Parr Oxygen Bomb Calorimeter.

Gas samples were collected in Tedlar gas sample bags and analyzed for their content of CO, CO_2, H_2, O_2, and total hydrocarbons using the process analyzer described in (14). Moisture content of the gas was determined by the condensation method (15).

EXPERIMENTAL RESULTS

To date, eight experimental runs have been conducted. Significant data from two of the runs, RUN 02-wood chips, and RUN 08-solid waste cubes, are shown in Tables I and II. The gas sampling equipment was not in operation for RUN 02, so a gas analysis is not available for that run; however, a good quality low-BTU gas was generated as evidenced by the clean burning gas flare. RUN 08, using solid waste cubes, is more completely documented.

COMPARISON OF SOLID WASTE CUBES WITH WOOD CHIPS

Since the experimental gasifier is a batch fed device, it must be periodically refueled. Due to the greater bulk density of solid waste cubes (495 kg/m^3) as compared to wood chips (231 kg/m^3), a longer run time between refuelings is possible. In a commercial sized gasifier, the greater density of cubed fuels would reduce the size of fuel storage and transport equipment.

A comparison can also be made of the operating temperatures of the reduction zone as measured by Thermocouple T2 (See Figures 5 and 7). Note that the gasifier reached steady state conditions faster with the cubed solid waste than with the wood chips. (Note: Flucuations in the reduction zone temperature for RUN 02 were caused by shutting down the gasifier for refueling. The large drop in reduction zone temperature for RUN 08 starting at T=140 minutes was caused by shutting down the gasifier to connect gas sampling equipment.)

TENTATIVE ECONOMICS OF A GASIFICATION SYSTEM

Gasification is an emerging technology, therefore, any cost estimates of full scale systems must be used with caution. However, since full scale gasifier systems are now commercially available in the United States (16 and 17), tentative cost estimates can be made.

Estimates were made of solid waste gasification systems for cities of 10,000 and 20,000 population. The estimates include capital and labor

Table I

FUEL AND OPERATIONAL DATA FOR
TWO EXPERIMENTAL GASIFICATION RUNS

Item	Run 02 Wood Chips	Run 08 Solid Waste Cubes
Fuel Summary		
Ultimate Analysis		
C,%	59.0	44.4
H,%	7.2	5.6
N,%	N/A	0.3
S,%	N/A	0.1
O,%	32.7	45.8
Residue	1.1	3.8
Proximate Analysis		
VCM,%	N/A	83.5
FC,%	N/A	7.9
Ash,%	N/A	3.1
Moisture,%	9	5.5
Energy Content, MJ/kg (Dry Basis HHV)	20.9	19.8
Operational Summary		
Net Run Time, min.	140	266
Gasification Rate, kg/hr	25.4	20.6
Char Production, kg/hr	2.7	2.5
Condensate Production, kg/hr	0.2	0.7
Air Input Temperature, °C	33	28
Reduction Zone Temperature, °C	900	940
Gas Outlet Temperature, °C	210	220
Fuel Weight Reduction,%	89	84

Table II

GAS ANALYSIS - RUN 08

Item	Value
Dry Gas Analysis, % by volume	
CO	16.5
CO_2	8.5
H_2	12.5
O_2	2.4
N_2	60.1
Gas Moisture Content,% by volume	13.6

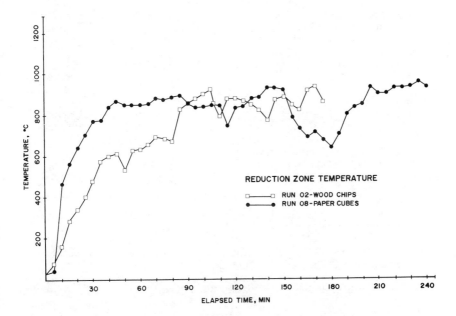

Figure 7. *Reduction zone temperature vs. elapsed time*

Table III

OPERATING AND CAPITAL COSTS FOR SOURCE
SEPARATION OF SOLID WASTE

Item	Population, persons	
	10,000	20,000
Combustibles Collected[a]		
Metric tons/year	1161	2322
Metric tons/day	4.4	8.8
Capital Costs[b]		
Total Cost	$40,000	$40,000
Annual Cost[c]	$ 7,830	$ 7,830
Operating Costs		
Maintenance, tires, oil, insurance	$ 7,440	$ 8,470
Gasoline ($0.26/liter)	$ 2,840	$ 5,220
Gasoline ($0.53/liter)	$ 5,680	$10,440
Labor		
Operator ($11/hr)	$12,430	$22,880
Supervision & Overhead[d]	$ 3,110	$ 5,720
Total Yearly Cost		
(Gasoline @ $0.26/liter)	$33,600	$50,100
(Gasoline @ $0.53/liter)	$36,500	$55,300
Total Unit Cost ($/metric ton)		
(Gasoline @$0.26/liter)	$ 28.94	$ 21.57
(Gasoline @$0.53/liter)	$ 31.44	$ 23.83

a. Assumes 50% participation, solid waste @ 40% combustible, and solid waste generation = 1.6 kg/cap-day.

b. Based on one 12.2 m , gasoline powered, side loading compactor truck, June 1979.

c. Six year life, 15% salvage value, 8% interest.

d. 25% of direct labor.

costs for a solid waste source separation system and capital costs only for a gasification system.

Costs for the Source Separation of Solid Waste

Costs were estimated for a source separation system as shown in Figure 3. It was assumed that 50 percent of the theoretically available combustible solid waste could be recovered. The total costs ranged from $28.94/metric ton for a city of 10,000 population (gasoline at $0.26/liter) to $23.83/metric ton for a city of 20,000 population (gasoline at $0.53/liter). The computations and all assumptions made are summarized in Table III. The computations do not include a credit for savings on tipping fees and the extension of landfill life as these costs are site specific. Gasification of 50 percent of the available combustible solid waste in a community could reduce landfill requirements by 20 percent.

Capital Costs of a Gasification System for Small Communities

A cost estimate was obtained for a 122 cm diameter downdraft gasifier with gas cleanup and a dual fuel diesel engine-generator of 400KW capacity. One such system, operating two shifts per day, could handle the source separated solid waste from a community of 10,000 population. Two such systems, operating two shifts per day, would be required for a city of 20,000 population. Shredding and densification equipment would also be required for each system. Capital costs of each system (as of June 1979) are summarized in Table IV. Construction, installation, and operating costs are not included.

CONCLUSIONS

Based on experimental work with a downdraft gasifier the following conclusions can be made:
1. Densified, source separated solid waste is an acceptable fuel for downdraft gasifier.
2. A source separation system can be operated by small communities to produce a combustible fuel in the range of $23.83 per metric ton to $28.94 per metric ton.
3. Capital costs for a gasification system for a communities of 10,000 and 20,000 population would be on the order of $305,000 and $555,000 respectively.

Table IV

CAPITAL COSTS - GASIFICATION SYSTEM

| Item | Population, persons | | | |
| | 10,000 | | 20,000 | |
	Quantity	Cost	Quantity	Cost
Gasifier-Engine-Generator[a]	1	$250,000	2	$500,000
Shredder-Conveyor System[b]	1	15,000	1	15,000
Densification System[c]	1	40,000	1	40,000
Total Capital Cost		$305,000		$555,000

a. MK VII Gasifier system (1.22 Meter Diameter Firebox) w/dual fuel diesel engine and 400 KW generator, Biomass Corporation. FOB Yuba City, California (June 1979).

b. Shredder-Conveyor System, Miller Manufacturing Co., 0.91 metric tons per hour capacity, FOB Turlock, California (June 1979).

c. John Deere Model 390 Stationary Cuber w/electric motor and switchgear, 2.2 metric tons per hour capacity, FOB Moline, Illinois (June 1979).

ACKNOWLEDGEMENTS

The assistance of J. Goss, B. Jenkins, J. Mehlschau, and N. Raubach of the Department of Agricultural Engineering, University of California, Davis, is gratefully acknowledged. The research discussed in this paper was supported by grants from the University of California Appropriate Technology Program and the U.S. Environmental Protection Agency (Grant No. R-805-70-3010). The contents do not necessarily reflect the views of the University of California or the Environmental Protection Agency, nor does mention of trade names or commercial products constitute endorsement or recommendation for use.

LITERATURE CITED

1. Tchobanoglous, G., Theissen, H., and Eliassen, R., "Solid Wastes Engineering Principles and Management Issues", McGraw-Hill Book Company, Inc., New York, 1977.

2. Jones, J., "Converting Solid Wastes and Residues to Fuel", Chemical Engineering, Jan. 2, 1978, McGraw-Hill Inc. New York, 1978.

3. Fisher, T.F., Kasbohm, M.L., and Rivero, J.R., "The PUROX System," in Proceedings 1976 National Waste Processing Conference, Boston, May 23-26, 1976, American Society of Mechanical Engineers, 1976.

4. Hollander, H.I. and Cunningham, N.F., "Beneficiated Solid Waste Cubettes as Salvage Fuel for Steam Generation," in Proceedings 1972 National Incinerator Conference, New York City, June 4-7, 1972, American Society of Mechanical Engineers, 1972.

5. Anonymous, "Strategy for Selling Dense RDF: 'Energy Factories'," Waste Age, July 1977.

6. Williams, R.O., Goss, J.R., Mehlschau, J.J., Jenkins, B., and Ramming, J., "Development of Pilot Plant Gasification Systems for the Conversion of Crop and Wood Residues to Thermal and Electrical Energy," in "Solid Wastes and Residues Conversion by Advanced Thermal Processes," Jones, J.L., and Radding, S.B., Editors, American Chemical Society, Washington, D.C., 1978.

7. Skov, N.A. and Papworth, M.L., "The PEGASUS Unit -Petroleum/Gasoline Substitute Systems," Pegasus Publishers, Inc., Olympia, Washington, 1974.

8. Nordstrom, O., "Redogorelse for Riksnamndens for ekonomish forsvar sberedskap forskningsock fursosverksamhet pa gengasomradet rid statens maskinpravnigar 1957-1963", Chapters 1-5, Obtainable from National Swedish Testing Institute for Agricultural Machinery, Box 7035, 75007 Uppsala, Sweden.

9. Jantunen, M., and Asplund, D., "Peat Gasification Experiments and the Use of Gas in a Diesel Engine," Technical Research Center of Finland, Fuel and Lubricant Research Laboratory, Espoo, Finland, 1979.

10. Brown and Caldwell, Consulting Engineers, "Central Contra Costa County Sanitary District Solid Waste Recovery Full Scale Test Report, Volume One", Walnut Creek, California, March 1977.

11. Jones, J.L., Phillips, R.C., Takaoka, S., and Lewis, F.M., "Pyrolysis, Thermal Gasification, and Liquefaction of Solid Wastes and Residues - Worldwide Status of Processes," Presented at the ASME 8th Biennial National Waste Processing Conference, Chicago, May 1978.

12. Eggen, A.C.W. and Kraatz, R., "Gasification of Solid Waste in Fixed Beds," Presented at the Winter Annual Meeting, ASME, November 17-22, 1974, New York City.

13. Williams, R.O. and Horsfield, B., "Generation of Low-BTU Fuel Gas from Agricultural Residues - Experiments with a Laboratory Scale Gas Producer", in Food, Fertilizer, and Agricultural Residues - Proceedings of the 1977 Cornell Agricultural Waste Management Conference, Ann Arbor Science Publishers, Inc., Ann Arbor, 1977.

14. Williams, R.O., and Goss, J.R., "An Assessment of the Gasification Characteristics of Some Agricultural and Forest Industry Residues Using a Laboratory Gasifier", Resource Recovery and Conservation, 3 (1979), pp 317-329.

15. Cooper, H.B.H. Jr. and Rosano, A.T. Jr., Source Testing for Air Pollution Control, McGraw-Hill Book Co., Inc., New York, 1974.

16. Biomass Corporation, "The Fuel Conversion Project", Sales Brochure, Yuba City, California, 1979.

17. Industrial Development and Procurement, Inc., "Moteurs Diesel Duvant - Gasifier-Diesel Engine Electric Generating Plants", Sales Brochure, Pittsburgh, Pennsylvania, 1978.

18. Williams, R.O., Personal Communication, Biomass Corporation, Yuba City, California, July 6, 1979.

RECEIVED November 16, 1979.

The Enerco Pyrolysis System

E. W. WHITE and M. J. THOMSON

Enerco, Inc., 126 Old Oxford Valley Road, Langhorne, PA 19047

Pyrolysis is the one process by which wood wastes and residues can be converted economically to transportable high grade fuels and liquid chemical feedstocks on a small scale consistant with the very dispersed nature of the forest resource.

Efficient pyrolysis can be achieved at atmospheric pressure with temperatures of only 350° to 500°C. These conditions lend themselves to simple and inexpensive system design. Pyrolysis can be an energy efficient process yielding a favorable energy balance because the process uses only a fraction of the energy contained in the wood.

Enerco's goal in the development of a mobile pyrolysis system has been to maximize the quality and yield of charcoal while making provision to optionally collect liquid distillates and to utilize the gases as an on-site fuel. Some markets for charcoal are well established but the uses of the char oil as fuel and especially as chemical feedstocks are not well defined at this time. With that in mind, provision is made to clean burn the uncondensed gases when the liquid fractions are not wanted.

This development was initiated in an effort to help stimulate improved forest management by making it possible to produce valuable products from the kinds of unused residues that result from intensive forest management programs. This argument has been discussed in considerable detail in two earlier publications (1,2). The first patent resulting from this development has recently issued (3).

Throughout this development, major emphasis has been placed on controlling the environmental problems of smoke and particulate emission that traditionally have been associated with charcoal production.

Construction of a 10 ton per day prototype unit

0-8412-0565-5/80/47-130-275$05.00/0

was begun in the spring of 1976. Tests were conducted
in fall 1976 and spring 1977. A variety of wood feeds
were run including wet (50% H_2O) and dry (6% H_2O) run-
of-mill sawdust, wet (50% H_2O) and air dry (21% H_2O)
chips and dry (10% H_2O) hogged wood. The prototype
unit was modified three times to evaluate different
reactor, combustion and condenser configurations.
     Design of the 24 t/d unit was initiated summer
1977, and serious discussions were undertaken with
potential customers. Two customers finally placed firm
orders with Enerco. These were Forest Energy Corp.,
King of Prussia, PA. and the Tennessee Valley Authority.
Mr. C. Robert Enoch of Forest Energy was interested in
the system primarily for the charcoal production capa-
bilities. His plans were to use the 24D unit at a
conventional sawmill site and to supplement the mill's
residue supply with residues from nearby sources.
     Dr. E. Lawrence Klein, who headed the TVA inquiry,
became interested in the 24D unit, not only for its
charcoal and charoil production but also for its potential to supply excess wood gas that could be used as a
substitute for fuel oil or natural gas while the charcoal was being produced.
     TVA was interested in being able to collect the
oils from a variety of wood feeds for evaluation and
use as a low sulfur fuel. TVA studies lead to formulation of the Maryville project.[4] At Maryville College,
the 24D unit will be used to produce charcoal from
purchased wood feed. During operation excess wood gas
will be burned in one of the college's two boilers to
supply steam. Oils will be collected and stored for
sale or use during pyrolyzer downtime or during peak
fuel loads. The charcoal will be sold to generate income to the college.
     The TVA unit was delivered early January 1979.
The Forest Energy unit was delivered early May 1979.
Both units have been undergoing shakedown tests and
evaluation. At present, the TVA unit is being fitted
with automated feed loading and charcoal handling
equipment. Additionally, TVA has reconfigured some of
the components to shorten wood gas runs and has added
additional controls to make the routine operation of
the system more straightforward.
     All components, including burner and duct insula-
tion, are ruggedly constructed and weather proofed to
withstand repeated site to site moves. Figure 1 shows
the TVA unit the day it arrived at Alcoa, TN. in a
snow storm. The trip originated the previous afternoon
at DuBois, PA. Temperatures during the first evening's
travel were below -16°C. The 24D unit with gross

*Figure 1.   Photograph of Enerco 24D pyrolyzer delivered to TVA, Alcoa, TN,
January 1979*

THERMAL CONVERSION OF SOLID WASTES AND BIOMASS

weight of approximately 12 ton rode smoothly at speeds
of up to 90 km/hr and arrived in less than 24 hours at
the end of the 1100 km trip with no apparent road
damage.

Design Concept

A schematic of the 24D system is shown in Figure
2. The title 24D is derived from the design capacity
of 24 ton per day charcoal output. The basic trailer-
mounted system is comprised of three subsystems: 1) a
reactor, 2) a recirculating gas loop to bring heat into
the reactor by indirect contact of the recirculating
gas with hot products of combustion in a heat exchanger
and 3) the combustion system. Ancillary subsystems
such as a condenser to obtain liquid distillates (shown
in sketch) and secondary heat exchangers (not shown)
are not critical to the operation of the 24D system.

The reactor accepts comminuted wood in the form of
run-of-mill sawdust, hogged wood, chips or shavings.
The feed enters the top of the static vertical bed
reactor. Level sensors automatically interrupt the
loading when the reactor is filled to the proper level.
Whenever the wood feed level drops below a second
selected level, the load auger automatically runs until
the full load level is reached.

Throughput rate of the reactor is governed by the
char off-loading rate. This is controlled by the
operating schedule of the off-load auger which can
either be run continuously or intermittently. At a
given throughput rate the quality of the char is dependent on 1) the dimension of the larger pieces of wood
in the feed 2) temperature and volume of the recirculating gas stream and 3) moisture content of the feed.
The reactor is built to accomplish a horizontal flow of
gases through the beds in such a way that the top and
bottom of the reactor are stagnant zones and the gases
exiting from the beds do so at a sufficiently low
velocity to prevent lofting and entrainment of fines.

A hot fan (F_1) in Figure 2 acts in a push-pull
mode to suck gases from the reactor and pressurize them
to move through the circuits. The volume of recirculated gas through the heat exchanger-reactor loop is
set by valve V_2. The off-gas valve V_1 operates automatically in response to a pressure impulse line in the
top of the reactor above the wood feed. This line
provides a pressure signal referenced to ambient baro-
metric pressure for adjusting V_1 to accomplish the
desired pressure differential. Typically this is set
at zero or minus 3.0 Pa to prevent wood gas escape

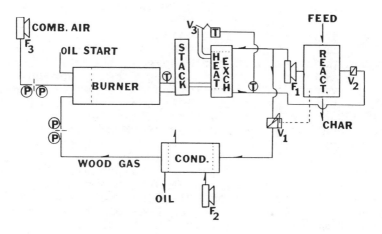

Figure 2.   Schematic of Enerco pyrolysis unit

while minimizing intake of ambient air through the feed
auger. Thus, action of the off-gas valve V_1 is the
dominant control over fugitive smoke emission from the
system. By using the pyrolysis gas as the heat carrier
to the reactor, nitrogen is essentially excluded from
the reaction. This is important in preventing formation of nitrogenous pyrolysis products including, of
course, the nitric acid that forms in air-using pyrolysis processes.
    The heat exchanger is operated in counter flow
with the wood gas entering at the top while the hot
products of combustion (POC) from the combustion chamber enters at the bottom. Temperature of the returning wood gas is controlled by automatically adjusting
the volume (V_3) of POC allowed to pass through the heat
exchanger. The temperature can be set anywhere in the
range of 300° to 500°C with a control precision of
±15°C.
    The combustion system uses an oil-fired startup
burner to initiate the pyrolysis reaction and to incinerate the early wood gas stream that is too deficient in fuel value to sustain its own flame. Once the
reaction produces burnable wood gas, the oil burner can
either be shut down or put on low fire. Because the
wood gas varies both in volume and fuel value, and because no gases are permitted to escape to the atmosphere, the gas burner must be quite flexible. A pre-
packaged ratio controller (not shown on sketch) senses
the wood gas flow using the pressure drop across an
orifice plate. Combustion air is automatically metered
in a fixed ratio to the burner. Compensation for gross
changes in wood gas fuel value are made manually by
changing the wood gas/combustion air volume ratio.
Temperature of the POC at the rear of the combustion
chamber (burner) is read out continuously. The POC
stack is equipped with an automatic back pressure device to give the heat exchanger first priority call on
POC. The base of the stack has a flanged port from
which POC can be ducted to driers or other equipment.
Similarily, a blanked off part on the wood gas run can
be used to direct some of the wood gas to an adjacent
burner if desired.

Results and Discussion

    Table I gives results of the proximate analysis of
charcoal and charoil obtained from the 10 t/d prototype.
    The charcoal is of excellent quality for most
uses. However, the value of 0.34% sulfur in this
sample was quite surprising. Analyses of three other

TABLE I. Analysis of Enerco Products From Prototype

	CHARCOAL	CHAROIL
Heat of Combustion MJ/kg (BTU/lb)	29.85 (12,834)	22.27 (9,573)
% ASH	4.74	1.14
% S	.35	.11
% VOLATILE MATERIAL	23.03	N/A
% FIXED CARBON	68.44	N/A

samples of the same run were made to check this result.
All three yielded values in the range of 0.07 to 0.10
wt% sulfur.

Table II summarizes results obtained from TVA for
their 24D system. Again, the charcoal is of high quality with sulfur results in the range one would expect.

TABLE II. Preliminary Testing of TVA Pyrolysis
Specimens

	Gross Heat of Combustion MJ/kg	(BTU/lb)
White Oak (Uncarbonized)	18.81	(8089)
White Oak (Partly Carbonized)	20.52	(8822)
Carbonized White Oak (No H_2O) Average of 2 Specimens	31.11	(13377)
Carbonized Wood (Possibly with H_2O Cool Down)	30.84	(13261)
Condensate (Solid)	28.39	(12207)
Condensate and Carbonized Wood (Solid)	32.04	(13774)
Condensate (Viscous)	25.83	(11106)

At this point in the development of the Enerco 24D
pyrolysis system the design concept and its advantages
have been demonstrated. In particular, the reactor
design successfully excludes nitrogen from the reaction
and does not require cyclones or other particulate
cleanup downstream. The combustion system cleanly

282 THERMAL CONVERSION OF SOLID WASTES AND BIOMASS

incinerates the low fuel value startup gases and takes over to cleanly burn the variable (quantity and quality) pyro gases. As of this writing, neither unit has confirmed all the economic assumptions that have to be finally met before the next units are to be built. In particular it is not yet certain just what the exact size of the reactor and heat exchanger needs to be to meet the 24 t/d production capacity for all the kinds of wood feed one would like to pyrolyze. It is, however, becoming increasingly apparent that definite advantages accrue from using bone dry wood feed. The higher the moisture content, the slower the pyrolysis proceeds and the oils appear to be of lower quality.

The original premise that charcoal would find markets as an alternative fuel for power generation and for home heating is now widely discussed. TVA has concluded that charcoal could be used in coal fired generating plants either alone or as an extender to be mixed with high sulfur coals. Two large electric utilities have indicated to the authors that they would purchase charcoal if it were to become available in quantities on the order of 10,000 t/y or more and assuming that no unusual dusting problems resulted. Recently Brooks and Field have reported on their analysis of the role charcoal could play as a domestic heating fuel in Maine. (5)

Acknowledgements

Mr. Fred Knoffsinger and Francis Gross collaborated in the development and early testing of the original prototype unit. Lee-Simpson Associates aided in engineering the first production scale unit. Mr. C. Robert Enoch of Forest Energy has been highly instrumental in supporting the commercial development of the 24D unit and has been very helpful in design and testing of his unit. Dr. E. Lawrence Klein of TVA has been particularily instrumental in promoting the pyrolysis concept. His insights into practical forestry and the role that wood energy should have in overall timber and wildlife management have been most instructive and stimulating. Mr. Robert Marshall of North American Mfg. has been invaluable by working out the combustion burner and control system as well as devoting a great deal of time to the entire 24D system. This development program has been supported by private individuals and by the sale of the two units.

Literature Cited

1. White, E. W., "Wood Charcoal as an Extender or
   Alternative for Coal: An Immediately Available
   Energy Source", Earth and Mineral Sciences
   Vol. 46 No. 4 (Publication of the College of
   Earth and Mineral Sciences, Penn State Univer-
   sity) 1977.
2. White, E. W., "Charcoal: Fuel for Thought",
   American Forests, p. 21, Oct. 1978.
3. White, E. W., F. M. Gross and F. E. Knoffsinger,
   "Continuous Drying and/or Heating Apparatus"
   U.S. Patent No. 4,165,216, 1979.
4. Klein, E. L., "Maryville College Pyrolysis Project",
   Southern Lumberman, Dec. 1978.
5. Brooks, D. J. and D. B. Field, "Potentials of
   Charcoal Production for Forest Stand Improvement
   and Domestic Space Heating in Maine" Cooperative
   Forestry Research Unit, University of Maine,
   Orono, Research Bulletin 1, March 1979, 52 pp.

RECEIVED November 20, 1979.

Development of Modular Biomass Gasification and Combustion Systems

JAMES FLETCHER and RICHARD WRIGHT

Industrial Combustion Inc., 4465 N. Oakland, Milwaukee, WI 53211

The need for retrofit devices to convert biomass materials to combustible fuel gas is well established. Several developers have built and operated successful biomass gasifiers, for example the systems reported by the University of California-Davis and Georgia Tech Engineering Experiment Station.

The objective of the development reported here is to reduce tradional gas producer technology to commercial practice so that unattended biomass utilization systems under completely automatic control can be available to boiler operators now burning fluid fuels in packaged steam and heat generators. The system is to be designed for maximum gas production with minimum carbon remaining in the ash and zero liquid wastes production.

The system must be highly reliable and simple enough to operate at optimum efficiency with a minimum of unskilled attention.

Defined specifications are:
1. Feed stock to be densified pellets.
2. Generator to be of crossflow type.
3. Gas combustor to be close coupled to generator.
4. Structure to be an integrated packaged unit.
5. Able to resist abnormal conditions.
6. Minimum power required to operate.
7. Combustion and system operating controls to include programming, combustion monitoring and operating rate regulation.
8. Gas combustor to have integral automatic ignitor and electronic flame sensor.
9. Capacity ranges to be from 2×10^6 to 30×10^6 B.t.u. per hour. (2.1 to 31.7 GJ/hr.).

Biomass gas producers can generate liquid, gaseous and solid end products. The desired end product of this unit is gas. Gas temperature is maintained high enough to inhibit condensation of low boiling point distillates.

Design and development tasks were separated into individual projects:
1. Feed stock supply.

0-8412-0565-5/80/47-130-285$05.00/0
© 1980 American Chemical Society

2. Feed stock surge reservoir.
3. Pyrolysis reactor.
4. Oxidation - Reduction reactor.
5. Refuse removal.
6. Forced draft air supply.
7. Gas combustor.
8. Generator controls.
9. System and combustor controls.

To verify design parameters, a series of structures designed for approximately 4×10^6 B.t.u. (4.22 GJ) were constructed and modified as improvements were indicated and better components became available. The first crude prototype might have been considered as commercially viable by some based upon an offer to purchase "as is" by a serious prospective customer.

Regulation of process rate in response to load demand has been accomplished. A modulating type actuator simultaneously positions valves separately regulating generator air and gas combustor air flow rates. The rate of ash removal is regulated in proportion to fuel and air supply. The heat output demand is therefore matched by the air delivery rate to the machine.

Controls for reactor temperature and gas temperature have been developed for both regulation and emergency high limit protection. Temperature in the reactor oxidizing zone is attenuated by water vapor introduced into the air flow. The optimum reactor temperature depends largely upon ash fusion temperature. Maximum reaction rate occurs at 2500°F (1370°C) or higher. With most fuels it is not practical to operate substantially above 2200°F (1200°C) because of ash and dirt contaminant fusion.

Air for gasification and for combustion of product gas is delivered by a common blower with individual regulation of separate steams to the reactor and to the combustor.

Hot gas is delivered to the combustor above atmospheric pressure. The gas combustor is designed to fire against negative, balanced or positive boiler firebox pressure to conform with the boiler requirements. The combustor is comparable in size and similar in configuration to other Industrial Combustion firing heads.

Reliability is enhanced by complete absence of gas valves. Maximum safety is achieved without depending upon pressure relief valves or explosion discs. Wide open burner design would relieve any abnormally elevated pressure directly into the boiler which in turn is directly vented to atmosphere.

The need for conversion of pyrolysis products to gas and vapors having boiling points lower than the gas temperature narrowed the design arrangement decision to either a downflow or crossflow reactor. Many other factors were considered in the selection of a crossflow reactor as most suited for the design objectives including feed stock and refuse handling, air pressure requirements, compact arrangement and process control characteristics.

Complete achievement of performance goals require control of feed stock properties. At the start it was assumed that commercially available densified wood pellets were substantially consistent in properties. For example, 7500 to 8500 B.t.u./lb. (17.5 to 19.7 MJ/kg) 1-4% ash and 7-12% moisture. Experience has shown that while these ranges are typical, available products may sometimes vary appreciably. This demonstrates the need for commercial fuel pellet standards. This need is recognized by both suppliers and users. Several agencies including the Forest Resources Department of North Carolina State University have expressed interest in preparing standard specification language. An ASTM subcommittee is looking at specifications for cellulose fuel pellets.

The chemical reactions and stoichiometry of the system described are basically the same as all other types of pyrolysis based gas generators, so are not discussed here. The schematic arrangement of the machine and location of the various reaction zones is shown in Figure 1.

Figure 2 shows the general arrangement of a gasifier/reactor/combustor applied to a typical heating or power boiler.

Many difficult problems were solved. Most were of an elementary nature not requiring new technology, but a great deal of genius was required to develop solutions.

The problems might be classified as material handling techniques, selection of commercially available components and application of suitable controls and regulating systems.

Arranging reliable components into acceptable compact, but dependable packages many times smaller than those previously used by other gasifier developers was a difficult assignment.

Probably the most trying task was arranging the flow path of volatile pyrolysis products through the char oxidation and reduction zones in such a manner that condensed tar would not render the reduction process ineffective.

Performance testing was conducted without sophisiticated gas analysis instrumentation. This might be best described as an engineering rather than scientific approach. Measured output divided by measured input is a practical means for determining system efficiency. This is entirely compatible with the objective of rapidly developing a practical machine for commercial utilization.

Historical practices once used with coke and anthracite coal gas producers do not apply to systems for gasifying high volatile biomass. The only similiarity between a biomass gasifier and a low volatile coal gas producer is that both convert solid fuel to a low heating value gas. The systems by nature are almost entirely different.

Published data for wood char shows reaction rates six to ten times as rapid as reaction rates for coke or anthracite coal. This might lead to conclusion that wood char gasification should be many times more rapid than that of coal based materials.

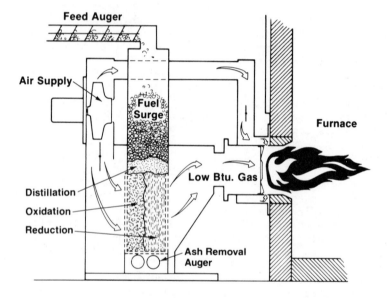

Figure 1.   Schematic arrangement of pyrolysis-based gas generator and location of reaction zones

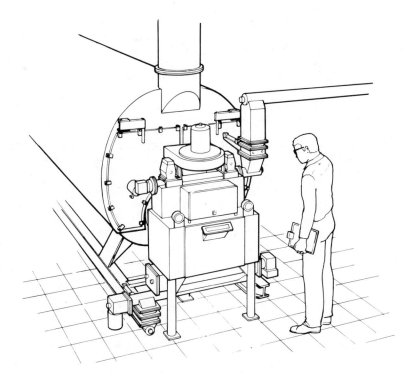

Figure 2. General arrangement of a gasifier/reactor/combustor applied to a typical heating or power boiler

While the well documented wood char reaction data is probably accurate, it does not specifically apply to this system. Precise laboratory determined reaction data does not necessarily apply to the wood gasifier because of possible incomplete pyrolysis to pure char or pollution of char within the reduction reactor with condensed tar.

In order to accurately test the system with a minimum of instrumentation, accurate analysis of the feed stock was necessary. This includes both proximate analysis to establish volatile fired carbon, moisture, ash and ultimate analysis for carbon, hydrogen and oxygen.

Calorimeter determined heating value is needed. Efforts to calculate heating value from ultimate analysis disclosed that the Dulong and similar equasions generally suitable for coal give excessive deviation from measured heating value.

Oxygen content is an important fuel property. This is a measure for the amount of chemically bound water in the dry fuel. This has an important bearing upon both the process operation and final gas composition.

The present state of development is the completion of several third generation production units for outside evaluation by qualified interested institutions and users. In house testing and development of larger capacity units is continuing at an increasing rate.

RECEIVED November 16, 1979.

Solid Waste Gasification and Energy Utilization

FRANKLIN G. RINKER

Midland-Ross Corporation, Thermal Systems Technical Center,
P.O. Box 985, Toledo, OH 43696

Until recently, solid waste has been considered only in the
context of a disposal problem. With the threat of possible fuel
curtailments and subsequent price increases, the energy content
of solid waste is being recognized as a viable energy resource.
The U. S. Bureau of Mines has forecast that the gross energy consumption will double by the year 2000 to approximately 193.4
quadrillion BTU's. Of this, approximately 41.2% will be consumed
by industrial users. In many cases, these same industrial users
are generating waste materials containing a significant amount of
recoverable energy.

Gasification of hydrocarbon bearing materials is not new.
Gasification of coal and wood was practiced from the mid-19th
century until the advent of large natural gas distribution systems. The gasification of solid wastes can be thought of as an
extension of this existing technology, using updated materials,
process control, and end use combustion techniques. In my discussion, I will attempt to answer the following questions concerning solid waste gasification:

1. What is gasification?
2. Why gasify solid waste?
3. How is gasification accomplished?
4. What is the present industrial application?
5. What is the future of solid waste gasification?

What is Gasification?

Technically speaking, gasification is the controlled thermal
decomposition of organic material, producing a gaseous fuel and
an inert solid ash residue. This controlled thermal processing
can be accomplished through two types of reactions:

1. Pyrolysis of the solid waste is generally a low temperature endothermic reaction yielding a fuel rich offgas
   and solid residue containing the inert gas and fixed
   carbon fractions of the original material.
2. Complete gasification consists of the pyrolyzer reaction
   and an additional exothermic reaction of the fixed car-

0-8412-0565-5/80/47-130-291$05.00/0
© 1980 American Chemical Society

bon with oxygen. The resultant products of this process are a fuel rich gas and a solid inert ash residue.

Pyrolysis has the advantage of not requiring an oxidizing agent such as air; thus the offgas has a high heating value and, in some cases, liquid hydrocarbon fractions can be separated and stored. The process is normally carried out under relatively low temperatures (800-1200°F), simplifying equipment design and material handling requirements. The main disadvantage of pyrolysis is that a significant portion of the chemical energy contained in the original material exits the pyrolysis process as solid char. The utilization of this char for end use heating applications involves special combustion and handling equipment. It is, therefore, usually landfilled as char with a resulting loss in its heating value.

Why Gasify Solid Waste?

Solid waste gasification offers the following benefits over conventional disposal means:
1. Gasification provides a captive energy source in a time of questionable energy availability and cost.
2. Gasification reduces the quantity of waste material to be landfilled and the associated costs.
3. The inert ash material discharged from the process is often more compatible with landill restrictions than the original waste.
4. The fuel gas can be used in conventional end use systems, such as steam boilers, process and liquid air heaters, and large industrial furnaces.
5. The low process velocities inherent in gasification minimize the need for offgas particulate cleaning devices required for air quality assurance.

From a fuel gas generation standpoint, the complete gasification of solid waste is desirable. Previously, the high temperatures resulting from the oxidation reaction of the fixed carbon contained in the original material limited the application of gasification to well defined feedstocks, such as coal, where the operating parameters were somewhat well defined. We have found that by using steam as an oxidizing agent in the gasification process, these reaction temperatures can be controlled, thus widening the application of gasification to the solid waste area.

How is Gasification Accomplished?

Gasification of solid waste can be accomplished in a variety of furnace designs. At Midland-Ross, we have selected the vertical shaft furnace primarily because of its uncomplicated material transport method--gravity--and its efficient counterflow of material and process gas. This furnace is shown on Figure #1.

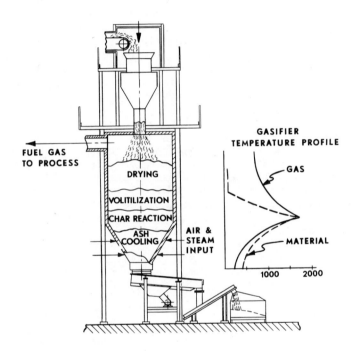

Figure 1.   Solid waste gasifier

Material enters the top of the furnace through a charging hopper and metering mechanism and falls through to the top of the material bed. It then moves down slowly through the four zones of reaction and is discharged for disposal at the bottom as ash. At the same time, air and steam are admitted at the bottom of the shaft in a manner designed to promote uniform distribution of these gases throughout the cross-section of the shaft furnace. These gases then travel upward through the ash cooling zone, cooling the ash and picking up super heat as they travel. As the air and steam enter the char reaction zone, two distinct and different chemical reactions occur. The oxygen contained in the air reacts with the carbon liberating heat and forming carbon monoxide and carbon dioxide. Concurrently, the steam reacts with the carbon, giving off carbon monoxide and hydrogen. This water gas shift reaction is endothermic and absorbs heat. By carefully controlling the balance of these two reactions, the temperature of the char reaction zone can be controlled producing optimum reaction. The determination of this balance is best accomplished by pilot scale testing on the actual solid waste to be gasified.

After passing through the char reaction zone, the steam and air have been converted to a high temperature gas stream consisting of mainly nitrogen, hydrogen, carbon dioxide, and carbon monoxide. This high temperature gas stream then passes through the volatilization zone where it transfers this heat to the waste material and causes a pyrolytic reaction, releasing the volatile hydrocarbons contained in the waste material. These hydrocarbons consist mainly of methane and ethane with heavier fractions occurring at lower reaction temperatures.

The final reaction is the forced drying of the incoming waste material through the transfer of heat from the hot volatile gases to the waste by convection. The now fuel rich gas then exits the bed and passes out of the gasifier for end use application.

What is the Present Industrial Application of Gasification?

The gasification of industrial solid wastes is not universal in application. The following points must be considered prior to application:

Material Considerations. The material should conform to certain criteria for the successful application of gasification. First, it should have a relatively low moisture content in the range of 0-50%. The addition of water to the incoming material causes a decrease in overall process efficiency as the excess moisture must be carried through the process and discharged from the process at elevated temperatures.

Second, the material should not melt or soften when exposed to gasification temperatures. Since these temperatures range from 1200-2200°F most plastic wastes would be eliminated for consideration. However, even these materials can be gasified using a pyrolytic process and a different furnace design.

Third, the material should have a relatively high heating value somewhere in the order of 5000-20,000 BTU/lb as received.

Fourth, the material should be able to be sized for proper flow through the shaft furnace. Sizing simplifies continuous charging of the material while still maintaining an atmosphere seal. Ideal particle sizes range from 1/2" to 6" mean dimension.

End Use Considerations. The main requirement for successful end use of energy generated by gasification is a sufficient and continuous capacity to utilize all of the energy produced. The most common end use is the production of process steam where the gasifier fuel gas is burned in the conventional water tube boiler and the steam produced is fed to the main plant steamline, relieving the load on fossil fuel fired boilers. Other end uses are process air heaters for large product dryers, industrial furnace firing, and electrical generation boilers.

Economic Considerations. Economics, of course, will justify the technique of solid waste disposal and fuel gas generation. Some important variables to be considered in the economic balance are as follows:

1. The quantity of solid waste generation should provide a sufficient load for maximum utilization of the size gasifier selected.
2. Solid waste generation should be consistent or sufficient storage capacity should be installed to provide consistent loading of the gasification system.
3. The present and future fuel cost and availability must be balanced against the capital and operating cost.
4. The present and future solid waste landfill cost and restrictions must be considered.

Taking the above considerations into account along with the cost of capital and return on investment goals, the economic justification of solid waste gasification can be evaluated. An example of the proper application of solid waste gasification is a project Midland-Ross is presently constructing for a major tobacco manufacturer. The solid waste consists of cigarette papers and filters, packaging paper, cardboard, used tobacco containers, and transport pallets. The average heating value of this material is 7500 BTU/lb as received, with an average moisture content of 10%. The waste is presently baled and transported to a landfill. A consistent end use heat requirement existed in the form of process steam generation for the manufacturing operation.

The gasification system selected for this application consisted of two (2) vertical shaft gasifiers, each capable of gasi-

fying 5000 lbs/hr of solid waste, one conventional water tube
boiler rated at 50,000 lbs of steam per hour at 175 psig and
equipped with a Surface rich fume burner unit designed to combust
the hot fuel rich gases produced by gasification; and finally,
a Surface Rich Fume Incinerator, used as a standby fume disposal
means in the event of a boiler shutdown.

Referring to Figure #2, the gasification system will be
charged with 10,000 lbs/hr of solid waste with the waste flow
being equally distributed to each gasifier. The steam and air
mixture will be proportioned to the waste flow to each gasifier.
The ratio of this air to steam will also be adjusted to provide
proper gasification temperatures. The ash discharge rate is adjusted to maintain a constant bed level at all times.   The
gas generated from the gasifier will amount to approximately
25,000 lbs/hr with a gross heat content of 68 MM BTU/hr. This
gas is then mixed with an appropriate amount of air in the rich
fume burner to fully combust the fumes at an excess air level
of 10%. The proportion of air and fuel is controlled to maintain this 10% excess air rate under various waste loadings thus
achieving a high boiler efficiency. All process variables are
monitored at a central system control panel operated by one man.
On weekends, or off shifts, when the waste generation is halted,
the system is put in a banked condition and operating temperatures are maintained until resumption of waste flow, thus minimizing restart time and maintaining system productivity.

The operating economics of gasification as compared to a
landfill operation break down as follows:

Gasification

Operating annual costs (4000 hrs/yr):
```
        Manpower . . . . . . . . . . . . . $110,000
        Fuel . . . . . . . . . . . . . . .   13,800
        Power  . . . . . . . . . . . . . .   10,000
        Maintenance  . . . . . . . . . . .   60,000
        Ash landfill . . . . . . . . . . .   25,000
                    Total Operating Costs $218,800
```

Income

```
Steam production @ $4/1000 lbs steam   $800,000
Net Income                             $581,200
```

Landfill

```
Total annual cost                      $300,000

Net savings through gasification       $881,200
```

This is shown graphically as a function of operating hours

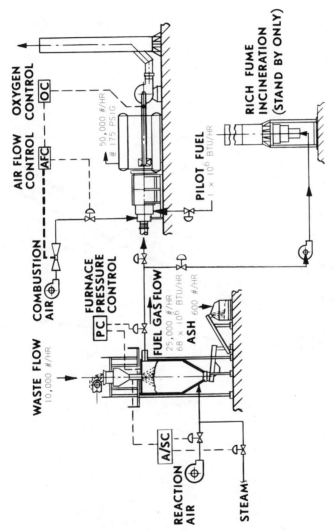

Figure 2. Solid waste gasification system

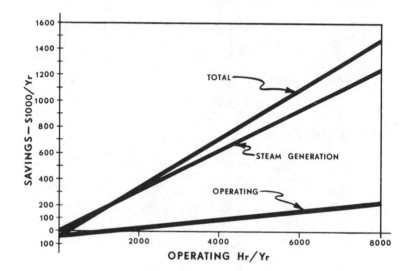

Figure 3.   Income generation curve

per year on Figure 3. As seen in this figure, the energy utilization provides the bulk of operating savings, although the landfill volume reduction of 30:1 and weight reduction of 15:1 provide appreciable savings.

What is the Future of Solid Waste Gasification?

During the next decade, at Midland-Ross we expect solid
waste gasification to grow technically in two areas: waste product
application and fuel product end use. In our pilot scale gasifier
we have demonstrated the successful gasification of many industrial waste materials, from rubber tires to forestry waste to
waste paper to pharmaceutical sewage sludge. Each application
has its unique operating parameters to define. As more applications are investigated, this parameter development will become
less tedious.

Fuel product end use is presently restricted to "same time"
energy users such as steam boilers, process air and liquid heaters, and large furnaces. Midland-Ross, through its Thermal
Systems Technical Center, is presently investigating and developing processes to store this energy in a readily transportable
and usable form. This would be either as a concentrated high
BTU gas or an oil-like product. This development, when it comes,
will enable industry to generate its own fuel and store it for
end use as economic or availability factors dictate.

In summary, solid waste gasification is a here and now technology for industry's profitable use. Using a process developed
over a century ago and applying today's method and materials, we
are able to relieve the burden placed on our landfills and produce additional income through an environmentally acceptable
process.

RECEIVED November 16, 1979.

Thermodynamics of Pyrolysis and Activation in the Production of Waste- or Biomass-Derived Activated Carbon

F. MICHAEL LEWIS

142 Waverly Place, Mountain View, CA 94040

C. M. ABLOW

SRA International, Menlo Park, CA 94025

The use of activated carbon, or activated charcoal as it is sometimes called, dates back as far as 2000 B.C. when it was first used by the ancient Egyptians. The term "activated" refers to the chemical or thermal treatment given to the carbon to increase its adsorptive capacity. Estimates for activated carbon demand in the U.S. vary but a nominal value for current U.S. usage would be 10^8 kg/yr.

Government regulations, such as the recent proposal to reduce trace organic compounds in drinking water with the use of activated carbon, if enacted, could substantially increase demand. However, the market growth anticipated when the Federal Water Pollution Control Act Amendments of 1972 were passed never came to fruition. With this degree of uncertainty, the best estimates for demand growth appear to be in the range of 5% - 11% per year.

Raw materials for the production of activated carbon have historically included coal, sulfite mill liquor, lignite, peat, coconut shell, and wood. Changing economics and a realization that some materials previously discarded represent a potential resource have stimulated investigations into using municipal solid waste and agricultural and forestry residues as feedstocks. In a study conducted by Stanford University, municipal solid waste was thermally processed into activated carbon and after removal of inert material, compared favorably with commercial grades of activated carbon for use in wastewater treatment. A commercial firm in Northern California is processing redwood sawdust to make activated carbon that can be used for cleaning and decolorizing food oils, water, fabrics, industrial gases and fluid wastes.

Practical processes for activating carbon include high temperature thermal processes, where the carbon is reacted with steam and carbon dioxide, and chemical processes where activation takes place in the presence of metallic chlorides. These processes were developed at the turn of the century and today variations of this basic technology are still used to manufacture nearly all activated carbons. In this paper conditions are presented for maximizing the thermal efficiency of the high temperature processes.

0-8412-0565-5/80/47-130-301$05.00/0

The combustible fraction of municipal solid waste and biomass consists primarily of carbon, hydrogen, and oxygen. The quantities of these elements vary slightly but for illustrative purposes we can treat both combustible fractions as pure cellulose $(C_6H_{10}O_5)$ with the following weight percents:

Carbon	44.4%
Hydrogen	6.2%
Oxygen	49.4%

The heating value of a typical biomass is sufficient to produce steam in excess of that required by the activated process if the system has been designed for maximum thermal efficiency. This can be especially important to developing countries who have large supplies of biomass such as rice hulls or coconut shells and who are currently contemplating the manufacture of activated carbon for export or local water treatment.

Designing a process to maximize energy recovery requires a fundamental understanding of the system thermodynamics. Unfortunately most commercial firms consider their processes highly proprietary and tend to treat the thermodyanmics as a kind of "black magic." As a result there is little or no technical data available.

A process that is similar to carbon activation is coal gasification. Here coal, consisting primarily of carbon, is reacted at high temperatures with various mixtures of air, oxygen, and steam to produce a fuel gas. We have applied some of the techniques used to analyze the thermodynamics of coal gasification systems to develop a technical data base.

Thermodynamic Fundamentals

The thermodynamic combustion/gasification processes involved can be analyzed on the basis of the First Law of Thermodynamics (Heat in = Heat out: Material in = Material out) and chemical equilibrium. The set of nine reactions listed in Table I describe the most important reactions for combustion/gasification processes. Thermodynamic data pertaining to these reactions can be found in engineering texts but K_9, the equilibrium constant for Reaction (9), is generally given for beta-graphite. Experimental data obtained from gasifiers operating on peat indicate that methane is found in the output gases at concentrations higher than that predicted by the beta-graphite equilibrium constant. In our analysis we have compared the output predictions for the beta-graphite equilibrium, the "textbook" value, and for the largest reported carbon equilibrium constant, the "experimental" value. From data reported by Feldmann (1975) we use $\ln K_9 = A/T-B$ where T is in degrees Kelvin and $(A,B) = (10885, 13.208)$ for the textbook value and $(18397, 17.445)$ for the experimental value.

The five product gases of the forward reactions in Table I are H_2O, CO_2, CO, H_2, and CH_4. Higher order hydrocarbons, sometimes called illuminants, occur in small quantities that can be neglected in a heat and mass balance analysis as can the small changes in nitrogen concentration. Carbon, hydrogen, and oxygen balances provide three equations for determining the equilibrium concentrations. Reaction (8), which is sometimes called the water gas shift, is a gas phase reaction that has been experimentally shown to approach equilibrium. The known value of the equilibrium constant gives us a fourth relation. Reaction (9) is a gas solid reaction that is probably not in equilibrium in a typical furnace. However, calculations based on equilibrium for reaction (9) give methane concentrations not far from experimentally determined values. The quantity of CH_4 that is produced under typical activation conditions is small enough that no significant error in the heat and mass balance is introduced.

Table I

Pyrolysis and Activation Reactions

1. $C + O_2 \rightleftharpoons CO_2$
2. $C + 1/2\ O_2 \rightleftharpoons CO$
3. $C + CO_2 \rightleftharpoons 2CO$
4. $CO + 1/2\ O_2 \rightleftharpoons CO_2$
5. $H_2 + 1/2\ O_2 \rightleftharpoons H_2O$
6. $C + H_2O \rightleftharpoons CO + H_2$
7. $C + 2\ H_2O \rightleftharpoons CO_2 + 2\ H_2$
8. $CO_2 + H_2 \rightleftharpoons CO + H_2O$
9. $C + 2H_2 \rightleftharpoons CH_4$

Reaction (6) plays a major role in the activation process but this gas-solid reaction does not approach equilibrium. Steam is the gaseous reactant in the forward reaction. A fraction of completion or Percent Steam Utilization (PSU) for this reaction is needed in order to calculate the mole percent of the different output gases. We have displayed the results of all calculations graphically with PSU as one of the variables.

Pyrolysis and Activation

The first unit operation in the production of waste or biomass derived activated carbon is pyrolysis, also known as baking or charring. In this process the material is heated in an essentially oxygen-free atmosphere to drive off the free moisture and volatiles. The material that remains is called char or fixed carbon. The char yield is dependent upon heating rate and in a study performed by Roberts et al. (1978) on the production of municipal solid waste derived activated carbon, the following equation was experimentally determined:

$$Y = -2.4 \ln\tau + C$$

Y = Percent char yield (dry basis)
τ = Heating rate, °C/min
C = Experimentally determined constant
   related to feedstock material and final
   temperature.

Figure 1 illustrates this equation for C = 26.5%, a value consistent with values reported for pure cellulose and municipal solid waste, and with data obtained in discussions with commercial firms. The importance of a slow heating rate to maximize the char yield is quite apparent in the figure. It follows that fluid beds and downdraft or updraft gasifiers, though excellent for maximizing gas yield, are not desirable for char production.

Pyrolysis is an endothermic process that requires heat. This heat can be generated by admitting a small quantity of air into the furnace where it hopefully consumes some of the volatile gases without burning any of the char. Indirect heating is attractive because it avoids any opportunity to lose the char.

To activate the char produced in the pyrolysis step, a portion of the carbon is reacted with steam as in Reaction (6) to develop the internal pore structure which gives the carbon its high adsorptive properties. The percent of carbon that is reacted will depend upon the final properties desired but 25% to 50% is a typical range. This reaction between steam and carbon is highly endothermic so that heat must be supplied to sustain the process. The following list represents alternatives for generating or supplying this heat.

1. Admission of a small quantity of air into the furnace: Carbon-Steam-Air system
2. Admission of a small quantity of oxygen into the furnace: Carbon-Steam-Oxygen system
3. Firing the furnace with a natural gas (CH_4) burner operating with stoichiometric air: Carbon-Steam-CH_4/Stoichiometric Air system
4. Heating the furnace indirectly: Carbon-Steam-Indirectly Heated

Figures 2 and 3 present our calculated equilibrium concentrations of the gaseous products of pyrolysis with 50% of the carbon reacted. The major effect of a temperature rise is to increase CO and decrease CH_4. The results for the two values of K_9 are also such as to balance a fall in CO concentration with a rise in CH_4.

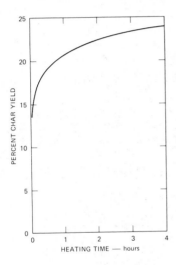

Figure 1.   Typical char yield from biomass by pyrolysis at 500°C

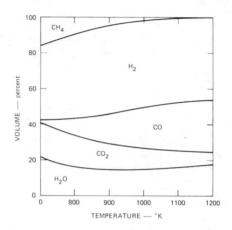

Figure 2.   Cellulose pyrolysis equilibria (textbook CH_4 output)

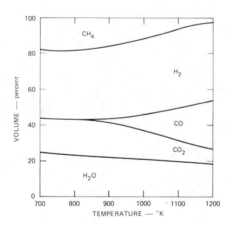

Figure 3.   Cellulose pyrolysis equilibria (experimental CH_4 output)

Thermodyanmics of Activation

To illustrate the effects of the different process parameters, we have developed computer programs to analyze various systems and plotted the results of these calculations graphically. All. graphs represent systems in heat equilibrium where the quantity of heat generated by combustion or applied indirectly, satisfy the system requirements. Figures 4 - 6 illustrate the changes in mole percent of the output gases as a function of PSU at temperatures of 1000°K, 1100°K, and 1200°K (1340°F, 1520°F, 1700°F) respectively. This represents the range of typical activation temperatures. The equilibrium concentration of steam necessarily falls as more steam is assumed to be reacted. Although the CO fraction of the product gases rises, the total of reactive gases falls with respect to the inert diluent N_2 since more air is needed to maintain the temperature. These effects are accentuated at the higher temperatures. The fraction that is CH_4 becomes less than 1% at the highest temperature.

The concentration of CH_4 in the output gases was calculated using the experimentally determined equilibrium constant for Reaction (9). Figure 7 gives the results of the calculation at 1100°K using the standard textbook value for K_9. Comparison with Fig. 5 shows the appearance of CH_4 as the only apparent difference in the two cases. Calculations neglecting CH_4 and omitting Reaction (9) give the same resulting equilibria within a fraction of one percent. Figures 8 and 9 indicate the quantity of air (shown as an O_2/H_2O ratio) that must be used to achieve heat equilibrium. From these graphs it is apparent that except for the lowest temperature, 1000°K, the presence of CH_4 in the output gas has negligible effect on the air or heat requirements.

From Figures 8 and 9 it is also apparent that for any given O_2/H_2O ratio the system can be in thermodynamic equilibrium at a variety of temperatures depending upon the actual steam utilization. Temperature increases tend to increase reaction rates which would indicate a higher PSU. However from these graphs PSU decreases with increasing temperature at a given O_2/H_2O ratio. This indicates that in an actual furnace, once an operating condition is reached, the furnace will be very stable.

Heat Source Considerations

Figures 10 - 12 present calculated equilibrium compositions at 1100°K neglecting CH_4 for the carbon-steam system with various ways of providing the needed heat. The reactants are indirectly heated for Fig. 10, heated by addition of a stoichiometric mixture of air and methane for Fig. 11, and by consumption by oxygen of some of the carbon for Fig. 12. The results for heating by an air-carbon reaction are given in Fig. 7.

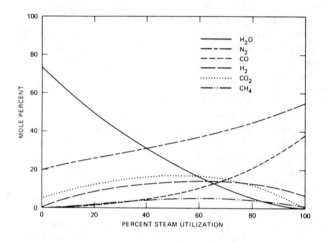

Figure 4.   Carbon–steam–air equilibria (T = 1000°K, experimental CH₄ output)

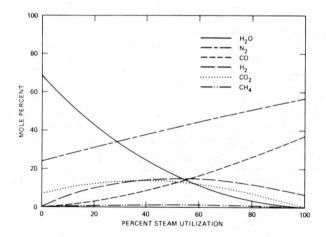

Figure 5.   Carbon–steam–air equilibria (T = 1100°K, experimental CH₄ output)

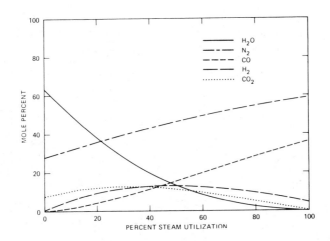

Figure 6. Carbon–steam–air equilibria (T = 1200°K, experimental CH₄ output)

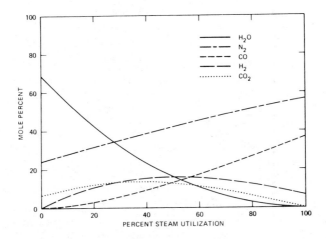

Figure 7. Carbon–steam–air equilibria (T = 1100°K, textbook CH₄ output)

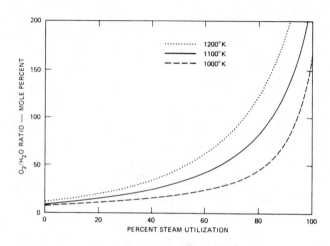

Figure 8.   Carbon–steam–air equilibria (experimental CH₄ output, oxygen demand)

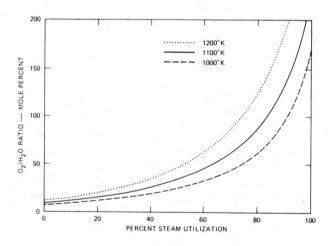

Figure 9.   Carbon–steam–air equilibria (textbook CH₄ output, oxygen demand)

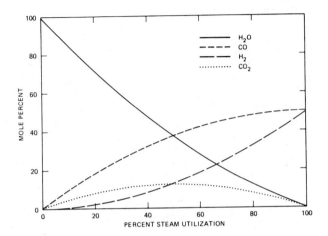

Figure 10.   Carbon–steam equilibria (T = 1100°K, no CH$_4$ output, indirect heating)

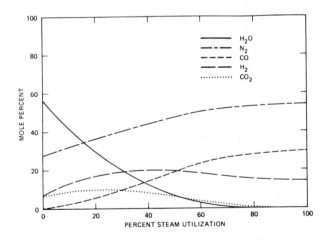

Figure 11.   Carbon–steam–CH$_4$/stoichiometric air equilibria (T = 1100°K, no CH$_4$ output)

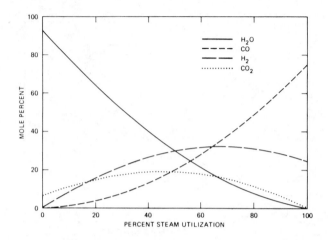

Figure 12.   Carbon–steam–oxygen equilibria (T = 1100°K, no CH$_4$ output)

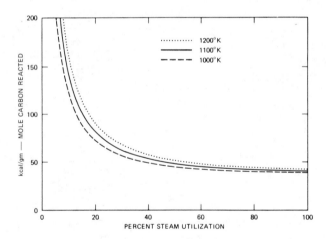

Figure 13.   Carbon–steam equilibria (no CH$_4$ output, indirect heating demand)

From a comparison of Figures 7, 10, 11, and 12 it can be seen that at a given PSU, the mole percent H_2O is greatest for the indirectly heated system with only an oxygen blown system approaching the same values. Oxygen blowing is common on large scale coal gasification units but is probably not feasible on the smaller scale carbon activation systems. At any given temperature, the higher the mole percent (partial pressure) of H_2O, the more quickly Reaction (6) proceeds which will give a higher PSU and therefore higher thermal efficiency. With proper modification, an indirectly heated kiln can represent an optimum configuration for activation.

Fig. 13 illustrates the quantity of heat that must be added to the indirectly heated system as a function of PSU to achieve heat equilibrium. Once this system reaches approximately 40% PSU the quantity of heat that must be added reaches an almost constant value indicating that high thermal efficiency is reached as soon as the system approaches this value.

Application of Theoretical Data to Real World Furnaces

Because commercial manufacturers of activated carbon consider their processes proprietary we have been unable to obtain any significant data on operating furnaces. Carbon regeneration furnaces which operate on similar thermodynamic principles generally use 0.6 - 1.0 kg steam/kg carbon feed but this number by itself is insufficient to determine a PSU. It is necessary to make an assumption as to the quantity of carbon that is actually reacted with steam and the average temperature at which this occurs to estimate PSU. Any use of auxiliary fuel further complicates this estimate. Data is available from coal gasification plants but these generally operate at much higher temperatures. The best assumption we can recommend, based upon a review of the literature and informal discussions with plant personnel, is that a design based upon 40% - 50% mole percent H_2O in the product gases is not unreasonable.

Conclusions

Based upon the analysis developed in this paper, the indirectly heated system can obtain approximately 45% PSU with mole percent H_2O equal to 40% as shown in Fig. 10 . From Fig. 13 we see that this PSU brings the process into the range of minimum energy requirements and therefore maximum thermal efficiency. On this basis we have recommended further research and development of indirectly heated reactors for the production of activated carbon from municipal solid waste and biomass.

Acknowledgement

This analysis was supported in part by Arcalon b.v., The Netherlands.

Cited Literature

1. Feldmann, H. F., "The Role of Chemical Reaction Engineering in Coal Gasification," Chemical Reaction Engineering Reviews, Advances in Chemistry, No. 148, pp. 132-155 (1975).

2. Roberts, P. V.; Leckie, J. O.; and Brunner, P. H., "Pyrolysis for the Production of Activated Carbon from Cellulosic Solid Wastes" Solid Wastes and Residues, J. L. Jones and S. B. Radding, eds. ACS Symposium Series No. 76, pp. 392-410 (1978).

Bibliography

3. Althouse, V. E. and Treskon, J. E., "Research Report on Activated Carbon," Chemical Economics Handbook, Stanford Research Institute, Menlo Park, California (1977).

4. Cheremisinoff, P. N., "Carbon Adsorption of Air and Water Pollutants," Pollution Engineering, Vol. 8, (July 1976).

5. Chiyoda Chemical Engineering & Construction Co., Ltd., Kawasaki Research Center, P. O. Box 10, Tsurumi, Yokohame, Japan 230, "New Water Treatment Using Granular Activated Carbon," presented at the Second National Conference on Complete Water Reuse, Chicago, USA (May 1975).

6. Johola, A. J., "Regeneration of Spent Granular Activated Carbon," NTIS PB-189 955, U.S. Department of Commerce (February 1969).

7. Klei, H. E.; Sahagian, J.; and Sundstrom, D. W., "Kinetics of Activated Carbon-Steam Reaction," Ind. Eng. Chem., Process Des. Dev., Vol. 14, No. 4 (1975)

8. Lewis, F. M. and Ablow, C. M., "Pyrogas from Biomass," Conference on Capturing the Sun Through Bioconversion, Washington, March 1976, Proceedings.

9. von Dreusche, C., Jr., "Process Aspects of Regeneration in a Multiple Hearth Furnace," AIChE 78th National Meeting, Salt Lake City, Utah (August 19, 1974).

10. von Fredersdorff, C. G. and Elliott, M. A., "Lowry's Chemistry of Coal Utilization" Supplementary volume, pp. 892-1022, Wiley, New York (1963).

RECEIVED December 10, 1979.

PYROLYSIS, GASIFICATION, AND LIQUEFACTION PROGRAMS IN CANADA

Canada's Biomass Conversion Technology R&D Program

RALPH OVEREND

National Research Council, Rm 206, Bldg M-50, Montreal Rd., Ottawa, Ontario, Canada K1A 0R6

Canada is a country of over 10 million km^2 in extent with a small population of about 23 million. Though possessing natural resources on a large scale; traditional sources of crude oil are declining, and as with the world as whole a transition is taking place in which alternative fuels are being examined. Already syncrudes are being produced from oil sands and heavy oils and these sources are predicted to be major sources of liquid fuels by the end of the century. Coal and renewable sources of energy are being studied as to their possible contributions as either substitutes for oil in non critical applications not requiring those attributes such as high energy density demanded of transportation fuels; or in the production of liquid fuels directly.

The R&D effort in renewable energy includes an extensive effort in the area of biomass fuels. After initial assessments of the potential of MSW, agricultural residues and crops and forestry (1) it has been recognised that while all three renewable resources will contribute to the near term substitution of liquid fuels in non critical applications, only forests and the recently dead form of biomass - Peat - can contribute liquid fuels on a scale close to that demanded which is to say at the Exajoule level. The Canadian forest covers almost one third of the land mass, while agriculture utilises less than 7 per cent. At least 10 per cent of the land area is underlain with organic terrain variously described as muskeg and peat. Peat which is not renewable in the sense that trees are; is; however a low grade resource with biomass characteristics such as heterogeneity, high moisture content and high combined oxygen content with low ash and sulphur contents. The mature forest has an energy density of around 5 Terajoule per hectare whilst the average peat energy density is around 16 TJ/ha, both of which might be compared to the Athabasca oil sands or coal which have a real energy densities of over 1000 times i.e. 10 - 50 Petajoules per hectare. Thus forests and peat not only share similar raw material characteristics but both have large and widespread implications with respect to land use and water resources.

0-8412-0565-5/80/47-130-317$05.00/0

The R&D program is intended to chart pathways through the matrix of possibilities illustrated in figure I. Obviously the system to be optimised is the pathway from resource to end use; however as can be seen the conversion technology plays a pivotal role in determining that optimum.

The Role of Conversion Technology in Biomass

Peat and forest resources are in general far from the population that they are required to serve so that the immediate goal of the conversion process can be seen to be that of providing a transportable commodity that maximises the energy transfer per unit distance.

The conversion technology also serves to match "impedances", it is the transformer that moves a wet, bulky, heterogenous, highly oxegenated substance into the applications that provide us with the goods and services we require.

Conversion Process Evaluations

While the topic of this paper is Canada's PTGL (Pyrolysis Thermal Gasification and Liquefaction) research program it is useful to review the current status of conversion technologies for biomass. The goal is to describe the characteristics of each technology so that efficiencies, process steps and environmental factors are well quantified and under these circumstances for a known cost and nature of feedstock an economic or social decision can be taken as to whether or not to implement the technology. With the exception of combustion technology the data base is far from this ideal; prospective processes range in development status from conceptual designs through 10 gram batch process scale to 20 tpd process development or demonstration units.

The cost of the feedstock is equally vague – the forest industry could in principle displace all non-transportation/fossil fuel needs by using relatively low cost self-generated residues in well established combustion technologies -- thus large scale use will require harvest or collection of the fuel. Forest residues - tops and branches - are numerically adequate though harvesting technology for this material is almost non existent; peat for dry thermal uses is expensive (2) costing around $1.3/GJ because of short seasons for drying with solar energy and concomitant storage costs; yet in principle, peat if useable in the wet state could be harvested by mining techniques on a year round basis.

To facilitate comparison of different technologies using widely different prices of feedstock a simple life cycle costing method is developed below for use with the conceptual supply curve for biomass (exclusive of peat) illustrated in figure II. The supply curve has been developed mainly from data contained in reference(1).

RESOURCE	HARVEST & TRANSPORT	CONVERSION TECHNOLOGY	TRANSPORT	END USE
1- FOREST Mill Residue Forest Residue Primary Harvest	ROAD RAIL BARGE PIPELINE	COMBUSTION eg: Steam Electricity	STEAM ELECTRICITY PIPELINE	PROCESS HEAT ELECTRICITY CHEMICAL SYNTHESIS
2- AGRICULTURE Field Residue Animal Residue Energy Crops		GASIFICATION eg: Low BTU Gas Synthesis Gas PYROLYSIS eg: Char Oil	TANKER ROAD RAIL	TRANSPORTATION FUELS eg: Methanol
3- AGRO-FORESTRY Energy Plantation		HYDROGENATION eg: Oil SNG		
4- MARICULTURE Fresh Water Salt Water 1-Estuaries 2-Open Ocean		FERMENTATION eg: CH_4 C_2H_5OH		

Figure 1. Biomass technology and resource assessment chart

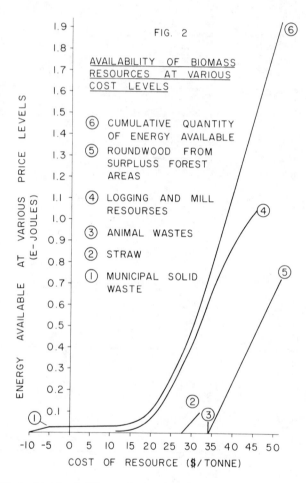

Figure 2.   Availability of biomass resources at various cost levels

Economic Comparison Model

The model is designed to identify either the range of feedstock prices appropriate for that technology or for a given feedstock cost the permissible capital intensity for the process.
The model is an economic model not a fiscal analysis of a plant
to be located at a specific site at a specific time. As such the
capital cost of land and services is excluded from the analysis
since those are either outside of or under the plant and will
endure through time and be reusable and cannot be counted as an
expense to social income.

Over the life of the energy conversion system a stream of
revenues from the sale of energy product are anticipated whilst
costs such as O & M, input fuel, capital charges and repayment can
also be envisaged. A life cycle methodology is used to relate
these streams so that the difference between revenues and costs
is used to compute a present value for the process plant. This
approach presumes that the alternative technology has relatively
minor market penetration such that its' output is "opportunity"
priced at the marginal rates that crude oil derived fuels or
electricity (the major energy forms anticipated over the next
30 - 40 years) will command.

The life cycle costs of the system are computed as follows:
revenue, fuel costs and O & M are assumed to occur at the end of
the year. The sum of these over the lifetime is discounted back
to the present and the difference between the present value of
the revenue and costs is identified as the capital expenditure
permissible for an "Instant" plant erected at t = 0.

Because the results of this are intended to be an economic
yardstick the arguments here are not financial arguments concerning the emplacement of a plant at a specific location and time
but rather concern social investment in alternative energy sources
such that the time value of money is set at a social discount
rate of:

$$\text{Discount Rate} = i = 10\% \text{ annum.}$$

With this value of discount rate a maximum plant life of 20 years
is the baseline case presented with little or no salvage value at
the end of that time.

$$\text{The plant life} = n = 20 \text{ years.}$$

The calculation at a social discount rate implies no capital
recovery since it represents an investment at the margin and in
any case is extraneous to the argument below. Because of the fuel
replacement characteristic of the new technologies the opportunity
cost of the energy sold is presumed to escalate at

$$f = 3\% \text{ per annum in real terms.}$$

Energy revenue = A = $/unit of energy.

Inflation of O & M costs and feedstock energy costs from say biomass (which is a regional rather than an international commodity) is not explicitely included except in so far as these terms are held to be in constant dollars throughout the life of the facility.

O & M = 0.2 C/annum where C = Capital Cost

C = $/capacity of 1 unit of Energy/year

fuel cost = B = $/unit of energy input

The efficiency of the conversion plant is specifically identified as a parameter which will be variable.

Efficiency = η = ratio ie E out/E in.

The present value of the fuel receipts can be shown to be

$$PV(revenue) = \frac{A(1-x^n)}{(i-f)} \text{ for } i \neq f$$

and $x = (1+f)/(1+i)$

The present value of the costs can be shown to be

$$PV(costs) = \left[\frac{B}{\eta} + 0.2C\right]\left[\frac{1-(1+i)^{-n}}{i}\right]$$

Since the capital cost at t = 0 is equal to the difference of these ie.

$$C = PV(revenue) - PV(costs)$$

the equations can be solved for the ratio C/A ie.

$$C/A = E/(1+0.2^{*}D) - \frac{B}{A} \cdot \frac{1}{\eta}\left(\frac{D}{1+0.2D}\right)$$

where C = Capital cost of facility to produce 1 unit output/annum
  A = value of 1 unit of fuel sold in the first year
  B = cost of 1 unit of feedstock in the first year
  η = process efficiency

$$D = \left\{\frac{1 - (1+i)^{-n}}{i}\right\} \quad \text{where} \quad \begin{array}{l} i = \text{discount rate \%/year} \\ n = \text{lifetime of facility} \end{array}$$

$$E = \frac{1 - x^n}{(i-f)} \text{ for } i \neq f \quad \text{where } f = \text{escalation of energy \%/year}$$
$$\text{and } x = (1+f)/(1+i)$$

$E = n/(1+i)$ for $i = f$.

* The factor 0.2 is an explicit statement that O & M cost = 20% of Capital Cost. The equation has the form $Y = MX + C$ where the units of Y are year and X is dimensionless, figure 3. Then the feedstock is of zero cost, e.g. WECS Solar and occasionally biomass the Capital cost becomes $E/(1+0.2D)$ and although efficiency is not explicit in the zero fuel cost solution it is still an economic consideration if not a thermodynamic fact since the size of the plant still has to meet a defined capital cost limitation. This model was used in table I which lists the characteristics of current biomass technologies and those under development. The unit of capacity C is defined here as the metric unit of

$$C = 1 \text{ TJ/annum capacity.} \quad 1 \text{ TJ} = 10^{12} \text{ Joule.}$$

for those not undergoing enforced metrication or familiar with other units the equivalences are:

$$C = 31.7 \text{ kW}$$

$$C = 0.45 \text{ Bbl/day}$$

$$C = 2.6 \ 10^3 \text{ ft}^3 \text{ methane/day.}$$

The maturity of the technology is indicated by a code:

A = commercial or proven

B = precommercial or demonstration

C = development or process development unit

D = bench scale process.

An index of the accuracy of the capital cost estimate is:

1 = commercial venture reliability

2 = detailed estimate

3 = conceptual estimate

4 = guesstimate.

TABLE I

TECHNO ECONOMIC ASSESSMENT FOR BIOENERGY CONVERSION PROCESSES

Process Technology	Product	Process Efficiency %	Maturity & Accuracy Index	Capital cost k$/TJ annum capacity @ 100% utilisation ie. C	Opportunity cost/$GJ ie. A	Feedstock price range $/GJ ie. B	Comment	Ref.
COMBUSTION								
Suspension	Hot air or stream	65-75	A – 1	1.29	3.84	3.22	Small scale 30x10⁶ Btu/hr or 366 KWt	
Fluid Bed			A – 1	1.57	3.84	3.16		9
Multi Chamber			A – 1	1.44	3.84	3.19		
Single Chamber			A – 1	2.68	3.84	2.89		
Package Boiler			A – 1	1.52	4.1	3.41		
ELECTRICITY GENERATION								
Steam Electric	Electricity	25	B – 2	51.2	6.94	–	14 MW	10
			B – 2	19.0	6.94	0.6216	50 MW	11
			B – 2	42.8	6.94	–	20 MW	12
			B – 2	32.4	13.89	1.691	50 MW	13
Retrofit Co-Generation	Steam/ Electricity	65	A – 2	9.6-16	6.94	3.56 –	5 MW & up	14
Combined-Cycle	Electricity	50	C – 3	32	13.89	2.24 3.44	300 MW	
Gasifier/ Diesel	Electricity	21	A – 2	31.7	41.7	8.64	1 MWe rating	

Table 1 – continued

DENSIFIED WOOD	Solid fuel	90	A – 1	0.614	3.84	4.07	300 t/d product	15
			A – 2	0.44	3.84	4.12	1000 t/d product	
GASIFIERS								
Updraught LBG	Low Btu gas	85	B – 2	2.54	3.1*	2.55	10 MWt	16
Downdraught LBG	6 MJ/m^3		B – 2	2.73	3.1	2.50	1 MWt	
Fluid Bed LBG	14 MJ/m^3	75	B – 2	0.54	3.1	3.09	10 MWt	5
Updraught MBG	SNG 37MJ/m^3	62	A – 1	5.1	3.84	2.32	350 MWt	13
Methanation			C – 3	20.9	3.84	–	10^{7}ft^3/day	
ANAEROBIC DIGESTION							*Most applications of LPG require derating of processes by up to 20%	
Piggery	SNG	50	C – 3	45.0	3.84	–	250 swine = 10^3 ft^3/day	17
Large Feed lot	SNG	50	B – 2	35.5	3.84	–	50,000 head feed lot = 10^{6}ft^3/day	18
LIQUID FUELS								
Chem System	Methyl fuel	57	C – 2	21.6	5.49	–	600 st/day	19
ICI	Methyl fuel	42	B – 2	18.8	5.49	0.32	1000 t/day	5
Wood Fermentation	Ethanol	31	C – 3	22.6	5.49	–	226x10^{6}l/year	5
Carboxolysis	Oxygenated #6 equivalent	55	B – 2	12.1	3.89	0.51	2000 bl/day	13
Wood Fermentation	Ethanol	31	A – 2	20	5.49	0.12	40x10^{6}l/year	19
Densified Wood & Gasifier	LBG/Combustion Process Steam	81	B – 2	1.2	3.1	2.75	65 MWt	

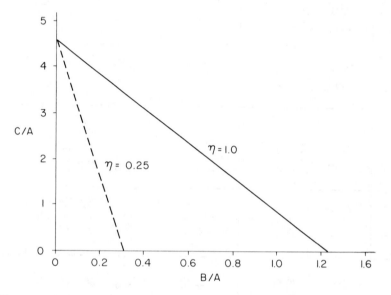

Figure 3.   *Equation* C/A = E/(1 + 0.15D) − B/A · 1/η (D/1 + 0.15D) *where* n = 20 *years,* i = 10%, *and* f = 3%

Reference prices for the opportunity cost "A" are:

   Oil   20$/Bbl (U.S.) or 23.50$/Bbl (Cdn)

      = 3.84$/GJ for both Oil and Gas

   Gasoline = 5.49$/GJ

         Electricity 2.5c/kWh = 6.94$/GJ eg nuclear

               5.0c/kWh = 13.89$/GJ eg fossil fueled

              15.0c/kWh = 41.67$/GJ eg remote northern
                                       communities.

N.N.   1 Btu = 1.055 kJ

       1 kWh = 3.6 MJ

Technology Assessment Conclusions

     The table I can be summarised into generic classes and using
a conversion of 18.6 GJ/tonne for wood biomass the data in table
II can be generated in which biomass costs technical risk and
likely market share are enumerated. Quick inspection of this
table shows that the "easy" low cost mill residues will likely be
consumed within the forest industries using well developed combustion technology. Conversely liquid fuels while having an extremely large potential market are constrained to a very low cost
requirement for their feedstock which cannot be met by current
harvesting technology. The supply curve of figure 2 indicates
that the likely average cost of the 500 PJ/annum increment beyond
the anticipated forest industries expansion will be around
30$/Odt for the supply.
     These figures may appear to be daunting economic goals for
biomass not to be restricted to essentially captive use within
the present biomass industries. An opportunity cost of 3.84 $/GJ
coupled to a 40% efficient process constrains the capital cost to
1.5 k$/TJ/annum output capacity. Only the densified biomass
option coupled with gasifiers at the point of use can meet this
cost criterion allowing that there will be prepared fuel transportation costs. If the liquid fuel opportunity cost of $5.49 is
used then the capital cost for the conversion has to be less than
7.9 k$/TJ/annum output capacity. Allowing that the usual scaling
law of an exponent to the 0.7 power is likely to apply to methanol
plants for example then a 4000 tpd plant would be feasible.
MacKay (3) has shown that scale increases effectively outweigh

TABLE II

TECHNOLOGY ASSESSMENT SUMMARY

TECHNOLOGY	BIOMASS COST $/ODt	TECHNICAL RISK	POSSIBLE END USE MARKET & SHARE
Direct Combustion	70	v - low	Forest Industries Self-Sufficiency present 300 PJ. 1985 600 PJ
Co-Generation in Forest Ind.	44-70	v - low	Large steam users in forest Ind. about 600 MW feasible. (20 PJ)
Densified Biomass	51	low	External to Forest Industries. Mining, Industrial process steam heating plants etc. Mainly to displace distillate fuels.
Gasification	55	moderate	High temperature processes and natural gas substitution. Retrofit potential for oil and gas boilers.
Electricity	12	v - low	Utility plants of greater than 50 MW. Very dependent on region. Market share likely to be low.
Remote Community and LDC applications of Gasifier/Diesel	160	moderate	Depends on region. Unit size is less than 250 kW in Canada. 50 MW market in northern regions.
Liquid Fuels	0-10	moderate to high	Potential market 1 EJ. End use compatibility problems for MeOH upgrading difficulties for hydrogenated wood products.

the incremental harvesting costs of increasing the supply base so
that this option is not entirely out of reach. It is worth examining not only the economic constraints but also some of the
chemical and end use constraints for biomass derived fuels.

Dulong's Formula and the Choice of Biomass Derived Fuel Form

Dulong's formula is used to predict the higher heating value
of fuels containing various percentages by weight of carbon,
hydrogen and oxygen.

$$Q_h = 338.3 \ C + 1442 \ (H-O/8)$$

where Q_h = higher heating value of fuel in J/g

  C = per cent of carbon

  H = per cent of hydrogen

  O = per cent of oxygen.

The percentages of C,H,O are by mass on a dry and ash free basis.
If the table of higher heating values for simple organic compounds
in Perry (4) is used as the true values then the calculated values
from the above formula fit a straight line of slope 1.04 with an
intercept of - 1502 J/g and a correlation coefficient of 0.99.
The relationship is evidently a good fit with at most 5 per cent
error in predicting the higher heating value of a compound of
specified C,H and O composition. The greatest error is obtained
for carbon monoxide where 4208 J/g is calculated when the true
value is 10110 J/g.

Since biomass is essentially free of nitrogen and sulphur
the three components control the heating value and satisfy the
relationship 100 = C + H + O. Thus all biomass derived compounds
can be assigned a unique coordinate on a triangular grid as is
indicated in figure 4, on which the Dulong heats of combustion
are also indicated. The triangular coordinates include all compositions unconstrained by chemistry so that heating values for
compositions containing more than 24 per cent hydrogen are not
feasible. Inspection however shows that the ideal conversion
technology for biomass will be one that loses oxygen while maxi-
mising the hydrogen to carbon ratio. Figure 5, is the same figure
as the previous one with the vectors shown for different techniques
of oxygen removal namely as water, carbon dioxide and carbon
monoxide. Ignoring for the present the precise chemistry by which
this is achieved some general points can be made
  i) Methanol is a compound that is less optimum in composi-
     tion than wood with the sole virtue of being a liquid
     that is just and so compatible with the gasoline system.
 ii) Water removal will produce a carbonaceous fuel.
iii) The most logical moiety to remove is carbon monoxide as
     the product vector is most close to a saturated hydro-
     carbon fuel.

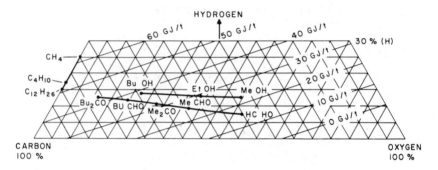

Figure 4.   Dulong diagram heat of combustion and CHO composition

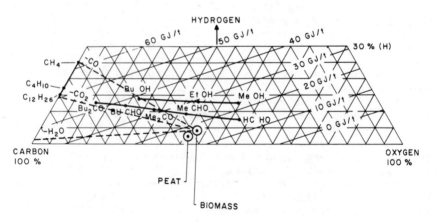

Figure 5.   Biomass conversion routes

The removal of all oxygen as carbon monoxide can be summarised as follows on a mole basis.

$$(CH_{1.45}O_{0.64})_n \qquad n\ 0.36\ CH_4 + n\ 0.64\ CO$$

The thermochemistry of the three vectors ranges from thermoneutral for water removal to 2.2 GJ/t endothermic for carbon dioxide removal and 4.5 GJ/t endothermic for carbon monoxide removal. The heat of combustion of the carbon monoxide is however 7.66 GJ so that it could be used either as a chemical reagent or as a heat source in an entirely biomass fueled process to produce fuels that most closely approximate todays preferred hydrocarbon fuels.

Canadian PTGL R&D Goals

Pyrolysis, Gasification and Liquefaction have high priorities within the Canadian R&D program with minor emphasis on direct combustion technology. Direct combustion is an old but well established technology with nevertheless some characteristics that still require development efforts. The present day systems used to burn bark and hog fuel use supplementary fuels such as fuel oil and natural gas to minimise emissions and the compensate for feedstock moisture variations. The net efficiency of burning wet wood with considerable excess air ratios is poor and leaves a lot to be desired. During the eco-excesses of the sixties this was acceptable because the large hog boilers were used as disposal units for excess residues. Since that time more residue has entered the process as fibre and of course the over fire fuels have become very expensive. Because of the forest industries' economic importance and the significant savings of fossil fuels that might accrue it is likely that direct combustion will become a higher priority. Already in the case of experimental work on fluid bed combustors for high sulphur coals provision is being made to study low grade fuels such as wet wood and dried peats.

In the Canadian context there need not be a restriction on biomass conversion processes to be self-sufficient and therefore be required to sacrifice carbon to drive the processes. Natural gas and hydro/nuclear electricity are likely to be available on a large scale through to the early decades of the next century. One or all of the following options have been discussed in the context of biomass conversion and in the specific case of methane additions to alter hydrogen to carbon monoxide ratios in syn gas work is presently under way to combine oxygen blown wood gasification with the reforming of natural gas, (5).

Table of Supplementary Sources for Biomass Conversion

Supplementary Source	Chemical Effect	Origin
External Heat	Eliminates need to sacrifice carbon	Thermal Power Stations eg fluid bed coal combustion or nuclear electric "Co-generated" thermal energy
Hydrogen	Chemical reagent for "CO" removal and hydrogenation	Steam reformed natural gas, electrolytic hydrogen generation from WECS, Hydro or nuclear
Electricity	Very high temperature plasma chemistry and or direct electrochemical reduction of biomass	WECS, Hydro or nuclear sources

Such complimentary synthesis routes or "HYBRIDS" can achieve remarkable reductions in capital cost and expensive biomass carbon utilisation. The methane and wood hybrid to produce methanol from the resulting syn-gas would use 722 m^3 of natural gas and 0.4 tonne of biomass to produce 1 tonne of methanol at a capital cost of around 10 k$/TJ/annum capacity on the 1000 tpd product scale. Though part of the feedstock is then priced at the international oil price (energy equivalent) a plant scale of 3000 tpd operating at 60 per cent efficiency will meet the economic criteria of the simple model used in this paper.

Hybrids can be viewed as "bridge" technologies which will smooth the transition from depletable resources to an era in which renewables and inexhaustibles such as coal and possibly nuclear fission (breeder) are the dominant resources, (6).

The Canadian Programme

The Canadian R&D program is described at the individual project level in Bioenergy Research and Development Program Review 1978, (7). From the preceeding discussion it can be seen that the major goal is to produce a liquid fuel or liquid fuel substitution product. A 1 EJ displacement of fossil fuels would be equivalent to a doubling of todays roundwood harvest, (8). Feedstock cost reduction is required and this is likely to be obtained by developing technology appropriate to biomass harvesting rather than the present roundwood systems.

A short term goal of the feedstock conversion programme is
to prove the gasification process step of synthesis gas production from biomass. This is being developed for both the methane
HYBRID and the biomass only case for both oxygen and air blown
gasifiers. The short term goal to produce synthesis gas as a
precursor for methanol acknowledges the possible wide-spread
adoption of methanol in the early nineties as a result of catas-
trophic shortfalls in the world petroleum supply. Preparation
of non-oxygenated or essentially hydrocarbon fuels is in an early
stage of development; however, provided that early adoption of
methanol does not occur, this route offers flexibility of choice
for what is after all an unknown vehicle power plant configuration of the 21. The current program is funded out to 1984 at
which time it is anticipated that there will be sufficient data
available to make a decision on whether or not to have a large
bioenergy contribution to the energy diet of the nineties.

REFERENCES

1. InterGroup Consulting Economists. "Liquid Fuels from Renewable
   Resources: Feasibility Study";

   Volume C: Forest Studies;
   Volume D: Agricultural Studies;
   Volume E: Municipal Waste Studies.

   Fisheries and Environment Canada: Ottawa 1978.

2. Montreal Engineering Co Ltd, "The Mining of Peat - A Canadian
   Energy Resource"; Energy Mines and Resources Canada: Ottawa
   1978.

3. Chemical Engineering Research Consultants Ltd, "The Production
   of Synthetic Liquid Fuels for Ontario"; Ministry of Energy
   Ontario: Toronto 1977; Volume 5.

4. Perry, R. H.; Chilton, C. H., Eds. "Chemical Engineers Handbook";
   McGraw Hill: New York 1963; 5th Edition.

5. InterGroup Consulting Economists, "Liquid Fuels from Renewable
   Resources: Feasibility Study"; Volume B: Conversion Studies;
   Fisheries and Environment Canada: Ottawa 1978.

6. Gander, J.E.; Belaire, F. W. "Energy Futures for Canadians,"
   Report EP 78-1; Energy Mines and Resources: Ottawa 1978.

7. Summers, B.A. "Bioenergy Research and Development - Program Review," NRCC 17671; National Research Council of Canada: Ottawa 1979.

8. Love, P.; Overend, R. "Tree Power: An Assessment of the Energy Potential of Forest Biomass in Canada," Report ER 78-1; Energy Mines and Resources: Ottawa 1978.

9. Levelton, B.H.; and Associates. "An Evaluation of Wood Waste Energy Conversion Systems," Report to the British Columbia Wood Waste Energy Co-Ordinating Committee: Fisheries and Environment Canada: Ottawa 1978.

10. Acres Shawinigan Ltd. "Hearst Wood Waste Energy Study: Summary Report," Ministry of Energy Ontario: Toronto 1979.

11. H.A. Simons (International) Ltd. "Hog-Fuel Co-Generation Study. Quesnel, British Columbia," Report to the British Columbia Wood Waste Energy Co-ordinating Committee: Fisheries and Environment Canada: Ottawa 1978.

12. S.N.C. Tottrup Services Ltd. "Energy and Chemicals from Wood," ENR Report No. 90; Department of Energy and Natural Resources Alberta: Edmonton 1979.

13. Kohan, S.M.; Barkhordar, P.M. "Mission Analysis for the Federal Fuels from Biomass Program, Volume IV: Thermochemical Conversion of Biomass to Fuels and Chemicals, "Report prepared by SRI International for the US DOE: Washington D.C. 1979.

14. Archibald, W.B.; Gabriel, J.A. "Economics of Co-generation in "Hardware for Energy Generation in the Forest Products Industry": Proceedings of the FPRS meeting January 1979 in Seattle: FPRS Madison, Wisconsin 1979.

15. T. B. Reed - Personal Communication.

16. Overend, R. "Gasification an Overview" in "Retrofit 1979" Edited by T. B. Reed and D. E. Jantzen SERI/TP-49-183: Golden Colorado 1979.

17. Biomass Energy Institute Inc. "Biogas Production from Animal Manure," Biomass Energy Institute: Winnipeg, 1978.

18. Burford, J.; Varani, F.T. "Design and Economics of Lamar, Colorado Bioconversion Plant," CSM Mineral Indusries Bull., V21 (no 4): Golden Colorado 1978.

19. Moon, G.D.; Messick, J.R.; Easley, C.E.; Katzen, R. "Technical and Economic Assessment of Motor Fuel Alcohol from Grain and other Biomass," Report to U.S. DOE - Contract EJ-78-C-01-6639; Washington, D.C. 1978.

RECEIVED December 18, 1979.

Pyrolysis of Agricultural Residues in a Rotary Kiln

R. H. ARSENAULT, M. A. GRANDBOIS, E. CHORNET, and
G. E. TIMBERS[1]

Sandwell Beak Research Group, 6870 Goreway Drive, Mississauga, Ontario,
Canada L4V 1L9

The conversion of agricultural residues to more amenable forms of energy such as gas, oil or char shows good potential both on the large and small scale. This paper is concerned with the pyrolysis of agricultural residues using a rotary kiln to potentially supply the energy needs of one or a few farm operations.

The rotary kiln pyrolysis unit used in these experiments was originally used for studies in peat conversion directed by E. Chornet at the University of Sherbrooke, Québec (Campion, 1978).

The scope of work in a first phase of the project included the adaptation of this peat pyrolysis system for conversion of agricultural residues, preliminary testing to identify operating problems, characterization of two feedstocks (oat straw and corn stover) and finally, experiments to determine product yield and composition at various operating conditions.

The work was performed by the Sandwell Beak Research Group and supported by Agriculture Canada under contract number AERD 8-1802. This first phase lasted one year and results are presented in this paper.

The second phase of the work is now underway detailing the pyrolysis study and some gasification and based on these two phases, design criteria for small scale pyrolytic conversion of agricultural biomass will be recommended.

Description of Process

A schematic diagram of the process is shown in Figure 1. The feed was chopped to 5-10 cm in a forage harvester and air dried to less than 10% moisture content before use. The feeder system consisted of a funnel-shaped 0.2 m^3 capacity hopper with chain-driven mixer arms, emptying onto a 5 cm diameter screw

[1] Current Address: Engineering and Statistical Research Institute, Agriculture Canada, Ottawa, Ontario, Canada

0-8412-0565-5/80/47-130-337$05.00/0

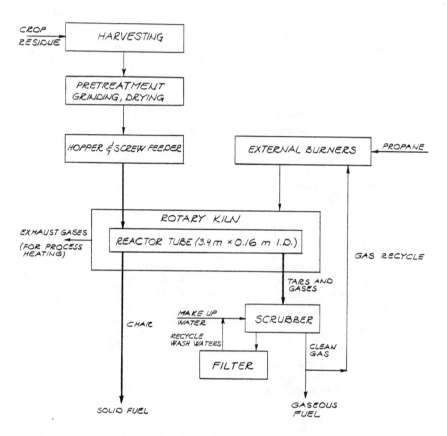

Figure 1.   Rotary kiln pyrolysis process

feeder. The reactor unit was a Bartlett-Snow Pacific experimental gas fired calciner with a 3.4 m by 0.165 m I.D. reactor tube. The furnace casing surrounding the tube is equipped with 14 burner ports. All process heating was done with propane during this phase. Residence time of the material within the reactor was controlled by varying the rotational speed and the angle of inclination of the tube. The product gas passed through a sedimentation cyclone and water scrubber, although this water scrubbing may not be necessary depending on product gas uses and pollution considerations. Wash waters were passed through an activated carbon filter to remove tars, and then recycled. The char was directed to a collector by gravity. A detailed schematic of the system is presented in Figure 2.

Experimental Procedure

The following reaction conditions were tested:

Temperature (K)	870, 1020, 1170
Residence time of feed in reactor tube (h)	0.25, 0.50, 0.75
Feed rates (kg/h)	Straw 1.5 to 2.2
	Stover 3.4 to 10.2

The reactor was purged with N_2 before and after each experiment. Reactor pressure varied between 5 and 10 kPa; at the base of the sedimentation cyclone, a collector tube was immerged to a depth of 40 cm in water, thus fixing the maximum operating pressure at 10 kPa. The reactor temperature was measured with a mobile thermocouple inside the reactor tube and 4 fixed external thermocouples. Temperature was kept uniform for a distance of 3 m along the tube and decreased linearly to about 400 K at the discharge end. The feed would thus reach the region of maximum temperature within 3 to 8 minutes after entering the reactor tube, depending on the residence time.

Once the desired temperature profile was obtained, the feeding was started. Steady state feeding lasted 1 to 4 hours.

The straw and stover feedstocks were analysed for C, H, O, N, S, moisture, ash and volatile matter content, as well as for several metals to gain insight into the potential catalytical activity of the ash. Product gas samples were analyzed for H_2, O_2, N_2, CH_4, CO, CO_2 and C_xH_y and the remainder of the gas was flared. Char samples were analysed for C, H, O, S, N and moisture content (moisture content by oven drying at 378 K); the remainder was stored in large plastic bags. The tars were not analysed since very small amounts were produced.

The feed and char were weighed and gas production was continuously monitored with a test meter to permit mass and energy balances. The water product was not determined, however, because of condensation problems in the scrubber.

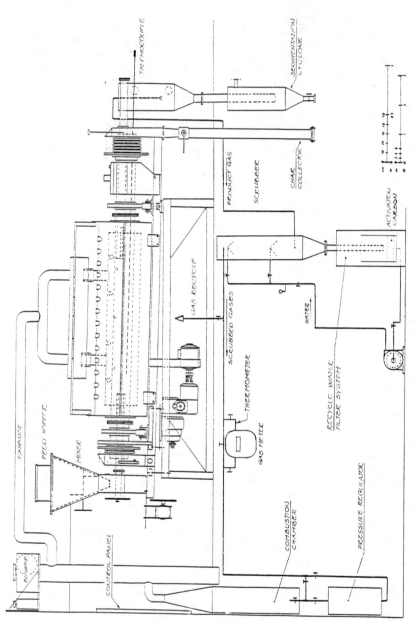

Figure 2. Detailed schematic of pyrolysis system

Heat inputs (propane) and outputs (combustion exhaust gases) were monitored.

Results

Feedstock Characteristics. Feedstock characteristics are presented in Table I. Results of the metal analysis will be considered in Phase II of the project. Calorific values were approximated using Dulong's formula and not actually measured. The oat straw had a lower ash content than the stover and a correspondingly higher calorific value. Calorific values, calculated on a dry basis and including the latent heat of water vapour in the products of combustion (high heat value), were 17.9 MJ/kg for the straw and 16.0 MJ/kg for the stover.

Table I: Feedstock Characteristics[1]

Type of Feed:	Straw		Stover	
	Mean	S.D.[5]	Mean	S.D.[5]
Ultimate Analysis[2]-% wt				
Carbon	48.1	0.7	43.2	0.4
Hydrogen	6.1	0.1	5.2	0.1
Oxygen	41.6	1.0	34.0	0.2
Sulphur	0.05	0.01	0.25	0.01
Nitrogen	0.4	0.1	2.8	0.1
Calorific value[3] (MJ/kg)	17.9	0.4	16.0	0.3
Proximate Analysis-% wt				
Moisture	5.4	0.7	7.8	1.4
Ash	3.8	0.4	14.6	0.6
Fixed carbon[4]	0.7	0.1	0.4	0.4
Volatile matter	90.1	0.6	77.2	0.9

1 From the analysis of 5 different samples for each feedstock
2 Moisture free
3 High heat value approximated using Dulong's formula
4 Calculated by difference (100%-%ash-%moisture-%volatile matter)
5 S.D. = Standard deviation

Product Yield and Composition. Product yield and composition did not vary significantly within the range of residence times studies. Feed rate did not have a noticeable effect either, although further experiments at 10 kg/h are required. The system output was limited by the feeding system used. It is likely that the kiln could convert more material since conversion was completed within 10 minutes. This was indicated by the fact that gas production fell to zero within 5 to 10 minutes after feeding was stopped. As well, during previous experiments with peat (Campion, 1978), feed rates up to 18 kg/h were used.

Product mass yields are shown in Figure 3. Char yields decreased almost by a factor of 2 from 870 K to 1170 K; converse-

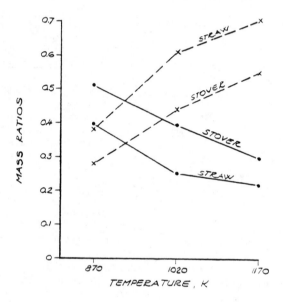

Figure 3.   Mass product/feed ratios ((———) char; (— — —) gas)

ly, gas yields doubled. Char yields were greater for stover
than for straw by about 20%; gas yields were lower by the same
amount. Gas production at 1020 K averaged 0.55 m^3/kg for straw
and 0.44 m^3/kg for stover. It was surprising to find that tar
yields were only around 1% or less. Apparently the large molecules constituting the tar are easily broken down in secondary
pyrolysis and gasification stages. The primary pyrolysis stage
where tars are produced will be studied in a bench scale retort
in Phase II to verify the secondary pyrolysis and gasification
assumptions.

Results of gas composition analysis are presented in
Figure 4 for the straw experiments and Figure 5 for the stover
experiments. As expected, CO_2 content fell considerably (by
about 1/2) when temperature was increased from 870 K to 1170 K.
CO and H_2 increased with temperature and CH_4 content was a maximum at 1020 K. Although the C_2 fraction was not analysed for
the individual components, approximate molecular weights and
calorific values were used to calculate the specific gravity
and calorific values of the gases. These were quite similar for
both product gases (straw and stover). Calorific values averaged 12 MJ/m^3. Higher calorific values around 14 MJ/m^3 were obtained at 1020 K where CH_4 production was greatest. Specific
gravity varied from 1.1 at 870 K to 0.8 at 1170 K.

Results of the char analyses are given in Table II. They
are only presented for two specific experiments since the characteristics did not vary considerably with reactor temperature
or residence times. The calorific value of the straw char was
higher than that of the stover char by about 20%. The calorific
value of the straw char on a unit weight basis was about 1.6
times greater than the straw feed and the calorific value of
the stover char was about 1.4 times greater than the stover
feed. Further characterization of the char will be carried out
in Phase II. Of particular interest are the burning characteristics, hygroscopicity, ash and acid insoluble ash content and
measured calorific value.

Table II: Results of char analysis for pyrolysis of straw
            and stover at 1020 K and 0.5 hour residence time

Feed	Elemental Analysis of Char (air dried) (% by weight)						Calculated Calorific Value[2] MJ/kg
	C	H	O	S	N	Ash[1]	
Straw	74.9	2.4	7.0	0.20	1.8	13.7	27.6
Stover	60.0	1.7	9.2	0.28	3.1	25.7	21.1

1 Calculated by difference (100% - %C, H, O, S, N)
2 Dulong's formula
  Average for all experiments: straw char - 27 MJ/kg
                               stover char - 23 MJ/kg

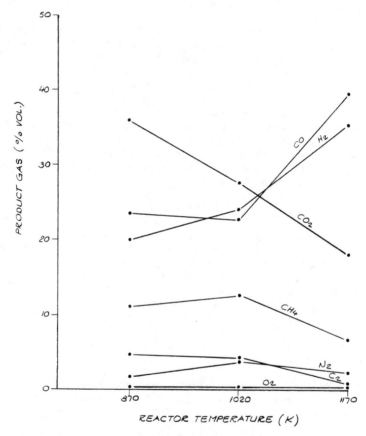

Figure 4.   Gas composition from straw pyrolysis—Series II (Note: heavy components not considered)

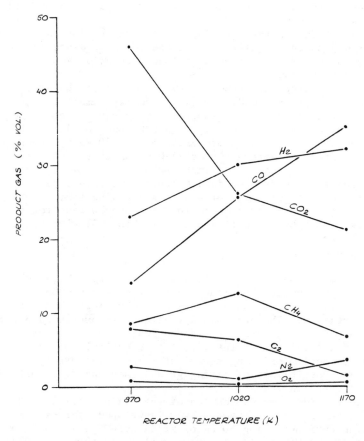

Figure 5. Gas composition from corn stover pyrolysis—Series III (Note: heavy components not considered)

Energy Data.   Figure 6 gives the energy product/feed
ratios, that is, the energy content found in the gas or char
for a unit quantity of feed. At 870 K there is almost 3 times
more energy in the char than in the gas. However, at 1170 K the
energy in the char and gas are about equal using stover feedstock and approximately twice as much in the gas as in the char
using straw feedstock. Therefore, if char is the desired product, reactor temperatures should be low (870 K), but if gas is
the desired product, reactor temperatures should be higher
(1020 K or 1170 K).

Mass balances averaged 85%. Losses in the operation were
mainly due to the inability to adequately condense and trap
pyrolysis waters. Experiments of longer duration would improve
the mass balances.

Energy in the char and gas averaged 77% of the gross calorific value of the feedstock. In these experiments, energy to
pyrolyse the feedstock was supplied by propane. About 100 MJ
were required to heat the kiln to 870 K. Another 121 MJ/h were
required to maintain this temperature and 189 MJ/h were required to maintain a temperature of 1170 K. About 90% of this heat
was lost in the exhaust and in an actual system most could be
recovered for process heating. Calculations were made to predict thermal efficiencies of a self-sustaining system where a
portion of the product is recycled and combusted to provide
process heat; thermal efficiencies of about 65% were predicted.

Product Uses

The most convenient application for the fuels obtained
would be to provide substitute heat for direct use on the producing farm or within short distances requiring a minimum of
transportation. The char has a similar heat output to bitumin-
ous coal on a mass basis, while the gas has about one-third the
value of natural gas. Relatively little capital would be required to modify existing equipment for burning pyrolysis products. The possibility of mixing ground char and fuel oil will
be considered in Phase II.

To provide some indication of the use potential in Eastern
Canada, the following example was developed for an Eastern
Canadian farm producing 50 ha of grain corn per year. Assume
the mean production given by Southwell and Rothwell (1977) of
5604 kg grain corn/ha at 28% moisture and 4700 kg stover/ha.
Assume as well that one-half the stover can safely be harvested
without damaging the soil or increasing winter runoff. Converting all of the harvested stover to pyrolysis products and
using a 10 kg/h, self-sustaining process would yield 1.5 x 10^6
MJ/yr. About 20% to 30% of the energy produced may be required
to dry the feedstock, depending on the year and harvesting conditions. About 20% of the output would supply adequate fuel for
an average farm residence, leaving about 50% of the output for

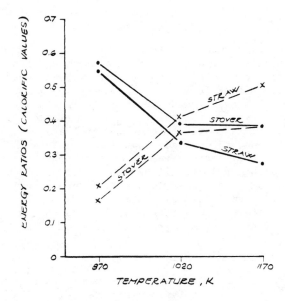

Figure 6. Energy product/feed ratios ((———) char; (– – –) gas)

other potential on-farm applications such as grain drying and farm heating.

Other uses may include replacement of hydrocarbon fuels with char in engines equipped with gasifiers. The gas may also have an application as an engine fuel although storage may be expensive.

Integration of crop-residue pyrolysis into the work routine on the farm would differ depending on the crop residue available, the end-use objective, the kind of farm, and the weather zone in which the farm is located. A 10 kg/h unit would require well over a year to process the crop residue from the 50 ha farm discussed earlier. A 50 kg/h kiln operating continuously would process the same amount in 100 days. The resulting char could be stored more easily than the feedstock.

Economic Analysis

An elementary economic analysis of the situation follows. We have seen that a typical 50 ha farm could yield about 10^6 MJ of energy in the form of high energy char or low energy gas. Compared to fuel oil at (U.S.) $0.11/litre (U.S. values $0.45/gallon) this would have a value of about (U.S.) $3,400.

Design criteria for the small scale pyrolytic conversion of agricultural biomass will be proposed in Phase II, keeping this figure in mind. Increased expenditures for harvesting, transport, storage of residue fuel, and equipment modifications required to use the product will have to be considered. There may also be economies of scale involved. A co-operative of a few farms within a small radius of each other may be more practical. Larger western farms (typically 500 ha) should also be considered - the scale up may prove advantageous.

Comparison with Peat Pyrolysis

It is interesting to compare these experiments with previous experiments on peat pyrolysis using the same unit. Ground sod peat with a degree of decomposition of 27% (Baturin formula - Baturin, 1975; H4 on the Von Post scale) was air dried to 12% moisture content and fed in at rates up to 18 kg/h (Campion, 1978). Yields and calorific values at reaction temperatures of 1120 K are compared with stover and straw pyrolysis at 1170 K in Table III.

Mass yields are similar for the different feedstocks but calorific values for both feed and products are somewhat higher for peat. Clearly, all three feedstocks show good potential. Depending on location, a farm could conceivably use peat as feedstock when agricultural residue was not available.

Table III: Comparison of straw, stover and peat pyrolysis

Type of Feed:	Oat Straw[1]	Corn Stover[1]	Sod Peat[2]
Feedstock			
Calorific value (MJ/kg)	17.9	16.2	20.1
Char			
Yield (%)	22	30	30
Calorific Value (MJ/kg)	26.0	24.0	26.9
Tar			
Yield (%)	< 1	< 1	4
Calorific Value (MJ/kg)	–	–	26.0
Gas			
Yield (m^3/kg feed)	0.795	0.582	0.465
Calorific Value (MJ/m^3)	11.5	11.0	15.3

1 Reactor temperature = 1170 K
2 Reactor temperature = 1120 K

Conclusions

These preliminary experiments indicate that pyrolysis of agricultural biomass appears to have a good potential for small scale production of fuels for farm operations. The kiln was easily started and operated and would require little supervision for continuous operation. Thermal efficiencies for a self-sustaining process are expected to be about 65%.

Abstract

Pyrolysis of agricultural residue was experimentally assessed as a fuel production process for farm applications. A rotary kiln (3.4 m by 0.165 m I.D.) was used due to its ease of operation, commercial availability, low operating costs and ease of start-up and shutdown. Ground oat straw and corn stover at less than 10% moisture were pyrolysed in an indirectly fired continuous-flow rotary kiln located at the University of Sherbrooke. The principle products were char and gas, less than 1% of the feed mass was converted to tar. Calorific values were about 17 MJ/kg for the feed, 26 MJ/kg for the char, and 12 MJ/m^3 for the gas. Calculations indicate that the thermal efficiency of a self-sustaining process would be around 65%.

Acknowledgements

Sandwell Beak Research Group is sincerely grateful to Agriculture Canada for their financial support of this research project.

Appreciation is expressed to the University of Sherbrooke for the use of their facilities during the experiments.

Thanks are due to the following for their assistance provided during the project: I. Coté, Technician (University of Sherbrooke); S. Jenkins, Research Chemist; M. Mislan, Project Scientist; J. Robinson, Professor (University of Guelph); R. Szawiola, Director of Laboratory Operations; and C. Roy, Research Engineer.

Literature Cited

Baturin, A.P., "Measures to Increase the Heating Value of Peat", The Symposium on the Combustion of Peat, Kuopio, Finland, 1975.

Campion, J.P., Masters Thesis, Université de Sherbrooke, 1978.

Southwell, P.H., and T.M. Rothwell, "Analysis of Output/Input Energy Ratios of Crop Production in Ontario", University of Guelph, Ontario, 1976.

RECEIVED November 16, 1979.

Fluidized-Bed Gasification of Solid Wastes and Biomass: the CIL Program

J. W. BLACK[1], K. G. BIRCHER[1], and K. A. CHISHOLM[1]

Eco-Research Limited, P.O. Box 200, Station A, Willowdale, Ontario, Canada M2N 5S8

Canadian Industries Limited (CIL) is a large Canadian chemical company which manufactures a range of heavy industrial chemicals. A significant volume of these products are based on natural gas feedstock. About six years ago; three circumstances evolved which prompted the Company to examine the potential of municipal refuse as a source of synthesis gas for chemicals manufacture. The events were,
1)   Threatened curtailment of natural gas supply,
2)   Requirement for additional ammonia capacity,
3)   Acquisition of a waste disposal company.

An initial economic study of the conversion of municipal waste to synthesis gas provided significant justification to proceed with a demonstration plant. The program was designed to demonstrate two new processes; 1) preparation of a solid fuel (RDF) from mixed municipal refuse 2) conversion of this fuel to a low calorific value gas. A demonstration plant was constructed in 1976. Evaluation of the first module, RDF preparation, was satisfactorily concluded in 1978. The second step, gasification of the RDF, will be completed this year. In late 1978, the gasification program was expanded to include other feedstocks primarily forest biomass and the use of oxygen. This latter program has received partial government support. Experiments are underway also to examine the gas processing steps subsequent to gasification i.e. high temperature gas cleaning and reforming.

Refuse as a Substitute Feedstock

Municipal refuse is a highly heterogeneous mixture of materials of which more than half are organic. Refuse contains significant amounts of water and has a comparatively low heating value. Garbage has, however, one redeeming feature - it is cheap. After separation of metals and glass, the combustible fraction can be thermally decomposed into three components; a high ash content char, a mixed condensible phase comprising water and liquid organics and a gas containing primarily

[1]Current address: BBC Engineering, 65 Pringle Avenue, Markham, Ontario, Canada L-P 2P5

carbon oxides, hydrogen and low molecular weight hydrocarbons.
Also produced are small but significant quantities of hydrogen
sulphide and hydrogen chloride. By contrast the requirements
for a chemical feedstock are continuous supply of unvarying
composition, minimal inerts and negligible amounts of particulates, corrosives or catalyst poisons.

To produce a feedstock from garbage therefore necessitates
the separation of a low inert,organic fraction with consistent
characteristics,followed by efficient conversion of this
material into a clean gas in a continuous, rapid response reactor. It was recognized that such a system would also produce
a good fuel. In fact this was to be the first commercial objective.

Description of the Prototype Unit. As proof of concept,
a facility was constructed in association with a conventional
refuse transfer station. The plant is comprised of two modules,
a 15 TPH RDF system in conjunction with a 1 TPH fluidized bed
gasifier. A flow diagram is shown in Figure 1.

The first module, the RDF preparation system, was designed
with sufficient capacity to separate all of the refuse from the
City of Kingston, Ontario where it is located. Conceptually
the system has been engineered to provide the following processing steps,

a)   Initial opening of all containers to release the refuse
     for subsequent processing but without significant size
     reduction,

b)   Size separation into two fractions, an oversize stream com-
     prised primarily of paper products and an undersize stream
     containing bottles, cans and wet refuse,

c)   Air classification of the undersize to remove the combus-
     tibles from the residue,

d)   Shredding of the organic fraction by a cutting action.

The basic system which is designed around low speed equipment can be upgraded to accommodate metals and glass recovery.
As consequence of the fact that only the organic materials are
shredded, both glass bottles and cans remain intact. Maintenance and power consumption are also significantly reduced. In
addition, the extremely low glass-content of the RDF minimizes
equipment abrasion and reduces potential slagging problems in
the thermal conversion process.

Conversion of the RDF into a low calorific value,fuel gas
is accomplished in a fluidized bed reactor. Both air and solid
fuel are fed into the reactor under positive pressure. The
RDF is immediately decomposed into gas, char and small amounts
of tar. The char is subsequently oxidized by air to provide
the heat necessary to sustain reaction conditions. Ash from
the system is removed by cyclone collectors for subsequent
disposal. The cleaned off-gas is then burned in a tangentially
fired combustor.

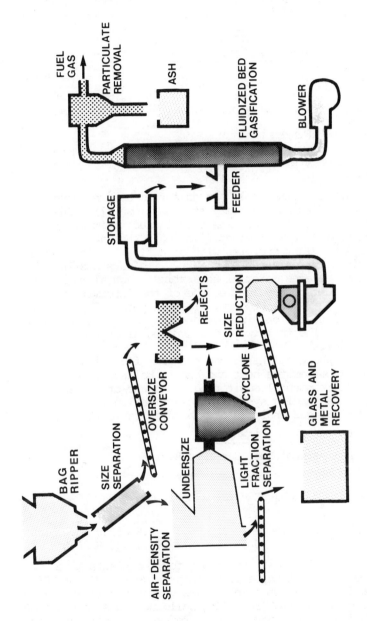

Figure 1. Resource recovery pilot plant—Kingston, Ontario

Results. The experimental program for the RDF module was designed to examine both separation efficiencies and total power requirements. Evaluation of this module is now complete and the data are summarized in Table I.

TABLE I

Summary of RDF Production Tests

Plant Location	CIL (Kingston)	Ames, Iowa
Average load Mg/h	20	30
Peak load Mg/h	40	45
Installed power kW	140	1790
Power consumption kW.h/Mg	2.7	26
RDF production % by wt of input MSW	77	84
Residue production % by wt of imput MSW (glass, metals, etc)	23	16

Also included in the Table are data from a more conventional garbage processing facility at Ames, Iowa (1). The power consumption results for Ames only include primary and secondary shredding and air classification. The Table shows that the Kingston Plant operates with significantly less power than conventionally designed RDF plants. i.e. systems in which size reduction precedes material separation. Actual maintenance costs are difficult to predict because of the intermittent nature of the Pilot Plant Program. To this point, 3 years after installation, the only component which has been repaired is the shredder. The teeth were resharpened after the second year of operation.

The second phase of the refuse program, fluidized bed gasification, incorporates heat and energy balances, gas analyses and environmental monitoring. Because of initial difficulties with feeding systems, the evaluation is not yet complete. Typical gas analysis are shown in Table II.

TABLE II

Analysis of Fuel Gas (Dry Basis)

Fuel	Gas Analysis								
	H_2	CO	CH_4	C_2H_2	C_2H_4	C_2H_6	CO_2	N_2	O_2
Refuse	12.5	11.6	2.5	---	3.2	---	15.4	45.1	0.8
Rubber	14.0	14.4	5.5	---	2.9	0.3	7.9	52.0	0.6
Wood	17.9	16.0	5.8	0.1	1.4	0.3	16.7	40.8	0.8

Commercial Activities. Proposals for two plants have been submitted. The first, a plant to provide 150 TPD of RDF for use as a fuel in a cement plant was guaranteed at an installed cost of $1 million including building and site preparation. A submission for conversion of refuse to steam has also been made. This plant is designed for 200 TPD of mixed municipal refuse at a steam capacity of 20 Mg steam/h.

Wood Gasification Program

Because of strong government interest and a more immediate market potential, CIL redirected the gasification program to incorporate forest biomass. It became apparent that there were three areas of interest. In order of increasing complexity these are,

1) An air fired, atmospheric pressure system suitable for small and medium scale energy producers such as industrial steam plants or hot gas generators,

2) An air fired, high pressure gasifier more appropriate to large power plants for high efficiency, combined cycle electrical generation,

3) An oxygen fired, high pressure gasifier necessary for chemicals and liquid fuel production.

A program was established to incorporate these three phases of development. We have completed Phase I and are planning for Phases 2 and 3. Part of the work has been supported by the Ontario Ministry of Energy and by the ENFOR program of the Environment Department of the Federal Government.

Program for the Production of a Fuel Gas From Wood. The refuse gasification unit was modified to facilitate wood handling and to incorporate a test unit for hot gas cleanup, reforming and condensation. The essential system components are shown in Figure 2. Various forms of forest biomass including wood chips, bark, sawdust and shavings have been evaluated in the gasifier.

The testing program was initially designed to evaluate the effect of parameters such as moisture content, bed temperature and the physical nature of the feed. Subsequently, experiments were conducted to examine long term effects, by simulating commercial operation under a variable load/demand situation for extended periods of time (2 periods of 670 hours continuous operation). Also included in the program were stack sampling tests, and tar production measurements.

Typical operating characteristics are presented in Table III.

TABLE III

Operating Characteristics During Wood Gasification Tests

Moisture in Wood	12-55%
Residue Production	0.017 kg/kg wood
Combustible Fraction of Residue	0.4 kg/kg residue
Turndown Ratio	5:1
Dust Loading in Stack (after cyclone)	$1.2 g/Nm^3$
Tar Loading in Gas	<0.1%
Organic Loading of Condensate	6000 ppm BOD_5

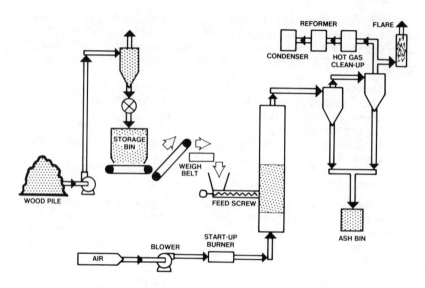

Figure 2.   Fluidized-bed system for wood gasification

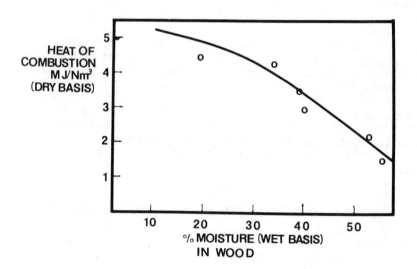

Figure 3.   Effect of wood moisture on heating value of fuel gas

The continuous testing program was concluded recently, August, 1979, and the data evaluation is only partially complete. It is obvious however that the gasifier can be operated successfully under the wide range of conditions that might be expected in a commercial situation. For normal operations, cooling and condensation of the gas is not recommended because of potential environmental problems with the condensate. The oxygen demand of the condensate is high and the liquor contains both polynuclear aromatics and phenols.

One of the more important parameters in the utilization of the gaseous fuel is the moisture content of the wood. It can be concluded from Figure 3 that an upper limit of 50% should be placed on the fuel moisture for stable combustion, particularly if it is necessary to avoid cooling the gas.

Synthesis Gas Production. As a preparatory step towards chemicals production from the gasification of wood, a series of experiments were conducted to evaluate the processing sequence required in the production of synthesis gas. Four key items were reviewed,
1) Reformer catalyst poisoning by sulphur in the wood,
2) High efficiency, hot gas dust removal,
3) Oxygen enrichment of the fluidizing gas,
4) The effect of reformation on the organic loading of the condensate.

The equipment was set up as shown in Figure 2. The hot gas filter, reformer and scrubber/condenser were installed on a side stream taken prior to gas combustion.

Results of the experiments have been very positive. The high temperature dust removal system, although overloaded, performed to expectations and no dust was observed on the reformer catalyst. The catalyst also retained its activity and successfully converted both tars and hydrocarbons to hydrogen by reaction with the water vapour present in the gas (Table IV).

TABLE IV

Reforming of Wood Gas

Gas	Composition Before Reformation - mol%	Composition After Reformation - mol%
H_2	5.0	14.4
CO_2	19.0	14.5
C_2H_4	.94	< 0.01
C_2H_6	.11	< 0.01
C_2H_2	.07	< 0.01
O_2	1.03	.88
N_2	64.9	59.7
CH_4	2.7	.068
CO	6.2	10.5

In addition the condensate showed significant improvements in clarity.

The oxygen enrichment experiments were performed by blending oxygen with the fluidizing gas. Figure 4 shows the effect of oxygen on the heating value of the product gas. The heating value increases significantly as the oxygen content becomes richer. Pure oxygen experiments are planned in the near future.

The overall results are very encouraging and the production of a synthesis gas may be cautiously projected from the experimental work. The processing steps from the synthesis gas to methanol, ammonia or hydrocarbons are relatively straightforward.

Commercial Activities. CIL has now begun marketing wood gasification technology for direct fuel firing applications. The capacity of the units being marketed are currently in the range of 20 - 400 TPD.

Process Economics

The gasification system must be tailored to suit each application, in particular with regard to consumption of the energy products and availability of feed. Cost of the final energy is very sensitive to the cost of waste or biomass. The price of $14/od ton used in these analyses is felt to represent the value of wood residues as an energy source once the market for these products has been established. Of course, in many instances these residues represent a disposal problem and an energy conversion facility would have to be built for them to realize this energy value.

The following represent some of the possible process applications. Capital costs incorporate engineering, profit, contingencies and interest during construction for a turnkey installation. Material handling, the gasifier and auxiliaries and tie-in to existing facilities are included where applicable.

Gas Generation. This is the simplest application for the fluid bed technology. The gas can be used as a substitute for natural gas in an existing boiler or for the drying of wood products in a kiln dryer etc.

As can be seen in Table V the cost of energy produced from such a facility is very low, the major portion being the cost of the wood residue or biomass fuel.

Steam Production and Co-generation of Electricity. For this application, the gas is fired in a boiler to produce high pressure steam that is subsequently expanded in a turbine/generator to produce electricity. Low pressure exhaust steam from the turbine is used for process heating.

An analysis of the costs is given in Table VI. Electricity and steam are available at very reasonable rates provided the full production of the plant can be utilized.

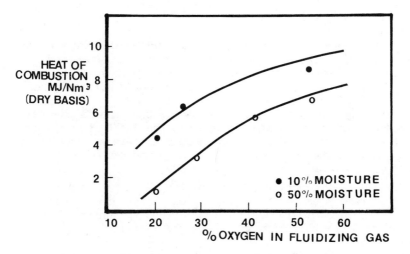

Figure 4.   Effect of oxygen enrichment on heating value of fuel gas

TABLE V

Gas Generation
(basis 250 odt/day facility)

Capital Cost $2,125,000

$/odt feed

Wood Residue Cost	14.00
Operating & Maintenance	1.90
Capital Charges 18% pa	4.64
	20.54
Energy in Wood	19.8 GJ/odt (8500 Btu/lb)
Energy Conversion Efficiency	70%
Energy Output	13.86 GJ/odt
Energy Cost	$1.48/GJ ($1.56/MM Btu)

Gas Turbines. The use of a pressurized gasifier and consequent firing of the gas in a gas turbine for the combined cycle generation of steam and electricity would increase the electricity produced per ton of feed when compared with co-generation.

The major problem in the use of gas turbines is the small amount of fine particulates present in the gas that would erode the turbine blades. Work is being done in this area and it is hoped that in the not too distant future the use of a biomass fired gas turbine will be demonstrated.

Methanol Production. Methanol from wood has been the topic of many studies in Canada, culminating this year in the full scale preliminary design of a 180,000 tpa unit and the design of a demonstration synthesis gas facility. (2,3).

TABLE VI

Co-generation
(basis 250 odt/day facility)

Capital Cost $10,000,000

$/od ton

Wood Cost	14.00
Operating & Maintenance	5.90
Capital Charges	21.82
	41.72
Electricity Production	2.4 GJ/odt (660 kW.h/odt)
Steam Production	5.04 t/odt wood
Electricity Cost	1.8¢/kW.h
Steam Cost	$5.90/ton steam ($2.67/1000 lb)

Pilot experiments described above for the production of a reformed synthesis gas have been incorporated into a conceptual process flowsheet for the production of methanol. Because of the high cost of transporting wood residues and biomass, the usual economy of scale does not apply and facilities using as little as 400 odt/day wood, producing 60,000 tpa methanol could be built costing around $41 million each. An economic analysis is given in Table VII based on a 500 odt/day facility. A larger facility would have lower capital charges per ton of feed but higher wood cost resulting in little change in the methanol cost. The cost of methanol at $219/ton is competitive on the open market in Eastern Canada and untaxed, as a gasohol would be competitive with a gasoline price around $1.20/gallon after allowing for marketing and distribution.

TABLE VII

Methanol Production
(basis 500 odt/day feed,
72600 tpa methanol production)

Capital Cost	$47.5 million
	$/odt
Wood Cost	21.00
Operating & Maintenance	21.50
Capital Charges 18% pa	51.82
	94.32
Methanol Production	0.44 t/odt feed
Methanol Cost	$ 214/t methanol (61¢/US gallon)

Conclusions

The status of CIL's development program is as follows,
1) Low power RDF preparation system - available for commercialization,
2) A fluidized bed system for gasification of RDF - in demonstration stage,
3) Fluidized bed wood gasifier - demonstrated at 25 TPD,
4) Synthesis gas production using enriched air - Pilot Plant tests partially complete.

References

1) Evaluation of the Ames Solid Waste Recovery System, Nov., 1977. EPA - 600/2-77-205.
2) Etude de l'implantation d'une usine de methanol-etape N°1 - Ministere des Terres et Forets, Quebec.
3) Etude de l'implantation d'une usine de methanol-etape N°2 - Ministere des Terres et Forets, Quebec.

RECEIVED November 16, 1979.

The Direct Liquefaction of Wood Using Nickel Catalysts

D. G. B. BOOCOCK, D. MACKAY, and H. FRANCO

Department of Chemical Engineering and Applied Chemistry,
University of Toronto, Toronto, Ontario, Canada M5S 1A4

Recent events have again emphasized the precarious nature of imported petroleum. Canada currently imports about 80,000 m^3/day (500,000 Bbl/day) into Eastern provinces and exports 9,000 m^3/day (56,600 Bbl/day) to the U.S.A. from western Canada. These exports will probably be phased out in the 1980's. At the present time natural gas exports to the U.S.A. result in an overall dollar trade balance for oil and gas in Canada. The continued development of the oil sands, the promising oil discoveries in the Beaufort Sea, and natural gas discoveries in the Arctic Islands and eastern Canada are key factors in Canada's energy future. However, it is evident that Canada for sometime will rely on oil imports for which the degree of future price escalation is unknown.

Renewable sources of energy are a desirable component of energy self sufficiency. The Ontario Ministry of Natural Resources has for some years been investigating plantations of fast-growing hybrid poplars (1). A reasonable average yield for these poplars is about 17,000 kg.ha^{-1}.y^{-1} (dry weight). We describe here procedures by which oil containing 10 to 14 per cent oxygen can be obtained from poplar wood in 30–40 per cent yield. This corresponds to about 35 barrels of oil per hectare per year. It has been calculated that plants producing 10^6 kg per day of oil would require a land area of 350 square miles assuming a 60 per cent use factor. To satisfy Canada's present petroleum requirements (1.8 x 10^6 Bbl/day) would require approximately 100,000 square miles of suitable land.

Poplar wood, like other woods, is made up of three major polymeric components which are cellulose, hemicellulose and lignin. Lignin comprises 18–23 per cent of hybrid poplar, and it is the most complex component, being non-linear in nature. The conversion of wood into a high energy liquid fuel requires depolymerization of the components and removal of as much oxygen as possible. The oxygen content of dry poplar is around 45 per cent and ideally this should be reduced to zero. However, we do not believe that the complete elimination of oxygen should be a dominant consideration particularly as the last few per cent of oxygen may require disproportional costs. We suggest that a unique refining procedure would be used to treat the oil.

Traditionally thermal liquefaction studies on biomass have been carried out in the presence of one or both of the reducing gases, hydrogen and carbon monoxide (2, 3, 4, 5, 6). Equation 1, in which cellulose has been used to approximate the elemental composition of wood, shows that theoretically a reducing gas is not required for wood liquefaction when internal carbon is used to remove the oxygen.

$$(C_6H_{10}O_5)_{2n} \longrightarrow nC_8H_{16} + 4nCO_2 + 2nH_2O$$

Equation 1

The removal of oxygen as carbon dioxide is desirable, but the formation of carbon monoxide is inefficient and, therefore, undesirable. When methane is not the major product some oxygen must also be removed in the form of water, but this mode must not be excessive as it would lead to hydrogen depletion of the products and encourage char formation. In the absence of hydrogen and carbon monoxide from other sources, the gasification of wood for the production of these gases is costly and should be avoided. The Bechtel Corporation estimated that gasification would account for 36 per cent of the capital cost of a plant based on the Albany process (7).

In most of our studies we have used hydrogen as a reducing agent and employed Raney nickel as catalyst, but in view of the low hydrogen uptake observed with mature catalyst some experiments were performed in the absence of hydrogen and these have yielded very encouraging results.

In an earlier paper (8) we described the complete liquefaction and gasification of fast-growing hybrid poplar (1). The wood (< 0.5 mm mesh) was suspended in water and treated with hydrogen in a well stirred and sparged autoclave using an initial pressure (24°C) of 10.3 MPa and reaction temperatures of 325-375°C. In all cases Raney-nickel (9) was used as catalyst. In these single batch reactions hydrogen uptake was high and considerable wood gasification occurred. For example, for the highest catalyst/wood mass ratio of 0.2, 50g of wood at 350°C consumed 3.8g of hydrogen and produced 1.3g of carbon dioxide, 16.0g of methane, 4.4g of saturated C_2 to C_4 alkanes and 8.0g of oil. No carbon monoxide was produced. The oil products contained 10-12 per cent oxygen and had viscosities of 700-2200 MPa.s at room temperature and heating values of 37-41 MJ.kg^{-1}.

We have now examined the beneficial effect of prolonged use of the catalyst in extended batch reactions. We report here some results from these extended batch reactions and from runs in which hydrogen was not used. We also address the problem of the wood feed system for possible commercial application.

Experimental

In a typical set of runs 150g of air-dried poplar, Populus x Euramericana, clone I-45/51 (8), which had been ground to less than 0.5 mm mesh, were added to a 2-ℓ magnedrive packless auto-clave (Autoclave Engineering) together with water (750 ml) and catalyst (20g). The autoclave was flushed with nitrogen and then hydrogen after which the hydrogen pressure was raised to 6.6 or 10.3 MPa. The autoclave temperature was then raised to 340–350°C over a period of 1 h with stirring and sparging and this tempera-ture was maintained for 1 or 2 h after which the autoclave was cooled to room temperature over a period of 2 h. Carbon dioxide carbon monoxide, methane, ethane, propane and butane in the gas phase were measured by gas chromatography as described previously (8). The autoclave was vented and the volume of the gas in the head space was measured. The sparging in the autoclave produced a foam which appeared to be stable after the autoclave was opened. Between runs a pipe, extending almost to the bottom of the auto-clave, was used to extract some of the liquid. This was accom-plished by pressurizing the autoclave and forcing liquid up the pipe. The oil product had a density greater than water, but because of the foam formation the removed liquid was mostly water. More wood (150g) was added and, in the relevant experiments, more catalyst was also added. The autoclave was sealed and the contents reacted as described previously. After the final run of a series, the autoclave contents were discharged as before. About 60 ml of liquid remained in the autoclave because the inlet of the discharge pipe was not located at the base of the lower internal surface due to the presence of the stirrer blades. This residual liquid was extracted several times with acetone, a procedure which also cleaned the inner surface of the autoclave. The acetone fraction was kept apart from the bulk liquid. In previous experiments, the product oil was extracted with chloroform. Introduction of a solvent presents many problems and a simple phase separation of oil and water was adopted. Catalyst was removed from the oil by centrifugation. Oil viscosities, heating values and elemental analyses were obtained as described previously (8). The carbon content of the aqueous phase was measured using a Beckman Total Carbon Analyser.

Results and Discussion

Table 1 shows results of seven sequential 2 h runs. A hydrogen pressure of 8 MPa was used in the first run to satisfy the demand for methane formation. Thereafter a lower hydrogen pressure was used. The liquid removed before each run was mostly water and considerably more was removed (except in run 5) than had been formed in the previous run. Thus the oil/water ratio was varied throughout this series. This ratio which varied from 0.3 to 2.0 had no effect on the course of the reaction. Gas

consumption, gas production and product gas composition were the
same for runs 3 through 7. At the end of each run only an oil and
water phase were present and no char was formed. The pH of the
aqueous phase was in the range 3 to 4 due to cleavage of the
acetate groups in the poplar hemicellulose and perhaps also due to
phenol formation.

The effects of prolonging catalyst use were all beneficial.
Hydrogen consumption decreased and carbon dioxide production
increased from run 1 to run 3. The organic product was an oil
which contained 13 per cent oxygen and was 95 per cent benzene
soluble. Approximately 55g of water was formed in each run. Thus
the molar ratio of carbon dioxide product to water product was
about 1:6 compared to an ideal ratio of close to 2:1. The oil
yield appeared to be 38 per cent but subsequent distillation showed
that the product contained about 9 per cent by weight of water.

Table 1
RESULTS OF SEVEN SEQUENTIAL RUNS, REACTION TEMPERATURE 350°C,
INITIAL PRESSURE 10.7 MPa RUN 1 AND 8 MPa IN SUBSEQUENT RUNS,
2 h REACTION TIME, 740 ml WATER, 150g WOOD ADDED AT EACH STEP

LIQUIDS REMOVED (ml)	H_2 (g) CONSUMED	PRODUCT GASES			
		CO_2 (g)	CH_4 (g)	C_2-C_4 (g)	CO (g)
Nil	11.6	9.7	36.0	2.9	0
200	3.1	13.1	2.4	1.4	0.2
200	2.0	23.0	0.4	0.7	0.4
200	2.0	22.0	0.4	0.8	0.5
0	2.2	20.1	0.3	0.6	0.4
100	1.9	22.3	0.4	0.7	0.1
200	2.1	23.3	0.4	0.8	0.6

OIL YIELD	398g (37.9%)
OIL ANALYSIS	C, 77.7; H, 8.4; O, 13.3
RESIDUE	0.5%
OIL HEATING VALUE	35.8 MJ.kg^{-1}
OIL VISCOSITY (24°C)	5075 mPa.s
AQUEOUS PHASE	390g
AQUEOUS PHASE CARBON	41.0g

The aromatic content of the oil was 33 per cent (10) as measured
by carbon-13 nmr. Crude distillation at atmospheric pressure
yielded a major fraction (42 per cent) which boiled in the range
190-350°C and unlike the crude product was completely miscible
with diesel fuel. A duplicate of this seven run series was carried
out and it yielded very similar results. Results for shorter
series in which different amounts of catalyst were added will be
reported elsewhere (11).

The relatively low consumption of hydrogen in this first

series prompted a series of runs in which no hydrogen was used, and except in the first and last runs the reaction was blanketed with nitrogen. The results of this series are shown in Table 2.

Table 2

RESULTS OF SIX SEQUENTIAL RUNS REACTION TEMPERATURE 350°C, NITROGEN BLANKET (ATMOSPHERIC PRESSURE EXCEPT RUN 6 (4.7 MPa H_2)), 2 h REACTION TIME, 750 ml WATER, 150g WOOD ADDED AT EACH STEP EXCEPT RUN 6, NO CARBON MONOXIDE DETECTED IN ANY RUNS

H_2 CONSUMED (g)	H_2 PRODUCED (g)	PRODUCT CASES		
		CO_2 (g)	CH_4 (g)	C_2-C_4 (g)
–	0.9	37.0	12.5	2.8
–	0.7	35.8	4.5	1.4
–	0.6	32.5	2.2	1.4
–	0.8	33.6	1.4	2.3
–	0.65	33.6	1.2	2.0
1.3	–	2.6	0.4	0.9

OIL YIELD   253g (33.7%)
OIL ANALYSIS   C, 77.5; H, 7.7; O, 12.2
AQUEOUS PHASE CARBON 40g

Again, no char was formed but, the amounts of carbon dioxide produced were greater than in the first series, and no carbon monoxide was detected. The oil product after each run (and before run 6) just flowed at room temperature. However, hydrogen treatment of this oil in run 6 caused solidification of the product. This suggested that the oil contained some alkene linkages which hydrogenated in the last run. It therefore appears that some forms of nickel metal have the ability to liquify wood at elevated temperature even in the absence of hydrogen.

Some preliminary runs using nickel carbonate in the presence of hydrogen have also yielded promising results. In these runs the nickel carbonate was reduced in situ to finely divided nickel which presumably functioned as the catalyst. Commercially available nickel powders did not appear to be as effective as Raney nickel or nickel produced in situ.

Wood Feed System

The introduction of solids into a high pressure system usually presents problems particularly if this is done on a continuous basis. Reducing wood to small particles for the purpose of producing slurries is not viable because of the large energy requirements. However, it appears that in the case of wood

368 THERMAL CONVERSION OF SOLID WASTES AND BIOMASS

this problem can be partially solved by a pretreatment of the wood with steam at much lower pressures than those used in the reaction vessel. This weakens the structure of wood by hydrolysis of the hemicellulose, and the treated wood suspended in water is easily macerated to a slurry. Some hydrolysis takes place in the reactor prior to liquefaction, and we have converted 1 cm cubes (size limited by the inlet port) of poplar to oil directly. No residue remained, and the oil was slightly more viscous than that formed from slurries of powdered wood. It also appears from previous results that in the reaction, simple sugars liberated by hydrolysis can be incorporated into the oil. Therefore, pre-hydrolysis of hemicellulose and even cellulose is not considered a problem in the wood liquefaction process.

We are currently constructing a laboratory scale semi-continuous unit which will use the prehydrolysis concept.

Acknowledgements

We would like to acknowledge financial support from the Department of Energy, Mines and Resources, and the Ontario Ministry of Energy. Current work is also funded by the ENFOR program of Environment Canada. We thank Dr. J. E. McCloskey of Canada Packer's for nmr measurements.

Literature Cited

1. Anderson, H.W. and Zsuffa, L., Forest Research Report No. 101, October 1975, Ministry of Natural Resources (Ontario), Division of Forests.
2. Appell, H.R. and Miller, R.D., Conversion of Cellulosic Wastes to Oil, Bureau of Mines Report of Investigation (1975).
3. Appell, H.R., Wender, I., and Miller, R.D., Conversion of Urban Refuse to Oil, Bureau of Mines Technical Progress Report - 25 May 1970.
4. Appell, H.R. and Wender, I., Converting Organic Wastes to Oil, Bureau of Mines Technical Report of Investigation 7560 (1971).
5. Weiss, A.H., Textile Res. J., 1972, 42, 526.
6. Weiss, A.H., Kranich, W.L., and Gupta, D.V., Ind. Eng. Chem., Proc. Des. Dev., 1976, 15, 256.
7. Lindemuth, T., Biomass Liquefaction Program, Experimental Investigation at Albany, Oregon, Second Annual Symposium on Fuels for Biomass, Troy, N.Y., June 1978.
8. Boocock, D.G.B., Mackay, D., McPherson, M., Nadeau, S.J., and Thurier, R., Can. J. Chem. Eng., 1979, 57, 98.
9. Covert, L.W. and Adkins, H., J. Amer. Chem. Soc., 1932, 54, 4116.
10. Schoolry, J.N. and Budde, W.L., Anal. Chem., 1976, 48, 1458.
11. Boocock, D.G.B., Mackay, D., and Franco, H., "The Production of Synthetic Liquids from Wood Using a Modified Nickel Catalyst", Paper accepted for presentation at 29th Canadian Chemical Engineering Conference, Sarnia, Ontario, 30 Sept. - 3 Oct. 1979.

RECEIVED November 16, 1979.

The Role of Catalysis in Wood Gasification

D. P. C. FUNG[1] and R. GRAHAM

Forintek Corporation Canada, 800 Montreal Road, Ottawa, Ontario, Canada K1G 3Z5

Early in 1978, a program was initiated at the Eastern Forest Products Laboratory of the Forintek Canada Corporation to perform research and development on wood gasification. This research includes the identification of those parameters which significantly affect the yield and distribution of products, the rate of reaction, and overall process efficiency. Long-term objectives of Forintek are to demonstrate a viable wood gasification technology and to encourage the forest product industries to use wood-bark residues as a substitute for rapidly depleting petroleum resources.

Coal gasification technology was developed prior to that of wood and is somewhat applicable to the gasification of wood. As suggested by Hoffman (1), the development of coal gasification technology may be grouped into three generations. The first generation technology was developed in Germany before World War II for producing liquid and gaseous fuels from coal. This included simple pyrolysis processes (for char & oil), rudimentary gas producers, fixed-bed hydrogenation, low and medium pressure oxygen gasification for synthesis gas production, and the development of the Fischer-Tropsch process for the production of hydrocarbons from synthesis gas. The second generation refers to the technical movements and modifications of coal gasifiers. The third generation can be regarded as the application of catalysis to gasification.

Catalysis research should permit a maximum yield of combustible or synthesis gas at the expense of less desirable char and liquid products (at lowest possible operating temperatures). In turn, this would reduce several technical problems and would increase the process efficiency and throughput. The use of catalysis could also be made to pre-determine the final composition of the

[1]Current address: CANMET, Department of Energy, Mines and Resources, 555 Booth Street, Ottawa, Ontario, Canada, K1A 0G1

gas product depending on intended end use.

Since catalysis is of such current interest, this paper presents a review of some of the literature which deals with the catalytic gasification of biomass. It also summarizes the pre-liminary results of a bench-scale fluid bed gasification study which employed Canadian hybrid poplar. The effect of such cat-alysts as potassium carbonate and calcium oxide on wood gasific-ation is reported. Hybrid poplar was selected due to its rapid growth and, therefore, its relevance to the concept of wood energy plantations.

Inorganic Salts as Gasification Catalysts

Alkali carbonates such as sodium carbonate and potassium car-bonate were studied intensively during early gasification resear-ch prior to World War II. White and Fox (2) noted that sodium carbonate increased graphite gasification rates, increased net gas yields and altered the gas product distribution in favor of carbon monoxide and hydrogen over carbon dioxide and methane. They presented a reaction mechanism for the carbonate catalyst in the optimum temperature range of 800 to 1000°C:

$$C + CO_2 \xrightarrow[Na_2CO_3]{} 2CO$$

Gas yields using sodium carbonate at lower temperatures are sim-ilar to those from the noncatalyzed reaction at higher temper-atures.

Subsequent graphite gasification research by White and his group (3,4) found potassium carbonate to be an excellent catalyst at an optimum proportion of 20 percent (by weight) of the carbon-aceous fuel feed. White and Weiss (3) noted that alkali carbon-ates also catalyzed the water-gas shift reaction when steam was injected during gasification:

$$CO + H_2O \xrightarrow[\text{Alkali Carbonate}]{} CO_2 + H_2$$

In 1953, Lewis, Gilliland and Hipkin (5) investigated the rate of gasification of wood charcoal catalyzed with potassium carbonate (10% by weight) at 650°C. The reaction rates at 650°C with the catalyst were approximately equal to the non-catalyzed rates at 870°C.

More recent research has recognized alkali carbonates as efficient catalysts in the pyrolysis and gasification of urban refuse, coal, and biomass.

Prahacs (6) observed that spent pulping liquors treated with additional sodium carbonate showed a significantly higher rate but not as rapidly as those liquors treated with sodium carbonate. Magnesium oxide was found to inhibit the catalytic effect of sodium carbonate.

Cox and co-workers (7, 8, 9) catalyzed the gasification of wood, coal, paper, sludge, manure and municipal refuse with potassium carbonate. They found that alkali carbonates promoted the steam-carbon and steam-hydrocarbon reactions and concluded:

   (i) conversion rates would be significantly less at the same temperature in the absence of potassium carbonate,

   (ii) the catalyzed products tend to be lower molecular weight hydrocarbons,

   (iii) the optimum alkali carbonate/coal ratio is about 27% by weight.

Rai and Tran (10) employed potassium carbonate as a catalyst for Douglas-fir bark and pulping liquor gasification, and found that the catalyzed gasification rate and gas yields were greater than for uncatalyzed reactions under otherwise identical conditions. Alkali carbonates exhibited selectivity towards the production of light hydrocarbons. They concluded that little had been done to assess the potential for alkali carbonate catalysis in large-scale gasification and encouraged further research in this direction.

Additional literature by Cox, (8), Appell (11), Hooverman (12), Cavalier (13), Knight (14) and Love (15) serves to reinforce the evidence that alkali carbonates increase reaction rates and overall gas yields in the gasification of biomass, coal and refuse. The general consensus is that alkali carbonates catalyze the water-gas shift and carbon-steam reactions.

Laboratory Studies at Forintek

The laboratory experiments were designed for preliminary work and for screening of catalysts to be used in our current catalytic gasification research program. Results from these studies are intended to identify those catalysts which best promote synthesis gas production (H_2/CO) and maximize reduction of char and liquid yields.

Feed Material

Five year old Ontario hybrid poplar was used during the experiments. It had a calorific value of 4700 cal/g and was hammermilled to a size range of 600 to 800 microns. The wood was then oven dried prior to gasification.

Reactor Assembly

The reactor assembly is depicted in Figure 1 and consisted of a screw feeder/hopper, the reactor proper, furnace, and a condenser train. The reactor proper was made of 316 stainless steel tubing (5 cm I.D. and 134 cm long) in three detachable segments. The top segment supported the gas exit line and contained a filtering medium (supported by a perforated plate) to restrict the movement of entrained fines from the reactor to the condenser train. The middle segment was equipped with the screw feeder and hopper. The hopper was designed to contain a 150 gm wood sample and could be readily sealed. Wood was added to the reactor at a steady rate for 20 minutes by means of the screw feeder. The base of the middle segment was connected to the bottom segment by flanges, between which a gas distributor grid was placed. The gas distributor had twenty holes (0.4 mm diameter) and was covered with a 74 micron wire mesh to prevent solid particles from plugging the grid holes. The grid served to distribute the nitrogen fluidizing gas evenly and also supported a 15 cm bed of commercial cracking catalyst (silica/alumina). The cracking catalyst was selected as a bed material due to its ideal fluidizing properties and not for its catalytic properties. The nitrogen gas was preheated to furnace temperature and was fed to the reactor through an inlet on the bottom segment. The reactor was heated by a tubular Lindberg furnace and the temperature was controlled by a thermocouple feedback mechanism connected to the oven control unit.

Gasification products were led along the exit gas line from the reactor to four consecutive sealed condensers in the recovery train. The first two condensers were immersed in an ice bath at $0°C$ and the last two were immersed in a $-70°C$ dry ice/acetone mixture. The gas line from the last condenser was fed to a series of sample bags where the gas could be stored for later chromatograph analysis.

Gas Analysis

The gas composition was determined with a dual column Hewlett Packard gas chromatograph equipped with a thermal conductivity detector. Chromasorb 102 and Molecular Sieve 5A columns with a helium carrier gas were employed for gas analysis.

Experimental Procedure

Wood samples were prepared by dry mixing 150g of oven dry wood with 5 g of potassium carbonate or 5 g of calcium oxide. Uncatalyzed wood samples were also used to establish baseline control data.

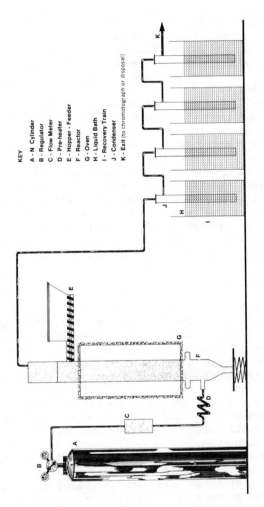

KEY

A - N Cylinder
B - Regulator
C - Flow Meter
D - Pre-heater
E - Hopper - Feeder
F - Reactor
G - Oven
H - Liquid Bath
I - Recovery Train
J - Condenser
K - Exit (to chromatograph or disposal)

Figure 1. The Forintek fluid-bed gasifier

The reactor was placed in the furnace and heating was initiated at least one hour before the run. Immediately after heating
was initiated, 300 grams of CBZ-1 cracking catalyst (Davison
Chemical Co.) was added to the reactor. Preheated nitrogen was
then introduced to the reactor at a volumetric flow rate of 1.5
liters per minute.

When the reactor and bed material reached the desired temperature (500°C), wood was fed continuously into the reactor for a
period of twenty minutes. Gas samples were collected at 4, 8 and
12 minutes after the feeding was initiated.

At the completion of the experiment (approx. 1.5 hr), the
char was allowed to cool and was then removed and weighed along
with the original bed material.

Four portions (200 ml each) of acetone were used to flush
and remove the liquid products from the condensers. The acetone
mixture was then filtered to collect any entrained charcoal/ash
fines which were then accounted for in the mass balance. The
acetone was evaporated off at reduced pressure and the remaining
liquid residue (pyroligneous acids and tars) was weighed. The
amount of gas product was then determined by difference. Where
wood was treated with potassium carbonate or calcium oxide, it
was assumed that these two catalysts could be accounted for in
the solid residue for purposes of determining the mass balance.
This method should give a good approximate mass balance of the
solid, liquid and gas products for comparative study purposes.

Results and Discussion

Table 1 summarizes the results of noncatalyzed hybrid poplar
gasification at 500°C in a fluidized bed reactor. Approximately
forty five percent (45%) of the wood was gasified. Twenty five
percent (25%) was converted to a liquid and thirty percent (30%)
remained as a solid product. The gasification reached a steady
state by at least four minutes as indicated by the chemical
composition of the gas. Samples collected at 8 and 12 minutes
had similar gas composition. No hydrogen was detected in the
gas product. Carbon monoxide and carbon dioxide were the major
components with a small amount of methane and low molecular
weight hydrocarbons.

Table II summarizes the catalyzed gasification results of
the hybrid poplar in the presence of potassium carbonate and
calcium oxide. Both of these catalysts increased the liquid
and gas yield at the expense of the solid. Both catalysts promoted a final product which was approximately fifty percent (50%)
gaseous. The remainder of the product was thirty percent (30%)
liquid and twenty percent (20%) solid.

It appears that potassium carbonate enhances the formation
of hydrogen and methane and reduces the formation of carbon
dioxide. Calcium oxide has a similar but more pronounced effect.

Table I

Gasification Results of Hybrid Poplar
at 500°C in a Fluid Bed System

Run No.	Solid %	Liquid %	Gas %	Gasific- ation Time, min.	Gas Composition,% by Volume[a]				
					H_2	CO	CO_2	CH_4	C_2-C_3 Hydro- carbons
1				4	0	56	38	4	2
				8	0	51	40	4	5
				12	0	51	40	6	3
	32	26	42	90					
2				4	0	54	38	5	3
				8	0	55	35	6	4
				12	0	55	30	10	5
	29	24	47	90					
3				4	0	53	39	5	3
				8	0	56	32	8	4
				12	0	55	37	5	3
	31	25	44	90					

[a] Nitrogen and oxygen free basis.

Table II

Gasification Results of Hybrid Poplar with
Catalysts at $500^{\circ}C$

Catalyst Used	Solid %	Liquid %	Gas %	Gas Composition, % by volume[a]				
				H_2	CO	CO_2	CH_4	C_2-C_3 Hydrocarbons
Control[b]	31	25	44	0	54	37	6	3
Potassium[c] Carbonate	19	29	52	14	36	31	15	4
Calcium[c] Oxide	19	30	51	20	43	20	13	4

[a] Nitrogen and oxygen free basis

[b] Average of 5 experiments

[c] Average of 3 experiments

Cox and his co-workers (9) reported that alkali carbonates increased the rate and yield of gas production of lodgepole pine at $550^{\circ}C$ using nitrogen as a carrier gas during gasification. The hydrogen composition remained unchanged at two percent (2%) in both catalyzed and noncatalyzed runs. Our results with potassium carbonate do not support this conclusion.

Conclusions:

The results of some preliminary research indicate that catalysts may be used to alter gas composition and gas yield. Specifically, potassium carbonate and calcium oxide appear to promote hydrogen production at the expense of the solid char product.

The literature indicates that gasification rates may be significantly accelerated by employing such catalysts as alkali carbonates and metallic oxides.

Results from the hybrid poplar gasification are encouraging and warrant additional research. Future catalytic work at Forintek will employ incremental temperatures ranging from 600 to $1000^{\circ}C$ and will include a variety of catalysts and catalytic combinations. An attempt will be made to promote the production of the following gases which enjoy different end uses:

(a) methane/hydrocarbon synthesis gas (H_2/CO)

(b) ammonia synthesis gas (H_2/N_2)

(c) substitute natural gas (CH_4)

(d) producer gas for combustion $(H_2, CO, CH_4, C_nH_m, CO_2, N_2)$.

Research will be conducted in the apparatus described in this paper, and also in larger fluidized bed (100,000 BTU/hr) and fixed bed downdraft (1×10^6 BTU/hr) reactors. These two larger reactors better approximate commercial gasification chemistry in that the process heat is supplied internally by partial combustion of the biomass feedstock.

Acknowledgement

The authors wish to thank Mr. E.E. Doyle of Forintek for performing the gas analysis.

References

1. E.J. Hoffman, "Coal Conversion", Modern Printing Company, Laramie, Wyoming, 1968, pp 11-15.

2. A.H. White, and D.A. Fox, Ind. Eng. Chem., 23 (3), 259 (1931).

3. A.H. White, and C.B. Weiss, Ind. Eng. Chem. 26 (1), 83 (1934).

4. A.H. White and Q.W. Fleer, Ind. Eng. Chem., 28 (11), 1301 (1936).
5. W.K. Lewis, E.R. Gilliland and H. Hipkin, Ind. Eng. Chem. 48 (8), 1697 (1953).
6. S. Prahacs, Advances in Chemistry Series 69, Div. Fuel Chem., 152nd National Meeting of the American Chemical Society, Chicago, September 1967.
7. J.O. Cox, G.W. Wilson and E.J. Hoffman, J. of Environmental Engineering Division, Amer. Chem. Soc., June 1974.
8. W.G. Wilson, L.J. Sealock, Jr., F.C. Hoodmaker, R.W. Hoffman, J.L. Cox, and D.L. Stinson, Preprints Amer. Chem. Soc., Div. Fuel Chem. 18 (2), 29(1973).
9. L.J. Sealock, Jr., R.J. Robertus, L.K. Mudge, D.H. Mitchell, and J.L. Cox, Papers Presented at the First World Conference on Future Sources of Organic Raw Materials, Toronto, July 1978.
10. C. Rai and D.Q. Tran, Amer. Inst. Chem. Eng., Symp. Series 157, 72, 100 (1976).
11. H.R. Appell, I. Wender, and R.D. Miller, U.S. Bureau of Mines, Tech. Prog. Report 25 (1970).
12. R.H. Hooverman, and J.A. Coffman, Symposium on Clean Fuels from Biomass and Wastes, Institute Gas Technol. Conference, Orlando, January 1977.
13. J.C. Cavalier and E. Chornet, Fuel, 56, 57 (1977).
14. J.A. Knight, "Thermal Uses and Properties of Carbohydrates and Lignins", F. Shafizadeh, E., Academic Press, New York, 1977.
15. P. Love and R. Overend, Energy, Mines and Resources Canada, Report ER-78-1, Ottawa, Canada, 1977.

RECEIVED November 16, 1979.

Wood Gasification for Gas and Power Generation

ARUN VERMA and G. A. WEISGERBER

Research and Development Centre, Saskatchewan Power Corporation,
2025 Victoria Avenue, Regina, Saskatchewan S4P 0S1

A study (1) was initiated by Saskatchewan Power Corporation (SPC) in 1976 to look at alternatives for heat and electrical power production in isolated northern communities. The present method of electrical power production at these communities is from diesel engines and fuel oil is largely used for space heating. The cost of these fuels has been going up recently, as well as the cost of transportation to these communities. The aim of the above mentioned study was to zero in on the most practical method for providing heat and power from locally available energy sources. The energy source was identified as wood.

The work identified and evaluated a number of technologies, their status and processes which could be used to go from wood to production of electricity or for space heating.

The appropriate technology for space heating was found to be the direct use of wood in high efficiency wood furnaces and for electricity production it was concluded that wood gasification followed by combustion of this gas in an internal combustion engine was the practical alternative.

Evaluation of high efficiency wood furnaces is being carried out under a separate project (2). For power production, a decision was taken to build a plant by scaling up the gasifier at B.C. Research Council (3, 4) which is said to be a fluidized bed gasification unit. A 4.22 GJ/h gasifier thus was fabricated with other auxilliaries. The Saskatchewan Forest Products plywood mill was chosen as the site for installation of the gasifier. The reason for that location was the easy availability of fuel as well as the ease with which the gasification technology could be transferred to the forest product industry in Saskatchewan for their own use of drying plywood. The plant was run for the first time in January 1979.

Parallel to the above work, a detailed resource, social and environmental impact study (5) was initiated to see the effect of power production via wood gasification plants on Northern Communities in the province.

Gasification Reactions

The term gasification as used here refers to the reaction of wood (wet) with air to yield a low heating value gas. The major components of the fuel gas produced are formed by various reactions between carbon and water in the fuel with oxygen in the air.

The first step towards gasification of wood after it has been introduced into the gasifier, is drying (100-200°C).

Wet wood + heat → Dry Wood + H_2O (Steam)

The second step is the pyrolysis step (200-500°C) which releases the pyroligneous acids, tar, etc.

Dry Wood + heat → Char + CO + CO_2 + H_2O

$$+ CH_4 + C_2H_4$$

$$+ \text{Pyroligneous acids}$$

$$+ \text{tar}$$

The oxidation or partial oxidation of char occurs next (300°C upwards).

Char + O_2 + H_2O (Steam) → CO + CO_2 + H_2 + heat

The first two processes are driven by the heat given out by the third process. The detailed thermo-chemistry can be summarized as follows:

$$C + O_2 \;\rightleftarrows\; CO_2, \text{Exothermic} \tag{1}$$

$$C + CO_2 \;\rightleftarrows\; 2CO, \text{Endothermic, Boudouard Reaction} \tag{2}$$

$$H_2O + CO \rightleftarrows CO_2 + H_2, \text{Endothermic, Water gas Shift Reaction} \tag{3}$$

$$H_2O + C \;\rightleftarrows\; CO + H_2, \text{Endothermic} \tag{4}$$

$$C + 2H_2 \;\rightleftarrows\; CH_4, \text{Exothermic} \tag{5}$$

The first reaction is the only one which provides the thermal energy for carrying out the other endothermic reactions. Reaction 5 is favoured only at lower temperatures and high pressures. Because of this, most of the methane formed in the product gas in gasifiers operating under atmospheric pressures, is a result of the pyrolysis step.

Hudson Bay Plant Details

Following are some of the details of the plant. Complete details may be seen elsewhere (6).

Gasifier Design Criteria. The gasification plant (Figure 1) has incorporated considerable design improvements and modifications over the smaller unit although the gasifier output target of 4.22 GJ/h and the grate heat release rates determined from B.C. Research unit served as the basis for determining the gasifier internal diameter. An expanded freeboard model was decided upon to reduce gas velocities in order to lower particulate entrainment in the gas exit stream.

Design parameters are listed below:

Net reactor output	4.22 GJ/h
Conversion efficiency	72%
Gas H.H. V.	5.59 MJ/m^3
Waste Wood Moisture Content	50% (wet basis)
Fuel Consumption	568 kg/h (as fired)

Gasifier. The gasifier is approximately 4 m in height having an internal diameter of 1.2 m in the lower section and 1.8 m in the expanded freeboard section.

The gasifier outer shell is composed of carbon steel. A layer of refractory forms the inner surface of the gasifier.

The windbox region of the gasifier is bolted to the bottom section of the gasifier to facilitate repairs to the ash removal system. Distribution of the incoming air is accomplished with a pinhole grate supported above the windbox. The sectional grate has refractory lining to protect the steel from the $1000^{\circ}C$ temperatures expected in the combustion zone immediately above.

Access to the gasifier is through a manhole. The windbox zone is also supplied with a clean out door.

The gasifier is equipped with 8 ports for thermocouple installation and 4 ports for pressure sensing. Two sets of ports at 180 degrees to each and capped with tempered glass are provided for the fuel height optical sensors.

Wood Feeding System. The feed system consists of a shift bin, drag conveyor, metering bin, air lock and screw conveyor.

The live bottom shift bin has a capacity of 6 m^3. A grid covers the bin to prevent large sized particles from damaging the system. An inclined drag conveyor transports fuel from the shift bin to the metering bin.

The metering bin has a capacity of 1.6 m^3 and has a live bottom discharge. A hydraulic motor drives the drag chain conveyor. A hydraulic power unit equipped with an electrically driven pump, flow regulator valve, pressure relief valve, and solenoid valve control the flow and pressure of hydraulic fluid

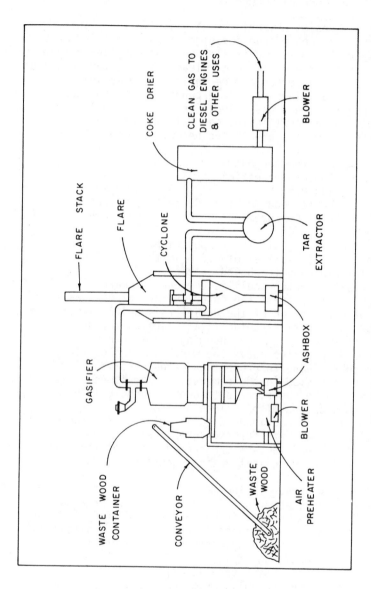

Figure 1.   Schematic of the wood gasification plant

to the hydraulic motor. When the solenoid valve which allows fluid to reach the hydraulic motor is closed the fluid circulates within the reservoir. The 12 position flow regulator valve directs the required flow of fluid to the hydraulic motor and bypasses the excess back into the reservoir.

A gear motor drives the rotary air lock and screw conveyor. The air lock provides a seal between the screw conveyor below and the small fuel hopper above. The 15 cm diameter screw conveyor has heavy duty 0.62 cm thick flighting. It extends to the inner refractory wall of the gasifier.

Air Feed System. A turbo pressure blower heat exchanger and burner compose the air feed system.

The blower is designed to deliver 12 m^3/min of air at 12 kPa guage pressure at 70°C. At 315°C this decreases to 6 kPa guage pressure. On the outlet of the blower is a manually controlled damper to regulate flow of air. Air is transported to the windbox area via a pipe equipped with an orifice plate.

The air preheater is a double shell direct-fired heat exchanger designed to deliver air at a maximum of 315°C. It is 1.2 m in diameter and is 2.3 m in length. A 0.45 GJ/h oil burner supplies the heat.

Ash Removal System. A rake, screw conveyor and ash box comprise the ash removal system. A rake mounted above the pinhole grate ploughs ash towards the centrally mounted ash discharge screw. A screw conveyor transports the ash via a gate equipped chute into the ash bin below.

Flare System. Following the gasifier is the flare system which is composed of a cyclone, cyclone hopper, butterfly valve, flare and stack.

The cyclone is 2 m tall with the diameter decreasing from 1.2 m at the raw gas entrance to 0.3 m at the particulate discharge point at the bottom. The 0.95 m^3 cyclone hopper is castor mounted and equipped with a cleanout door.

On top of the cyclone is located a cross-tee which is equipped with two manually operated, and one motorized butterfly valve. The manual valves are employed to direct the gas to a particular end use. The motorized valve controls the gas entering the flare.

The flare chamber is refractory lined and has an adjustable louvre at the bottom to regulate combustion air. Ignition is provided with a 0.20 MJ/h oil burner. A totally enclosed hood captures the combustion products and directs them to the 0.30 m diameter stack which extends above the hood.

Gas Sampling System. The gas sampling system consists of a knockout pot, filters, gas partitioner, integrator and pressure-vacuum pump.

The pump draws gas from the gasifier outlet through a knock-out pot where moisture, tars and dust are removed. Between the knockout pot and the gas partitioner are a couple of filters which ensure the gas is clean and will not damage the analyzer. The gas is drawn through a Fisher gas partitioner and is exhausted outside. The integrator then provides the gas composition on a volume per cent basis.

Wood Gas Cleaning System. The wood gas cleaning system consists of a cyclone, centrifugal tar extractor, drier and blower. In the cyclone hot raw gas enters tangentially with ash and carbon particles dropping to the bottom into a hopper and the gas exiting from the top. The hot gas is directed from the cross-tee to the mechanically driven tar extractor in which water is injected to cool the gas and condense the tars which are then separated from the gas by the centrifugal force. The blower draws the cool gas through the column filled with dry wood blocks which remove moisture. The water and tar mixture is taken to a separation tank where the water and tar are separated by gravity. The water is recirculated back to the tar extractor.

Diesel Generator System. The cool clean gas can be combusted in a naturally aspirated 250 HP Deutz diesel engine (Model F12L413F) which has been modified for burning wood gas. This has been purchased from Imbert Co. in Germany. The diesel oil injection system has been modified to allow approximately 10 per cent of the normal flow to be injected. This amount of oil is ignited in compression which in turn ignites the wood gas and air mixture. The wood gas is fed into the engine intake system, mixed with air and sucked into the engine. A Brown Boveri generator rated at 150 kW is frame mounted with the diesel engine. Three banks of 75 kW electric furnaces act as the loading units.

Burner System. A modified 5.28 GJ/h burner is also on site to combust the low heating value gas. The burner consists of a 15 cm special wide range burner, pressure pilot, flame rod, combustion air blower with motor, micro ratio valve assembly and an emergency shut off valve. The burner size has been increased compared to a normal burner operating with natural gas. The pilot is operated on propane to ensure adequate ignition. Because of the low heating value of the gas, the air enters the normal gas inlet and the wood gas enters through the normal air inlet.

Process Control. The control panel contains the start-stop locations and indicating lights. Temperature, pressure and flow parameters are also displayed.

A bindicator on the metering bin and a timer control the wood feed addition to the metering bin. The bindicator shuts off the conveyor from the shift bin when the metering bin is full. A timer which can be varied from 0 to 30 minutes automatically starts the shift bin conveyor after the selected time interval.

The rate of addition of solid fuel is controlled from the metering bin. A hydraulic system drives the metering bin conveyor. When the metering bin conveyor is stationary the electrically driven pump on the hydraulic power unit recirculates the hydraulic fluid to the reservoir. To start the metering bin conveyor a solenoid valve is opened allowing hydraulic fluid to turn the hydraulic motor. The flow and pressure of fluid to the hydraulic motor and ultimately the conveyor speed is controlled by a 12 position regulator valve on the hydraulic power unit. A single motor drives both the screw conveyor and airlock. The motor is equipped with reversible switch gear. Both the solenoid valve on the hydraulic power unit and the screw conveyor motor are actuated by one start location on the control panel.

Flow of combustion air to the gasifier is controlled by a manually adjustable damper at the outlet of the turbo blower. An orifice plate in the air line to the gasifier indicates the rate of air addition on the control panel. Hot combustion air for start-up of the gasifier is obtained by operating the oil burner on the air preheater.

Ash removal is on an intermittent basis with time between removal and amount removed at a single time determined by operating experience.

The gasifier is equipped with chromel-alumel Type K thermocouples which are tied into a multipoint temperature chart recorder housed in the control panel. Temperatures of the combustion air, fuel bed at various heights, exit gas and combustion products in the stack are monitored. Ports in the gasifier are connected to manometers to provide pressure data. Two sets of optical sensors provide fuel bed height determination by operating indicating lights on the control panel. Fuel bed height is maintained between the two sets of sensors. The operating pressure of the gasifier is controlled between 5 and 16.5 cm water column.

The pilot oil burner on the flare can be operated manually or automatically with it starting and stopping as the flare valve opens and closes.

Several safety features have been included in the control system. The exit gas thermocouple is connected to a variable temperature controller to maintain gasifier temperatures below safe operating limits. A pressure switch is connected to the outlet point of the gasifier which will shut the combustion fan down if the pressure exceeds 120 cm water column. Various pieces of equipment are also interlocked to protect the system.

Results and Analysis

The Gasifier. Table 1 shows a typical analysis of wet spruce chips of nearly 58% moisture content. Data for the gasification of these is shown in Table 2. The gas compositions are only preliminary and do not represent gas quality when the gasifier is operated under optimized conditions. Even in these preliminary runs the gas heating value has been determined to be in the range 4.4 - 5.07 MJ/m3. The gasifier output was 4.27 GJ/h. Optimization of the gasifier operation is proceeding.

Figure 2 shows the temperatures of the various points in the gasifier over a two hour period while gasifying wet spruce chips. The temperature in the combustion zone varied between 600 - 700°C with less than 1% oxygen showing in the exit gases. Other temperatures were considerably lower.

Large differences between thermocouple points 1 and 5 indicate that the bed is not in a fluidized state. This was confirmed by further experimental tests carried out on a three dimensional cold model. It was observed that the chips tend to lock with one another and cannot be fluidized under the design flow conditions in the reactor. It was also observed that some channeling also occurs in the bed. Further work is presently under way to determine the fluidizing conditions for other types of waste wood fuels.

The Burner. The burner to date has been run effectively with clean wood gas without any problems and it has been demonstrated without a doubt that the clean gas can be easily utilized in a modified burner or a burner designed to burn wood gas. The hot raw gas also burns effectively in the burner.

The Engine. The diesel engine, by the time of writing, had run for eight hours total under part load, using the wood gas from this gasification plant. However, the same engine was run on wood gas from a down draft gasifier having similar quality gas. Data from this run is shown in Table 3.

As may be seen, with 11% diesel and balance wood gas, the output of the engine drops by 20%. The rest of the engine performance is similar to that when operating with diesel fuel alone.

Environment, Resource and Social Impact of Gasification. Here, the forest resources and the social - environmental consequences of power generation via wood gasification were evaluated on 15 northern communities in Saskatchewan which are serviced by diesel electric generators. A general assessment was undertaken for 14 communities and a detailed one for 'Pinehouse'. These communities are listed in Table 4. The table shows the size and the electrical load at these communities. The general assumptions of this study were that forest

Table I

Typical Analysis of Spruce Wood Chips

1. Moisture (as received), %   57.74

2. Proximate Analysis (dry basis)
   Ash, %   0.79
   Volatile Matter, %   79.40
   Fixed Carbon, %   19.82
   Heating Value   20,099 J/g

3. Ultimate Analysis
   Carbon, %   44.11
   Hydrogen, %   5.59
   Nitrogen, %   0.08

4. Ash Analysis, %
   Al_2O_3 0.21
   SiO_2 1.34
   CaO   39.72
   MgO   5.56
   Na_2O 17.27
   K_2O 11.59
   P_2O_5 5.78
   Fe_2O_3 0.01

Table II

Composition of gas produced from 57.74%
Moisture Spruce Chips

Analysis #	CO_2	H_2	C_2H_6	O_2	N_2	CH_4	CO	H.H.V.* MJ/m^3 Dry	Sat.
1	12.08	11.89	0.501	0.880	54.79	1.95	17.91	4.69	4.62
2	12.24	12.05	0.525	0.898	55.35	1.63	17.3	4.53	4.46
3	12.21	12.37	0.605	0.908	53.54	1.86	18.5	4.86	4.77
4	11.55	11.97	0.604	1.065	55.69	1.712	17.39	4.62	4.54
5	11.77	11.74	0.56	0.99	56.79	1.661	16.49	4.43	4.36
6	11.42	9.93	0.61	1.37	55.74	1.91	19.03	4.68	4.60
7	9.62	9.84	0.507	.991	54.3	2.05	22.7	5.07	4.98

* At 14.73 psia & 60°F.

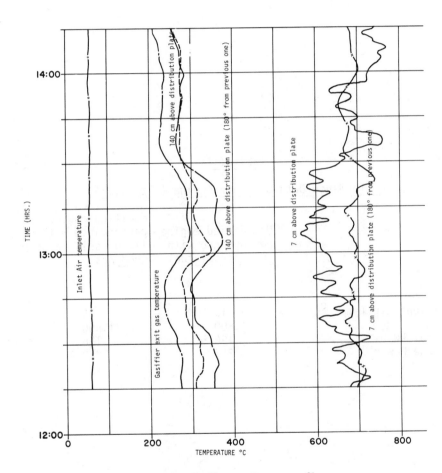

Figure 2.   Gasifier temperature profile

Table III

Performance Data of Deutz Engine (F12L 413F)

	Unmodified engine performance with 100% diesel fuel	Modified engine performance with 11% diesel, balance wood gas
Ambient Air Temp. °C	15	15
rpm	1,800	1,800
Total kW	196.5	156
Horse Power	267	212.4

Table IV

Communities in Northern Saskatchewan
Using Diesel for Power Generation

Community	Total Capacity (kW)	Average Load (kW)	Peak Load (kW)	Diesel Fuel Consumption (litres) 1977	No. of Electrical Connections 1978	No. of Persons per Connection
Camsell Portage	100	30	45	75,009	24	5
Fond du Lac	500	100	150	272,760	92	6
Stony Rapids	500	110	180	270,487	95	2
Black Lake	500	90	130	264,577	94	6
Wollaston	600	120	200	268,214	92	5
Kinoosao	100	20	35	75,009	20	4
Southend	350	80	130	183,972	87	7
Deschambault	425	50	95	184,113	77	6
Sturgeon Landing	125	25	45	102,740	40	5
Stanley Mission	500	100	175	233,033	121	9
Pinehouse	600	100	175	206,566	82	7
Patuanak	500	100	175	225,695	111	6
Dillon	550	110	185	220,936	80	5
Michel	100	15	25	73,645	14	8
Brabant	125	25	45	75,009	17	3
TOTAL				2,731,764		

resources would have to be harvested on a sustained yield basis
and use of waste wood could be made in the gasification plant.
Wood reserve data was found adequate for 9 of the 15 communities.
The remaining 6 communities which are located north of 57 degrees
latitude did not have sufficiently detailed forest volume
information to allow an accurate assessment of its wood reserves.
Nevertheless, the findings indicate that forest resources are
adequate to supply a small wood gasification plant for at least
50 years in each community. Such a time frame work for logging
ensures a renewable timber resource even without an active
reforestation program although it is desirable.

One of the major findings was that waste wood could provide
an adequate supply of fuel for a few communities. Examples of
wood waste could be found at local sawmills, right of way clear-
ing for roads, and wood damaged by forest fires.

The social consequences of wood gasification on a community
are positive. Though the scale of the gasification project is
small, it may require local people to operate the gasifier and
to harvest an adequate supply of wood. With the use of local
labour, the construction of the project in the community would
also have mainly positive implications for the community.

The environmental consequences based on emission data could
not be completed as the data is being collected but the effect
of gasification on vegetation, wildlife and aquatic life, etc.
was estimated to be minimal. Overall, the project would offer
permanent local employment and ensure a greater electrical power
capacity. Earned income would be increased and spin-off benefits
could result in the establishment of small businesses. The
utilization of waste wood is another positive element associated
with this proposed project. As a result, the gasification
project is likely to be viewed positively in northern communities.

Conclusions

  i)   The scale up version of the B.C. Research gasification
       unit did not perform as a fluidized bed gasifier when
       operating on wet spruce chips. It essentially operated
       in a fixed updraft flow mode.
  ii) The fuel gas produced from wood can be easily combusted
       in a modified burner or a burner built to combust this
       type of gas.
  iii) The naturally aspirated modified diesel engines can be
       run successfully with wood gas while using 10% diesel
       fuel for ignition purposes. The drop in output is 20%.
  iv) Power generation via wood gasification in Northern
       Communities has a minimal impact on the environment and
       resources while it has a net positive social impact.

Abstract

A 4.22 GJ/h waste wood gasification plant is being evaluated at Hudson Bay, Saskatchewan, Canada. The demonstration scale gasifier is a scale up of a smaller unit at the B.C. Research Council, Vancouver.

The gasifier is designed to convert minus 1 inch wood waste with less than 50% moisture into a low heating value gas of approximately 4.5 - 5.6 MJ/m^3. In operation wood waste is metered out from a live bottom storage hopper into the gasifier via a feed screw. Air is fed into the windbox under the pinhole grate. Hot raw gas exits from the expanded freeboard gasifier at the top. The gas is cleaned in a 'Crossley' type cleaning train consisting of a cyclone, centrifugal extractor and a drier. The clean gas is fed to a modified diesel engine or a burner. The aim of the work is to evaluate the gasification plant for production of fuel gas suitable for use in a burner and a modified engine for electrical power production.

Plant details and preliminary performance data are discussed in the paper, along with environment, social and resource impact of wood gasification on northern communities in the province.

References

1. Verma, A., Stobbs, R.A., "Wood for Power generation in isolated Northern Communities." Saskatchewan Power Corporation report, 1977.

2. Verma, A., Weisgerber, Gordon, "Thermochemical Conversion and other biomass related work at Saskatchewan Power Corporation", Proceedings of "Energy from Biomass seminar", Ottawa, March 1979.

3. "Engineering Feasibility Study of the British Columbia Research Hog Fuel Gasification system", H.A. Simons (International) Ltd., May 1978.

4. Liu, M.S., Serenius, R., "Fluidized Bed Solids Waste Gasifier", Forest Products Journal 26a, 56-59, 1976.

5. "The Social, Environmental and Resource Impact of Wood gasification on Isolated Northern Communities, Part I"; March 1979, Report by Saskatchewan Power Corporation for the Federal Government of Canada under contract. DSS File No. 07SB.KL016-8-0060.

6. "Evaluation of the Saskatchewan Power Corporation Wood gasifier at Hudson Bay, Saskatchewan, 1978-79". May 1979, Report by Saskatchewan Corporation for the Federal Government of Canada under contract. DSS File No. 07SB.KL016-8-0061.

Acknowledgement:

This project is a joint project of Saskatchewan Power Corporation, Saskatchewan Forest Products Corporation and the Federal Government of Canada.

RECEIVED November 16, 1979.

PYROLYSIS, GASIFICATION, AND LIQUEFACTION PROGRAMS IN EUROPE

Basic Principles of Waste Pyrolysis and Review of European Processes

A. G. BUEKENS and J. G. SCHOETERS

Department of Chemical Engineering and Industrial Chemistry, Vrije
Universiteit Brussels, Pleinlaan 2, B-1050 Brussels, Belgium

1. Introduction

In the early fifties more and more Municipalities in W. Europe
were confronted with a declining availability of suitable landfill
sites. This resulted in the large scale introduction of incinerator plant during the period 1960-75. Initially quite a few of
these units showed mechanical deficiencies in the handling of refuse and clinker as well as corrosion problems in the boiler and
gas cleaning plant ; some even suffered from gross design errors.
As years went by more experience was gained and through the combined skills of designers, manufacturers and operators refuse incineration attained its present status of a well proven though expensive method of waste disposal.
        Gradually the densely populated regions of W. Europe were
amply equipped with incinerator plants. The latter complied with
the air pollution regulations of that time by using highly efficient electrostatic precipitators. Thus they could be situated in
the very centre of the refuse generating territory reducing the
cost of collection and transportation. In large plants the heat
of combustion was recovered under the form of medium to low pressure steam, used for power generation and district heating. The
low volume, sterile residue was a suitable substitute for gravel,
provided the material is graded and the unburnt and magnetic substance is removed.[1,2,3]
        This explains why the progress of PTGL-technology was slower
in W. Europe than in the U.S.A. or Japan. Even now, the only
process that attained commercial operation in Europe is of American origin. All other processes only exist at pilot scale, although a few have been conceived at a larger scale and participate
in public tenders. Unfortunately innovation is both difficult and
dangerous. Municipalities are very well aware of this fact and
will not embark on a yet unproven demonstration project unless very
important incentives are provided by the Government. The spectacular failure of a couple of large scale American PTGL-units
could only corroborate this tendency.

0-8412-0565-5/80/47-130-397$06.25/0

TABLE 1a : PTGL-Systems for municipal refuse.

NAME	LOCATION	SIZE	T_{MAX}	DATE[*]	CHAR	OIL	GAS	STEAM	REMARKS
ANDCO-TORRAX	Luedelange Grasse Frankfurt Creteil	200t/a 170t/a 200t/a 2x200t/d	1500°C	1976	-	-	-	X	Slagging gasifier. Original U.S. design.
PYROGAS	Gislaved	50t/a	1500°C	1977	-	X	X	-	Countercurrent vertical shaft. Gasification of refuse/coal mixture with air and steam.
SAARBERG-FERNWARME	Velsen	24t/a	1000°C	1977	-	X	X	-	Countercurrent vertical shaft. Gasification with pure oxygen. Low temp. seperation of gases.
DESTRUGAS	Kalundborg	5t/a	1050°C	1972	X	-	X	-	Cocurrent vertical shaft. Indirectly heated.
WARREN-SPRING	Stevenage	1t/d	800°C	1975	X	X	X	-	Cross-flow vertical shaft. Direct heating with recycled pyrolysis gases.
T.U. Berlin	Berlin	.5t/d	950°C	1977	X	X	X	-	Indirectly heated cocurrent vertical shaft.
SODETEG	Grand-Queville	12t/d			X	-	X	-	Indirectly heated vertical shaft. Project inactive.
KRAUSS-MAFFEI	Munchen	12t/d	600°C	1978	X	-	X	-	Indirectly heated rotating drum. Heavy hydrocarbons decomposed in thermal afterreactor
KIENER	Goldshöfe	6t/d	500°C		X	-	X	-	Indirectly heated rotating drum. Thermal afterreactor with hot cokes. Gas drives gas engine.
UNIVERSITY EINDHOVEN	Eindhoven	.5t/d	900°C	1979	X	X	X	-	Indirectly heated spout-fluid bed reactor.
D. ANLAGEN LEASING	Mainz						X	-	Indirectly heated rotating drum.

[*] Start-up date of the unit.

TABLE 1 b : PTGL – Processes for industrial wastes and tires.

NAME	LOCATION	SIZE	T_{MAX}	DATE*	CHAR	OIL	GAS	STEAM	REMARKS
HERKO/KIENER	Goldshöfe	6t/d	500°C		X	X	X	X	Same as Kiener. Uses no afterburner. For tires.
BATCHELOR-ROBINSON	Stevenage	6t/d	800°C	1975	X	X	X	–	Warren Spring process for tires.
FOSTER-WHEELER	Hartlepool	1t/d	800°C	1976	X	X	X	–	Same as Warren Spring.
HERBOLD	Meckesheim	2.5t/d	500°C		X	X	X	X	Indirectly heated screw conveyor. For tires.
G M U	Bochum	5t/d	700°C	1977	X	X	–	–	Indirectly heated rotating conveyor. For tires, cable, plastics.
UNIVERSITY HAMBURG	Hamburg	.5t/d	800°C	1976	X	X	X	–	Indirectly heated fluidized bed. For tires.
UNIVERSITY BRUSSELS	Brussels	.2t/d	850°C	1978	X	X	X	–	Indirectly heated fluidized bed. For plastics, tires, wood waste.
RUHRCHEMIE	Oberhausen	1t/d	450°C		–	X	–	–	Stirred, indirectly heated retort. For polyethylene waste.
PPT	Hanover		430°C		–	–	–	–	Indirectly heated fixed bed. For recovery of metals and heat from cable, painted metals...
RAMMS	Essen		500°C		–	–	–	–	See PPT.
GUILINI	BRD				–	–	–	–	Vertical shaft gasifier. For tires.

TABLE 1c : Biomass Gasification Processes.

Developer	
A.T.F. (France)	Gas producer driving dual fuel engine
Distibois (France)	Gas producer manufacturer.
Duvant (France)	Gas producers for wood waste, and coconut. Drives diesel engine for lean gas.
Ets. Lambiotte (Belgium)	Manufactures furnaces for charcoal production. Testprogramme(in coöperation with VUB) on wood waste gasifier driving diesel engine.
National Swedish Testing Institute (Sweden)	Tests on downdraft gasifier. Test with turbo charged diesel engine.
Pillard (France)	Gas producer for straw and coconut.
Royal Institute of Technology (Sweden)	Steam gasification in fluid-bed (1-20 kg/hr).
Secfi (France)	Gas producer for straw and coconut.
Sté. Alsacienne de Construction Mécanique	Manufactures updraft gas producers.
T.H. Twente	Develops a downdraft gasifier for Third World countries.
Wellman Incandescent (England)	Manufactures updraft gasifiers.

For about ten years now private companies and research institutions have undertaken R & D-work in the field of pyrolysis or gasification of refuse, sewage sludge, plastics and rubber, or biomass. (table I). The potential advantages of PTGL-systems that stimulated this work are :
- the simplicity and potential lower cost of some of the systems
- their ability to deal with special wastes, such as sludge, plastics and rubber, or waste oil
- the reduction of the amount of gas to be cleaned
- the production of valuable products or of storable fuel
- the potential recovery of unoxidized metal and unmelted glass from the residue.

2. Survey of European PTGL-Technology

PTGL-processes (table Ia,b,c) can be characterized by their operating principle (pyrolysis, gasification or liquefaction) and reaction conditions, their type of reactor and method of heat supply and the possible presence of auxiliary systems. The selection of a particular process will ultimately be based on its proven reliability on one hand and on the quality of the process output on the other [4,5].

Pyrolysis processes generate oil or a limited quantity of a rather rich gas (Table II), which can be purified, stored and conveyed over a large distance. Co-disposal of plastics, rubber and waste oil is often possible. On the other hand the condensating liquors are grossly polluted and the pyrolysis coke is difficult to valorize. The pyrolysis process is complicated by the necessity of indirect heat transfer and the condensation of tar often leads to operating problems.

Gasification with oxygen yields a moderate quantity (Table II) of a rich gas, which can be cooled, purified and used as a synthesis or a fuel gas. The use of oxygen increases the operating cost and creates an explosion hazard. Gasification with air yields a large quantity of lean gas, which is often directly burned. In that case the process can be considered as a two-step incineration.

TABLE II

Typical yield and calorific value of product gas (s.t.p.)

	Volume m^3/t of refuse (dry)	Calorific Value kJ/m^3
Pyrolysis	200 - 500	12500 - 19000
Gasification with oxygen	600 - 1100	7500 - 15000
air	1000 - 2500	3000 - 8000

Pyrolysis processes can be subdivided according to their <u>operating temperature</u> :
- low temperature processes (below 550°C) yield mainly oil, tar and carbonized residue. Indirect heat transfer is no problem. Heat losses are low and thermal yield is high.
- medium temperature processes (550 - 800°C) give a high yield of gas, which has a high calorific value due to the presence of methane and some higher hydrocarbons. Indirect heating becomes more difficult.
- high temperature processes (above 800°C) give the highest volume of gas. The additional gas arises by the secondary cracking of the primary products. This explains the lower calorific value of the gas. Most tar is converted into gas and carbon so that obstructions formed by condensing tar are rarely a problem. On the other hand the weakening and melting of glass and ash are responsible for difficulties in the elimination of residue and for a reduced lifetime of the refractory lining.

Gasification processes normally operate at higher temperatures than pyrolysis processes. The highest temperatures are attained in the zone in which carbonized residue is combusted. When using oxygen or preheated air as a gasifying medium it is often possible to operate in the slagging mode, which in principle simplifies the problem of ash removal. The slag can be tapped continuously or not ; it is either granulated by quenching in a water bath or solidified to large blocks. Auxiliary burners and addition of flux may be required to regulate and enhance the fluidity of the slag. Heavy metals and chlorides evaporate in the hearth zone and are collected together with fly-ash.

The fluidized bed gasifier operates at the other extreme of the temperature scale. Low operating temperatures (550- 850°C) are the rule because the combustion of carbon occurs throughout the bed and heat exchange is fast. Moreover the operating temperature has to be limited in order to avoid agglomeration and clinkering of the bed material.

<u>Countercurrent operation</u> of shaft and rotary kiln furnaces leads to extensive heat transfer between feedstock and outgoing products. The sensible heat of the products remains within the system which enhances its thermal yield. Cocurrent operation shows none of these advantages ; yet it is a simple and efficient method to avoid tar problems. Indeed the tar is extensively cracked while moving through the hearth zone. Also the problem of wastewater is somewhat simplified. Crosscurrent operation is only seldom used.

The required process heat can be supplied by 3 different methods :
- <u>indirect heating</u> through a metal or a refractory wall
- <u>direct heating</u>   by means of a separately heated circulating heat carrier, e.g. hot sand, steel balls, glowing carbon particles, hot flue gas, melted salts or metals

- <u>internal heating</u> by partial combustion of pyrolysis coke, gas
or auxiliary fuel.

Indirect heating through a refractory wall, as in the
Destrugas process, is very slow. It takes about 20 hours to heat
a slowly descending layer of refuse up to the desired temperature
of 950 - 1050°C. This results in a low capacity per unit reactor
volume. Even inductive heating of steel balls, admixed to the
refuse, as in early WSL-experiments, does not sufficiently accelerate the heating process. Better results are obtained by forced
circulation of pyrolysis gas, as in the Lambiotte wood carbonization furnace and in the WSL cross-flow reactor.

Heat transfer can also be activated by the periodic contact of the material to be heated with an externally heated wall.
This method is extensively used in Europe by tumbling the feed
material in an externally heated metallic rotary kiln. Another
method of activating the wall-to-feedstock heat exchange is to
suspend the feed-stock in a fluidized bed of sand, coke or ash.
External heating of fluidized bed reactors is often used at bench
scale, two-bed systems being preferred at an industrial scale.

Molten metals, salts and slags received little attention
in Europe. It is recognized that such melts are corrosive and
difficult to regenerate.

The most popular type of gasification furnace in Europe
is the vertical shaft gasifier, used in the Andco-Torrax, Saarberg
Fernwärme and Pyrogas process. Pyrolysis processes are often conducted in an indirectly heated rotary kiln reactor, e.g. in the
Kiener, GMU or Krauss-Maffei process. Fluidized bed reactors are
used at the universities of Hamburg, Eindhoven and Brussels and
thus seem more popular in academic than in industrial spheres.

3. Status of European PTGL-plants

The only PTGL-process currently operating at a commercial scale
is the American-developed <u>Andco-Torrax</u> process, pioneered in Europe by S.A. Ets Paul Würth (Luxemburg). Since its start-up in 1976
the Leudelange plant has operated intermittently, the frequent
periods of shutdown being used for modifying various parts of the
system. At present the reliability of the process has considerably improved but the nominal design capacity (200 t/day) will probably not be attained. [6].

One of the main problems common to all large diameter
shaft reactors operating on a heterogeneous material is that of
channeling. Channeling of rising gas causes local overheating of
the reactor wall and also occasional small explosions. Part of
the descending refuse reaches the heart at too low a temperature,
causing unmelted pieces of metal to float upon the slag being discharged. The channeling problem was tackled by systematic modifications of the reactor profile, by diminishing the diameter of the
reactor and by improving the cooling of the reactor wall.

In another problem area the slag tapping system was sim-

plified and rendered more accessible. Also the removal system for granulated slag was changed several times ; the quenching trough was enlarged and the original pumping of granulated slag was first replaced by a chain conveyor, later by a crane and grapple system. Other modifications include :
- improvements in the refuse hopper and feeding system, giving a better tightness and backfire control
- adding a mechanical level controller to the existing γ-ray device
- improving the refractory masonry around the tuyeres
- addition of a generator powered by an internal combustion engine.

The generator in case of power failure serves to maintain overpressure in the blast system to avoid the entrance of slag.

At present the actual factor limiting plant capacity is the unstable combustion of the product gas ; this causes untolerable vibrations of the waste-heat boiler when operating near full capacity.

Other Andco-Torrax plants have been built in France (Grasse, Créteil) and W. Germany (Frankfurt). The last plant is not operating yet because of excessively stringent local regulations on permissible noise levels.

From European experience it can be concluded that the Andco-Torrax system is a feasible PTGL-process but that it takes much time and trouble before such a new process can be operated with a sufficient degree of reliability. It is still too early to compare the process to conventional mass-incineration. At present the consumption of auxiliary fuel is still very high at 5 to 12 GJ/h (Luxemburg). The power consumption, at 60 - 80 kWh/t of refuse is about twice that of a conventional incinerator. The consumption of cooling water and the composition of the flue gas are roughly comparable.

The Pyrogas process has been developed by Motala Werkstatt a Swedish company that formerly constructed two-stage coal gasifiers. The feed is first carbonized in the upper part of the gasifier and then descends into the gasifier proper, which yields a tar-free gas.

After preliminary test runs on an old gas generator in Oaxen a larger plant was built at Gislaved, with an hourly capacity of 2 t of refuse, 0.2 t of rubber waste and 0.55 t of coal. It was claimed that all process wastewater could be disposed of by evaporation in a special boiler for generating gasifying steam. The present status of the project is unknown, but Motala seems to have stopped its activities in this field [7].

Saarberg Fernwärme utilizes a rotary-grate vertical shaft gasifier operating on a mixture of oxygen and steam. The refuse is pretreated by shredding and magnetic removal of ferrous metal. [8,9]. After some preliminary testing at a small 1.2 t/day scale, a 50 t/day was constructed and operated. Some problems were encountered with the rotary grate and the breakthrough of oxygen. A remarkable feature is the cryogenic separation of the product gas, which has not operated yet.

The Destrugas process is the oldest European PTGL-process on the market. Original testing work took place in Denmark but was later halted. Prof. Thomé-Kozmiensky built a smaller test unit at the T.U. Berlin, which operates according to the same principle . The process is currently being marketed by Wibau Matthias & Co ; the full scale unit would consist of a number of 5t/day modules, placed side by side as in a conventional coking chamber battery [10-15]

An externally heated vertical shaft has also been used in the now discontinued Sodeteg process.

An indirectly heated rotary drum was developed by the firm Kiener in Goldshöfe. The process was evaluated by Prof. Tabasaran of the T.U. of Stuttgart. The tar laden pyrolysis gas is partially oxidized and cracked on a bed of glowing coke, the temperature of which is controlled by suitable addition of combustion air. [16,17,18,19] .

Kraus Maffei has adopted the same principle by using a thermal afterburner for heating and cracking of the pyrolysis gas. Afterburning also reduces the contamination of the gas scrubbing liquor. After test runs at a 12 t/day scale on refuse, sludge, wood waste and waste oil a 120 t/day unit is in the planning stage [20, 21] .

Veba Öl (originally GMU), Rütgerswerk A.G. and Eisen und Metall A.G. collaborate in the development of a heated rotating drum furnace for the recovery of metals and oil from special wastes, such as tyres, cable wastes and acid tars with a high calorific value . The oil is separated into benzene, toluene, naphtalene and heavier fractions by distillation [22,23,24,25] .

Deutsche Anlagen Leasing designed a process with the same type of pyrolysis reactor.

Refuse pyrolysis can be conducted in an externally heated fixed barrel provided with a screw conveyor. This process has been proposed in a recent patent application of Mr. Greul (Frankfurt).

The development work of Warren Spring Laboratory (WSL) mentioned before was the starting point of 2 industrial ventures. Foster Wheeler continued WSL-testing at a 1t/h plant situated in Hartlepool and offers commercial scale units (2 - 5t/h) for specific industrial wastes. [27,28] .

Batchelor & Robinson also ordered test work at WSL. Later they built an improved pyrolysis plant, the oil product of which is fired in a boiler. The company claims to have solved the technical problems inherent to this technique but does not at present offer the process for licensing.

The fluidized bed pyrolysis process developed at the Hamburg University aims at the recovery of products from plastics and tires. A pyrolysis unit for unshredded tires has been constructed. It produces gas, a carbon black, steel cord and an oil that has a

high content of aromatics because of the recirculation of pyrolysis gases and the specific reaction conditions that are used. [29,
30,31,32].

At the university of Eindhoven, a spout-fluid bed reactor
is being developed for the pyrolysis of crushed and classified refuse. The feed is suspended in the fluidizing gas and blown in
the reactor through an inlet pipe in the bottom of the reactor.
Mathematical models for the pyrolysis of refuse particles have
been developed and start-up of the reactor is planned for 1979[33].

Much research on the pyrolysis of special wastes has been
done in Poland. These tests were usually performed on a small
scale and very little of their results has been published.

Ruhrchemie uses a stirred tank reactor, heated externally
with molten salt for converting polyethylene waste to a fuel oil.
[34].

The indirectly heated screw conveyor developed by Herbold
for pyrolyzing shredded tires has been tested on a 2.5 t/d pilot
installation. The main products are oil and a carbonaceous residue. The research project seems to be inactive at this moment[35].

P.P.T. and Ramms both use a pyrolytic process for recovering metals from cable wastes and painted metals. The reactors
are externally heated by combustion of the pyrolysis gases [36,37].

Guillini designed a vertical shaft gasifier for tires.
The produced gases are flared.

Work on the gasification of wood and agricultural residues
is performed by different private companies making gas generators
(Table I.c.).

Various universities are also active in this field. At
the university of Twente, The Netherlands, a downdraft gasifier
is being developed for Third World countries[38].

4. Fundamentals of Pyrolysis and Gasification

4.1. Introduction

PTGL-technology can only be succesful if practical
solutions are found for problems such as refuse handling and feeding, maintenance of stable operating conditions with a constantly
variable feedstock, smooth elimination of residue and proper
treatment of reaction products. Quite normally process developers
center all their energy on the design,construction and demonstration of their plant and equipment.

Household refuse by itself shows such a diversity in the
properties and composition of its components that it does not lend
itself to modelling, not even to substitution for experimental
purposes by synthetic refuse. Its tendency for bridging or solidification under its own weight as well as the possible presence
of contraries such as barbed wire, massive iron, gas bottles or
ammunition make it a good candidate for beating even the very
toughest handling system.

Studying the fundamentals is not a proper way for developing a PTGL technology operating on refuse. On the other hand a vast number of industrial and agricultural waste have well established dimensions and physical and chemical properties. In such cases experimental work at a bench scale can be used to anticipate the problems of a full scale disposal or recovery unit and to optimize its design for a maximum value of the reactor output. A study of the process fundamentals is a useful guide in such an enterprise.

4.2. Thermodynamic Aspects

1. Pyrolysis

Organic compounds decompose upon heating. This tendency is easily explained by the increase in entropy which accompanies the fragmentation of the original molecule into simpler products. In a system of organic compounds there will be a steady evolution towards more stable end products.

For hydrocarbons the thermodynamic stability can be measured by means of the free energy of formation from the elements, i.e. carbon and hydrogen. All hydrocarbons become unstable above $500^\circ C$. The mechanism of conversion is quite complicated and can be modified by the presence of catalysts. In the case of thermal decomposition the primary products are converted into compounds of increasing stability [39] : (Figure 1)

paraffins → olefins → diolefins → aromatics →

polycyclic aromatics (tar) → carbon + hydrogen.

At lower temperature the olefins and diolefins tend to polymerize to tar with a highly complicated, aliphatic structure.

The thermal decomposition of oxygenated compounds yields simpler and stabler compounds, such as formaldehyde, acetone, acetic acid, etc. Final products are often CO, CO_2, H_2O, $CH_2 = CO$..

Van Krevelen and Hoftyzer [40] characterized the relative stability of various polymers and polymerization initiators by means of the temperature $T_{1/2}$, at which the substrate is half decomposed after a reaction period of 30 minutes. These authors found a linear correlation between this parameter $T_{1/2}$ and the energy of the weakest bond of the decomposing molecule.

Thermodynamic considerations thus allow one to predict the relative stability of different reacting compounds or to explain the evolution in a reacting system. In the Hamburg pyrolysis process (29) for example, a high yield of aromatic compounds is obtained by combining a high pyrolysis temperature with the recirculation of less desirable and less stable gaseous reaction products.

Another very important factor in pyrolysis is the heat of

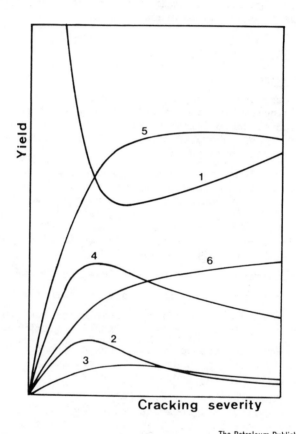

The Petroleum Publishing Company

Figure 1. Evolution of the product distribution with cracking severity (Naphta cracking): (1) C_{5^+}; (2) $C_4H_8 + C_4H_{10}$; (3) C_4H_6; (4) C_3; (5) C_2; (6) $H_2 + CH_4$

<u>reaction</u>. The latter can easily be calculated from the heat of formation of the reacting material and the reaction products. Unfortunately these data are not always known with precision for polymeric or biological material and the product analysis is often incomplete. In the first case the heat of reaction can be determined experimentally by means of bomb calorimetry. The heat of pyrolysis can also be obtained by differential thermal analysis (D.T.A.). In this technique the temperature of the product sample is compared to that of a reference sample ; integration of the temperature differential vs. the operating time yields a quantity related to the heat of reaction.

2. Gasification

Both the thermodynamic equilibrium and the heat of reaction are reasonably well known for the heterogeneous gasification of carbonized residue with oxygen, steam, carbon dioxide and hydrogen. The outlet gas composition is normally computed by a trial and error procedure ; for each operating temperature the resulting composition is calculated, then the attained temperature is determined from an enthalphy balance and the temperature correspondence is checked.

The most important factor affecting equilibrium gas composition is temperature. Deviations from equilibrium can be explained by the presence of pyrolysis gas, arising during thermal decomposition, by uneven gas and solid distribution, due to channeling, baking, clinkering or formation of blowholes, or by inadequate rates of reaction.

Increasing the operating pressure has an unfavorable influence on the gasification equilibria of steam and carbon dioxide, but favors the formation of methane. In practice all processes operate at substantially atmospheric pressure.

4.3. Kinetics and Mechanism

1. Pyrolysis

The distribution of primary reaction products is mainly determined by the structure of the raw material and by the reaction conditions. At low temperatures reactive primary products may immediately polymerize to a tar, obscuring the primary reaction steps. At high temperatures the primary fragments are sufficiently stable or rearrange to stable products, so that the first reaction steps are often easily recognizable.

The reaction patterns in the <u>thermal decomposition of polymers</u> are fairly well known. Generally one of the following mechanisms dominates [41,42] :

- thermal depolymerization of a polymer into a monomer

-stochastic fragmentation of a polymer chain into smaller molecules of various length
-elimination of side chains or adjacent substituents yielding
products such as H_2O, CO_2, NH_3, HCl, H_2S, CH_4 and higher paraffins.

It is possible to predict the relative importance of the
several modes of decomposition from an analysis of the polymer
structure[41] . Van Krevelen[40] even showed that the tendency for
coke formation is an additive property, which can be computed from
the monomer structure.

The pyrolysis of cellulose is of outstanding importance in
the study of PTGL-processes involving refuse, wood waste or agricultural waste. Unfortunately the products and mechanism of thermal decomposition are extremely sensitive to a number of physical
and chemical factors. Even the decomposition of the purest α-
cellulose has led to conflicting opinions regarding kinetics and
mechanism[43].

It is generally accepted now that there are 2 competitive
methods of breakdown :
- carbonization, proceeding by a ring opening of individual monomer units with a loss of CO, CO_2, or H_2O, but with conservation of
a polymeric carbon-chain
-depolymerization to levoglucosan tar after complete "unzipping"
of cellulose crystallites.

The relative importance of the two modes of decomposition
depends on temperature, structure, moisture and ash content and
presence of additives. Low temperatures and the presence of flame
retardants favor carbonization.

Paper, board, agricultural, garden and wood waste largely
consist of cellulosic material. Hemicellulose, a low molecular
weight polysaccharide present in vegetable material and in paper,
gives a higher yield of methanol and acetic acid than cellulose.
Lignin is a fairly stable, carbon-rich aromatic compound present
in wood. It yields mainly tar and carbon.

The kinetics and mechanism of thermal decomposition have
often been studied by methods such as thermogravimetric analysis
(TGA), differential thermal analysis (DTA), pyrolysis gas chromatography, or - by means of small externally heated retorts, agi-
tated vessel or fluidized bed reactors. The usefullness of the
results of these studies is often limited because :
- kinetics and mechanism can be vastly different from one temperature domain to another or be profoundly influenced by the presence of foreign matter (ash, moisture, catalysts)
- even in one and the same temperature domain the rate controlling
step may differ depending on the size of the particle or on the
flow conditions.
- generally the kinetic data show so much spread that data selec-

tion and interpretation becomes a problem. In the case of cellulose pyrolysis activation energies ranging from 70 to 230 kJ/mole have been reported.

Such differences can be attributed to differences in starting material and experimental conditions and techniques. Further it should be born in mind that thermal decomposition processes occur along complex mechanisms that cannot adequately be described by a simple rate law. This explains why apparent energies of activation or reaction orders can vary almost continuously with experimental conditions.

A number of models that take account of some of the physical phenomena occuring during pyrolysis have been developed for a single wood particle [44,45]. The relevant phenomena are :
- heat transfer by convection and radiation from the surroundings to the wood particle
- internal conductive and convective heat transfer.
- drying and thermal decomposition of the material.
- inner diffusion and outer convection of the pyrolysis products.

A practical difficulty in the application of such models to real reactors is the poor knowledge of the numerical values of the parameters involved (porosity, heat of reaction, thermal conductivity, heat capacity, kinetic parameters).

Yet from these studies it follows that heat and mass transfer phenomena by no means can be ignored. Kung [46] showed that among wood particles of 2, 0.2, 0.02 cm only the smallest ones showed a uniform temperature distribution. Hence in actual pyrolysis reactors the rate of conversion will usually be controlled by heat transfer. When the latter is very slow, as in the Destrugas process, plant operating capacity may be strongly affected by the moisture content of the refuse.

2. Gasification

Gasification is a heterogeneous gas-solid reaction, composed of three elementary steps :
- adsorption of the gaseous reactants onto the carbon surface.
- chemical reaction between adsorbed gas and solid.
- desorption of the gaseous product.

The chemical rate of reaction is determined by the slowest of these 3 steps. A simplified rate low is given by

$$r_c = k.A.m_c.p_G$$

where
r_c = rate of carbon consumption
k = rate parameter, following an Arrhenius law
A = specific surface of the carbonized residue
m_c = amount of C
p_G = concentration of the gasifying agent

The reaction with oxygen is extremely fast. The relative rates of the 4 heterogeneous gasification reactions at 800°C and 10^4 Pa (0.1 atm) are :

carbon-oxygen	:	10^5
carbon-steam	:	3
carbon-carbon dioxide	:	1
carbon-hydrogen	:	3.10^{-3}

Chemical reaction is only rate determining in a first low temperature domain, situated roughly below 1000°C. In a second, medium temperature domain the gaseous reactant is gradually depleted inside the porous particle, so that the reaction proceeds at the rate at which internal diffusion supplies new reactants. In this internal diffusion controlled region the apparent energy of activation falls off to half its initial value.

At very high temperatures, i.e. above 1200°C, chemical reaction is so fast that the gaseous reactant is immediately consumed upon contact with carbon. At that moment the rate is controlled by external diffusion. The activation energy becomes almost zero.

In each of the 3 domains the rate of gasification is proportional to the quantity of fuel and to the partial pressure p_G (all other factors remaining equal). It is also proportional to the specific surface A, except at high temperature when external diffusion is the controlling step. Smaller particles have a relatively larger outer surface and a shorter pore length ; they react faster in the domains of external and internal diffusion control, but not in the low temperature domain. A high gas velocity (u) stimulates external transfer ; it thus enhances the overall rate of reaction at the highest temperatures (Figure 2).

In practice the gas composition can often be approximated by the actual equilibrium composition. The deviation can be corrected for by assuming a somewhat lower gasifier temperature than the real one. Another correction method defines an apparent equilibrium parameter determined experimentally :

$$K_{app} = x.K_{theory}$$

In the updraft gasifier x essentially equals 1 for gasification by steam and carbon dioxide, but attains only 0.1 -0.6 for the methanation reaction, the exact figure depending on the reactivity of the coke [48].

Fluidized bed and suspended particle gasification are characterized by much shorter reaction times. The equilibrium approach is incomplete even for gasification with steam or carbon dioxide so that more elaborate computation methods are required[48].

Gasification of a single particle has also been the subject of a number of mathematical models, that take into account part of the heat and mass transfer aspects of the problem [49,50] Here also the lack of hard data and the mathematical complexity

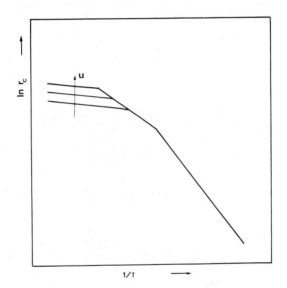

Figure 2.   Influence of gas velocity (u) on reaction rate

hinter the practical application of such models.

    In case chemical reaction is fast compared to internal diffusion or when the porosity of the solid is very low, conversion starts at the outer surface of the particle. As reaction proceeds further, a reacting boundary layer progressively moves inwards. The particle thus consists of an unreacted inner core surrounded by an outer ash layer (provided the particle contains ash). This concept is the basis of the well known shrinking core model ; it yields some practical conversion (X)-time (t) relations.[51]. Assuming a steady-state and isothermal conditions the shrinking core model gives for spherical particles [52] :

Controlling step	Conversion/time relation
external gas film diffusion	$\frac{t}{\tau} = X$
ash layer diffusion	$\frac{t}{\tau} = 1 - 3(1-X)^{2/3} + 2(1-X)$
chemical reaction	$\frac{t}{\tau} = 1 - (1-X)^{1/3}$

(τ = time required for complete conversion ; function of density of the particle, bulk gas concentration and mass transfer or chemical reaction parameters).

4.4.Chemical Reactor Engineering Aspects.

    When the behavior of a single particle in a pyrolysis or gasification reactor is studied, the particle is generally assumed to be at a given temperature and surrounded by a gas of well known composition.

    In the design of an actual reactor a more comprehensive model is needed to account for the spatial distribution of temperature, concentration or extent of conversion. The reactor model describes how gas and solid reactants flow and are contacted and how heat and mass are transferred. The simplest types of flow pattern conceivable are : perfect mixing and plug flow.

TABLE III

Type of reactor	Flow type gas	solid	deviations
vertical shaft	plug	plug	channelling, radial flow due to uneven porosity
rotating kiln	plug	plug	radial concentration patterns in gas, axial dispersion
fluidized bed	plug	well mixed	two phase gas flow, backmixing of gas, slugging, channelling,
entrained reactor	plug	plug	uneven mixing of gas-solids, axial dispersion

Several examples of reactor modelling have been described elsewhere [53, 54, 55].

The gasification of coal has been modelled several times [55, 56] for a vertical shaft reactor. The models assume plug flow for both the gaseous and solid reactants, and homogeneous or shrinking core behavior for the individual particles. The value of the various models is difficult to evaluate, because the detailed computational results can only be compared with experimental data at the outlet of the reactor. The models of coal gasification do not consider the effect of deviations from idealized flow. The phenomena taking place during refuse gasification and the effect of the deviations from an ideal packing have been discussed elsewhere [4, 57].

The modelling of a fluidized bed reactor is rather difficult because of the complicated flow of gas and solids in the reactor. In a well-fluidized reactor it is often possible to assume perfect mixing of the solids, since only large differences in density and size of the particles lead to segregation. But the gas flows through the reactor according to two different mechanisms. A first part creeps slowly upward through the dense phase and has a good contact with the bed material. The remainder flows through the bed under the form of fastly rising bubbles and has but a poor contact with the bed material.

It follows that the concentration of the gaseous reactant is different in the "dense" and in the "bubble" phase and also varies with height.

Hence there is no simple or direct relation between the observable gas outlet concentration and the kinetic parameters of the gas/solid reaction. The conversion also depends on the rate of mass transfer between the two phases, which varies in a complicated manner with bubble size and rise velocity, particle size distribution, diffusivity, bed geometry, the presence of bed inserts, etc. A number of empirical correlations are available to estimate these parameters. In most modelling studies however, adjustable parameters such as the bubble size are used to fit the experimental data.

The safest method of designing a full scale reactor is largely based on experiments at successively larger scales of capacity. The use of a comprehensive mathematical model as a guide in the design is still out of question because of the mathematical complexity of such a model, the lack of accurate physical and chemical data, and the necessity of mastering a number of technical and operating problems. On the other hand it is useful to try and understand the operating principles and the behavior of such a reactor and the main features of the controlling mechanism. Spectacular errors in the scaling-up of PTGL-reactors may thus be avoided.

5. The PTGL-Project of the University of Brussels (V.U.B.)

Since its foundation the Department of Chemical Engineering and
Industrial Chemistry of the V.U.B. acquired considerable experience in the field of high temperature processes, with studies on
steam-reforming of natural gas, pyrolysis of hydrocarbons and catalytic combustion of hydrocarbons. The Department conducted fundamental studies as well as contract work for industry, e.g. in
the domain of fluidized bed techniques, incinerator grate mechanisms and small waste-fed boilers. An assessment on current thermal disposal techniques was prepared on behalf of E.E.C.[5,57,58] .

In autumn 1976 a contract was signed with the Ministry of
Science Policy for a study on the pyrolysis of waste. At that
moment it seemed impossible to catch up with the work of excellent
companies such as Union Carbide, Monsanto or Occidental Petroleum
and develop a proprietary process. Furthermore the amount of financial funds provided and the extent of facilities available at
the University precluded such an approach.

So it was decided to concentrate on more fundamental aspects of pyrolysis and gasification and to limit large scale work
to collaboration projects with industry, such as the development
of small boilers for firing miscellaneous wastes, moving bed carbonization of wood slabs, fluidized bed gasification and pyrolysis
and moving bed gasifiers for various applications of the product
gas [59].

The fundamental research work was subvided into 3 parts
[60]: - thermogravimetric analysis (TGA) and differential thermal
analysis (DTA) of minute samples (5-50 mg) of the material to be
studied. Both techniques yield information on the rate of thermal
decomposition, the heat of reaction, the kinetic parameters of
this process (pseudo-order and activation energy) and the amount
of residue. The analysis of the evolving product has been monitored by means of gas chromatography. High temperature oxidation of
the residue allows to compare the reactivity of the carbonized residue.

- pyrolysis gas chromatography (Py.G.C.) of microsamples (10
-200 µg) rapidly yields information on the rate and products of
decomposition in an extremely broad range of reaction temperatures.
Precise measurement of the transient temperature profiles of the
sample holder and of the evolution volatile products as a function
of time allowed to obtain a much better insight of the physical
and chemical phenomena occuring during Py.G.C. This insight is
obviously required for the comparison with or the extrapolation to
real scale experiments.

- pyrolysis in a fluidized bed bench scale reactor. The test
unit is composed of a 15 cm I.D. fluidized bed reactor, a hopper
and a screw feeder mounted on top of the expansion section of the
bed, a tubular furnace preheater for the fluidizing agent, a quench
cooler, cyclone, tubular condensor and oil mist scrubber (Fig.3).

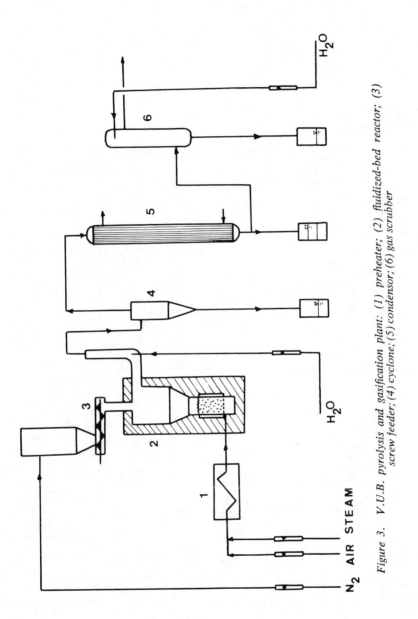

Figure 3. *V.U.B. pyrolysis and gasification plant: (1) preheater; (2) fluidized-bed reactor; (3) screw feeder; (4) cyclone; (5) condensor; (6) gas scrubber*

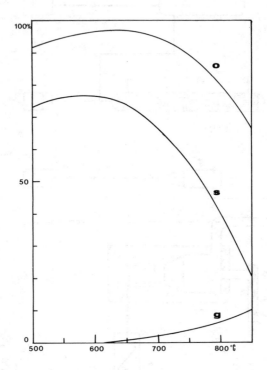

Figure 4.   Polystyrene pyrolysis: (O) oil; (S) styrene; (G) gas

Pyrolysis tests have been made on polyethylene and poly-styrene with steam as the fluidizing agent. Polyethylene yielded a gas and a wax. Their relative amounts could be adjusted by varying the operating temperature [61].

Polystyrene pyrolysis yielded an oil with a high content of styrene monomer (Fig.4). The reaction conditions were optimized to maximize the yield of styrene. A maximum of 76% (by weight) was obtained. The oil is suitable for reuse in the styrene manufacturing process.

Presently, tests are performed on various agricultural and wood wastes such as bark, soy hulls and wood shavings.

ACKNOWLEDGEMENTS

The research work described here was carried out as part of the R & D Programme "Economy of Wastes and Secondary Raw Materials", sponsored by the Prime Minister's Services -Science Policy.

LITERATURE CITED

1. Buekens,A; Schoeters,J 2nd World Recycling Conference, Manilla, Exhibitions for Industry Ltd, Oxted,Engl.,1979.
2. Martin,W; Weiand,H 1978 National Incinerator Conference, Chicago, Ill.,ASME, 1978, p.83.
3. Mutke,R."Conversion of Refuse to Energy",Montreux, 1975,p.105.
4. Buekens,A. Müll und Abfall, 1978, 10(12), 353.
5. Buekens, A. and Schoeters,J. "Assessment of Current Technology of thermal processes for Waste Disposal",E.E.C. Contract 282 - 76-9,ECIB, Nov. 1977.
6. Mélan, C. Müll und Abfall, 1978,10(12),363.
7. Roslund,J. in ref. 63,136.
8. Besch,G.M. in ref. 64,196.
9. Besch,G.M. Müll und Abfall,1978,10(12),384
10.Müller,H. in ref. 64,154.
11.Rymsa,K.H. in ref. 63,87.
12.Rymsa,K.H. Müll und Abfall, 1978,10(12), 377.
13.Heil, J. in ref. 64,112.
14.Segebrecht, J. in ref. 64,92.
15.Link, K. in ref. 64, 100.
16.Tabasaran,O. To be presented at this meeting.
17.Tabasaran,O; Besemer,G; Thomanetz,E. Müll und Abfall, 1977, 9 (10), 293.
18.Nowak,F. 1978, Natl. Waste Conf., Chicago,ASME, 1978, 29.
19.Lenz,S. Müll und Abfall, 1978, 10(2), 371.
20.Willerup,O.C.R.E.Conference, Montreux, 1975, 235.
21.Schmidt,R.Müll und Abfall, 1978, 10(12), 375.
22.Bracker,G.P. Metall, 1977, 31(5), 534.
23.Collin,G;Grigoleit,G; Bracker,G.P. Chem. Ing. Techn.; 1978, 50,(11) ,836.
24.Collin, G; Grigoleit,G; Michel,E. Chem. Ing. Techn., 1979, 51 (3), 220.

420 THERMAL CONVERSION OF SOLID WASTES AND BIOMASS

25. Bracker, G.P. Müll und Abfall, 1979, 11(4), 96.
26. Douglas,E; Webb,M; Daborn,G.R. Presented on "Treatment and Recycling of Solid Wastes", Manchester, Jan. 1974.
27. Wilson, H.T., Fletcher, R. Presented on "2nd World Recycling Conference", Manilla, Exhibitions for Industry Ltd, Oxted, England, 1979.
28. Anon. "Pyrolysis Review", Foster Wheeler Power products Ltd.
29. Kaminsky,W. to be presented at this meeting.
30. Sinn, H. Chem. Ing. Techn. , 1974, 46, 579.
31. Kaminsky, W, Menzel,J, Sinn,H. Conserv. and Recycling, 1977, 1 , 91.
32. Sinn, H. Kautschuk. + Gummi Kunststoffe,1979, 32 (1), 23.
33. Kox, W. Thesis, T.H. Eindhoven, 1975.
34. Speth, S. Chem. Ing. Techn, 1973, 45 (6), 256.
35. Besemer, G. "Disposal of Industrial Wastewater and Wastes", Symposium, Wroclaw, 1977, 282.
36. Anon.Metalloberfläche,1978, 32(8), 357.
37. Anon.Wasser, Luft und Betrieb, 1978, 22(8).
38. Groeneveld, M. to be presented at this meeting.
39. Zdonik,S.B., Green, E.J., Hallee, L.P. "Manufacturing Ethylene" Petr.Pub. Co. 1970.
40. Van Krevelen, D.W., Hoftijzer, P.J. "Properties of Polymers", Elsevier Scient. Pub.Co, Amsterdam, 1976.
41. Cameron, G.G. "The mechanisms of Pyrolysis, Oxidation and Burning of Organic Material", N.B.S. special pub. 357, June 1972, 62.
42. Madorsky,S.L. "Thermal Degradation of Organic Polymers", Interscience Publishers, 1964.
43. Alger,R. "The Mechanisms of Pyrolysis, Oxidation and Burning of Organic Material",N.B.S. Special pub. 357, June 1972,171.
44. Maa, P.S., Bailie, R.C. "Combustion Science and Technology", 1973, 7, 257.
45. Miyanami, K. Fan, L.S., Fan,L.T., Walawender, W.P. Can. Ind. Chem. Engineering, 1973, 55, 317
46. Kung, H.C. Combustion and Flame, 1972, 18, 185.
47. Walker, P.L., Rusinko, F. Austin, L.G. Advances in Catalysis, 1959, 11, 133.
48. Rammler, E."Technologie und Chemie der Braunkohleverwertung", 1962, 296.
49. Szekely,J. Evans, J, Sohn, H.Y. "Gas-Solid Reactions", Academic Press, N.Y., 1976.
50. Petersen, E. A.I. Ch. E. J., 1957, 3, 443.
51. Smith, J.M. "Chemical Engineering Kinetics", McGraw Hill, N.Y. 1970.
52. Levenspiel, O. "Chemical Reaction Engineering", J.Wiley & Sons, N.Y., 1972.
53. Yoshida, K. Kunii, D.,J. Chem. Eng. Japan,1974, 7,(1) 34.
54. Ubhayakar, S.K., Stickler, D.B., Gannon, R.E. Fuel, 1977, 56, 281.
55. Amundson, N.R., Arri, L.E. A.I. Ch. E. Journal,1978,24(1),87.

56. Klose, E, Toufar, W. Energietechnik, 1976, 26(12), 546.
57. Eggen,A.C., Kraatz, R. Winter Annual meeting, Power Division of ASME, N.Y., Nov. 1974, 17.
58. Buekens, A., Schoeters, J. "W. European Experience on Small Waste fed Incinerators with Heat Recovery", Preliminary Report prepared for S R I, 1978.
59. Buekens,A.G. Conservation and Recycling, 1977, 1, 247.
60. Buekens,A.G.; Masson, H. 2nd World Recycling Conf.Manilla 1979.
61. Buekens,A.;Mertens,J.;Schoeters,J.Steen,P. 1st World Recycling Conference, Basel, 1978.
62. Schoeters,J.;Buekens,A. MER/CRE Conference, Berlin, Oct.1979.
63. "New Technologies in Waste Disposal" (in German), Symposium at T.U. Berlin, Erich Schmidt Verlag, 1977.
64. "Thermal Treatment of Household Refuse" (in German), Symposium at T.U. Berlin, Jurgen Kleindienst, 1978.

RECEIVED November 30, 1979.

Pyrolysis of Plastic Waste and Scrap Tires Using a Fluidized-Bed Process

WALTER KAMINSKY and HANSJÖRG SINN

Universität Hamburg, Institut für Anorganische und Angewandte
Chemie, Martin-Luther-King-Platz 6, 2 Hamburg 13, West Germany

Increasing numbers of old tyres and waste plastics are creating problems regarding their disposal. Alternatively, there is a
continuing search for new sources of hydrocarbons as a result of
increasing energy and raw material needs. Old tyres and waste
plastic can be converted by pyrolysis into almost residue free
organic raw materials, to meet this need (1, 2, 3, 4, 5). About
340.000 t.a. of scrap tyres arise out of Western Germany's total
production of 6.4 million tons p.a. of plastic and rubber which
are dealt with in the following ways (6).

Table I
                    Removal or Reuse of Old Tyres.

Methods of Removal or Reuse	Scrap Tyres t/a.	%
Regeneration of Rubber	20.000	6
Recycling	30.000	9
Pyrolysis	–	0
Burning	7.000	2
Remoulding	90.000	27
Controlled Dumping	85.000	25
Uncontrolled Dumping (whereabouts unknown)	100.000	30
Others (e.g. boat-fenders) circa	6.000	1
Total	338.000	100

About 180.000 t/a. of scrap tyres remain to be processed, for
instance by pyrolysis. In comparison the amount of plastic waste
is about $1.4 \cdot 10^6$ t/a which is significantly higher. The main part
(about $1.1 \cdot 10^6$ t/a) is mixed up with household waste (7), and
therefore only a few hundred thousand tons are easily available from
the producing and processing industries.
Two different procedures exist for dealing with the latter

0-8412-0565-5/80/47-130-423$05.00/0

waste. Through Re-use the macro-molecular structure stays intact in contrast to other methods (Pyrolysis, Hydrolysis, Burning) which change the chemical structure to produce raw materials or energy.

Fifty percent of the total energy needed to produce plastic materials is used in the polymerization process. Thus at first sight, it seems sensible to re-use plastic and rubber waste materials (8). But Re-use processes have a negative influence on the macro-molecular structure which is mainly a chain depolymerisation effect (9). Therefore Re-use by itself is not a satisfactory process. In addition, a large proportion of the plastic waste is polluted and mixed with other types of waste so that Re-use is impossible.

Therefore, the way is open for pyrolysis processes. These take place either in the total absence of oxygen or with a lack of a stochiometrically needed amount of oxygen. These processes aim to maintain the valuable C-H-bonding of the macro-molecules to obtain smaller molecules, which can be used as energy recources or chemical raw materials. These processes take place in rotary kilns, screw-kilns, melting vessels, or fluidized bed reactors. The problem lies in transferring the heat to solid materials in such a way that the conditions of pyrolysis are evenly distributed to produce the desired product spectrum. Using rotary or other indirectly heated kilns with their varying temperature zone results in an uneven decomposition of the material. However by using a quartz sand or carbon black fluidized bed with its even heat transfer properties, uniform conditions for decomposition can be reached (10, 11).

At the University of Hamburg we have been developing a fluidized bed process for the pyrolysis of plastic waste and scrap tyres since 1970. We used three stages of up-scaling - 0.1 kg/h; 10 kg/h; 100 kg/h.

Laboratory scale experiments

Figure 1 shows a flow diagram of the plant for laboratory experiments with a continuous throughput of 0.1 kg/h (12):

A screw conveyor feeds through a cooled downpipe the electrically heated quartz-reactor which has a diameter of 5 cm and a fluidizing zone of up to 8 cm. The fluidizing gas is about 500 l/h of either nitrogen or circulated cracker gas. A cyclone separates solids from the hot pyrolysis gas stream; an electrostatic precipitator and a system of intensive coolers and hydrocyclones condense the liquid portions of the cracked products. The non-condensable pyrolysis-gases are measured.

The first column in Table II shows the composition of the pyrolysis products from an input of polyethylene. Beside the gaseous products, i.e. mainly methane, ethane, ethylene and propene, the liquid products consist of up to 95 % benzene, toluene, styrene and naphthalene. Only a small amount (1.0 wt-%) of carbon black is produced.

The product spectrum can be influenced to a certain degree by changes in temperature. Figure 2 shows the pyrolysis of granulated

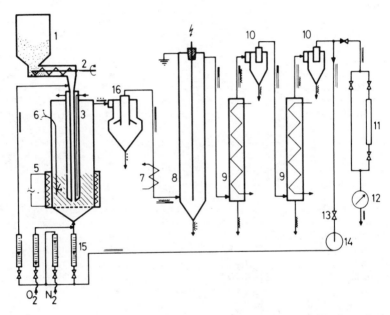

Figure 1. Flow diagram of the laboratory test plant of fluidized-bed pyrolysis.

(1) plastic feed hopper; (2) screw conveyor; (3) downpipe with cooling jacket; (4) fluidized-bed reactor; (5) heater; (6) cyclone; (7) cooler; (8) electrostatic precipitator; (9) intensive cooler; (10) cyclone; (11) gas sampler; (12) gas meter; (13) throttle; (14) compressor; (15) rotameter.

Table II.

Products of fluidized bed pyrolysis of
plastic waste and scrap tyres (wt-%)

Reactor	LWS	TWS-1	TWS-1	TWS-1	TWS-2
Feed material	PE	PE	used Syringes	tyre pieces	whole tyres
Temperatur °C	740	780	720	750	700
Hydrogen	0,5	1,17	0,49	1,30	0,42
Methane	16,1	20,27	19,09	15,13	6,06
Ethane	5,3	4,33	6,64	2,95	2,34
Ethylene	25,4	16,89	15,44	3,99	1,65
Propane	+	0,80	0,12	0,29	0,43
Propene	9,3	5,35	9,93	2,50	1,53
Butene	0,5	0,08	3,03	1,31	1,41
Butadiene	2,8	1,28	1,38	0,92	0,25
Isoprene	+	0,09	0,31	0,34	0,35
Cyclopentadiene	1,0	2,63	2,08	0,39	0,25
Other aliphatic compounds	13,3	0,69	3,13	0,36	1,07
Benzene	12,2	24,75	13,62	4,75	2,42
Toluene	3,6	5,94	4,20	3,62	2,65
Xylene	1,1	+	+	+	+
Styrene	1,1	1,46	0,45	0,17	0,35
Indan, Indene	0,3	1,27	0,46	0,31	0,48
Naphthalene	0,7	3,73	2,48	0,85	0,42
Methylnaphthalene	0,15	0,84	0,92	0,83	0,67
Diphenyl	0,02	0,34	0,33	0,49	0,39
Fluorene	0,01	0,29	0,15	0,16	+
Phenanthrene	0,02	0,59	0,47	0,29	0,19
Pyrene	+	0,22	+	0,21	0,06
Other aromatic compounds	5,1	5,40	8,24	8,50	13,67
Carbon monoxide				3,80	1,48
Carbon dioxide				1,95	1,74
Water				0,10	5,11
Hydrogen Sulfide				0,23	0,02
Thiophene				0,15	0,25
Carbon soot, fillers	0,9	1,50	5,80	40,59	40
Steel cord				1,62	11,30
Balance	99,4	99,91	98,76	98,10	96,96

LWS: laboratory scaled reactor, TWS-1: pilot plant, TWS-2: pilot
plant for whole tyres, PE: polyethylene, +: trace detection

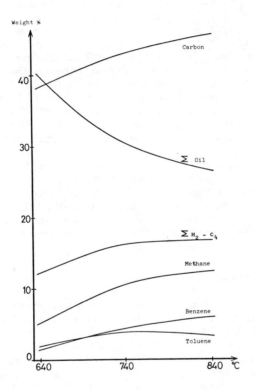

Figure 2.   Product composition from fluidized-bed pyrolysis of granulated rubber tires, plotted as weight percent vs. temperature

scrap tyres. By increasing the temperature, the yield of carbon
black, hydrogen, methane and benzene rises, and the yield of oil
falls. Because a high percentage (about 35 wt-%) of the original
feed material is carbon black, it is evident that it will form a
large part of the pyrolysis products (over 38 wt-%). The liquid
fraction has also a high aromatic content.

The bound chloride in plastics containing PVC is recovered as
hydrogen chloride gas from pyrolysis. Considerable amounts of carbon soot are produced in this reaction. By adding superheated
steam and hydrogen, the amount of carbon black produced can be reduced from 8.8 to 2.1 wt-% (13).

For the pyrolysis of plastic waste with a high PVC content,
it is probably better to first dehydrohalogenize it in a reactor
with added sand. Tests have shown that hydrogen chloride gas can
be produced from PVC at temperatures of between 350 and 400°C,
which is up to 99 % pure and, after adsorption of the residual hydrocarbons, may be used to make very pure hydrochloric acid. After
treating PVC for 20 minutes at a temperature of 350°C, more than
90 % of the bound chloride has split off. This time is reduced to
less than 10 minutes if the temperature is increased to 400°C
(13).

If the PVC content of the plastic waste is low, the hydrogen
chloride produced can be absorbed either directly in the fluidized
bed to which calcium oxide or magnesium oxide is added, or in a
separately connected fluidized bed. This method has proved satisfactory, at least for the absorption of hydrogen fluoride in the
pyrolysis of PTFE-containing plastic wastes and for hydrogen sulfide in the pyrolysis of rubber.

Pilot plant experiments

These results led to the design of a pilot plant with a
throughput of 10 kg/h whose overall layout is shown in Figure 3.
The project was financed by the Association of the Plastic Producing Industry (Verband Kunststofferzeugende Industrie, VKE) and
the Federal Ministry of Research and Technology (BMFT).

Plastic material is pyrolyzed in a fluidized bed reactor utilizing quartz sand heated to about 800°C. The fluidizing medium is
preheated pyrolysis gas.

To observe the principle of indirect heating and to avoid any
pollution of the products, we used fire tubes inside the sand bed,
in which either propane (for the starting period) or pyrolysis gas
can be burned.

The feed enters the sand bed either at the top of the reactor
through an air lock (ZR 1) or through a water-cooled screw feeder
(ZR 2 with S 1) at the side. A water-cooled screw conveyer (S 2)
is installed as an outlet for the quartz sand. At the top of the
reactor is a safety bursting disc. The lower half of the reactor
is bricked with fire-proof mortar. The reactor is built up with
three parts, each of a diameter of 0.5 m and different lengths of

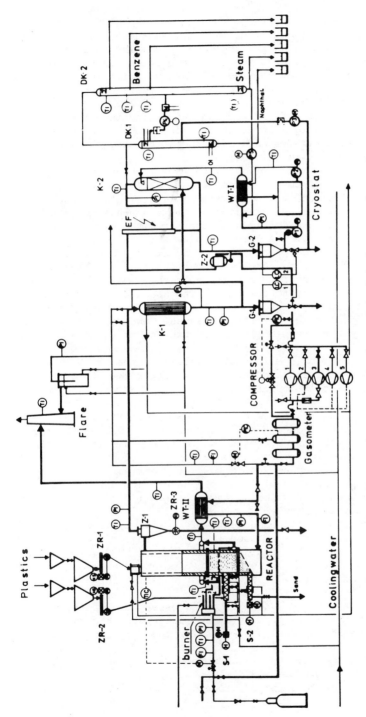

Figure 3.   Flow diagram of the pilot plant for the pyrolysis of plastic waste and scrap tires.

(PI) Pressure indication device; (PIC) automatic pressure control; (TI) temperature indicating device; (TIC) automatic temperature control; (FI) flow-meter; (LC) level indicator; (R) reactor; (M) motor; (WT) heat exchanger; (ZR) bucket wheel lock; (KM) membrane compressor; (KR) cryostat; (K) cooler; (S) screw feeder; (G) container; (Z) cyclone; (P) pump; (TG) dip pipe; (EF) electric separator; (DK) packed distillation column

1.0 m, 0.5 m and 0.3 m. The upper part carries the feeding tunnel
for coarse material (e.g. rubber pieces) and sand, the safety disc
and the connection flange for the gas outlet. The feed screw conveyor and the four P-shaped burners - two of each are respectively
above and below the conveyor level - are installed in the middle
part. The power output of the four burners is regulated by varying
the pressure of the burning gas.

The bottom of the fluidized bed has an incline of about 15°
and carries the bent gas inlet tubes. These tubes are movable ver-
tically so that their distance from the bottom of the fluidized
bed can be varied. With this arrangement there is a variable settling zone for small metal pieces such as will be deposited on pyrolytic stripping of copper-cored wiring. A flanged inspection
cover is fitted to the bottom of the gas distribution zone.

A picture of this pyrolysis reactor is shown in Figure 4. The
pyrolysis gas produced is cleaned together with the fluidizing gas
in a cyclone (Z 1); if the feed contains scrap tyre material, char
and zinc oxide will be precipitated. The cleaned gas is led to the
condenser (K 1) which is actually a vertical tubular heat exchanger. In the case of condenser failure, a three way cock allows the
gas to be fed directly to the flare. The gas leaves the condenser
(K 1) at 20 to 50°C, or at 100°C if the cooling capacity is reduced to avoid clogging by paraffin waxes. Condensed products are
collected in a receiver (G 1). In a counter-current packed washer
(K 2) the gas is then cooled to -20°C. The condensed products are
dissolved in a scrubber oil pump. For cooling the scrubber oil to
-20°C a counter-current tubular heat exchanger (WT 1) is used. The
liquid products collected are processed together with the scrubber
oil first in a 0.1 m diam packed glass rectification column, in
which high boiling (b.p. above 150°C) and low boiling products are
separated. Gaseous products are recirculated to become the pyrolysis gas. The low boiling fraction is refined again in a second
8 cm diameter packed column to give four fractions: a low boiling,
a high boiling, a benzene, and a toluene fraction. The high boiling fraction (i.e. xylene) is pumped back continuously into the
washing tower, any excess being taken off. All products are drawn
off and separately analyzed.

The non-condensable components of the pyrolysis gas are compressed to 2 to 3 bar and stored in a gasometer. Fluidizing gas as
well as fuel gas for the radiating fire tubes is drawn off as required. The fluidizing gas is preheated to 400°C in a countercurrent heat exchanger (WT 2) by the exhaust gas of the four fire
tubes.

Because of safety reasons it is necessary to avoid any air or
oxygen within the whole product cycle. Therefore in case of a drop
below atmospheric pressure, a by-pass for the compressor allows
product gas to flow back from the gasometer into the system.

The plant is equipped with numerous data-collecting instruments. Because we have deliberately separated the hot and cold
parts of the plant, the latter, together with the processing of

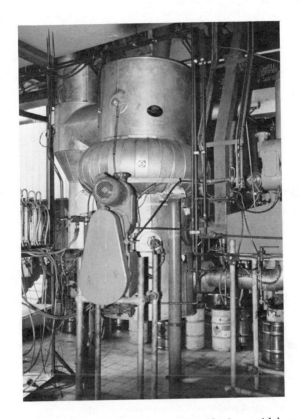

Figure 4.   Pyrolysis reactor with a feed screw for plastic, a withdrawal screw for solid materials, and a cyclone

the pyrolysis oil, are situated in an explosion proof room. Surveillance is carried out by continuous analysis of the room air as well as by explosion limit controls. The pyrolysis gas is analyzed automatically by a gas chromatograph.

All data collected are graphed to achieve energy- and mass-balances.

Some basic components are continuously monitored by infrared spectroscopy, i.e. ethylene in the pyrolysis gas, sulphur dioxide and oxygen in the exhaust gas.

The described pilot plant has been operating for more than 700 running hours. Within these runs, it has been proved that the process is self-sufficient in its energy needs (Table III).

Table III

Energy balance for the pyrolysis of 19,1 kg/h Polyethylene/Polypropylene (ratio 1:1) at 710°C . Pyrolysis gas (produced): 8,0 m^3/h

Heat supplied (burning of 4,6 m^3/h Gas) $209,8 \cdot 10^6$ J

Dissipated heat

Burner gases	$77,0 \cdot 10^6$ J
Pyrolysis gases	$2,1 \cdot 10^6$ J
Crack energy	$47,3 \cdot 10^6$ J
Cooler (K 1)	$46,9 \cdot 10^6$ J
Water cooling	$14,7 \cdot 10^6$ J
Insulation losses (remainder)	$21,8 \cdot 10^6$ J

Almost half the amount of pyrolysis gas produced was burnt in the exhaust gas flare; the other part was sufficient to support the energy demands of the process, being burnt in the radiation fire tubes. Waste heat from the reactor heating (cooler) was sufficient to heat the distillation column.

Analysis of the product spectrum shows that between 40 and 60 wt-% of the feed is gained as worthwhile liquid products (Table II).

Table IV shows the particular fractions which are obtained from the process.

In the pyrolysis of polyethylene at 810°C, the low boiling fraction contains, apart from benzene, more than 10 wt-% of cyclopentadiene. If benzene-rich first runnings and last runnings (toluene-rich) are removed, one can obtain a fraction containing about 98 wt-% of benzene. The main constituents of the high boiling fractions are naphthalene, methylnaphthalene and indene, as well as highly condensed aromatics.

As a general example for contaminated plastic scrap we pyrolyzed shreddered syringe materials - containing polyethylene, polypropylene and rubber (Table II). The pyrolysis of these syringes is without doubt hygienic and gives nearly the same products as

Table IV

Composition of various product fractions
obtained through the pyrolysis of polyethylene at 810°C

Substance	gas vol %	low boiling wt-%	benzene- rich wt-%	toluene- rich wt-%	high boiling wt-%
Hydrogen	20,7				
Methane	40,8				
Ethane	7,1				
Ethylene	22,2				
Propene	5,9				
Cyclopentadiene	0,2	10,6	0,1		
Benzene	1,0	76,9	97,9	47,6	
Toluene	0,2	2,0	0,9	51,9	0,1
Styrene				0,2	2,3
Naphthalene					24,5
Others	1,9	10,5	1,1	0,3	73,1
Boiling range (°C)		50-70	70-80	80-100	above 150

from an input of polyethylene. The products show only traces of
water, carbon dioxide and monoxide, and acetonitrile; the percentage of carbon soot can be up to 5.8 wt-% because of the rubber
content.

These results prove that pyrolysis is an adequate process to
reprocess pure plastic materials as well as contaminated plastic
fractions of household waste.

Prevailing conditions in West Germany mean that aromatics are
to be favoured as products.

We have now optimized the process to achieve the maximum
yield of benzene possible, by varying the temperature of the fluid
bed, the feed and the flow of the fluidizing gas. Statistical
planning of test runs led to the following effects and regression
equations. The yields of benzene and ethylene can be calculated
from the following equation, using percentage weight in relation
to the feed.

$$\text{wt-\% benzene} = 21.095 + 2.323\ t - 1.381\ d + 0.731\ w - 0.680\ td + 0.581\ dw + 0.104\ tw - 0.377\ t^2 - 0.910\ d^2 - 0.457\ w^2$$

$$\text{wt-\% ethylene} = 18.733 - 1.195\ t - 0.132\ d - 1.785\ w - 2.436\ td + 2.643\ dw - 0.341\ tw - 1.618\ t^2 + 3.240\ d^2 + 1.263\ w^2$$

in which

$$t = \frac{temperature - 750^\circ C}{30^\circ C}$$

$$d = \frac{throughput - 14,18 \ kg/h}{5,03 \ kg/h}$$

$$w = \frac{fluidizing \ gas/h - 32,85 \ m^3/h}{365 \ m^3/h}$$

Figure 5 shows the graphing of the flow of fluidizing gas against the feed; it will be noticed that a maximum in the yield of benzene almost corresponds to a minimum in the yield of ethylene. From these results and from experiments with single particles, it can be concluded that aromatics are produced from olefins such as ethylene, propene and butadiene. The slight difference between the above maximum and minimum yields, and the different gradients of the yield curves indicate the existence of successive and parallel reactions: the pyrolysis reaction has a two-stage mechanism. In the first stage, macromolecules are decomposed into small, mostly olefinic compounds. During the second stage, aromatics are produced from the inter-reaction of these compounds, and hydrogen and methane are given off. By raising the temperature, the flow of fluidizing gas and the residence time, the yield of aromatics increases. Larger amounts of benzene were produced by the pilot plant than by the laboratory scale plant, as a result of a longer residence time in the reactor.

Prototype for Whole-Tyre Pyrolysis

Fluidized sand beds are surprisingly insensitive to the unit size of the feed material. Pieces of scrap tires up to a weight of 2.7 kg/each were fed and quantitatively pyrolyzed. These results proved the feasibility of pyrolytic processing of scrap tyres without prior size reduction.

Most pyrolysis processes use crushed feed down to between 200 and 20 mm in size which causes considerable expense (10, 14, 15).

The following companies have carried out pyrolysis experiments in indirectly heated rotary kilns using such feed material:

Kobe Steel (16) in Japan
Goodyear and TOSCO (17) in the U.S.A.
GMU (18) and Herko-Kiener (19) in West Germany.

In cooperation with the Hamburg company C.R.Eckelmann and promoted by the Federal Ministry of Research and Technology a pilot plant fluid bed reactor for a 1.5 to 2.5 t/d throughput of scrap tyres

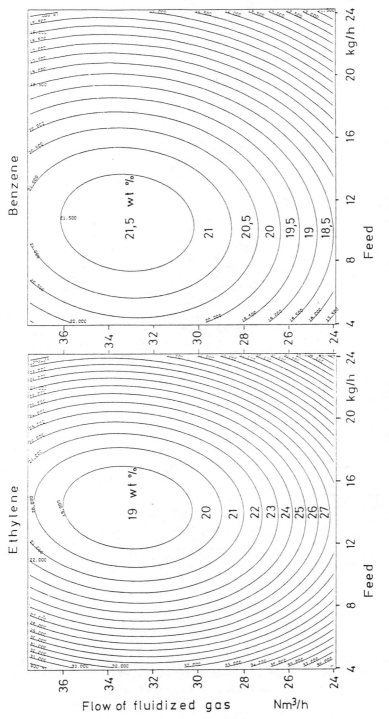

Figure 5.   Optimization process to maximize benzene production; the flow of fluidized gas vs. feed showing the yield of ethylene and benzene

was built at the Hamburg institute (20, 21). The fluid bed has internal dimensions of 900 x 900 M.M. so that individual unshredded
tyres of up to 20 kg in weight can be processed. The high cost of
shredding can therefore be dispensed with (Fig. 6).

The pyrolysis zone, a fluidized sand bed or carbon black bed
is indirectly heated up to between 650 to 850°C by seven radiating
fire-tubes, arranged in 2 layers. One part of the product gas is
used to fluidize the bed, the other being burnt to heat the process.

The whole tyres roll through a gastight lock into the reactor
(Fig. 7). Observable events can be described as follows. The tyre,
landing on the fluid bed gradually sinks into the sand. The material heats up and softens, its surface becoming covered with hot
sand grains.

Through the shearing forces of the fluid bed - increased because of the reduction of the free cross section of the bed - an
"exchange" of sand grains takes place at the surface of the
softened material; the abrasion of small particles and their decomposition begins.

A methane and ethylene rich gas as well as a condensable
liquid part with a high percentage of aromatics are produced (see
Table II). The main part of the filler materials - carbon soot and
zincoxide - is blown out of the fluid bed and can be separated in
a cyclone. The carbon soot is dry and trickles out of the cyclone.

After 2 or 3 minutes, the tyre is completely pyrolyzed. What
remains in the sand bed is a mass of twisted steel wires, which
are removed by a tiltable grate extended into the fluid bed.

The pyrolysis products together with the fluidizing gas (i.e.
the non-condensable pyrolysis gases) leave the reactor via a
cyclone, where dry carbon soot and filler materials are precipitated.

The hot gases are cooled to room temperature by an oil scrubber and then refined in the washer and rectification unit used in
the smaller pilot plant described above.

The pilot plant has been running since late 1978. It was
proved in a test run with a feed of more than 150 car tyres, that
the process is self-sufficient in its energy needs, producing at a
temperature of 720°C, even excess pyrolysis gas to supply other
heat necessary. Waste heat from the reactor heating is sufficient
to heat the distillation column.

The balance of the products is as follows (compare with
Table II):

        22 wt-% gas
        27 wt-% liquids
        39 wt-% carbon soot
        12 wt-% steel cord

The sulphur content of the aromatic fraction is less than
0.4 wt-% and that of the gas less than 0.1 wt-%. Carbon black containing zinc oxide and sulphides can be separated into re-usable

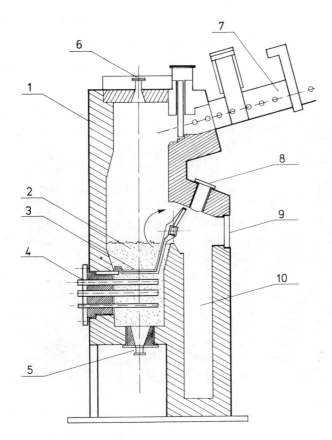

Figure 6.   Flow diagram of the prototype reactor for whole-tire pyrolysis.

(1) steel wall with fireproof bricking; (2) fluidized bed; (3) tiltable grate; (4) radiation fire tubes;
(5) nozzles to remove sand and metal; (6, 8, and 9) flange for observation and repairs; (7)
gastide lock; (10) shaft for steel cord.

Figure 7.   Placement of whole tires into the lock of the prototype reactor

H.A.T.-quality and waste carbon black by an air classifier. The waste carbon black is an interesting source of zinc (4-5 wt-%).

If the price we can obtain for pyrolysis oil stays above 580 DM/t (this corresponds to a benzene price level of 1100 DM/t) and that of carbon black over 300 DM/t, then operating costs and capital costs of a plant with a capacity of 7500 t/a will show a profit of 7 % p.a. At present, the price of benzene amounts to 900 - 1000 DM/t, and prices of up to 600 DM/t for carbon black are possible.

As a result of our work, the firm C.R. Eckelmann are planning to build a commercial plant with a capacity of 6.000 to 10.000 t/a of unshredded scrap tyres and hydrocarbon-containing waste.

In addition, we have succeeded recently in treating aliphatic waste oil and the liquid extracts of oil-sands to produce aromatic compounds.

In conclusion,it may be possible to gain aromatic liquids from oil-sands in a one-stage process, using the sand in the feed as the medium for the fluid bed.

Literature Cited

1. Sanner, W.S. Bureau of Mines TN 23. U7 No. 742862206173
2. Kaminsky, W.; Menzel, J.; Sinn, H. Int. J. Conservation & Recycling 1976, 1, 91
3. Tsutsumi, S. CRE Conference Paper, Montreux 1976, 567
4. Beckmann, J.A.; Crane, G.; Kay, E.L.; Laman, J.R. Rubber Chem. Technol., 1974, 47, 597
5. Spendlove, M.J. Umschau, 1973, 73, 364
6. Umweltbundesamt der BRD, Materialien zum Abfallwirtschaftsprogramm '75, Materialien 2/76, Berlin 1976
7. Glenz, W. Kunststoffe, 1977, 67, 165
8. Data received from Stanford Research Institute, April 1973
9. Ranby, B. Kunststoffe German Plastics, 1978, 68, 10
10. Sinn, H.; Kaminsky, W.; Janning, J. Angew. Chem. Ed. Engl., 1976, 15, 660
11. Kaminsky, W.; Menzel, J.; Sinn, H.; Plastic and Rubber, Processing June 1976, 69
12. Menzel, J. Chem.-Ing.-Techn., 1974, 46, 607
13. Kaminsky, W.; Sinn, H. Kunststoffe German Plastics, 1978, 68, 14
14. Schnecko, H.W. Chem.-Ing.-Tech. 1976, 48, 443
15. Luers, W. Gummi-Asbest-Kunstst., 1975, 28, 770
16. Takamura, A.; Inone, K.; Sakai, T. CRE Conference Paper, Montreux 1976, 532
17. Ricci, L.J. Chem. Eng. (Aug. 1976) 52
18. Collin, G.; Grigoleit, G.; Bracker, G.-P. Chem.-Ing.-Tech., 1978, 50, 836
19. Tabasaran, O.; Besemer, G.; Thomanetz, E. Müll Abfall, 1977, 9, 293
20. Kaminsky, W.; Sinn, H.; Janning, J. European Rubber J., 1979 161, 15
21. Kaminsky, W.; Sinn, H.; Janning, J. Chem.-Ing.-Tech., 1979, 51, 419

RECEIVED November 16, 1979.

Two–Stage Solid Waste Pyrolysis with Drum Reactor and Gas Converter

O. TABASARAN

University of Stuttgart, Bandtaele 1, 7000 Stuttgart - 80, West Germany

1. Introduction

Results of experiments conducted so far indicate that pyrolysis offers a number of basic advantages when compared with thermooxidative waste treatment. Examples of said advantages are:

- Relatively good adaptability towards fluctuations in quantity and quality of input material.

- Possibility of treating different wastes that are normally classified as being problematic.

- Simpler and more efficient gas purification.

- Low fresh water requirements.

- Small amounts of waste water.

- High energy yield.

- Principle possibility of storing energy.

- Relatively quick starting-up and shutting-down operations.

- Simple separation of glass, inert material and metals from residues.

- Direct product utilization in gas engines.

- Economical operation, also in connection with decentralized systems.

The author knows of several pilot plants in Europe with technical dimensions, e. g. the waste degasing plant Destrugas in Kalundborg, Denmark, with a throughput of 5 t/24 h; the Kiener System pyrolysis plant in Goldshöfe/Aalen, West Germany, with a capacity of approximately 6 t/24 h; a gas generator for 50 t/24 h for used tyre treatment in Gislaved, Sweden.

Further systems are being developed in West Germany by Kraus-Maffei, GMU and Saarberg-Fernwärme. The first high-temperature pyrolysis plant with liquid slag removal, Andco-Torrax-System (capacity: 120 t/24 h), has already been put to work in Luxembourg. It has been followed by a similar plant working under the same principle in Frankfurt, West Germany.

The Destrugas System pilot plant has been thoroughly examined by the author on order of the Minister of the Interior of the Federal Republic of Germany, whereas the Kiener System experimental plant was subject to examination by order of the Minister für Wirtschaft, Mittelstand und Verkehr and the Minister für Ernährung, Landwirtschaft und Umwelt, both in the State of Baden-Württemberg, furthermore by order of the Umweltbundesamt (German Environmental Protection Agency). In particular this investigation was to take place with a view to data that ally to environmental considerations. The following is a presentation of the problem of harmful substances in pyrolysis gases and some of the results of the work done at the plant mentioned last.

2. Harmful Substances in Pyrolysis Gases

The types of waste primarily considered for pyrolysis treatment, such as municipal refuse, plastics or used tyres contain a number of chemical substances which react to new products during degasing as well as during combustion, e. g. sulphur, chlorine and nitrogen compounds. Such compounds may pollute the atmosphere and must therefore be removed in their gaseous phases. S u l p h u r   c o m p o u n d s   are present primarily in the form of hydrogen sulphides which are split off at relatively low temperatures of about 250° C. By thermal dissociation of the hydrogen sulphide on hot or ovenheated surfaces elemental sulphur is produced. It may react with carbon to form carbon disulphide and with carbon monoxide to form carbon oxysulphide (carbonyl sulphide). Organic sulphur compounds such as carbon disulphide, carbon oxysulphide, hydrogen thiocyanate, thiophenes and ethyl hydrosulphide are usually formed in small quantities only. The sometimes unpleasant odour of the oil derived from carbonization must be attributed primarily to the presence of sulphur compounds, traces of which are noxious.

N i t r o g e n   c o m p o u n d s   are present partly in the form of ammonia and hydrogen cyanide. The intensity of ammonia formation is high if the pyrolysis (carbonization) temperature is low and if the residence time of the carbonization gas at high temperature levels is short. Here, the equilibrium

$$2 \ NH_3 \ \rightleftharpoons \ N_2 + 3 \ H_2$$

must be considered. It will be shifted to the right at high temperatures.

The higher the temperature the more hydrogen cyanide will be present in the gas, and the ammonia will react with the carbon in accordance with the equation:

$$C + NH_3 \ \rightleftharpoons \ HCN + H_2$$

As a rule approx. 4 % of the total ammonia will be converted into ferrocyanide at a temperature of about 800°C. At a temperature of 1000°C the portion may amount to 24 %.

The formation of   n i t r o g e n o u s   g a s e s is not to be expected to an extent worth mentioning during pyrolysis alone. During hybrid processes or as a result of partial or complete combustion these gases are formed compulsorily:

$$N_2 \quad + O_2 \ \rightleftharpoons \ 2 \ NO \ (T > 1000°C)$$
$$4 \ NH_3 + 5 \ O_2 \rightleftharpoons \ 4 \ NO + 6 \ H_2O$$
$$(2 \ NO \ + O_2 \ \rightleftharpoons \ 2 \ NO_2)$$

The pyrolysis of material containing   c h l o r i n e   such as PVC produces primarily hydrochloric acid which in the presence of ammonia is transformed into ammonium chloride as the gases cool down.

In contrast to oxidative thermal treatment low-temperature pyrolysis shows that sodium chloride does not react to yield hydrochloric acid, and therefore the formation of volatile heavy metal chlorides is negligibly small. Chlorine salts (chlorides) as well as heavy metals leave the system mainly with the solid residues.

H y d r o f l u o r i c   a c i d   is found in small quantities in pyrolysis gases. It is likely that the contents are similar to those obtained from burning.

Heavy metal vapours or heavy metal
dusts are produced by the high-temperature pyrolysis
process, whereas they remain in the solid residues
-save for mercury- in connection with the low-
temperature process.

Carbonization gas usually contains considerable
quantities of long-chain hydrocarbons or higher-con-
densed aromatic and phenolic hydrocarbons which condense during cooling to oily, tar-resembling substances or which contaminate the gas scrubbing water
considerably. Therefore a number of low-temperature
pyrolysis processes have an integrated high-temperature stage through which the carbonization gas circulates to accomplish the cracking of such substances
to lower-molecular compounds. The function and design
of the cracking zones determine to what extent remains
of the tarry substances reach the scrubbing water of
the on-line gas purification unit and thereby charge
it with substances that are at times difficult to
decompose biologically in sewage treatment plants.
Dusts contained in pyrolysis gas may cause difficulties in pipes and equipment because of their tendency
to settle there. It is, however, no problem to remove
them from the exhaust gas.

3.   Kiener Waste Pyrolysis System

As far as the Kiener pyrolysis system is concerned
there is experience available from pilot plants. In
February 1979 the formal official planning procedure was concluded with the objective of setting up
a refuse treatment plant with a capacity of approx.
140 t/d in the city of Aalen, West Germany.

Figure 1 is a flow diagram of the plant.

The system offers in principle two modes of
operation. One mode emphasizes the recovery of raw
materials whereas the other emphasizes the production
of energy. The type of operation chosen depends on the
input material and on the economic expectations involved.

3.1 Operation with Emphasis on Material Recovery

Figure 2 shows this type of operation. The input
material consists primarily of substances with relatively high energy contents, such as used tyres, plastics,
rugs, wood and special organic industrial wastes.

The continuously fed pyrolysis reactor consists
of a rotating steel drum which is heated indirectly
with the exhaust gases of a gas engine or the firing
gases of a gas or oil burner by passing these hot
gases through a system of specially designed, internally mounted heat exchanger tubes. Gastight air-locks

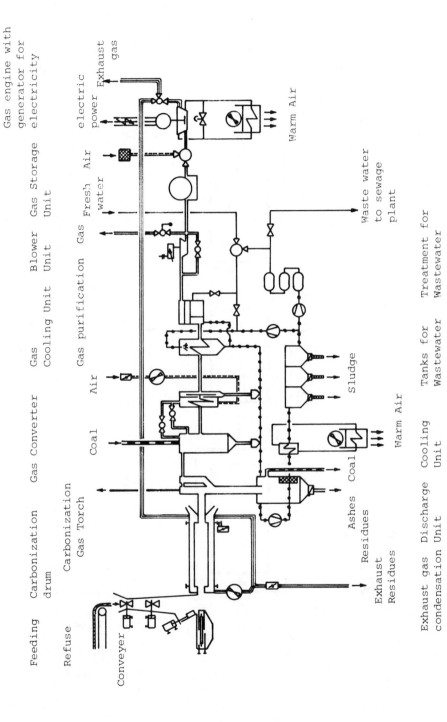

Figure 1.   General outline of the Kiener System two-stage refuse plant with drum reactor and gas converter

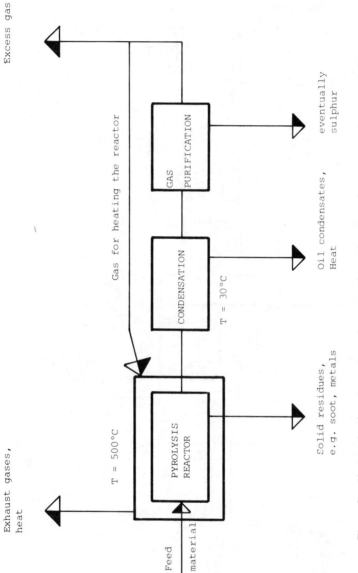

Figure 2. Variant 1—degasing and condensing with emphasis on the recovery of raw materials

provided at both ends of the drum make continuous
feeding as well as continuous slag and soot removal
possible. The tubing not only offers optimal heat
transfer but also rolls the material around and advances it subject to uniform drum rotation. Operation
according to variant "1)" comprises a condenser succeed-
ing the drum. It is cooled with air and is thus independent of water.

Type, quantity and quality of the reaction products naturally depend upon the properties of the input material. The following are examples of products
obtained by the pyrolysis of used tyres (Figures 3 and 4).

Gas:
The gas which is cleaned by scrubbing can be used
advantageously for the aspired, self-sufficient
operation of the plant in respect to energy. Excess
quantities of the gas may be flared off. The composition
of the high-calorific gas obtained from the pyrolysis
of old tyres is shown in table 1.

As can be seen the pyrolysis gas contains mainly
hydrogen, methane and other hydrocarbons. In our case
the yield was 167 kg for each ton of used tyres. The
average calorific value was determined to be approx.
34.500 kJ/m^3. The odour was specific and intense.

Oils:
The complex-compounded oily products obtained through
condensing can be used principally as engine fuel or
as fuel for oil burners.They may also be used as
additives or basic materials for rubber production
or in the chemical industry.

On the average 1 ton of used tyres yields 468 kg
of condensate having a typical and intense odour and
a calorific value of 40.000 kJ/kg. The flash point
is near 8°C. For further details please refer to
table 2.

Solid Residues:
These consist primarily of material high in carbon
content, which is practically odourless. Because of
their high calorific values of more than 30.000 kJ/kg
it is possible to use them as fuel or as raw material
for gas generators. If properly treated, they can
also be used as filler materials by the rubber industry
or as fillers or dyes by the chemical industry (see
table 3). In our case the pyrolysis of 1 ton of used
tyres produced 365 kg of solid residues, most of
which, namely 49.7 % had a grain size between 0.1 and
0.5 mm.

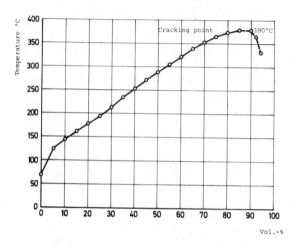

Figure 3. Boiling behavior of the condensate from the pyrolysis of old tires
(determined by DIN procedure 51751)

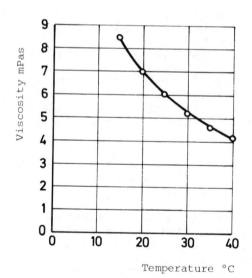

Figure 4. Viscosity of the condensate from the pyrolysis of old tires (determined
by DIN procedure 51562)

Table I. Mean Values of the Parameters Characterizing the Process Gas Obtained Through the Pyrolysis of Old Tires

Parameter	Method of Determination	Unit of Measurement	Value
Yield (based on tyres)	gravimetric	% by weight	16.7
Odour	Nose		intensive, specific
Net calorific value H_u	from GC	kJ/m_n^3	34 580
Gross calorific value H_o	from GC	kJ/m_n^3	46 180
Relative density d_v	from GC		0,959
Wobbe number	from GC		11 260
Carbon dioxide CO_2	GC	% by volume	4,6
Oxygen O_2	GC	% by volume	0,31
Hydrogen H_2	GC	% by volume	16,36
Nitrogen N_2	GC	% by volume	15,66
Carbon monoxide CO	GC	% by volume	4,66
Methane CH_4	GC	% by volume	21,86
Ethane and Ethene $C_2+H_6+C_2H_4$	GC	% by volume	13,9
C_3-hydrocarbons	GC	% by volume	9,53
C_4-hydrocarbons	GC	% by volume	6,7
C_5-and C_6-hydrocarbons	GC	% by volume	5,32
Other hydrocarbons	GC	% by volume	1,1
Hydrogen sulphide before purification	H_2S Draeger	ppm	6 000
Hydrogen sulphide after purification	H_2S Draeger	ppm	n. d.
Hydrogen cyanide before purification	HCN Draeger	ppm	6
Hydrogen cyanide after purification	HCN Draeger	ppm	4
Ammonia NH_3	Draeger	ppm	n. d.
Hydrogen chloride HCl	Draeger	ppm	n. d.
Nitrogen oxides NO_x	Draeger	ppm	n. d.
Dust	VDI 2066	mg/m_n^3	n. d.

GC   : Gas chromatography
n. d. : not detectable

Table II. Mean Values of the Parameters Characterizing the Condensate Obtained Through the Pyrolysis of Old Tires

Parameter	Method of Determination	Unit of Measurement	Value
Yield (based on tyres)	gravimetric	% by weight	46.8
Colour	Eye		black
Odour	Nose		intensive, specific
Net calorific value H_u	calorimeter	kJ/kg	40 161
Gross calorific value H_o	calorimeter	kJ/kg	42 406
Water content	DIN 51 582	% by weight	depending on moisture in the tyres up to 7
Density at 15° C	DIN 51 757	g/ml	0.937
Viscosity at 15° C			8.42
Viscosity at 20° C			7.09
Viscosity at 25° C	DIN 51 562	m Pas	6.03
Viscosity at 30° C			5.21
Viscosity at 35° C			4.55
Viscosity at 40° C			4.13
Coke residue acc. to Conradsen	DIN 51 511	% by weight	2.7
Closed-cup flash point	DIN 51 758	° C	+ 8
Combustion point	DIN 51 356	° C	+ 35
Setting point, lower	DIN 51 583	° C	- 23
Setting point, upper	DIN 51 583	° C	- 30
Ash content	DIN 51 575	% by weight	0.016

Table II.   Continued

Neutralization number	NZ	DIN 51 558/1	mg/KOH/g	2.4
Saponification	VZ	DIN 51 559	mg KOH/g	5.8
Bromine-number	BZ	DIN 51 774	% by weight	54
Carbon	C	EA	% by weight	86
Hydrogen	H	EA	% by weight	10.3
Oxygen	O	EA	% by weight	0.4
Nitrogen	N	EA	% by weight	0.7
Sulphur	S	EA	% by weight	1.3
Chlorine	Cl	EA	% by weight	1.3
Zinc	Zn	AAS	ppm	26
Lead	Pb	AAS	ppm	0.1
Cadmium	Cd	AAS	ppm	0.1
Mercury	Hg	AAS	ppm	0.1
Tin	Sn	AAS	ppm	2
Manganese	Mn	AAS	ppm	0.31
Iron	Fe	AAS	ppm	3.5
Chromium	Cr	AAS	ppm	0.14
Nickel	Ni	AAS	ppm	0.39
Copper	Cu	AAS	ppm	0.7
Aluminium	Al	AAS	ppm	2
Specific heat at 20° C	Cp	calorimeter	kJ/kg · K	1.96

EA   :   Elementary Analysis
AAS   :   Atomic Absorption Spectralphotometry
DIN ....:   German Industrial Norm

Table III. Mean Values of Parameters Characterizing the Solid Residues Obtained Through the Pyrolysis of Old Tires

Parameter		Method of Determination	Unit of Measurement	Value
Total yield (based on tyres)		gravimetric	% by weight	36.5
Soot fraction (based on tyres)		gravimetric	% by weight	30.5
Steel fraction (based on tyres)		gravimetric	% by weight	6
Soot fraction (based on solid residues)		gravimetric	% by weight	83.5
Steel fraction (based on solid residues)		gravimetric	% by weight	16.5

All further data given below are valid only for the soot fraction of the solid residues.

Parameter		Method of Determination	Unit of Measurement	Value
Bulk density		gravimetric	t/m^3	0.42
Colour		Eye		black
Odour		Nose		odour-less
Net calorific value H_u		calorimeter	kJ/kg	30 355
Gross calorific value H_o		calorimeter	kJ/kg	30 600
Annealing loss		DEV	% by weight	88.3
Extractable organic substance			Tetra/Chloro-form	0.8
Extractable water-soluble substance		Eurocop.Cost.	% by weight	0.6
Bulk density		gravimetric	g/ml	0.42
Carbon	C	EA	% by weight	85.8
Hydrogen	H	EA	% by weight	1.1
Nitrogen	N	EA	% by weight	0.4

Table III. Continued

Sulphur	S	EA	% by weight	2.6
Chlorine	Cl	EA	% by weight	0.7
Zinc	Zn	AAS	ppm	61 680
Zinc oxide	ZnO	Calculated from Zn	% by weight	7.7
Lead	Pb	AAS	ppm	229
Cadmium	Cd	AAS	ppm	22
Mercury	Hg	AAS	ppm	
Tin	Sn	AAS	ppm	400
Manganese	Mn	AAS	ppm	40
Iron	Fe	AAS	ppm	3 800
Chromium	Cr	AAS	ppm	28
Nickel	Ni	AAS	ppm	31
Copper	Cu	AAS	ppm	157
Aluminium	Al	AAS	ppm	3 400
Tellurium	Te	AAS	ppm	

DEV	:	German Standards Methods for the Examination of Water
EA	:	Elementary Analysis
AAS	:	Atomic Absorption Spectralphotometry

Metals:
The solid residues contain metallic substances which
can be removed with ease from the soot after pyrolysis because they are the shredded steel components of
tyres.

Harmful Gaseous Substances:
The sulphur compounds contained in tyre material enter
the gas phase in the form of hydrogen sulfide, which
can be brought to react with pelleted ferric oxide in
a dry cleaning stage. The purifying agent will become
saturated after a certain period of time and can be
regenerated by exposing it to air.

3.2 Operation with Emphasis on Energy Yield

All substances with larger proportions of inert components or lower energy contents are suitable as input
materials for this second process variant. Such substances are, for example, municipal refuse and sewage
sludge. In principle the same drum reactor is used as
explained for the first process variant. However the
carbonization gases are not condensed but thermally
cracked in a gas converter. The gas converter is a
solid-bed generator filled with charcoal or coke having
a low ash content and a certain size. In contrast to
conventional solid-bed generators its purpose is not
to gasify solid fuel but much rather to convert the
carbonization gas and to initiate the hydrogen gas and
methane-forming reactions. A hot coal bed acts partially as a catalyst. The hot carbonization gas leaves
the reactor at a temperature between 340 and 450°C and
is mixed with a certain stoichiometrically insufficient quantity of air to achieve partial oxidation,
thus causing the temperature of the gas to rise to
approx. 1100°C. This also heats up the coal bed where
the subsequent reactions are initiated. The gas mixture thus produced leaves the generator at a temperature of about 450°C because of thermochemical and
mainly endothermic processes and is passed through a
cyclone to separate fine-grain solids, through a gas
cooler to reduce the temperature, through a gas
scrubber for cleaning and eventually to the gas engine
after it has been mixed with air (Figure 5).

The gas engine is coupled directly to an electric
generator. Having reached a uniform speed, the operating temperature that has been set inside the drum
reactor can be kept constant with the aid of the hot
exhaust gases from the gas engine.

The gas converter is shown in Figure 6. Tables 4
and 5 show the mean values of the results of experi-

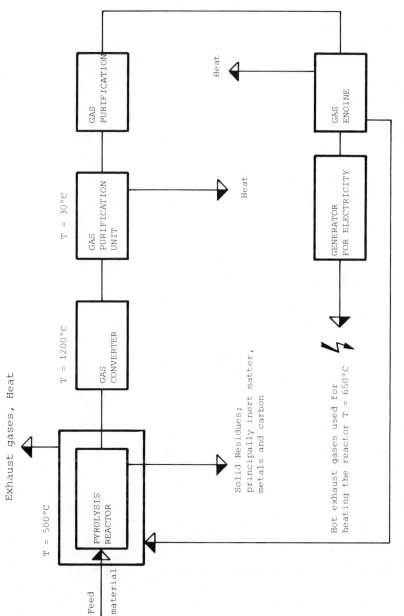

Figure 5. Variant 2—degasing and cracking with emphasis on the production of heat and electrical energy

Table IV.   Mean Values of Parameters Characterizing the Raw Material and the Reaction Products in Experiments with Municipal Refuse

Raw Refuse

Water content after pre-drying	18	% by weight
Annealing loss	44	% by weight
Organic carbon content	40	% by weight
Net calorific value	8 370	kJ/kg

Net calorific value	5 700	kJ/m³
Density	1.1	kg/m³
H_2	20	% by volume
CO	14	% by volume
CH_4	3	% by volume
C_nH_m	1	% by volume
O_2	3	% by volume
CO_2	8	% by volume
N_2	51	% by volume
H_2S	10	ppm
SO_2	20	ppm
HCN	30	ppm
NO_x		n.d.*
HCl		n.d.*
Hg		n.d.*
NH_3	120	n.d.*
Dust content		n.a.*

Solid Residues

Annealing loss		15	% by weight
Organic carbon content		12	% by weight
Bulk density		1.00	kg/dm³
Apparent density		1.15	kg/dm³
Net calorific density		2500	kJ/kg
Heavy metals:	Cr	2.9	ppm
	Mn	8.0	ppm
	Ni	1.6	ppm
	Co	2.2	ppm
	Zn	41.0	ppm
	Cd	0.2	ppm
	Cu	5.4	ppm
	Pb	20.0	ppm
	Hg	1.0	ppm

Gas Scrubber Water

Total solids	8	g/l
Annealing loss	35	% by weight
COD	1 700	mg/l
BOD_5	160	mg/l
N (Total)	700	mg/l
N (NH_3)	650	mg/l
NO_x	3.2	mg/l
Chlorides Cl^-	290	mg/l
Cyanides CN^-	7.4	mg/l
Sulfates SO_4^{--}	400	mg/l
Sulfides S--	14	mg/l
P (Total)	4	mg/l
Total phenols	22.5	mg/l
Tars	5.2	mg/l
pH-value	7.7	mg/l

*	n.a. =	not assayed	
*	n.d. =	not detectable	

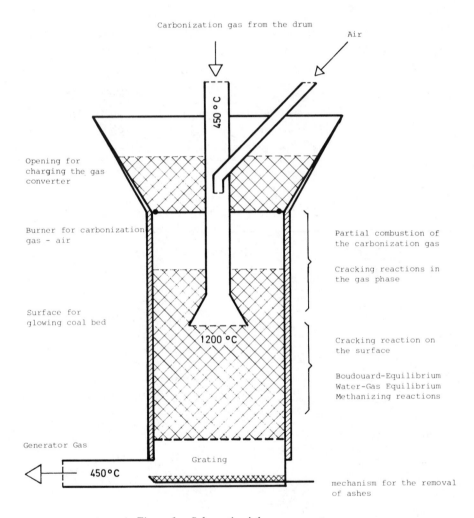

Figure 6.   Schematic of the gas converter

Table V.   Composition of the Eluates of Pyrolysis Slag from Experiments with Municipal Refuse

Kind of substance		Eluate of the pyrolysis slag
$N_{(Total)}$	mg/l	3
BOD_5	mg/l	10
COD	mg/l	450
CN^-	mg/l	0.06
Cl^-		900
SO_4^{--}	mg/l	650
PO_4^{---}	mg/l	0.50
Tars	mg/l	1
Phenoles	mg/l	Traces
pH-value		7.1

ments during a measuring period for "Municipal Refuse".
Figures 7 and 8 illustrate material and energy balances.

Material balance:

In short it can be stated that the pyrolysis of 100 kg
of municipal refuse, predried to a water content of
18 % by weight and having a calorific value of 8370 kJ/kg
and to which 77.6 kg of air have been added, produces
in the two-stage plant an average of

> 106.9 kg of pure gas and
> 56.8 kg of solid residues.

The balance of 13.9 kg can be explained by gas and
water losses that cannot be accounted for.

Substances such as glass and metals can be separated with relative ease. Provided the residues are
deposited in a landfill this process results in a
space-saving of approx. 60 % when compared with the
sanitary landfilling of raw wastes.

Solid residues:

In our case the solid residues contained approx. 12 %
organic carbon. The heavy metal concentrations were
low, probably and primarily due to the low quantities
of such substances in the input material. Leaching
experiments showed that the drum reactor residues
yielded mainly chloride-and sulphate-containing
extracts. The chemical oxygen demand of these solutions
was relatively high when compared with the BOD.

Scrubber Water:

In the experiments involved the amount of water from
the gas scrubber related to the amount of raw input
material was about 0.18 m³/t. The pH-value of the
water was mildly alkaline, the chloride concentration
was about 290 mg Cl⁻/l and the total amount of phenols
about 23 mg/l. Annealing loss of the scrubber water
sediment was 2800 mg/l, and the annealing residue amounted to 8000 mg/l. Sulphur was found mainly in an oxidized form. Results of initial experiments using an apparatus for the determination of the biochemical oxygen demand allowed the conclusion that an undisturbed decomposition of the organic components can take place.

Gas:

The pure gas with a specific gravity of 1.1 kg/m³ had
a net calorific value of 5700 kJ/m³ or 5180 kJ/kg. If
an excess factor of n = 1.5 is attributed to the
combustion of the pure gas at a theoretical air
demand of 1.5 m³ air/m³ gas the specific exhaust gas
production will be 2700 m³/t of waste. The low harmful substance concentrations in the pure gas allow the
expectation that the emission values will be low
which means that the clean air requirements of environ-

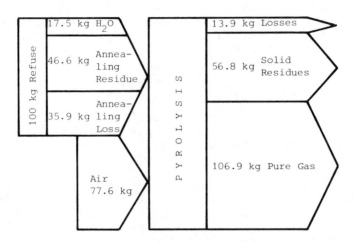

Figure 7.   Outline of the material balance for the pyrolysis of municipal refuse in the two-stage Kiener System

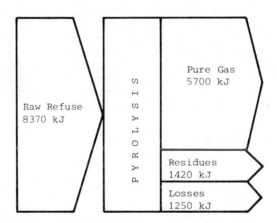

Figure 8.   Simplified energy balance for the pyrolysis of municipal refuse in the two-stage Kiener System

mental protection agencies can be complied with without problems.

Energy balance:

It is not possible to set up an exact energy balance because such functions as material feeding, removal, transport and the like were performed manually in the experimental plant. An estimate shows, however, that the pyrolytic treatment of the available energy in the waste transformed

> 68 % into the gas and
> 17 % into the solid residues.

If a gas engine is operated, the energy content of the exhaust is sufficient to bring the drum to the required reaction temperature.

4.   Summary

Pyrolysis as a waste treatment method has made considerable progress during recent years. After laboratory and semi-technical scale testing, several demonstration plants have been set up. Others are in the planning or construction stage.

As regards the economy of the different systems there exist only theoretical elaborations which still must be verified by practical experience.

There are still a number of questions to be answered that refer to the technology, the operating technique and the operational safety. These gaps in our knowledge must be closed during the years to come. In light of the encouraging experimental results achieved so far and in the interest of a further positive development of waste management public agencies should be well-advised in sponsoring efforts aiming at the realisation of more research and demonstration plants.

RECEIVED November 16, 1979.

The Design of Co-Current Moving-
Bed Gasifiers Fueled by Biomass

MICHIEL J. GROENEVELD and W. P. M. VAN SWAAIJ

Department Chemical Engineering, Twente University of Technology,
P.O. Box 217, 7500 AE Enschede, Netherlands

Co-current moving bed gasifiers have been selected for the gasification of biomass during one century and even now this reactor-
type seems to be attractive for this purpose in comparison to
other possible types like counter-current, fluid bed and entrained
bed [1]. The main advantages are that a) the reactor is simple
to construct, b) a tar free gas is produced and c) if required,
chlorine and sulphur components can be removed during gasification
by means of e.g. dolomite addition. A disadvantage is the high
outlet temperature (900 K).
For the classical design the use of feedstocks with a small particle size or a high ash content and scaling up of the reactor to
capacities larger than 250 kg/h seem to be difficult if a tar
free product gas is required. The reactor know-how is largely
empirical and dates mainly from 1925 - 1945 when the gasification
of dried woodblocks (0.04 m) for traction of cars and lorries was
practiced in Europe. An extensive survey of the various co-current
moving bed gasifier designs and their design rules has already been
presented by Schläpfer and Tobler [2].
The present revival of interest in biomass gasification envisages
the utilization of difficult feedstocks varying in particle size,
moisture and ash content etc.
To use these feedstocks properly, an analysis of the physical and
chemical phenomena taking place during gasification is necessary.
To this aim we studied the gasification of wood varying in particle size and moisture content, utilizing models for different
parts of the gasifier. From this analysis existing empirical
design rules and process limitations can be understood and an
improved design suitable for many types of biomass and wastes has
been derived.

Description of the co-current moving bed gasifier. Gasification is a complex process during which the solid fuel successively dries, pyrolyses (organic gases and liquids are distilled
off, consequently the solid residue achieves a higher carbon
content) and reacts with O_2, CO_2, H_2O in a complex manner given

below in a simplified form:

$$C + O_2 = CO_2$$
$$C + CO_2 = 2\ CO$$
$$C + H_2O = CO + H_2$$
$$CO + H_2O = CO_2 + H_2$$

In the gasifier (see Fig. 1) these processes can be found in different zones which show some overlap, however.

The solid feedstock is introduced at the top of the gasifier and has to flow downwards. Drying and pyrolysis take place due to heattransfer from the oxydation zone close to the air inlet. In the oxydation zone the reactions with oxygen result in a sharp rise in temperature upto ± 1600 K. An important function apart from this heat generation of this zone is to convert all condensable organic products (tar) from the pyrolysis zone. In a properly designed gasifier the products of the oxydation zone are charcoal and a hot gas rich in CO_2 and H_2O but not containing tar. In the reduction zone the sensible heat of the hot char and gases are absorbed in the endothermic reactions of CO_2 and H_2O with the carbon. Consequently a gas containing N_2, H_2, CO, CO_2, H_2O and some CH_4 is produced. This gas can be used for energy production in furnaces, turbines or internal combustion engines, as well as for further synthesis.
Deal wood being taken as a model feedstock, the different zones will be analysed.

Analysis of a co-current moving bed wood gasifier.

The influence of geometrical factors, the specific reactorload and the feedstock properties on the fuel gas production including the amount of tarry components will be analysed. No atttention will be paid to the possibility of in situ sulphur or chlorine removal. Earlier attempts to predict the product gas composition were based on thermodynamic models i.e. a combination of mass and energy balances assuming for one or more reactions chemical equilibrium at an empirically determined temperature e.g. outlet temperature.

Thermodynamic models. To calculate the product gas composition from the fuel characteristics the models assume either the heterogeneous char H_2O, CO_2 and H_2 reactions at equilibrium (Gumz model) [3] or the homogeneous watershift reaction at equilibrium and a fixed amount of methane produced as a pyrolysis product (Schläpfer model) [2].
In Fig. 2 -a ternary CHO diagram- free oxygen and solid carbon boundaries, and the dry wood composition are shown. By gasification a certain amount of oxygen is added so that about all the solid carbon is converted. The Gumz model is applicable in the region between corner C and the solid carbon boundary, the Schläpfer model between the solid carbon and the free oxygen boundaries. Both models can easily be solved using the element balances, the energy balance and the equilibrium constants. In the energy balance a fixed emperically fitted amount of heat losses has to

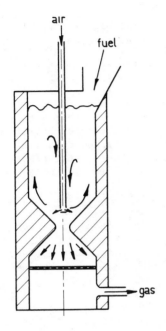

Figure 1.   Co-current moving-bed gasifier

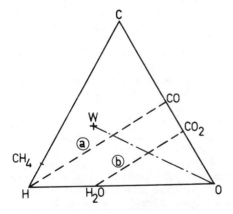

Figure 2.   Ternary CHO diagram for T = 1500 K: (a) solid carbon boundary; (b) free oxygen boundary; (W) dry wood

be introduced. In Table 1 an example of experimental and model
results is presented.

Table 1 Example of our experiments and thermodynamic model calculations (wood composition dry wt% C = 49.1; H = 6.1; O =
44.6; ash = 0.6; H_2O = 12.7) [5].

| | gas composition (vol%) | | | | | | m^3 gas/ | ΔH_{chem}/ | equil. |
	CO	H_2	CO_2	H_2	CH_4	H_2O	kg wood	ΔH_{in}	temp. K
experiment	17	16	11	47	1	8	2.98	79%	—
Gumz model	17	16	12	48	1	6	2.66	69%	910
Schläpfer model	16	15	12	45	2	10	2.71	73%	1073

It can be concluded that the gas composition and production for
the gasification of a feedstock can be estimated from simple
thermodynamic calculations. However, it is not sure whether equilibrium is reached and no information is obtained about the dimensions required for the reactor. To obtain these data, a more complex reactor model will be necessary, the various chemical and
physical steps being taken into account.

Reaction engineering model of the reduction zone. By this
model the conversion is calculated of the solid fuel and the gas
composition as a function of the distance from the airinlet, where
the reductionzone is supposed to start. It is assumed that the
pyrolysis and oxydation zone produce only char, CO_2, H_2O and some
CH_4. The inlet conditions of the reduction zone are obtained from
mass and energy balances over the previous zones only.
The sensible heat of the inlet flows, char and gases, is absorbed
in endothermic reactions of C + CO_2 and C + H_2O producing H_2 and
CO while CH_4 and N_2 remain unchanged. The rate of these processes
can be limited by 1) mass and heat transfer from the gas phase to
the particles; 2) diffusion of the gaseous reactants into the
porous particle; 3) chemical reaction at the active surface of the
particle; 4) diffusion and convection of the gaseous products out
of the particle to the gasphase.
The last three steps have been analysed in a particle model which
was tested on single char particles [4]. With a local volumetric
rate model it was possible to predict the measured carbon conversion profiles within the particle based on separately measured
chemical reaction kinetics and diffusivities.
This particle model is combined with the mass and heat transfer
equations, mass and heat balances and flow equations for the gas
and solid phase [5], the latter flow being governed by the
shrinking of the particles and the ash removal. In this reaction
engineering model the reactor geometry is simplified to a cylinder
with the throat diameter. It was found that under practical conditions about 2% of the carbon was removed together with the ash.
Heat leakage occurred and was taken to be proportional to the
reactor length and temperature difference.
In Table 2 calculated and typical experimental data are presented,
more experiments at different conditions can be found elsewhere [5].
In Fig. 3 conversion and temperature profiles are given showing

Table 2.
Typical experimental data and results from the reaction
engineering model for the reduction zone

		exp	calc	z↓	d_p↑	d_p↑	wood ↑	T air↑	moisture↑
In: wood	kg/s/m²	.225	.225	.225	.225	.225	2.25	.225	.225
moisture	kg/s/m²	.036	.036	.036	.036	.036	0.36	.036	.248
air	kg/s/m²	.596	.596	.536	.596	.596	5.96	.504	.724
d_p 10^3	m	24.8	24.8	24.8	10	50	24.8	24.8	24.8
z	m	0.45	0.53	4.38	0.44	0.44	3.74	1.6	1.0
T air	K	300	300	300	300	300	300	750	300
Out: gas	kg/s/m²	.846	.852	.794	.852	.852	8.52	.757	1.17
N_2	vol%	49.0	47.5	44.9	47.5	47.5	47.5	43.6	39.7
CO	vol%	15.0	15.3	15.9	15.4	15.4	15.4	18.2	2.3
CO_2	vol%	11.3	12.1	12.7	12.0	12.0	12.0	11.1	15.1
H_2	vol%	13.4	12.4	15.7	12.3	12.3	12.3	16.6	8.4
CH_4	vol%	0.8	0.8	0.8	0.8	0.8	0.8	0.8	1.0
H_2O	vol%	10.5	11.9	10.0	12.0	12.0	12.0	9.6	33.5
ΔH_{in}	kW/m²	4068	4068	4068	4068	4068	40680	4068	3739
$\Delta H_{chem}/\Delta H_{in}$	%	67	66	70	66	66	66	75	50
$\Delta H_{sens}/\Delta H_{in}$	%		27	21	27	27	27	16	42

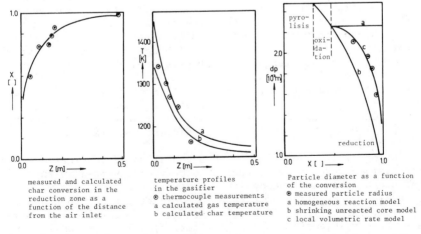

measured and calculated
char conversion in the
reduction zone as a
function of the distance
from the air inlet

temperature profiles
in the gasifier
⊙ thermocouple measurements
a calculated gas temperature
b calculated char temperature

Particle diameter as a function
of the conversion
⊙ measured particle radius
a homogeneous reaction model
b shrinking unreacted core model
c local volumetric rate model

*Figure 3.   Experimental and calculated conversion temperature and char particle
diameter for Experiment 1*

that the model describes the experiments well. From the table
the influence of the reduction zone length, particle size,
moisture content and heat recovery of the fuel gas can be seen.
In a longer reduction zone some more sensible heat of the gas is
consumed for CO and H_2 production, lowering the outlet temperature
to ± 950 K. In this final stage of the process no mass and heat
transfer limitations occur, so that the length as well as the
diameter could be increased to obtain the required volume. From
the model calculations and the experiments the particle size
appears to have hardly any influence on the performance of the
reduction zone. A larger capacity needs a longer reactor or more
precisely a larger reduction zone volume. Experimentally specific
reactor loads between 0.05 and 0.7 $kg/s/m^2$ (throat area) have
been used succesfully. A higher moisture content of the feedstock
makes the amount of water in the product gas increase and thus
its sensible heat at the cost of the chemically bounded energy.
It was found, however, that with a feedstock with 50% moisture the
reduction zone model is not valid any more. As under these conditions tars have been found in the gas, the assumption that all the
pyrolysis products are converted in the oxydation zone is incor-
rect. More chemically bounded energy in the fuel gas can be obtained if some sensible heat of the gas is used for preheating of
the air or preferably for drying of the fuel. Due to this extra
input of energy in the gasifier less air is required and the Joule
value of the gas is increased. The reduction zone model can thus
be used to calculate the required dimensions together with the
product gas temperature and composition, in case no pyrolysis
products are present there.

   Analysis of the oxydation and pyrolysis zone. The most critical function of these zones are the complete devolatilization
and subsequent conversion of the tars. Tar is produced during
the pyrolysis of the wood particles and consists of many organic
components. Quantity and composition of the pyrolysis products
highly depend on the pyrolysis conditions such as heating rate.
temperature and particle size [6]. Although pyrolysis is a very
complex process for large dry particles at high heating rates, the
rate of pyrolysis is mainly determined by the heat conduction into
the solid particle [5,7]. Therefore, the residence time for complete devolatilization can be estimated from heat penetration
times. The time needed for a particle to reach a centre temperature of 750 K, which is usually sufficient [8], is given in
Table 3.
If incompletely devolatilized char enters the reduction zone, tar
may evolve there and will remain largely unconverted in the gas.
The second reason for undesired tar in the product gas may be
incomplete conversion of tar in the oxydation zone. To study this
problem, three types of experiments have been carried out: a) methane tracer injection in the gasifier; b) cold flow model tests;
c) tar production measurement at different geometries.
   Methane injection. Methane was chosen as a tracer, because

it is the only pyrolysis product which is never completely converted. In Fig. 4 the injection places and the conversion of the
additionally injected methane are presented. It is clearly shown
that the place of injection in the throat section makes hardly
any difference. Moreover, the methane concentration in the bunker
was found to increase for all cases. Apparently there is a considerable mixing in the throat section and methane flows upwards
into the bunker.

   Cold flow models. In a two dimensional cold model the flow
patterns could be made visual by using a gas flow with char powder
or water and ink [5]. The patterns observed are sketched in Fig. 5.
The double circular flow patterns prevent the pyrolysis products
flow along the cold wall directly into the product gas. The circulary flow patterns induced by the horizontal air jet should
reach the wall, however. This is an important design criterium for
the throat geometry. Similar patterns can be expected if the air
is introduced through the side walls. Penetration of an air jet
in a packed bed was studied by measuring the radial velocity
distribution (see Fig. 6). As an example a typical velocity distribution is given in Fig. 7, together with the velocity in the
centre of the jet as a function of the length (z) for different
inlet flows and packing sizes (see Fig. 8). It is found that for
the 0.02 m packing and different inlet tubes increasing of the
inlet velocity at the same mass flows did not influence the
velocity distribution. From Fig. 9 it can be seen that the penetration length of the jet defined as the distance at which the
centre velocity is 20% of the centre velocity at z = 0.02 m (first
measurement point), is independent of the mass flow. For the
0.0037 m packing the jet reached 0.045 m and 0.095 m for the
0.02 m packing. Although the penetration of the jet under reaction
conditions might be different, it is clear that the penetration
of the jet and thus the dimension of the circulary flow patterns
are mainly determined by the particle size. This explains the
empirically found fact that for most co-current gasifiers a
minimum particle size is required for tar-free operation.

   Tar production at different geometries. Experiments during
gasification confirmed the importance of correct positioning of
the air inlet, with its related double circulary flow patterns,
in the throat. While in normal position no tar, e.g. ± 250 mg/m^3,
was produced, a sharp increase in tar production was observed if
the air inlet was moved upwards thus increasing the distance
between the inlet and the throat wall. It was also found to be
preferable to have the air horizontally flow into the reactor.
No influence on the reactor performance could be observed for
different inlet tube sizes, which is in accordance with the cold
jet experiments. The air inlet being placed up to a position
of 0.12 m above the throat, tar-free gasification was possible,
with 0.01 m woodchips and a throat diameter of 0.25 m.

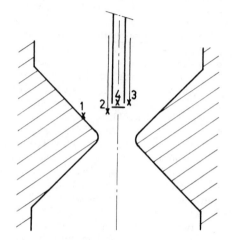

Figure 4. Methane tracer injection; conversion of additionally injected methane in Point (1) 0.36; (2) 0.24; (3) 0.35; (4) 0.38

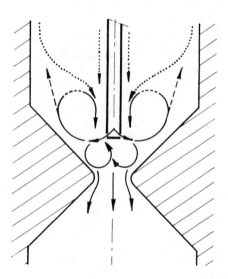

Figure 5. Double circular flow patterns occurring in the gasifier

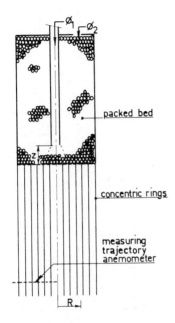

Figure 6.   Experimental set-up for the behavior of a jet in a packed bed

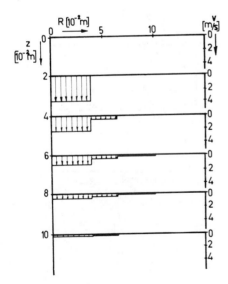

Figure 7.   Radial velocity distribution: (ϕ_1) $4.2 10^{-3} m^3/sec$; (ϕ_2) $5.6 10^{-3} m^3/sec$; (d_p) $0.02m$

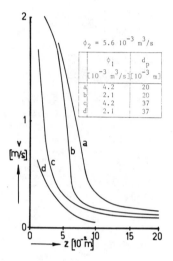

Figure 8. Velocity in the center of the jet as a function of the distance to the inlet

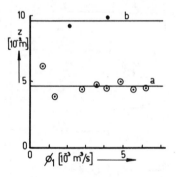

Figure 9. Penetration depth of the jet as a function of ϕ_1: (a) $d_p = 0.0037$ m; (b) $d_p = 0.02$ m

The distance air inlet - throat wall was then 0.17 m. Addition of large amounts of sawdust (up to 75%) did not influence this behaviour.

   Discussion. To prevent tar occurrence in the product gas it is necessary to have the particles completely devolatilized on passing the throat. Tar produced above the throat should not slip along the walls but should be mixed up in the circulary flow system close to the air inlet. In spite of the gas mixing the temperature close to the air inlet was found to be much higher (T = 1600 - 1800 K, for deal woodblocks 15% moisture) than the temperatures close to the throat wall (T = 800 - 1500 K, depending on the capacity). At a low capacity of 0.1 kg/s/m^2 throat area tar-free operation was possible at wall temperatures of 800 K. At this low temperature none or little tar conversion can be expected [9], this showing clearly that tars have been transported to the air inlet where temperatures are high and free oxygen exists. The zone of free oxygen is limited, however. On experimental combustion of char coal with air it was found that after two particle diameters (d$_p$ = 0.02 m) all the oxygen was consumed [5]. In the gasifier oxygen reacts still faster with the gaseous components in the order of: hydrocarbons, H$_2$ > CH$_4$ > CO > solid C. [10]. Because even pre-mixed methane in the air inlet is not completely converted (see Fig 4.), H$_2$ and tars must be responsible for most of the oxygen consumption. In some gasifier designs pyrolysis products are sucked out of the reactor, mixed with the air and combusted prior to its introduction in the gasifier. In a well-designed oxydation zone this process takes place within the reactor itself.

Design considerations

From the previous analysis three main design factors for co-current moving bed wood gasifiers are distinguished:

   a). Heat penetration for complete devolatilization the residence time in the pyrolysis zone should be equal to or longer than the Fourier time for heating up. This leads to a maximum reactor load which can be estimated in the following way. Due to radiative and convective heat transfer the pyrolysis zone is assumed to start roughly three particle diameters above the air inlet, and to end at the smallest cross-section of the throat. In Table 3, the heat penetration times and their related maximum reactor loads are presented.

Table 3. Heat penetration times and maximum (specific) reactor
loads. Assumptions: Fo = 0.1; a = 2 10^{-7} m^2/s; inlet height =
throat diameter, throat angle 45°

d_p m	t s	maximum reactor load kg/s d_{throat}: 0.1 m	0.3	maximum specific reactor load kg/s/m^2 0.1 m	0.3 m
0.1	5000	(3 10^{-3}) *	0.02	(0.4)*	0.3
0.05	1250	(4 10^{-3}) *	0.04	(0.5)*	0.6
0.02	200	0.01	0.2	1.3	2.8
0.01	50	0.03	0.7	3.8	9.9
0.005	12.5	0.1	(2.4)**	12.7	(34)**

* particles too large for the throat area
** air jet will not reach the wall see b).

For an IMBERT gasifier (throat diameter 0.15 m, d_p = 0.05 m) a
maximum load of 0.016 kg/s is so calculated, which is in agreement with the experimentally determined maximum load of 0.014 kg/s
[2].

b). Penetration length of the air jet. This criterium limits
the area which can be covered by one air inlet point. The particle
size is the most important factor here, but no quantitative relations can be given yet. If a large capacity per unit is required,
a large cross-sectional area with many inlet points is necessary.
However these air inlets should not hinder the solids flow in the
reactor.

c). Length of the reduction zone. The length depends on the
fuel characteristics, specific capacity, heat losses and gas
quality required. It can be calculated by utilizing the reaction
engineering model for the reduction zone.
Many additional factors may influence the final design: pressure
drop (especially with small particle diameters); solids flow (for
some feedstocks and at high specific capacities); cinder or slag
formation; entrainment of fine char particles; heat recovery of
the fuel gas, etc.

Design of a 100 t/d gasifier.

From the previous analysis it is concluded that the classical
throat design is not suited well for scaling up because of the
limited penetration of the air jet. Introducing more inlet points
over the throat area would easily hinder the downward solids flow.
A solution may be the use of an annular throat with a ring-shaped
air distributor in the centre and/or air inlet points along the
side walls [11]. An example will be given here for a 100 t/d
reactor gasifying biomass (d_p = 0.002 - 0.03 m, moisture content
14%, bulk density 400 kg/m^3). A sketch of the design is given in
Fig. 10. The volume of the pyrolysis zone is chosen such so as
to be sufficient for complete devolatilization of the 0.03 m
particles (residence time ≥ heat penetration time = 450 sec.)

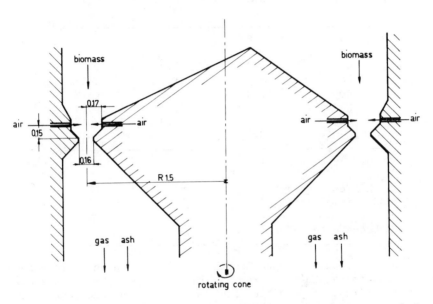

Figure 10.   Sketch of a 100 t/d co-current moving-bed gasifier fueled by wood
(type Twente)

The specific reactor load is 0.77 $kg/s/m^2$ (throat area).
Air is introduced through both side walls, so that the penetration depth of 0.17 m expected for the air jet will be sufficient for tar-free operation. On the basis of the model the required reduction zone and the mass and energy balances are calculated. Two cases are presented, the first without any heat recovery from the fuel gas, the second with air preheated to 700 K. (See Table 4.)

Table 4. Mass and energy balances for a 100 t/d gasifier, fueled by wood.

In: wood	1.16	1.16	kg/s
moisture	0.19	0.19	kg/s
air	2.49	2.00	kg/s
T air	300	913	K
Out: gas	3.83	3.34	kg/s
composition			
N_2	43.2	37.9	vol%
CO	17.5	21.1	vol%
CO_2	11.0	10.2	vol%
H_2	16.6	21.5	vol%
CH_4	0.9	1.0	vol%
H_2O	10.8	8.3	vol%
In: Energy	21	21	MW
Out:			
Chemical energy	73	82	%
Sensible Heat	22	12	%
Heat losses	2	3	%
Unconverted char	3	3	%

The required reduction zone length is ± 3 m based on a cross-section equal to the annular area. As increasing of the cross-section below the throat is possible, too, the reduction zone length can be reduced. The height of the bunker can be chosen based on refilling and fuel spreading arguments. The inner side wall of the annulus can be shaped as a slowly rotating cone to overcome solids flow problems. The top of the cone is then used for fuel spread and enhancing the bunker flow properties, the lower parts of the rotating cone for crushing cinders or slags formed from the ash.

Summary and conclusions

The physical and chemical phenomena occurring in a co-current moving bed gasifier have been analysed from experimental results and model calculations for different zones of the gasifier. A number of previously known empirical design rules can be understood and possible process limitations are predicted. With the reaction engineering model for the reduction zone the gas composition can be related to the fuel characteristics, reduction zone dimensions, heat losses and recovery. From the analysis of the pyrolysis/

oxydation zone it can be concluded that its dimensions are limited by the time to complete the devolatilization and the penetration length of the air jet. For larger capacities an annular throat is thus preferred over the classical throat design. An example is given for a 100 t/d gasifier.

List of symbols

a	thermal diffusivity	m^2/s
d_p	particle diameter	m
d_{throat}	throat diameter	m
Fo	Fourier number	-
R	radius	m
t	heat penetration time	s
T	temperature	K
v	velocity	m/s
X	conversion	-
z	distance from the air inlet	m
ΔH	specific enthalpy flow	kW/m^2
ΔH_{in}	ΔH of the fuel	kW/m^2
ΔH_{chem}	chemically bounded ΔH of the gas	kW/m^2
ΔH_{sens}	sensible heat of the gas	kW/m^2
ϕ	flow	m^3/s

Literature Cited

1. Groeneveld, M.J.; Van Swaaij, W.P.M., Applied Energy 1979, 5, 165-179
2. Schläper, P.; Tobler, J., "Theoretische und praktische Untersuchungen über den Betrieb von Motorfahrzeuge mit Holzgas", Bern 1937.
3. Gumz, W., "Vergasung fester Brennstoff", Springer-Verlag, Berlin 1952.
4. Groeneveld, M.J.; Van Swaaij, W.P.M., Sixth International Symposium on chemical reaction engineering, Nice, 1980.
5. Groeneveld, M.J., Ph-D-thesis, Twente University of Technology, Enschede, Netherlands.
6. Murty Kanury, A., Combust. Flame, 1972, 18, 75-83.
7. Hsiang-Cheng Kung, Combust. Flame, 1972, 18, 185-195.
8. Rensfelt, E., Symposium papers "Energy from biomass and wastes", I.G.T., Washington D.C. 1978.
9. Antal, M.J., Symposium papers "Energy from biomass and wastes" I.G.T., Washinton D.C., 1978
10. Field, M.A.; Gill, D.W.; Morgan, B.B.; Hawksley, P.G.W.; "Combustion of pulversised coal" B.C.U.R.A. Leatherhead, England, 1967.
11. Dutch Patent application, 79 03893.

RECEIVED November 20, 1979.

Pyrolytic Recovery of Raw Materials from Special Wastes

GERD COLLIN

Rütgerswerke AG, D-4100 Duisburg-Meiderich, West Germany

Summary

Pyrolysis can be used for the thermal decomposition
of waste materials that are predominantly organic in
nature, e.g. scrap tyres, scrap cables, waste
plastics, shredder wastes, and acid sludge. Rotary
kilns are particularly suitable as universally
applicable pyrolysis units for continuous operation.
Highly aromatic pyrolysis oils for use as chemical
raw materials are obtained at reactor temperatures of
about 700 °C. Such pyrolysis oils could form the basis
for the production of aromatics such as benzene,
naphthalene, and their homologues, thermoplastic
hydrocarbon resins and precursors of industrial
carbon, when the proven processes for the refining
of coal tar and crude benzene are applied.

Numerous projects in industrial countries with waste
and raw materials problems, including the Federal
Republic of Germany, have been concerned for some
years with the technical possibilities and the
economic viability of recovering materials from waste
(1, 2). Suitable conversion technologies, specifically for solid wastes of mainly organic nature
include pyrolysis processes, i.e. the thermal
decomposition in the absence of oxygen, leading to
the decomposition of organic substances into char
and low molecular gaseous and liquid chemicals (3).
Can pyrolysis be considered as a universal conversion
technology for the recovery of raw materials from
such wastes and what possibilities are there for the
utilization of the pyrolysis products?
To clarify this question, a R and D project

for the pyrolysis of different special wastes, carried
out jointly by the companies Eisen und Metall AG,
Rütgerswerke AG, and Veba Oel Umwelttechnik GmbH is
being dealt with among others. This project is
sponsored by the Federal Ministry of Research and
Technology and the Ministry of Food, Agriculture
and Forestry of the Land Northrhine Westphalia. The
laboratory scale tests began in 1976, and the semi-
technical trials will be concluded by the end of
1979. The development costs have so far amounted to
some 10 million DM.

Within the scope of this project the pyrolysis
of differing special wastes was investigated for the
first time under identical conditions. The results
reveal possibilities of how to produce highly
aromatic pyrolysis oils by pyrolyzing hydrocarbons
containing special wastes, such as scrap tyres,
scrap cables, waste plastics, shredder waste from
used cars, and acid sludge in an indirectly heated
rotary drum reactor at a reaction temperature of
some 700 °C in yields up to above 50 % in relation
to the organic substance. From these the chemical
raw materials benzene, toluene, xylene, naphthalene
and its homologues can be manufactured. In addition,
thermoplastic resins can be produced by polymerization of fractions containing styrene, indene and
their homologues (4).

In Fig. 1 the concept of the pyrolysis plant
is depicted; Fig. 2 shows the scheme of the pilot
plant. The semi-technical continuous reactor consists
of a rotary kiln furnace D 1 with a total length of
7.7 m, a diameter of 0.8 m, and a heating zone of
4.75 m. The throughput amounts to roughly 200 kg/h
when the reactor is in operation at a wall temperature of 700 °C. Solids are discharged via a solids
settler unit. Condensable parts of the raw pyrolysis
gas are separated by quenching with pyrolysis oil
in Q 1, indirect cooling in W 4/5 and counter-
current scrubbing with methylnaphthalene in K 3. The
acid noxious gases HCl, SO_2 and H_2S are removed in
K 1 and K 2 by countercurrent scrubbing with water
and caustic soda lye. The purified pyrolysis gas is
used for autothermal indirect heating of the reactor.

The pyrolysis oils are separated from the
quench and scrubber oils in the distilation units
K 4 and K 7/8; they can be further refined into
chemical raw materials by the application of processes used in upgrading coal tar and coke oven
benzene.

Fig. 3 shows the yields which can be obtained

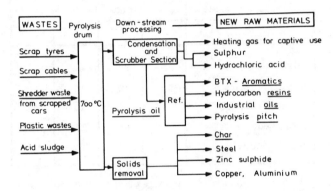

Figure 1. Recovery of raw materials by pyrolysis

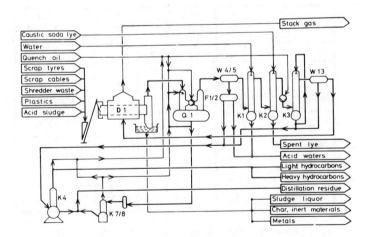

Figure 2. Schematic of the semitechnical rotary drum pyrolysis plant

TYPES OF WASTE	YIELDS [WEIGHT %]			
	Pyrolysis oil	Pyrolysis gas	Pyrolytic char	Metals
Scrap tyres	23	17	44	12
Waste plastics	53	30	15	-
Shredder wastes	13	12	33	40
Scrap cables	9	14	31	43
Acid sludge	13	37	31	-

Figure 3.   Main product yields of pyrolysis of special wastes

COMPONENTS [%]	TYPES OF WASTE				
	Scrap tyres	Waste plastics	Shredder wastes	Scrap cables	Acid tar
Aliphatic Hydrocarbons	0.5	trace	1	2	trace
Benzene	22	22	24	40	5
Toluene	18	23	19	20	18
$C_{8/9}$-Benzene Homologues Indane + Homologues	8	5	8	5.5	10
Styrene, Indene and Homologues	6	10.5	6	4.5	10
Naphthalene	10	10	9	4	14
Naphthalene Homologues	6.5	3.5	5	4	6
Higher bicyclic to tetracyclic Aromatics	3	6	8	7	5
Pyrolysis Resins (bp >450 °C)	21	20	20	13	41

Figure 4.   Pyrolysis oils from special wastes (700°C)

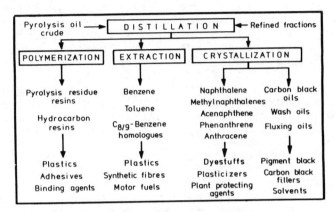

Figure 5.   Pyrolysis oil processing

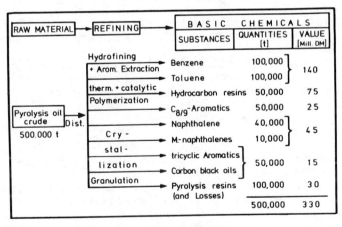

Figure 6.   Pyrolysis oil refinery

as main products; a breakdown of the chemical composition of the pyrolysis oils is given in Fig. 4.
The refining of the pyrolysis oil and the downstream
processing into chemical products is schematically
presented in Fig. 5; a balance of products from a
large industrial refinery is shown in Fig. 6 (5).
  The refining of highly aromatic pyrolysis oils
from waste yields the following range of products:
- Benzene for further processing into styrene, phenol,
  cyclohexane and aniline as intermediate products
  for thermoplastics and thermosets, synthetic
  rubber, synthetic fibres and dyestuffs,
- Toluene for further processing into nitro-, chloro-
  and isopropyl toluenes for polyurethane plastics,
  explosives, dyestuffs, antioxidants and plant-
  protecting agents, and if required, for dealky-
  lation to benzene,
- $C_{8/9}$-benzene homologues as high-octane blending
  components for low-lead and non-leaded supergrade
  gasolines,
- Condensed aromatic hydrocarbons such as naphthalene,
  methylnaphthalenes, acenaphthene, phenanthrene
  and anthracene for the synthesis of textile and
  pigment dyestuffs, and of aromatic carboxylic
  acids for PVC-plasticizer esters and polyester
  resins,
- Thermoplastic hydrocarbon resins and residue resins
  for the formulation of plastics, adhesives and
  binding agents,
- Highly aromatic oils as primary products for the
  production of carbon black fillers and pigment
  black for automobile tyres, other rubber
  articles, printing inks and paints.

References

1. Bundesminister des Innern: Abfallwirtschaftsprogramm '75 der Bundesregierung. Bonn, 22.10.75
2. Umweltbundesamt: Materialien zum Abfallwirtschaftsprogramm '75. Materialien 6/76,
   Berlin 1976
3. Collin, G.; Grigoleit, G.; Bracker, G.-P.:
   Chem.-Ing.-Techn. 50 (1978) No. 10, pp. 836/841
4. Collin, G.; Grigoleit, G.; Michel, E.: Chem.-
   Ing.-Techn. 51 (1979) No. 3, pp. 220/224
5. Collin, G.: Paper presented on the CRE/MER
   Conference, Berlin, October 1 to 3, 1979

RECEIVED November 20, 1979.

Gasification of Solid Waste in Accordance with the SFW–FUNK Process

HARALD FUNK and HORST HUMMELSIEP

Saarberg-Fernwärme, GmbH, Sulzbachstrasse 26, 66 Saarbruecken 3,
West Germany

The Saarberg-Fernwärme in Saarbrücken, West Germany,
decided to build a pilot plant to demonstrate the feasibility of solid waste gasification after having gained long range experience in operating incinerators.
The Federal Ministry of Research and Technology of West
Germany agreed to support this project because the two
committees, one for energy conservation and the other
for environmental protection, recommended it for implementation.

Incentives for gasification

The reason for developing such a gasification process
was a certain shortcoming of the incinerator.
     The incinerator with its moving grates cannot
burn industrial waste or toxic waste or used lubricating oil since the generation of excessive heat will
cause some damage. Furthermore the incinerator, although equipped with waterwalls, secondary combustion
chamber, electric precipitator, large scrubbing facility and a tall stack, still is an open system. In
spite of all the extra equipment, a tall stack is required to distribute impurities widely. Furthermore,
some components of the flue gas are corrosive and the
results are costly repairs. Another disadvantage is
the large amount of water required for scrubbing which
in average amounts to at least 120 gallons per ton of
solid waste burnt. After all, the volume of flue gas is
about 10 times that of the gas generated in a gas producer. The steam yielded from an incinerator has to be
used right there since the distance of steam transport
is rather limited in contrast to piping of gas. These
shortcomings of the incinerator gave an incentive to
compensate in a gasification unit.

0-8412-0565-5/80/47-130-485$05.00/0

Process description

The gasification unit is basically a simple design in the form of a shaft furnace with revolving grates. The solid waste is fed through locks from the top and then the ash is discarded in charges from the bottom, avoiding the escape of gas or harmful components. The gas discharged overhead from the reactor is cooled down to ambient temperature, thus condensing tars, light oils and steam.

The tars and oils after their separation from water are recycled to the reactor to enrich the gas stream by means of a carburetting effect. A portion of the water recovered in the condensation phase is used for cooling the gas and the excess is discarded after treatment. The ash discharged is inert and can be discarded for landfill operations. The primary object of this system is to yield gas at a high thermal efficiency (Figure 1).

The process is carried out in phases:
  a) Raw material preparation
  b) Gasification
  c) Condensation
  d) Gas purification

House-hold and industrial waste is shredded in a hammer mill where pieces of 4 inch maximum lump size are carried by a conveyor to screens through which glass, ceramics, slag, ash and other inerts smaller than 3/8" are discarded. Pieces larger than 4" are separated by gravity.

Then the solid waste is charged to the reactor through locks. While the material slides down through the various zones as:
          Drying,
          Devolatilizing or carbonizing,
          Reducing,
          Reaction or partial oxidation.

The gas discharged from the top of the reactor has a temperature of 500 to 600 K and after condensibles are settled out, it is compressed and charged to the purification section. This gas purification system works on a physical principle, since the heavier and harmful components are frozen out - either condensed or sublimed - in the regenerator by cooling down the steam to 170 or 130 K. Then the impurities retained in the regenerator are recovered by switching to a vacuum cycle, while the pure gas containing mainly H_2, CO, CH_4 and some CO_2 is yielded at ambient temperature. These 3 cycles, the

          loading phase (retaining heavier
              components)

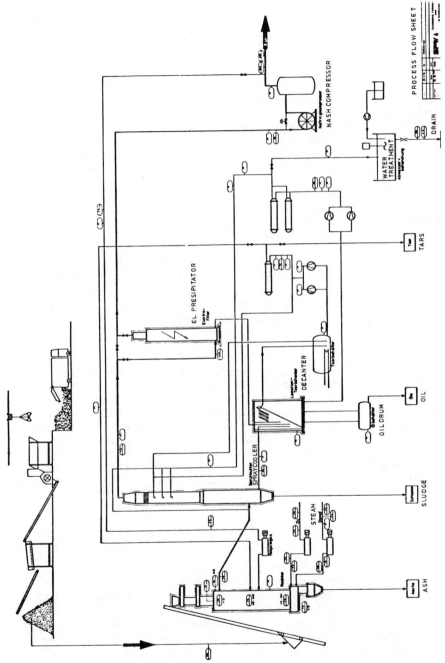
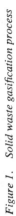

Figure 1.   Solid waste gasification process

recovery phase (vacuum cycle)
cooling down phase (yielding pure gas)
are switched every 6 or 8 minutes.

Product gas

The raw gas after condensation of liquids at ambient
temperature is recovered at a volume of 500 to 1000
normal cubic meters per ton of solid waste (about
19 000 to 37 000 SCF/t) containing up to 80 % of the
heating value charged with solid waste.

It can be used for heating, power generation or as
reducing gas for metallurgical processes or as a synthesis gas for the production of methanol. Methanol in
turn is a clean fuel and can be charged to a gas turbine for power generation or as additive or blending
component to gasoline. It can be stored more cheaply
than LNG (liquid natural gas) at less of a risk, to
make it more suitable for peakshaving, since it can be
converted to pure methane in a matter of hours. Methyl
alcohol is used as a base product for formaldehyde
(for plastics) or acetic acid (pharmaceutical industry)
or for the production of proteins and even for high
octane gasoline (via Mobil process).

Earlier test runs

Prior to the erection of the pilot plant at Velsen, a
small test reactor consuming up to 60 kg/h of solid
waste was installed at Neunkirchen next to an incinerator, which was to prove feasibility of solid waste
gasification. During 2 1/2 years of test runs amounting
to about 10 000 hours, all kinds of materials were
charged to this reactor. For instance: rubber, plastics, wood, used lubricating oil mixed with household
waste. A reasonable gas composition could be maintained
for any length of time:

H_2 :   30 Vol.-%

CO   :   20

CO_2 :   36

O_2 :   2

plus CH_4 and some higher hydrocarbons with nitrogen.

As a rule about 10 % of oxygen was added to the
charge and 0.25 to 0.40 kg of steam per kg of solid
waste to keep the reaction temperature at a level between 1100 and 1200 K. The results were encouraging
and therefore it was decided to proceed with the latter
approach to solid waste gasification.

Sampling of raw material and product

The solid waste from household was sampled and analysed frequently and varies in composition as follows:

Moisture	22 - 36 WT.-%
Inerts	20 - 47
Carbon	20 - 43
Hydrogen	3.0 - 6.0
Oxygen	20 - 42 (by difference)
Nitrogen	0.34 - 0.75
Sulfur	0.30 - 0.45
Chlorine	0.36 - 0.53
Fluorine	0.014 - 0.02

Upper heating value: as delivered   9500-12000 kJ/kg
                     dry   13500-15500 kJ/kg

Lower heating value: as delivered   8500-10500 kJ/kg
                     dry   12000-14500 kJ/kg

Shredded and screened:

Lower heating value: moist   9500-12500 kJ/kg
                     dry   12500-17000 kJ/kg

The lower heating value of the product gas prior
to gas purification was measured at 7500 to 9500 kJ/
normal cubic meter while a volume of gas of 700 to
1100 m^3_n per ton of gas was produced at normal operating
conditions keeping the reaction temperature at about
1300 K.
The composition of the gas varies as following:

H_2 :   25 - 35 Vol.-%

CO :   13 - 40

CO_2:   18 - 40

CH_4:   5 - 10

plus some illuminants.

Conclusion

The Saarberg-Fernwärme Co., is the only company in
Europe which is engaged in all kinds of solid waste
utilization, i.e. incineration, recycling operations
and gasification, to conserve raw materials and energy.
The gasification of solid waste as a supplement to
operation of incinerators is interesting enough to continue the development of such an approach.
The intermediate step for commercialization is the design of a gas producer at a capacity of 3 to 5 tons per
hour of solid waste which is now in the planning stage.

RECEIVED December 3, 1979

PYROLYSIS, GASIFICATION, AND LIQUEFACTION PROGRAMS IN JAPAN

Overview of Pyrolysis, Thermal Gasification, and Liquefaction Processes in Japan

M. HIRAOKA

Department of Sanitary Engineering, Kyoto University, Kyoto, Japan

The system of treatment and disposal of municipal refuse in Japan since 1963, is made up of collection and transportation followed by the incineration and disposal of the residues or direct landfilling. A large number of incineration facilities have been constructed in many cities, with grant-in-aid from the government. Indeed, the incineration process is excellent in the respect that perishable organic refuse can be converted to stable inert material, reduced in volume and disposed under good sanitary conditions, and for recovering energy from refuse. But it is inevitably apt to cause some pollution problems such as air and water pollution and leaching of heavy metals from ash. The facilities for air pollution control, water treatment, stabilization of ash in the incineration process and residues disposal, have become more costly as new standards are promulgated.

On the basis of these circumstances, the need for a better solid waste disposal system has stimulated a great deal of interest in the application of pyrolysis to solid wastes instead of the traditional incineration system in our country. Since the Agency of Industrial Science & Technology, MITI had launched the national project on development of technology and system of resource recovery from refuse in 1973, more than a dozen processes of PTGL* are being developed at the national level, municipal level, and the level of private companies as shown in Table I.

On the other hand, the spread of the activated sludge method to treat sewage has created problems of sludge disposal. The incineration process has been increasingly adopted as the method for reducing the volume and residues. But it has been recently pointed out that incineration of sewage sludge causes air and water pollution and leaching of heavy metals (especially Cr^{+6} compounds) from ash, the same as in incineration of domestic refuse. To cope with this situation, the drying-pyrolysis process has been developed as

*PTGL = pyrolysis, thermal gasification and liquefaction

0-8412-0565-5/80/47-130-493$05.00/0

Table I. Pyrolysis Process Developing in Japan

No.	Process Name	Company or Organization	Type of Reactor	Phase of R & D	Process
1	Dual chamber fluidized bed system	AIST (MITI) & Ebara MFG. Co. Ltd.	Dual chamber fluidized bed	5 t/d pilot plant test, 100 t/d DP is under construction	Pyrolysis/ Gas
2	A fluidized bed system	AIST & Hitachi Ltd.	Single fluidized bed	5 t/d pilot plant test	Pyrolysis/ Gas
3	Pyrox-system	Tsukishima Kikai Co. Ltd.	Dual chamber fluidized bed	150 t/d practical plant is being planned	Pyrolysis/ Gas, Oil
4	Incineration system	IHI Co. Ltd.	Single fluidized bed	Pilot plant 30 t/d	Incineration/Steam
5	Waste melting system	Nippon Steel Co. Ltd.	Moving bed shaft kiln	150 t/d practical plant is under construction	Pyrolysis/ Gas
6	Magma bed process	Shinmeiwa Industry Co.	Fixed bed electric furnace	Bench-scale test	Pyrolysis/ Gas
7	Shaft Kiln pyrolysis system	Hitachi Ship-building Co. Ltd.	Moving bed shaft kiln	20 t/d pilot plant test	Pyrolysis/ Gas
8	Destra-gas process	Hitachi Plant Const. Ltd.	Moving bed shaft kiln	Pilot plant test	Pyrolysis/ Gas
9	Purox system	Showa Denko Co. Ltd.	Moving bed shaft kiln	75 t/d practical plant is being planned	Pyrolysis/ Gas
10	Torrax system	Takuma Co. Ltd.	Moving bed shaft kiln		Pyrolysis/ Gas
11	Landgard process	Kawasaki Heavy Industries Ltd.	Rotary kiln	30 t/d pilot plant	Pyrolysis/ Gas/Steam
12	M. - Gallet system	Mitsubishi Heavy Industries Ltd.	Flash type reactor	Bench-scale test	Pyrolysis/ Oil
13	Pyrolysis of scrap tires	Kobe Steel Ltd.	External heated rotary kiln	23 t/d practical plant is under construction	Pyrolysis/ Gas/Oil
14	Pyrolysis of sewage sludge	NGK. Insulators Ltd.	Multiple hearth furnace	40 t/d practical plant is installed	Pyrolysis and Combustion

the most feasible method of sludge treatment and disposal which
minimizes the secondary pollution problems as discussed by Hiraoka
et al. (1).

The aim of this paper is to describe the present status of
development and application, and evaluation of PTGL processes in
our country.

Characteristics of Municipal Refuse (MR) and Sewage Sludge (SSL) in Japan

The general characteristics of MR and SSL in Japan are shown
in Table II and Table III. The Japanese MR has, on the average,
a heating value of 1,300 Kcal/Kg, generally ranging from 800 to
1,800 Kcal/Kg, and has a water content ranging 40-60%. Although,
the calorific value of MR tends to increase, this relatively lower
heating value and high moisture content compared with that of the
United States must be considered in the design of resource recovery systems from MR in Japan.

Heat Recovery System from MR. Since the Japanese government
adopted incineration and landfill systems for treatment of MR in
1963, the technology of European incineration has been used mainly
for constructing the large scale incinerator; except for a few
incinerators, the Japanese incineration system does not have the
heat recovery subsystem as in Europe. Since the Arab Oil Embargo
in 1973, the need for heat recovery from incinerators in large
cities is increasingly emphasized. As the main cities in Japan
are located in the temperate climate region, district heating is
not practical except in Hokkaido. Electrical power generation is,
in principle, quite reasonable for recovering the energy from MR
in Japan. The Nishiyodo incineration plant of Osaka City has a
processing capacity of 400 tons of MR per day, and is equipped
with two 2,700 KW generators, with output voltage of 6,600 V. In
Tokyo, Katsushika incineration plant which is in operation generates electricity of 12,000 KW, with a refuse feed of 1,200 tons
per day.

Application of PTGL Processes as an Alternative to Incineration for Treatment of MR. As stated above, as a first stage,
various PTGL process systems offer the possibility of decreasing
several pollution problems caused in the incineration process.
The pyrolysis and melting process developed by Nippon Steel Co.,
Ltd. was adopted at Ibaraki City to treat 450 tons/day (150 tons/
day X 3) of MR, and is under construction. Funabashi City has
decided to use the dual fluidized bed reactor system (Pyrox process) to treat 450 tons/day (150 tons/day X 3) of MR, and Chichibu
City has decided to use the Purox process developed by Showa Denko
Co., Ltd. and Union Carbide Co., Ltd.. The design and construction of these plants were started this April.

Table II (a). General Characteristics of MR

1. Physical Properties

Garbage	15–30%
Combustibles	20–50%
Uncombustibles	10–20%
Moisture	40–60%

2. Chemical Properties (WB)

C	10–25%
H	1.5–3.0%
O	10–20%
N	0.5–1.0%
Cl	0.2–1.0%
S	0.2–0.3%

3. Calorific Value   800–1,800 Kcal/Kg

Table II (b). General Characteristics of Sewage Sludge

1. Physical Properties

Combustibles	15–50%
Uncombustibles	10–20%
Moisture	60–85%

2. Chemical Properties (DB)

C	15–25%
H	1–5%
O	15–20%
N	2–7%
S	0.2–1%

3. Calorific Value   0–300 Kcal-Kg

Table III. An Example of Composition of MR

	Wet basis content (%)	Moisture (%)	Combustible (%)	Uncombustible (%)
1. Garbage	25.3	71.3	19.1	9.6
2. Paper	40.6	38.0	53.4	8.6
3. Plastics	10.8	25.6	63.3	11.1
4. Wood	5.0	50.0	47.0	3.0
5. Cloth	2.5	40.0	55.2	4.8
6. Ferrous metals	2.3	0.0	0.0	100.0
7. Glass, brick & pebbles	5.5	4.5	0.0	95.5
8. Miscelleneous & uncategorized	8.0	50.3	19.7	30.0
Total	44.0		38.6	17.3

Resource Recovery Process System Developed by National Project. Phase I of the national project was carried out from 1973 to 1976 for developing the feasible technology and systems of resource recovery from solid wastes, and the demonstration plant (100 tons/day) for the material - reclamation system in phase II is under construction. The pyrolysis processes which have been developed in phase I of the national project are as follows:

  a) Dual Fluidized Bed Reactor System for Fuel Gas Recovery
   (Ebara Corporation)

  b) Fluidized-Bed Pyrolysis for Oil Recovery (Hitachi Ltd.)
These reclamation systems may be suitable for the characteristic Japanese MR. As the MR in Japan has a relatively high moisture content, direct air classification of the crushed refuse as seen in the United States is hard to apply. A drying process may be necessary prior to the air classification. In this respect, a selective pulverizing classifier has been developed in the national project which is quite suitable to separate garbage of high moisture content and paper in Japanese MR. This demonstration plant with a capacity of 100 tons/day will be in full operation on November of 1979.

Application of PTGL Processes to Industrial Wastes. Industrial wastes which have a high calorific value such as scrap tires and plastics are suitable for PTGL processes. Though the many processes as shown in Table I can be applied to treat industrial wastes, a PTGL process using a external-heated rotary kiln has been developed by Kobe Steel Co., Ltd. to decompose the scrap tires and to get a liquefied fuel. The Clean Japan Center which was established to promote the resource recovery effort sponsored by MITI and the related private companies, are planning to construct the commercial plant which will treat the scrap tires by use of this technology and use the liquefied oil as the fuel for a cement kiln.

Evaluation of PTGL Processes

  Various PTGL processes have been developed and are in demonstration and commercial stages. From a comparison of municipal refuse handling options, we will determine if pyrolysis can be competitive with conventional handling options and how recovered material and energy values affect the operation costs. Three municipal refuse handling options were conceptionally designed and compared in this study. These are:

  Option 1, Incineration System,
  Option 2, Pyrolysis-Melting System,
and Option 3, Material Recovery and Pyrolysis System.

  Option 1 is a conventional stoker incineration system with heat recovery. Electric power will be generated with the steam from the waste heat boiler. To meet the air pollution standards, exhaust gas has to be cleaned by an electrostatic precipitator and a scrubber. Caustic soda solution will be applied in the scrubber to absorb hydrogen chloride and sulfur oxides. Residual ash will be disposed of by landfilling.

In option 2, municipal refuse will be size-reduced by a crusher and fed into the converter. Oxygen will be supplied to the converter and municipal refuse will be dried, pyrolyzed and melted at high temperature. Pyrolysis gas will be directly combusted in the combustor, where waste water will be evaporated and oxidized. Energy will be recovered in the form of steam by a waste heat boiler and electric power will be generated. Lime will be applied to absorb hydrogen chloride and sulfur oxides in the flue gas. Slag may be used for the construction of roads, etc..

Option 3 includes both material and heat recovery. A semi-wet pulverizing classifier developed by the Agency of Industrial Science and Technology, MITI is applied to recover garbage which has low calorific value and high moisture content. The highly calorific part of municipal refuse which mainly contains paper and plastics will be supplied to pyrolysis reactor. The pyrolysis reactor consists of two fluidized beds; cracking reactor and regenerator. Pyrolysis gas will be cleaned through the scrubber and stored in the storage tank. The cleaned gas will be combusted to produce high pressure steam with which electric power will be generated efficiently. Compost will be obtained from garbage through a process containing fermentator, dryer and separators. Electricity, ferrous metal, and compost will be recovered in this system.

Basis for Plant Design. The composition of municipal refuse is assumed as shown in Table III. The municipal refuse has the lower calorific value of ca. 1,500 Kcal/Kg. Plant size of 600 T/D is assumed. The capital investment costs, utilities, etc. were calculated using contacts with equipment vendors. Cost for repairs are assumed to be two percent of the plant construction cost per year. Unit costs of utilities and unit prices of recovered energy and material are assumed, based on the actual prices in 1979. Ash and other residues disposal cost is assumed to be 2,450 Yen/T, taking note of the representative cost data of large cities in Japan. The grant available to a municipality is assumed to pay up to fifty percent of the capital investment. The remaining investment cost must be amortized in fifteen years with the interest rate of six percent.

Incineration System (Option 1). The system flow diagram is shown in Figure 1. About 70.5 percent of the heat of the municipal refuse is recovered in the form of steam (265°C). A portion of the steam is used in the auxiliary equipment in the incineration system such as the air heater. The efficiency of the low pressure turbine generator is about 0.21. Utilities for the operation of the system and operating costs are listed in Table IV. The caustic soda solution costs to meet the air pollution standards are high.

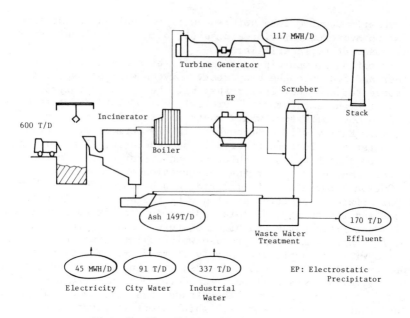

Figure 1.   Flow diagram of incineration option

Table Ⅳ . Utilities and Operating Cost (Incineration)

	T/D	$\times 10^6$ Yen/year
City Water	91	5.9
Industrial Water	337	6.1
NaOH (48 wt%)	7.3	65.8
other Consumption Material		151.7
Repairs		183.3
Ash Disposal	149	131.4
Total		544.2
Revenue (Electric Power)	72 MWH/D	129.6

As this option includes no material recovery process, the residual ash which amounts to about 25 percent of the raw municipal refuse in weight has to be disposed of. The generated electricity amounts to 117 MWH/D, and the plant operation needs 45 MWH/D of electricity. The outcome of this is that 72 MWH/D of electricity can be delivered.

Pyrolysis-Melting System (Option 2). The system flow diagram is shown in Figure 2. The heat recovery efficiencies of the converter and combustor are 0.81 and 0.71, respectively. About 60 percent of fed heat is recovered in the form of steam. The efficiency of the low pressure turbine generator is about 0.21 as in option 1. The 104 tons per day of slag are assumed to be taken by the construction firms. Utilities for the operation of the system and operating costs are listed in Table V. Auxiliary fuel cost takes the greater part of the total operating cost. As the oxygen generator needs much electricity (about 41 percent of the total usage of electricity), deliverable electricity is only 17 MWH per day. The recovered energy is used for the improvement of the characteristics of the residues in this system. Thus, ash disposal cost is small; on the other hand, revenue from the sale of electricity is small.

Material Recovery and Pyrolysis System (Option 3). The system flow diagram is shown in Figure 3. A semi-wet pulverizing classifier and a magnetic separator recover garbage and ferrous metal. Material fed to the pyrolysis reactor will have a calorific value of about 2,200 Kcal/Kg. About 54 percent of the heat of the feed material is recovered in the form of gaseous fuel. The conversion of the fuel gas to steam (390°C) is about 65.6 percent. About 29 percent of steam is used for drying the fermentated garbage. The efficiency of the intermediate high pressure turbine generator is about 0.46. The cleaned gas fuel through the scrubber make it possible to raise the steam temperature to 390°C. The strength of the waste water quality is offered up in the interest of the high efficiency of the turbine generator in this system. The utilities for the operation of the system and the operating costs are listed in Table VI. The consumptive materials take the greater part in the total operating cost, reflecting much requirement for the maintenance of the complex system. The recovery in this system includes electricity, ferrous metal and compost. The total revenue is the maximum of the three systems.

Comparison. The total operating costs are compared in Figure 4 for the aforementioned three options. Revenues from recovered materials and energy are also shown in the figure. Gross operating cost is lowest for Pyrolysis-Melting Process (Option 2). Gross operating cost minus revenue, or net disposal cost is lowest for Incineration Process (Option 1). However, differences in the net disposal costs are small for these three

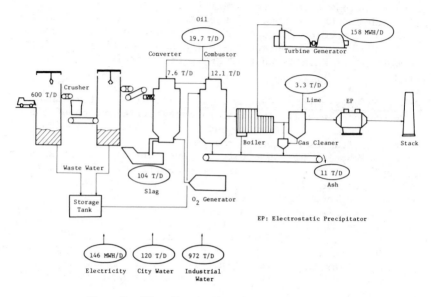

Figure 2.   Flow diagram of pyrolysis-melting option

Table V. Utilities and Operating Cost (Pyrolysis-Melting)

	T/D	$\times 10^6$ Yen/Year
City Water	120	7.8
Industrial Water	972	17.5
Oil	19.7	212.8
Lime	3.3	16.6
Other Consumptive Material		17.3
Repairs		176.2
Ash Disposal	11	9.7
Total		457.9
Revenue(Electric Power)	17MW	30.6

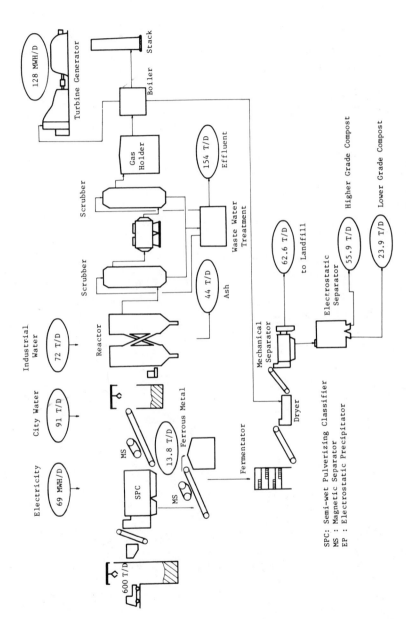

Figure 3. Flow diagram of material recovery and pyrolysis option

Table VI. Utilities and Operating Cost (Material Recovery
+ Pyrolysis)

	T/D	
City Water	171	11.1
Industrial Water	268	4.8
Other Consumptive Material		301.0
Repairs		216.4
Residue Disposal	107	94.3
Total		627.6
Revenue		
Electric Power	68 MWH/D	122.4
Ferrous Metals	13.8 T/D	23.0
Higher Grade Compost	55.9	30.2
Lower Grade Compost	23.9	4.3
Total		179.9

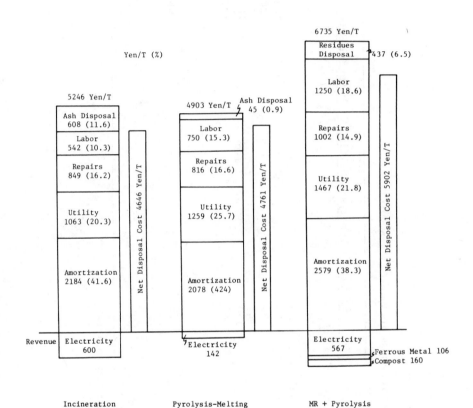

Figure 4. Total operating costs for MR handling options

Table VII. Comparison of Effluence and Recovery

	Incineration	Pyrolysis-Melting	Material Recovery + Pyrolysis
Effluence			
Flue Gas Volume (X10^6 Nm^3/D)	3.39	3.30	0.91
NO$_x$ as NO (Kg/D)	364	308	103
SO$_x$ as SO$_2$ (Kg/D)	155	131	20
HCl (Kg/D)	88	75	11
Waste Water (T/D)	170	nearly zero[a]	154
Residue (T/D)	149	11	107
Recovery			
Electricity (MWH/D)	72	17	68
Ferrous Metals (T/D)	—	—	14
Compost (T/D)	—	—	56 (Higher grade)
			24 (Lower grade)

a) Almost all water is used for cooling

options. For Pyrolysis-Melting System and Material Recovery and Pyrolysis (Options 2 and 3), the shares of the labor cost and utility cost are higher than Incineration System (Option 1). The high percentages of labor cost in option 2 and 3 are due to the increase of the number of unit operations. Those of utility costs for option 2 and 3 are due to the expenditure for auxiliary fuel and deodorizing chemicals, respectively. The inflation of the selling prices of electricity and other recovered materials makes option 3 to be competitive to other options. The shortage of landfilling site makes option 2 relative to others. A comparison of effluents and recovery from the systems is shown in Table VII. The table shows Pyrolysis-Melting System, and Material Recovery and Pyrolysis System (Option 2 and 3) are favorable in the respect of the air and water pollution protection.

Acknowledgment

The authors wish to acknowledge Dr. N. Takeda for his assistance of a portion of the work presented in this article.

References

1. Majima, T.; Kasakura, T.; Naruse, M.; Hiraoka, M. Prog. Wat. Tech., 1977, 9, 381.
2. Hiraoka, M.; Kawamura, M. The Third Japan - U.S. Conference on Solid Waste Management - Special Subject 2, 1976, May, 12, at Tokyo, Japan.
3. Hiraoka, M.; Takeda, N.; Fujita, K. 2nd Recycling World Congress, 1979, March 19-22, at Manila, Philippines.
4. Hiraoka, M. The International Congress of Scientist on The Human Environment, 1975, November 17-26, at Kyoto, Japan.

RECEIVED November 16, 1979.

Pilot Plant Study on Sewage Sludge Pyrolysis

T. KASAKURA
NGK Insulators, Ltd., Nagoya, Japan

M. HIRAOKA
Kyoto University, Kyoto, Japan

Sludge incineration process has been increasingly adopted by many municipalities as one of the effective methods for reducing the volume and stabilizing the organic materials of sewage sludge. But recently the incineration process has been pointed out to have several serious problems, namely higher energy consumption, and secondary pollution such as Cr^{+6} formation in ash and NOx formation in exhaust gas. Some investigators had proposed that "pyrolysis" would have a possibility to solve these problems, and have studied pyrolysis of sewage sludge.

The authors had already conducted the laboratory scale study and the preliminary pilot plant study, and proposed that "drying-pyrolysis process" (pyrolysis followed by indirect steam drying of dewatered sludge cake) (Fig.-1) could be one of the most economical and feasible alternatives for conventional incineration process. The authors have further conducted the feasibility study on a continuous system of "drying-pyrolysis process" to evaluate the performance of the process in pilot scale, and to demonstrate its effectiveness as a thermal processing of sewage sludge. This paper presents the results of this pilot plant study.

The purpose of this study can be summarized as ;
(i)   To evaluate the performance of drying-pyrolysis process in pilot scale
(ii) To demonstrate the effectiveness of drying-pyrolysis process as a thermal processing of sewage sludge
(iii) To compare drying-pyrolysis process with other thermal processes of sewage sludge
(iv) To show the feasibility of the practical plant of drying-pyrolysis process

PILOT PLANT and EXPERIMENTAL PROCEDURE

The pilot plant for this study consists of a four shaft indirect steam dryer of paddle type, a four hearth furnace for

0-8412-0565-5/80/47-130-509$05.00/0

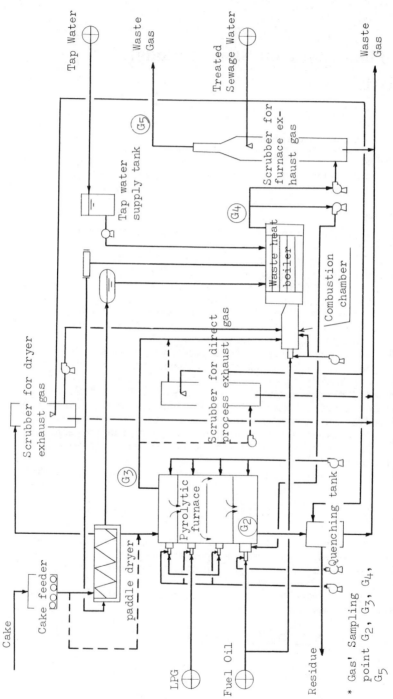

Figure 1. Schematic of drying-pyrolysis process pilot plant

pyrolysis, a combustion chamber for pyrolytic gas (after burner), and a heat recovery boiler for combustion chamber exhaust gas. The overall loading capacity of this plant is about 4 to 5 tons per day as dewatered cake with water content 75 %. (Fig-1) Five different kinds of dewatered cakes were selected for the pilot test, water content and higher heating value of sample cakes were varied from 46 to 76 %, from 1,500 to 2,960 kcal/kg D. S. respectively. (Table-I)

Cake A; Sludge from a combined sewerage system sewage treatment plant of a representative city in Japan. The cake was dewatered by vacuum filter after lime and ferric chloride were added to mixed primary and excess sludge.

Cake B; Sludge from a regional sewerage system sewage treatment plant containing industrial waste water. The cake was dewatered by filter press after an organic polymer was added to mixed primary and excess sludge.

Cake C; The same sludge as cake B, but the cake was dewatered by filter press after low temperature (150 - 160 °C) thermal conditioning.

Cake D; Sludge from separated sewerage system sewage treatment plant for domestic waste. The cake was dewatered by filter press after conventional (180 - 200 °C) thermal conditioning of the mixed sludge.

Cake E; Sludge of chromium tanneries' waste. The cake was dewatered by vacuum filter after an organic polymer was added to raw sludge.

The dewatered cake was dried to the water content of 30 to 40 % in the indirect steam dryer, and fed into the pyrolytic furnace through the screw conveyor. Combustion air less than the theoretical amounts for sludge is supplied into the furnace for partial combustion of combustible pyrolitic gas. That is, pyrolysis was carried out by "Direct Heating" method.

Off gas from furnace is introduced into the combustion chamber to burn its remaining combustibles, and to decompose its pollutant and other odor components. Excess air is automatically controlled by the oxgen concentration at the exit of the combustion chamber. Heat of off gas from the combustion chamber is recovered by the waste heat boiler. Steam generated in the boiler is supplied to the indirect steam dryer. When the amounts of steam generated is not sufficient for drying dewatered cake, auxiliary fuel is used to generate the additional steam.

In this study, to make a comparison with drying-pyrolysis process, the experiments on direct feed processes, i.e., direct pyrolysis process and incineration process, were conducted. In these processes, the dewatered cake was directly fed into the furnace. The plant was operated at steady state for at least 5 or 6 hours in all runs.

Table I. Analytical result of fed cake.

Cake	A	B	C	D	E
Run No.	803	816	922	819	823
Moisture content (wt%)	75.8	74.0	45.7	55.3	76.4
Ignition loss (wt% on dry basis)	59.5	56.7	32.0	83.6	64.3
Higher heating value (kcal/kg on dry basis)	2,960	3,290	1,500	4,310	4,000

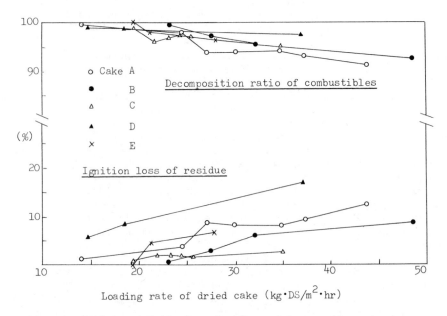

Figure 2. Variation of ignition loss of residue and decomposition ratio of combustibles with loading rate of dried cake

EXPERIMENTAL RESULTS AND DISCUSSION

Performance of Drying-Pyrolysis Process   Indirect steam
drying operation was available to all kinds of dewatered cakes
tested and its efficiency (overall heat transfer coefficient U)
varied with the kinds of cakes.

$U(Kcal/m^2 \ hr \ {}^{\circ}C)$	Cake A	Cake B	Cake C	Cake D	Cake E
	140-150	160-170	275	300-360	220-280

The overall heat transfer coefficient (U) based on the input heat
was calculated from the amount of condensed drain which released
from the dryer by the following equation.

$$q = U \cdot A \ (\varDelta t)$$

where q is supplied heat and A is total heat transfer area.
The temperature difference ($\varDelta t$) was obtained from the difference
between cake temperature ($95^{\circ}C$ constant) and steam temperature.
    The operating conditions in this study were selected on the
previous work (Majima et al., 1977) and the preliminary study by
the pilot plant. Pyrolytic furnace was mostly operated under the
following condition; air ratio for combustibles in sludge 0.6,
pyrolytic temperature $900^{\circ}C$, and retention time 60 minutes.
    In this study, to evaluate the weight reduction effect of
thermal processing, "Decomposition Ratio" difined as follows was
introduced.

Decomposition Ratio of Combustibles (%)

$$= (1 - \frac{100 - lc}{lc} \cdot \frac{lr}{100 - lr}) \cdot 100,$$

lc; ignition loss of fed cake (%)
lr; ignition loss of residue (%)

    When the loading rate to furnace was taken under 25 kg
$D.S./m^2 \cdot hr$, more than 95 % of combustibles in sludge cake was decomposed by pyrolysis regardless of the kinds of cakes. Thus,
pyrolysis is estimated to be as effective as incineration in
weight reduction of sewage sludge. The relationship between
loading rate and decomposition ratio or ignition loss in solid
residue is shown in Figure 2.

Heavy Metals   Table 2 shows the behaviour of chromium VI
(Cr+6) in pyrolytic operation. The behaviour of Cr^{+6} in incinerating operation on cake E is also shown in the table. The
ratio of Cr^{+6} to total chromium (Cr^{+6}/T-Cr) is an indicator of
the behaviour of chromium compounds in thermal processings. As
indicated in the table, a part of chromium III (Cr^{+3}) in fed cake
is oxidized into Cr^{+6} during incinerating operation, whereas a
considerable portion of Cr^{+6} in fed cake is reduced during pyrolytic operation.

Table II. Behaviour of Cr^{+6} in pyrolytic operation

cake (run No.)		T-Cr (mg/kg D.S.)	Cr^{+6} (mg/kgD.S)	Cr^{+6}/T-Cr (%)	Dissolved Cr^{+6} (mg/)
A (run 803)	Cake	300	8.6	2.6	–
	Residue	590	5.9	1.0	N.D.
E (run 823)	Cake	14,700	150	0.98	–
	Residue	56,000	16	0.03	N.D.
E (run 826)*	Cake	14,700	150	0.98	–
	Residue	40,900	4,150	10.1	140

* Incinerated at air ratio of 2.4

Table III. Dispersion ratio of heavy metals.

	Run No.	Cd	Pb	As
Drying-pyrolysis	803	86.4 %	84.1 %	43.6 %
Direct pyrolysis	903	85.9	84.4	34.6
Incineration	904	87.4	84.6	45.0

Table IV. Characteristics of exhuast gas

Process	Drying-pyrolysis			Direct pyrolysis		Incineration	
Cake (run No.)	A (803)			A (903)		A (904)	
Sampling point	G3	G4	G5	G3	G4	G3	G4
Gas temperature (°C)	760	215	49	290–345	–	390–410	–
Moisture (%)	24.6	20.8	10.3	45.6	6.5	32.3	7.1
Dry gas volume(Nm^3/hr)	200	330	280	75	111	134	248
Dust load (g/Nm^3)	8.40	3.10	0.34	0.76	0.04	0.60	0.02
HCN (ppm)	240	<0.1	<0.1	430	<0.1	320	<0.1
NH_3 (ppm)	2,080	1.0	0.50	1,970	48	170	4.5
NO_x (ppm)	54	210	190	70	58	140	160
SO_x (ppm)	130	230	130	83	84	12	19
HCl (ppm)	6.0	15.0	1.4	11	14	3.3	11
Cl_2 (ppm)	<0.2	<0.2	<0.2	<0.2	<0.2	<0.2	<0.2
H_2 (vol.%)	1.34	ND	–	1.63	ND	ND	ND
CO (vol.%)	1.64	ND	–	1.23	ND	ND	ND
CH_4 (vol.%)	0.34	ND	–	0.27	ND	ND	ND
C_2H_4 (vol.%)	0.11	ND	–	0.08	ND	ND	ND
O_2 (vol.%)	0.17	1.2	–	0.43	2.1	9.7	8.1

G3, G4, G5   see Fig.-1

During pyrolytic operation, it was possible not only to prevent perfectly the oxidation of Cr^{+3} into hazardous Cr^{+6} but also to reduce Cr^{+6} in fed cake. On the contrary, during incinerating operation, the oxidation of Cr^{+3} into Cr^{+6} occured and there was a danger of secondary pollution by dissolution of Cr^{+6} from ash. This fact was obviously confirmed by the test on tanneries sludge cake which contains 1.5 % chromium (Table II).

Table III is a summary of the dispersion ratios of heavy metals in three thermal processings in this study. They are very similar with each other. This suggests that there is no essential difference between pyrolytic and incinerating operations as for dispersion of heavy metals and that temperature is a dominant factor in dispersion.

Exhaust Gases   Table IV shows the analytical results of exhaust gas from drying-pyrolysis process and from direct pyrolysis process and incineration process on cake A. 43 to 83 % of S in the fed cake remains in the residue, 17 to 51 % of S converts into SOx and the remainder (0 to 7%) into H_2S. 55 to 87 % of $C\ell$ in the fed cake remains in the residue and the remainder converts into $HC\ell$.

HCN and NH_3 are typical thermally-decomposed products of nitrogen compounds in fuels and important intermediate products into NOx or N_2. The formation of HCN is considered to increase with the temperature rise, so that larger amount of HCN than usual incineration run would be attributed to higher decomposing temperature (exhaust gas temperature of 480-490°C) comparing with usual incinerating operation (exhaust gas temperature of 200 to 300°C). The concentration of NH_3 in pyrolytic operation with poor O_2 is of the 10^3 order in ppm, where of the 10^2 order in incinerating operation. Both HCN and NH_3 are almost completely decomposed in the combustion chamber and are in very low concentrations at the outlet of the boiler (G4).

Behaviour of NOx in drying-pyrolysis process is shown in Table V. Data of run 803 and 922 indicate that NOx is reduced in the furnace at air ratio of 0.61 and 0.71, whereas NO is formed in the furnace at high air ratio of run 823. In such a case as pyrolytic operation for sewage sludge where a large amount of NH_3 and HCN is produced, production or reduction of NOx in the furnace is determined by the balance of competitive reaction between converting reaction of NH_3 and HCN into NOx and reducing reaction of NO by these gases. Therefore, it is desirable to keep the air-fuel ratio of pyrolytic operation at about 0.6 from the view point of the prevention of NOx.

Residual NH_3 and HCN are decomposed by afterburning under atmosphere of low oxygen concentration, but if the structure of the combustion chamber is unsuitable, a large amount of NOx will be formed here. Table V indicates that an amount of NOx is formed by afterburning in each run. Further experiments for decreasing NOx formation at afterburning of exhaust gas has been conducted

Table V. Behaviour of nitrogen oxide in exhaust
gas of drying-pyrolysis process.

Cake	A	C	E	A
Run No.	803	922	823	904*
Air ratio	0.61	0.71	0.83	2.0
Sampling point G_2 NOx (ppm)	170	180	110	–
Gas volume (Nm³/hr, d.b)	110	135	60	–
NO (g/hr)	25.1	32.6	8.8	–
Sampling point G_3				
NOx (ppm)	54	120	46	140
Gas volume (Nm³/hr, d.b)	200	185	235	590
NO (g/hr)	14.5	29.8	14.5	110
NH₃ (ppm)	2,080	390	2,800	170
HCN (ppm)	240	200	230	320
Sampling point G_4				
NOx (ppm)	210	200	91	160
Gas volume (Nm³/hr, d.b)	330	410	480	1,090
NO (g/hr)	93.0	109	58.5	234

* Incinerated at air ratio of 2.0
Dry gas volume is corrected to the same feed amount
of dry solid.

Table VI. Experimental results of NOx prevention at afterburning
by the multiple combustion method

		1	2	3	4	5	6	7	10
Distribution of	A_1	100			50	50	50	50	33
Air to each air	A_2				50				
Port (%)	A_3		100			50			33
	A_4								
	A_5			100			50	50	33
Concentration of O_2 at outlet of the boiler	(%)			1.0-2.0			0.7-1.0		
Concentration of NOx at outlet of the boiler	(ppm)	410	380	150	230	180	160	90	85

Concentration of CO (%)   inlet 1.7 - 2.1   outlet 0
Temperature at inlet of the boiler (°C) 950

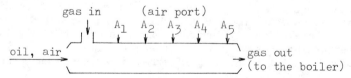

with the rebuilt combustion chamber. Experimental results are shown in Table VI. In single stage combustion, the amount of NOx is smaller in the case that air is fed from the position far from the auxiliary fuel burner than in the case that air is fed from the position near the auxiliary fuel burner.

However, as shown in Table VI multiple stage combustion is preferable in order to decrease NOx formation without fail. The factors which influence the behaviour of NO_x are temperature of the chamber, numbers of stage, feeding position of air, overall air ratio and distribution of feed air in multiple stage combustion method. It was found possible to minimize the formation of NO_x by multi-stage combustion under the condition of total air ratio of about 1.1 and temperature at the outlet of the chamber of below 950°C.

Comparison of Drying-Pyrolysis Process, Direct Pyrolysis Process and Incineration Process   Table VII shows the comparisons of experimental results for the three processes of drying-pyrolysis, direct pyrolysis, and incineration. In practical direct feed processes, however auxiliary fuel consumption would be lower than this experimental data, because in the experiment of both direct feed processes, exhaust gas temperature at the outlet of the furnace was considerably higher than that in practical plant, and because waste heat from the combustion chamber was not recovered in the experiment, whereas recovered in practical plant. Although there is a difference in loading rate, it has been confirmed on both drying-pyrolysis process and direct pyrolysis process that they can decompose combustibles to the same degree of incineration process, and prevent the formation of pollutant such as Cr^{+6} and NO_x. Comparing exhaust gas volume and auxiliary fuel consumption for the same amount of cake drying-pyrolysis process is superior to the other processes. Especially, auxiliary fuel consumption is remarkably small in drying-pyrolysis process.

PROCESS ANALYSIS

When utilization of a process is considered, quantitative evaluation under the practical conditions is required. Based on the data obtained from this experiment, simulating calculation for the three processes was conducted to know how the evaluating measure of each process (auxiliary fuel consumption) follows when characteristics of fed cake (water content and higher heating value) fluctuate. Water content and higher heating value were varied from 50 to 80 %, from 1,500 to 3,500 kcal/kg on dry solid base, respectively. According to the simulated result, auxiliary fuel consumption is the smallest in drying-pyrolysis process while it was the largest in incineration process, in the whole range of the cake characteristics. (Fig. 3)

Of the operation cost for each process, dominative factor is

Table VII. Comparison of drying-pyrolysis process, direct pyrolysis process and incineration process.

Process	Drying-pyrolysis	Direct pyrolysis	Incineration
Run No.	803	903	904
Cake	A	A	A
Moisture of cake (%)	75.8	77.8	77.5
Ignition loss of cake (%.DS)	59.5	66.0	66.0
Higher heating value(kcal/kg.DS)	2,900	3,300	3,300
Feed rate of cake (kg/hr)	200.1	50.0	50.0
Cake moisture at furnace inlet(%)	35.1	77.8	77.5
Air ratio	0.61	0.60	2.0
Exhaust gas temperature ($^\circ$C)			
at furnace outlet	700 – 790	290 – 345	390 – 410
at combustion chamber outlet	1,100	800	800
Oil consumption (ℓ/hr)	14.7	11.7	18.1
Cr^{+6} content in residue (mg/kg)	5.8	2.7	44.0
Dissolved Cr^{+6} (mg/ℓ)	ND	ND	0.4
Loading rate to hearth(kg.DS/m^2.hr)	29.9	6.8	6.8
Exhaust gas volume (Nm^3/ton cake)			
at furnace outlet	1,326 (1,000)*	2,748 (1,495)	3,949 (2,673)
at combustion chamber outlet	2,082 (1,649)	2,381 (2,281)	5,350 (4,968)
Oil consumption (ℓ/ton cake)			
at furnace	50	145	128
at combustion chamber	24	89	234
total	74	234	362
Decomposition ratio of combustible (%)	94.0	99.3	98.1
(Cr^{+6}/T-Cr) cake/(Cr^{+6}/T-Cr) residue	2.6/1.0	1.3/0.27	1.3/4.7

Experimental Data

* Dry gas volume

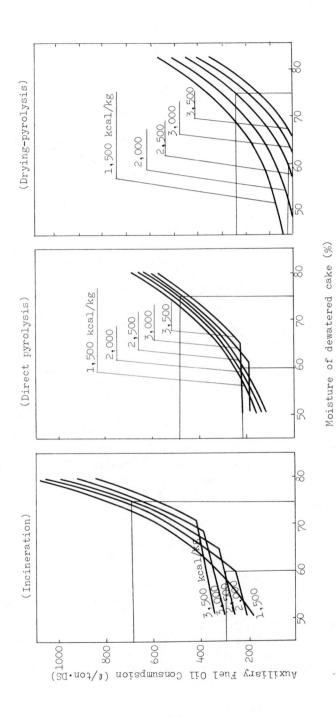

Figure 3. Simulation results for auxiliary fuel consumption

the fuel cost, so that as proved from the results of simulation, operation cost of drying-pyrolysis process is the least of all, and operation cost of incineration process is the most of all. Economical comparison among the three thermal processings were made based on the results of simulation. Table VIII shows the result. Construction cost of each process was estimated to be the same. The overall economics for the three thermal processings can be estimated from operation cost because construction cost is much the same among three thermal processings, and drying-pyrolysis process is the most economical and incineration process is the most expensive.

SUMMARY

A pilot plant study was carried out on "drying-pyrolysis" process in order to put the sludge pyrolysis process with multiple hearth furnace to a practical use. Additional studies were carried out on direct pyrolysis process and incineration process in order. to compare these three thermal processings. The samples in this study were 5 dewatered cakes obtained from various sewage treatment plants and from various dewatering methods. The pilot plant for the experiments of this study consisted of a four-shaft indirect steam dryer with a total heating area of 10.1 m^2, a four hearth furnace with a total hearth area of 1.63 m^2 and its incidental facilities.

The results of the study can be summarized as follows;
(1) The overall heat transfer coefficient (U) in drying sludge cakes by the indirect steam dryer varies from 140 Kcal/m^2hr°C to 360 Kcal/m^2hr °C with the kinds of cakes.
(2) Pyrolytic furnace was mostly operated under the following condition; air ratio for combustibles in sludge 0.6, pyrolytic temperature 900°C, and retention time 60 minutes. When the loading rate to furnace was taken under 25 kg D.S./m^2·hr, more than 95 % of combustibles in sludge cake was decomposed by pyrolysis regardless of the kinds of cakes. Thus, pyrolysis is estimated to be as effective as incineration in weight reduction of sewage sludge.
(3) During pyrolytic operation, it was possible not only to prevent perfectly the oxidation of Cr^{+3} into hazardous Cr^{+6}, but also to reduce Cr^{+6} in fed cake. On the contrary, during incinerating operation, the oxidation of Cr^{+3} into Cr^{+6} occured and there was a danger of secondary pollution by dissolution of Cr^{+6} from ash. This fact was obviously confirmed by the test on tanneries sludge cake which contains 1.5 % chromium.
(4) There is no essential difference between pyrolytic and incinerating operations as for dispersion of heavy metals and that temperature is a dominant factor in dispersion.
(5) Considerable amount of HCN is detected in furnace exhaust gas in every thermal processings. Concentrations of NH_3 in

Table VIII. Comparison of economics of three thermal processings

(Terms of comparison)	Case I	Case II
Treating sludge weight (kg.DS/hr)	1,000	1,000
Moisture of cake (%)	75	60
Higher heating value (Kcal/Kg·DS)	2,500	2,500
Plant scale (ton cake/d)	100	60
(Comparison of runing cost)		
Drying-pyrolysis process	0.38	0.12
Direct pyrolysis process	0.76	0.39
Incineration process	1.00	0.45
(Comparison of construction cost)		
Drying-pyrolysis process	1.0	0.8
Direct pyrolysis process	1.0	0.8
Incineration process	1.0	0.8

exhaust gas depends on the process and is in the order of 10^3 ppm for pyrolytic operation and of 10^2 ppm for incinerating operation. They are almost completely decomposed by afterburning.

(6) When pyrolytic operation is performed at air ratio for combustibles of about 0.6, it is possible not only to prevent formation of NOx, but also to reduce NOx by reducing reaction by NH_3 or HCN in the furnace. In afterburning process of exhaust gas containing NH_3 and HCN, it is possible to prevent formation of NOx by two stage combustion under the condition of overall air ratio of about 1.1 and temperature at outlet of the chamber of below 950°C.

(7) The two direct feed **processes (direct pyrolysis process and incineration process)** were experimented, and the result of comparison with drying-pyrolysis process was obtained. This result suggests that pyrolysis processes, particulary drying-pyrolysis process can solve the problems of secondary pollution by Cr^{+6} or NOx and of large energy consumption for conventional incineration process.

(8) According to the simulated result, auxiliary fuel consumption is the smallest in drying-pyrolysis process while it was the largest in incineration process, in the whole range of the cake characteristics. According to economical comparison among the three thermal processings, the overall economics for the three thermal processings can be estimated from operation cost because construction cost is much the same in the three thermal processings, and drying-pyrolysis process is the most economical and incineration process is the most expensive.

(9) As a conclusion from the above summaries, this study achieved the purposes. Namely, drying-pyrolysis process was proved to be the most cost effective and environmentally acceptable process of the three processes and its technical feasibility was verified. Moreover, direct pyrolysis process was ascertained to be slightly inferior to drying-pyrolysis process in economical point, but to be the process which can improve conventional incineration process.

ACKNOWLEDGEMENT

One part of this study was conducted under the disbursement of the public works technology and development subsidy of the Ministry of Construction of Japan for fiscal 1976. Further, this study included one part of the study which was conducted by the studying group of Osaka Prefecture, Japan. The authors thank many co-workers.

LITERATURE CITED

1. Arai, N.; Kawamura, S.; Kagaya, M.; Sugiyama, Y.; Matsuhiro, N.; Kudo, I. Study on prevention of fuel NOx formation. (in Japanese) <u>Proceedings of the 13th Japan Heat Transfer Symposium.</u>, 1974 442-444
2. Bartok, W. <u>Monthly report, EPA Contract</u> 1969 No. CPA70
3. Kajitani, Y. Sewage sludge incineration and behaviour of chromium. (in Japanese) <u>PPM</u>., 1977 No. 9, 37-42
4. Kashiwaya, M. Studies on Sewage sludge pyrolysis. <u>Papers presented for the fifth US-JAPAN Conference on Sewage Treatment technology.</u>, 1977 No. 3, 73-104
5. Majima, T.; Kasakura, T.; Naruse, M.; Hiraoka, M. Studies on pyrolysis process of sewage sludge. <u>Progr. Water Tech.</u>, 1977 <u>9</u>, 381-396
6. Matsui, S.; Hiraoka, M. Air pollution problems produced by incineration of sewage sludge. (in Japanese) <u>J. of Japan Sewage Works Ass.</u>, 1974 <u>11</u>, No. 124, 13-22
7. Morita, Y.; Kimura T. Catalysis of basis meterials for gasification of carbon. (in Japanese) <u>J. of the Fuel Soc. Japan.</u>, 1977 <u>56</u>, No. 598, 77-83
8. Takahashi, S.; Narita I. Researching report on air pollution of sewage sludge incineration. (in Japanese) <u>Japan Sewage Works Agency Report.</u>, 1977 51-013
9. Takeda, N.; Hiraoka, M. Combined process of pyrolysis and combustion for sludge disposal. <u>Env. Sci. & Tech.</u>, 1976 <u>10</u>, 1147-1150
10. Takeda, N. <u>Doctor thesis</u> (Kyoto University) 1978

RECEIVED November 16, 1979.

Gasification of Solid Waste in Dual Fluidized–Bed Reactors

M. KAGAYAMA, M. IGARASHI, M. HASEGAWA, and J. FUKUDA

Tsukishima Kikai Co., Ltd., 17-15, Tsukuda 2-Chome, Chuo-Ku, Tokyo 104, Japan

D. KUNII

Tokyo University, Tokyo, Japan

In Japan, pollution emanating from conventional municipal refuse incinerator has become a problem these past ten or more years. The problem centers on air pollution by HCL and SOx in flue gas which has worsening as a result of the increased plastic content in refuse and water pollution of land-fill sites by disposed incinerator ash. We have developed a new solid waste treatment process to solve these problems. It took about seven years to develop the system which involves circulating sand particles between two fluidized bed reactors. In 1972, after performing a survey of research activities concerning pyrolysis for more than one year, TSK (Tsukishima Kikai Co., Ltd.) made the decision to develop the new solid waste pyrolysis system which is comprised of two fluidized bed reactors. This system has been applied to a cracking process as Kunii-Kunugi Process (1), which has been under development as a national project in Japan, for production of olefins from heavy oil. Fundamental pyrolysis test using a small, single fluidized bed reactor was performed in the first stage in 1973. Various fundamental data of thermal destruction of solid waste were obtained by this test. In the same year, a pilot plant consisting of dual fluidized bed reactors and auxiliary equipment was constructed. I.D. of the cracking reactor and the regenerator is 150mm and 200mm. About 10 kg/hr of solid waste was fed and pyrolized continuously. Using this plant, a continuous pilot test of the process has been made since 1974. Beside the test using these small units, a mock-up plant test on a larger scale was also made to obtain the engineering know-how for the design of dual fluidized bed reactors. The mock-up plant was operated under normal temperature and the sand was fluidized by the compressed air. The scale up factor from the pilot plant to the mock-up plant was 3 times. Through these test the feasibility of this system was confirmed, so a demonstration plant was constructed in 1975 nearby a pulp and paper mill in Miyagi Prefecture. The Ministry of International Trade and Industry (Japanese Government) granted financial support for the demonstration plant as a very important technology to be materialized promptly.

The plant has been operated successfully for about 7000 hours
since April 1976, disposing sludge from the pulp and paper mill,
municipal refuse, waste plastic and blocks of spent tires.

Description of Process

The pyrolysis equipment of the process comprise of the cracking reactor and the regenerator, as shown in Fig-1. These two
reactors are filled with sand which is used as a heat transfer
medium. Superheated steam is blown into the reactors through
nozzles located at the bottom. The sand goes up through the
reactors with steam and forms a fluidized bed zone. Then it goes
down from one reactor to the other through the circulation pipes
mainly by gravity. Thus the sand circulates between the two
reactors. Disintegrated solid waste is fed to the cracking
reactor where it is mixed with the hot sand to be dried and
cracked. By the cracking reaction, organic matter in the solid
waste is pyrolyzed into three components: fuel gas, tar and char.
Produced gas and tar are taken out of the top of the cracking
reactor with steam. Char overflows from the cracking reactor to
the bottom of the regenerator with the circulating sand through
the circulation pipe. In the regenerator, char is carried up by
steam from the bottom to the fluidized bed zone, where it comes
in contact with air and burns. If the amount of char is not sufficient to maintain cracking and combustion reaction, (a phenomenon which occurs when the solid waste has low calorific value),
auxiliary fuel such as oil or produced combustible gas must be fed
to the regenerator. Incineration flue gases come out from the
regenerator and provide heat to the combustion air through the
heat exchanger. After heat exchange, they enter the heat recovery
process and the gas cleaning process. Hot sand overflows from the
regenerator to the cracking reactor, and then again heats up the
solid waste. The circulating sand is cooled by cracking reaction
and reheated by char (and auxiliary fuel) combustion. As
mentioned above, since the cracking zone is separated from the
combustion zone, pyrolysis is made under the optimum oxygen-free
condition and high calorific fuel gas can be produced.
Inorganic coarse material goes down and accumulates at the bottom
of the reactors, then is discharged periodically through the
bottom valves. Fine inorganic particles such as ash are exhausted
with the flue gas from the top of the regenerator and are caught
by the multi-cyclone and electrostatic precipitator.

Actual Demonstration Facility

In 1975, a 40 ton/day demonstration plant was constructed
nearby a pulp and paper mill in Miyagi Prefecture. The main
purposes of the demonstration plant were to check the durability
of the construction materials, to establish the method of operation
and to study some other questionable items which could not be

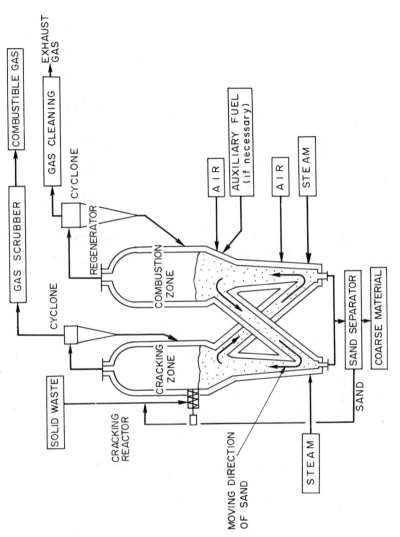

Figure 1.   Two fluidized-bed reactors

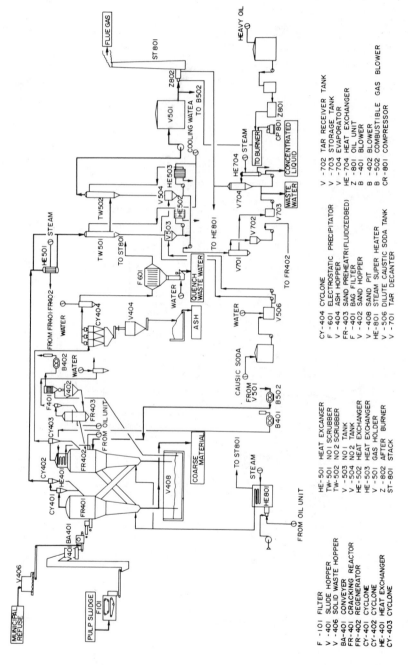

F - 101   FILTER
V - 401   SLUDE HOPPER
V - 406   SOLID WASTE HOPPER
BA-401   CONVEYER
FR-401   CRACKING REACTOR
FR-402   REGENERATOR
CY-401   CYCLONE
CY-402   CYCLONE
HE-401   HEAT EXCHANGER
CY-403   CYCLONE

HE-501   HEAT EXCANGER
TW-501   NO1 SCRUBBER
TW-502   NO2 SCRUBBER
V - 503   NO1 TANK
V - 504   NO2 TANK
HE-502   HEAT EXCHANGER
HE-503   HEAT EXCHANGER
V - 501   GAS HOLDER
Z - 802   AFTER BURNER
ST-801   STACK

CY-404   CYCLONE
F - 601   ELECTROSTATIC PRECIPITATOR
V - 404   ASH HOPPER
FR-403   SAND PREHEATR(FLUIDIZEDBED)
F - 401   BAG FILTER
V - 402   SAND HOPPER
V - 408   SAND PIT
HE-801   STEAM SUPER HEATER
V - 506   DILUTE CAUSTIC SODA TANK
V - 701   TAR DECANTER

V - 702   TAR RECEIVER TANK
V - 703   STORAGE TANK
V - 704   EVAPORATOR
HE-704   HEAT EXCHANGER
Z - 801   OIL UNIT
B - 401   BLOWER
B - 402   BLOWER
B - 502   COMBUSTIBLE GAS BLOWER
CR-801   COMPRESSOR

Figure 2.   Flow sheet of the demonstration plant

checked by the small unit test. The plant consists of the two
reactors of 2000mm I.D., feeding system, gas handling system and
other auxiliary equipment.   The plant has been operated for
about 7000 hours and necessary data obtained. During the operation
period, continuous operation for 1200 hours was twice accomplished
successfully. The flow sheet of this plant is shown in Fig-2.
Municipal refuse is taken from the practical incineration plant
and crushed by the hammer-type mill. Then it is stored in the
feed hopper. Pulp and paper sludge is dehydrated to a moisture
content of about 80% and stored in the sludge hopper. From the
hopper, crushed solid waste or sludge is taken out continuously
and supplied to the cracking reactor through the conveyor scale
where it is weighed. In case of the test for mixture of sludge
and plastic waste, crushed plastic waste is stored in the feed
hopper and is taken out and mixed with pulp and paper sludge at
the inlet of the conveyor scale.   After being supplied to the
cracking reactor, where temperature is around 700°C, all kinds
of solid wastes are treated in the same way. Produced combustible
gas is discharged from the top of the cracking reactor and passes
through the cyclones in which most of the particles are caught.
Residual particles contained in the gas are washed out at the
scrubbers where the gas is cooled at the same time and moisture
vapor is condensed. This condensed water is used as scrubbing
liquid and then is taken out continuously as process waste water.
The scrubber circulating liquid is cooled by cooling water at the
heat exchangers condensed water which is taken out from the
scrubbers is introduced to the tar decanter where tar is separated
from water by decantation. Tar-free water is concentrated in the
evaporator and concentrated liquid is treated in other equipments.
    Clean combustible gas which still contains hydrogen sulfide
is stored in the gas holder and is used as the fuel gas for the
regenerator and the after-burner.   Incineration flue gas which
is at a temperature of around 800°C is cooled by air for incineration at the heat exchanger which is installed on top of the
regenerator. After heat recovery, the flue gas goes through the
cyclone where large particles are caught, then passes through the
multi-cyclone and the electrostatic precipitator. In these
facilities almost all fine particles are caught. Particle-free
flue gas is discharged from the stack.   When heat is insufficient because of low calorific value of the solid waste, heavy
oil is supplied for the regenerator as auxiliary fuel. Steam for
fluidization is supplied from an off-site source. It is superheated by the superheater.

Operation Data of Demonstration Plant Test

    Municipal Refuse. In Table-I, average values of component
of municipal refuse are shown. Table-II shows the values of
chemical analysis which were made after pulverization. As shown
in Table-II, moisture content is lower than that of usual average

figure for Japanese urban refuse. This is due to the fact that water was evaporated during crushing and storage.

Table-I
Analysis of municipal refuse
(before pulverization, weight ratio on dry basis)

Items Percentage	% by weight
Wood waste	4.9
Fibers and cloth	8.1
Paper	48.6
Garbage	15.1
Iron	11.0
Nonironic metal	1.1
Glass and soil	4.9
Plastics	6.3
Total	100.0

Table-II
Chemical and physical analysis of
pulverized municipal refuse

Items		% by weight
Moisture		45.4
Combustible		37.6
Incombustible		17.0
	Total	100.0
	C	33.40
	H	4.42
Chemical	N	1.26
analysis	O	28.05
(dry	Total-S	0.47
basis)	Total-CL	1.00
	P	0.30
	Inorganic	31.10
	Total	100.00
	Bulk specific gravity (W)*	0.34
	Net specific gravity (D)	0.84
	Moisture (W)	45.4
	Ash (D)	31.1
	Ignition loss (D)	68.9
	Fixed carbon (D)	9.4
	Calorific value	3625 Kcal/kg

*W : Wet base, D : Dry base

Material Balance and Heat Balance. Required heat was mainly supplied by incineration of char, and some amount of produced combustible gas was fed as auxiliary fuel to the regenerator, as the amount of char was not sufficient for continuous thermal cracking. The material balance around the reactors is shown in Table-III and heat balance in Table-IV. Radiation and convection loss in Table-IV is larger than that of usual incinerators because of the thin refractory. It can be decreased in case of commercial plants. Energy balance of the total plant is shown in Fig-3.

Table-III

Material balance of municipal refuse pyrolysis

Input			Output		
Items		Kg/Hr	Items		Kg/Hr
Solid Waste	Moisture	726.4	C. gas*	Dry gas	393
	Combustible	601.9		Steam	1374
	Incombustible	271.7		Sub total	1767
	Sub total	1600	F. gas*	Dry gas	2770
Air feed		2650		Steam	721
Combustible gas		150		Sub total	3491
Steam		1136	Ash		278
Total		5536	Total		5536

* C. gas : Combustible gas
  F. gas : Flue gas from the regenerator

Table-IV

Heat balance of municipal refuse pyrolysis

Heat in			Heat out		
Item	x 10⁶ Kcal/Hr	%	Item	x 10⁶ Kcal/Hr	%
Municipal Refuse	3.182	67.4	Combustible Gas	1.812	38.3
Air	0.035	0.6	Steam 1	1.272	26.9
Combustible Gas	0.652	13.8	Flue gas	0.471	10.0
Steam	0.863	18.2	Steam 2	0.658	13.9
			Ash	0.054	1.1
			Heat loss	0.465	9.8
Total	4.732	100	Total	4.732	100

Steam 1 : Steam from the cracking reactor
Steam 2 : Steam from the regenerator

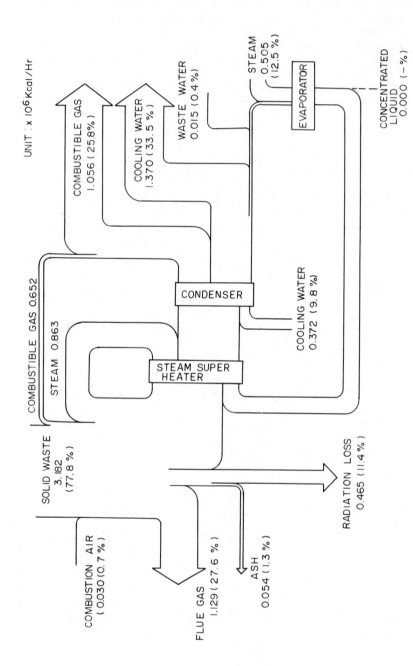

Figure 3.   Energy balance of municipal refuse pyrolysis unit ($\times 10^6$ Kcal/hr)

Combustible Gas. Component of the combustible gas which is produced by pyrolysis of municipal refuse is shown in Table-V. The produced combustible gas of the demonstration plant has a different composition from that of the small pilot plant. The former includes more carbon monoxide and less carbon dioxide than the latter. The difference in gas composition between the two tests seems to come from the fact that in case of the larger reactor, supplied solid waste is heated up promptly and over-cracking is prevented. Because of the high carbon monoxide content, the calorific value of the demonstration plant is higher than that of the small reactors.   Sampling of the combustible gas for chemical analysis was made at the outlet of the gas scrubber. The gas was not desulfurized, so the concentration of hydrogen sulfide is high as shown in Table-V. When combustible gas is supplied for off-site plants as clean energy, desulfurization (H_2S recovery) may be necessary, but in case of that gas is used on-site, either flue gas desulfurization after incineration or H_2S recovery can be applied depending on the situation.   Further clean gas can be converted to methane rich gas (town gas) as a substitute natural gas. In this case, the amount of town gas will be about half that of raw gas, and its calorific value is about 8500 Kcal/Nm^3. For example, the amount of town gas which could be produced from a 450 ton/day municipal refuse pyrolysis plant would be equivalent to the gas consumed by 100,000 people in Japan.

Table-V

Chemical analysis of combustible gas

Component	% by volume	
H_2	19.58	
CO	35.84	
CO_2	16.73	
CH_4	14.35	
C_2H_4	5.68	
Other hydrocarbons	3.40	
H_2S	0.34	
NH_3	0.75	ppm
HCL	15	ppm
HCN	0.5	ppm
N_2	4.08	
Total	100	
Calorific value	4716 Kcal/Nm^3	

Flue Gas. Incineration flue gas was continuously analyzed by an automatic analyzer. The average value of analysis are shown in Table-VI. From oxygen concentration it is clear that char incineration can be made under the condition of very low excess air.

NOx concentration is considerably low comparing with conventional incinerator. Particles in flue gas can be caught easily by the electrostatic precipitator, and it thus easily passes strict regulations for particles emission.

Table-VI

Analysis of incineration flue gas

Component	% by volume
O_2	1.9
CO_2	17.8
N_2	80.3
SOx	35 ppm
NOx	40 ppm

Ash. Two kinds of ash are discharged by this process. One of them is coarse inorganic matter which is taken from the bottom of the reactors, the other one is fine particle ash discharged from the multi-cyclone and the EP. The coarse material contains very little organic material because it remains in the reactors for a long time. Organic or combustible matter in fine ash is also small. It is less than two-percent as shown in Table-VII. Since all heavy metals are fixed in ash and are hardly soluble in water, the ash can be landfilled without any additional treatment. But it is preferable to solidify the ash for easy handling and minimizing the amount of soluble heavy metal in case of commercial plants. Solubility test was made for ash by the routine method stipulated by the Japanese Environment Protection Agency. The result is shown in Table VIII.

Table-VII

Analysis of ash

Items	
Ignition loss	1.83%
Net specific gravity	2.51
Bulk specific gravity	0.53
Mean particle size	18.1 μ
pH value*	9.83 [-]

* Measured as 10% slurry (It was stirred for one hour)

Table-VIII

Solubility test of heavy metal in ash

Heavy metal	Concentration ppm
Hexavalent chromium	< 0.025
T-cadmium	< 0.005
T-lead	<0.1
T-mercury	ND

Pyrolysis of Pulp and Paper Sludge. The filter cake containing about 80% moisture was supplied for the cracking reactor without predrying. Heavy oil was fed to the incinerator as the auxiliary fuel. This is different from the case of municipal refuse, but the combustible gas composition and calorific value, flue gas composition and ash were similar to that of municipal refuse. The chemical analysis of combustible gas and flue gas are shown in Tables IX and X.

Table-IX

Chemical analysis of combustible gas*
from pulp sludge

Component	% by volume
H_2	14.61
CO	40.26
CO_2	4.82
CH_4	14.58
C_2H_4	6.18
Other hydrocarbons	5.32
NH_3	1.35
H_2S	0.93
HCL	2.42
HCN	0.009
N_2	9.521
Total	100.0
Calorific value	4827 $Kcal/Nm^3$

* Gas was sampled at the inlet of the scrubber.

Table-X

Analysis of incineration flue gas from pulp sludge

Component	Content %
O_2	2.69
CO_2	13.82
N_2	83.47
SO_x	227 ppm
NO_x	33 ppm
Total	100.0

Pyrolysis of Mixture of Sludge and Plastic Waste. Large
amount of auxiliary fuel is necessary to pyrolyze the solid waste
of low calorific value such as pulp and paper sludge which contains a considerable of water. Since plastic waste has very high
calorific value, it will be used as a substitute for oil or gas
when it is pyrolyzed together with sludge. For the purpose of
confirmation of the above-mentioned idea, pulp sludge was
pyrolyzed with plastic waste. Operation was very successfull and
little auxiliary fuel was required. Combustible gas of high calorific value was produced. Analysis and material balance are shown
in Tables XI through XV.   Pyrolysis of plastic waste only was
also made and no trouble was encountered. Cracking and incineration temperature could be controlled easily. Chemical analysis
of the produced combustible gas were almost the same as that of
the sludge-plastic mixture.

Table-XI

Analysis of plastic waste and pulp sludge

			Plastic	Sludge	Mixture
Component %	Mixing ratio	D.B.	54.5	45.5	100
		W.B.	26.7	73.3	100
	Combustible		76.2	16.4	32.4
	Incombustible		0.4	6.9	5.2
	Moisture		23.4	76.7	62.4
	Total		100.0	100.0	100.0
Chemical analysis (dry basis) %	C		75.4	31.0	55.1
	H		12.2	4.3	8.6
	O		9.7	34.2	20.8
	N			0.4	0.2
	S		0.1	0.4	0.3
	CL		2.1	0.1	1.2
	Incombustible		0.5	29.6	13.8
	Total		100.0	100.0	100.0
	Bulk specific gravity	(W)	0.06	0.76	0.57
	Net specific gravity	(D)	0.53	1.35	0.90
	Moisture	(W)	24.3	76.7	62.4
	Ash	(D)	0.5	29.6	13.8
	Ignition loss	(D)	99.5	70.4	86.2
	Volatile matter	(D)	99.1	58.5	80.6
	Fixed carbon	(D)	0.4	11.9	5.6
Net calorific value			9435 Kcal/Kg	2886 Kcal/Kg	6450 Kcal/Kg
Bulk calorific value			6584 Kcal/Kg	158 Kcal/Kg	1828 Kcal/Kg

W: Wet basis   D: Dry basis

Table-XII

Material balance of the pyrolysis
of pulp sludge and plastic waste

Input			Output		
Items		Kg/Hr	Items		Kg/Hr
Mixed waste	Moisture	455.9	C. gas	Dry gas	150.4
	Combustible	236.4		Tar	30.5
	Incombustible	37.7		Steam	1054.0
	Sub total	730.0		Sub total	1234.9
Feed air		2933.4	F. gas	Dry gas	2944.9
Combustible gas		150.4		Steam	852.4
Tar		61.0			
Heavy oil		10.0		Sub total	3797.3
Steam		1190.0	Ash		42.6
Total		5074.8	Total		5074.8

Table-XIII

Heat balance of mixed waste pyrolysis

Input			Output		
Items	x 10^6 Kcal/Hr	%	Items	x 10^6 Kcal/Hr	%
Solid waste	1.777	40.1	Combustible gas	1.669	37.7
Feed air	0.044	1.0	Steam 1	0.962	21.8
Combustible gas	1.290	29.2	Flue gas	0.506	11.4
Tar	0.305	6.9	Steam 2	0.782	17.7
Heavy oil	0.100	2.3	Ash	0.003	0.1
Steam	0.906	20.5	Radiation	0.500	11.3
Total	4.422	100.0	Total	4.422	100.0

Steam 1 : Steam from the cracking reactor
Steam 2 : Steam from the regenerator

Table-XIV

Analysis of incineration flue gas

Component	% by volume
O_2	3.7
CO_2	13.6
N_2	82.7
SO_x	20 ppm
NO_x	54 ppm
Total	100

Table-XV

Chemical analysis of combustible gas
produced from the mixed waste

Component	% by volume
H_2	18.7
CO	17.6
CO_2	8.3
CH_4	18.9
C_2H_4	16.1
C_2H_6	5.0
C_3H_6	6.8
C_3H_8	0.7
C_4H_8	5.4
NH_3	1.0 ppm
H_2S	1700 ppm
HCL	150 ppm
HCN	0.5 ppm
N_2	2.3
Total	100
Calorific value	8,795 Kcal/Nm^3

Process Economics

The first commercial plant of the dual fluidized bed pyrolysis
process with a capacity of 450 ton/day (150 ton/day x 3 series) is
now under construction and is scheduled to commence operation in
October 1980 for disposal of municipal refuse in Funabashi City,
near Tokyo. The outline of this plant is as follows.

1) Composition of municipal refuse (wet base component).

Moisture	38 - 57%
Combustible	46 - 27%
Ash	16%
Total	100%

2) Operation time   24 hours/day (continuous)

3) Estimated conversion ratio of the combustible material

Gas	67.5%
Char	23.5%
Water	9.0%

4) Recovery of energy   Some of the steam generated by
gas-fired boilers and waste heat-recovery boilers is used
for the reactors to fluidized sand via waste water evaporator. The remains are used to generate 1500KWH of
electric power.

5) Flue gas

        SOx : Less than 370 ppm
        NOx : Less than 100 ppm
        HCL : Less than 10 mg/Nm^3
        Gas volume   Less than 3000 Nm^3/ton of municipal refuse

6) Ash   Ignition loss : Less than 1%

7) Process waste water. Not discharged from the plant. Recycled water is used in the process.

8) Utilities (Per ton of municipal refuse)

        Water   0.9 ton
        Electricity 154 KWH (Max.)
        Chemicals   2.8 dollars

9) Construction cost. $35,000,000 including cost of buildings for the plant, office and other civil works.

10) Operating cost ($ per ton of municipal refuse)

Labor	2.92	(29 persons)
Electricity	6.88	(11.8 yen/KWH in Japan)
Water	1.09	(250 yen/ton in Funabashi
Chemicals	2.80	City)
Maintenance	3.02	
Disposal cost	1.14	
Total	17.85	

        Estimation base : $1 = 200 yen

Conclusion

The outstanding features of this system are as follows.

1) High calorific value combustible gas can be produced, since produced gas is not diluted by flue gas in the reactor.

2) By means of these reactors, cracking occur under perfect oxygen free condition, and incineration also takes place under optimium condition.

3) Heating up of solid waste is achieved swiftly by a large amount of sand. It is preferable to prevent over-craking, so high calorific gas can be produced.

4) Char produced in the cracking reactor is burned in the regenerator and generated heat is used efficiently for cracking
reaction.

5) Since reaction is made at relatively low temperature (less
than 850°C), metals do not sinter and inorganic salts do not
melt.

6) Small amount of NOx and HCL is discharged.

7) The volume of flue gas is considerably smaller than that of
the conventional continuous incinerator.

8) The process has wide flexibility for variation of feed solid
wastes. It can be operated smoothly under stable condition
for various solid wastes.

Acknowledgements

The authors would like to express their appreciation to the
Ministry of International Trade and Industry for exceeding financial support to develop this process, and to Messrs. Y. Okino,
M. Hosono, Y. Nagayoshi and U. Ishii and others engaged in the
development of the process, for their sincere efforts to ensure
the success of the present project.

Literature Cited

1. Kunii, D.; Kunugi, T., paper presented on PD 19,
"9th World Petroleum Congress", May 12th, Tokyo.

RECEIVED November 16, 1979.

Disposal of Municipal Refuse by the Two-Bed Pyrolysis System

N. ANDOH, Y. ISHII, Y. HIRAYAMA, and K. ITO

Ebara Corporation, Fujisawa 4720, Fujisawashi, Japan

In the "two-bed pyrolysis system", fluidized fine solids circulate between a reactor to pyrolyze municipal refuse and an incinerator to burn up the produced char.
This system may have the following potentials;
(1) Gas of high heat value can be obtained due to little influence of incineration on the generated gas.
(2) Smoother fluidization can be expected with clean up effect by the incineration process.
(3) Less flue gas from the system.
(4) Similar system has been widely used and proved its reliability in the oil cracking industry.
When applying this system to the treatment of solid waste, however, several problems such as elimination of inorganics from the reactor and steadily feeding of the solid without gas leakage etc., as well as the control of fine solids circulation and gas contamination between reactor and incinerator must be solved. The development has been committed by the Agency of Industrial Science and Technology, MITI..

Plant Description.

Figure 1 indicates the schematic flow of two-bed pyrolysis plant. This plant consists of the pyrolysis reactor and the regenerator through which fluidized medium circulates, the pre-combustion burner, the raw material feeder, the cyclones, the heat exchangers, the vapor-liquid separator, the gas recycle blower, the air feeder, the gas cleaner and the produced-gas combustion device. The plant also includes various auxiliary equipment, the devices for measuring and recording flow rate, pressure and temperature, as well as the automatic analysis for gas.
Regarding design of this apparatus, the pyrolysis gas product recirculation system is adopted for fulidization to provide increased thermal efficiency, easy vapor-liquid separation and increased calorific value of gas product. The pipe which connects the two beds to each other is not provided with any moving valve;

0-8412-0565-5/80/47-130-541$05.00/0

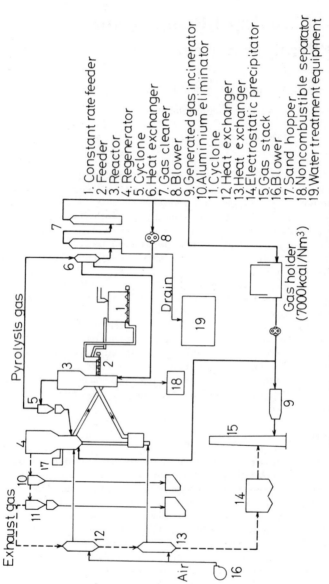

1. Constant rate feeder
2. Feeder
3. Reactor
4. Regenerator
5. Cyclone
6. Heat exchanger
7. Gas cleaner
8. Blower
9. Generated gas incinerator
10. Aluminium eliminator
11. Cyclone
12. Heat exchanger
13. Heat exchanger
14. Electrostatic precipitator
15. Gas stack
16. Blower
17. Sand hopper
18. Noncombustible separator
19. Water treatment equipment

Figure 1. Flow diagram of the two-bed pyrolysis system

rate of circulation of fluidized sand is controlled by a specifically developed controller adjusting flow rate of a very small
quantity of air flowing into the vertical conveyance portion. Two
connecting pipes are connected tangentially with pyrolysis reactor, so that circular motion created in the bed can promote dispersion of raw materials, thermal conveyance and inorganics elimination. To bring gas pressure inside the beds close to atmospheric pressure, the char and grain catching device with dual exhaust
type valve is designed with special care against possible pressure
loss. The same consideration is paid also to design of the heat
exchangers and the gas cleaner so that pressure loss by them can
be minimized. In the raw material feed, the gas sealing property
of which is largely affected by the level of material in the hopper, the solid material overflow system is adopted so as to maintain the proper level of raw material, and the gas sealing portion
is separated from the fluidized bed. All conveyors which handle
fluidized sand and raw material are fully enclosed to prevent sand
and raw material from scattering. Operation of the plant is automatically controlled at the central control room. Because this is
a gas producing facility, it is equipped with an abnormality
watching device, alarm device and emergency stop device.

Basic Study

The development of the system with basic study has been performed
since 1973.

(1) Experiments with single fluidized bed reactor of 300mm
in diameter. To clarify various operational influences on pyrolytic reaction, it is necessary to keep refuse composition constant. Since the composition of municipal refuse excluding ash is
similar to that of paper, preliminary study concerning influences
of the operating conditions such as temperature, gas velocity,
etc., on the generated products has been performed with use of
paper instead of refuse. In the next step, actual refuse experiments revealed the composition and heat value of the generated
gas, contamination of hazardous gases, heavy metals balance and
sintering problems etc., thus various input materials classified
by a pre-treatment sub-system were evaluated as an energy source
by pyrolysis.

(2) Cold model tests. The following techniques have been
developed by the aid of two cold models of 100mm and 400mm in each
diameter.
  (a) Techniques specifically associated with two-bed system, such
      as circulation control of fluidized medium, protection of
      gas leakage between two beds, etc..
  (b) Inorganics elimination from the reactors.
  (c) Optimum configuration of two reactors.

Figure 2.   Pilot plant, 5 t/d in capacity

(3) Feeder. Continuous and stable feeding of solid refuse without gas leakage has been established by developing a refuse feeder.

Pilot Plant of Two-bed Pyrolysis System.

Based on the above studies, the two-bed pyrolysis pilot plant of approx. 5T/D in capacity, as shown in the Fig. -2, was designed and constructed in 1975. Since then, its operational research has been performed for three years.

The plant was operated between 650°-750°C in the pyrolysis temperature under stable controls of pressure, temperature and sand circulation. Since the fluidized bed has an effect of heat accumulation, it is easy to control the operating temperature, though the heterogeneous refuse is fed to the bed. Consequently, no damage of reactor materials due to excessive temperature were recognized.

Table-1 and -2 indicate compositions of input refuse and produced pyrolysis gas components, respectively. Since no flue gas from combustion bed is mixed into the pyrolysis bed, calorific value of the pyrolysis gas shows extremely high. When using the group-III material, plastics-rich group classified by a pre-treatment sub-system, its calorific value is approx. 2.93×10^4 KJ/Nm^3 (7000kcal/Nm^3). In the steady state of the operation, the gas components showed almost constant.

Pyrolysis gas yield is 0.5 to 0.6 Nm^3/1kg of raw refuse, and the higher temperature yields more gas with less tar and char. Fig.-3 and -4 indicate the gas yields vs. temperature and the gas components vs. temperature, respectively. The energy recovery rate is highest between 650° to 750°C of temperature, where recovered energy rate shows 50 to 60% which is deemed to be considerably higher than that of a conventional incinerator. Due to low production of the char and the tar which is generally difficult to treat, the gas treatment and other maintenance are easy (Fig.-5).

As to material balance based on carbon, carbon content in raw material is distributed by pyrolysis process into gas, tar and waste water. Ca. 55 to 60% of total carbon is yielded into pyrolysis gas while ca. 30% of total quantity is yielded into char and exhausted from the regenerator as flue gas. The remainder forms tar, etc.. The pyrolysis gas contains several hazardous gases such as about 2,000 ppm of hydrogen sulfide, 200 to 400 ppm of hydrogen cyanide, and 600 to 2,000 ppm of hydrogen chloride. Referring to the gas of HCl, about 70 to 80% of produced HCl gas can be caught away in reactors when the additives (Ca(OH)2) is mixed in the refuse. Constituents of these hazardous gases can be reduced within the regulated values by alkali cleaning and water cleaning before collecting the produced gas. In the regenerator, scarcely any hazardous gases are generated because of combustion of only char produced in the pyrolysis reactor. A compact gas treatment equipment is effective enough to deal with such a small quantity

Table-1 Composition of Input Material

(%)

Component	$(II + III)_G$	III_G
C	44.4	51.6
H	6.32	7.51
N	1.14	1.05
O	40.68	33.08
S	0.14	0.17
cℓ	1.02	0.88
Ash	6.3	5.71
Calorific Value KJ/kg (Kcal/kg)	16300 (3890)	20800 (4960)

Table 2 Produced Pyrolysis Gas Components

Material Component	Total refuse	Classified refuse $(II + III)_G$	Classified refuse III_G
H_2 %	30.0	20.2	15.8
O_2	0.9	0.9	0.1
N_2	2.5	2.1	4.8
CO	34.7	26.8	21.1
CO_2	11.2	16.9	11.2
CH_4	12.7	16.7	23.8
C_2H_4	5.4	6.7	11.7
C_2H_6	1.1	3.6	4.7
C_3H_m	1.0	4.5	5.8
C_4H_n	0.5	1.6	1.0
Calorific Value (Net) KJ/Nm^3 ($Kcal/Nm^3$)	17200 (4100)	23700 (5670)	29300 (7000)

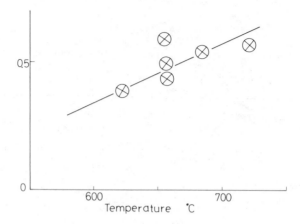

Figure 3.   Gas yields vs. temperature

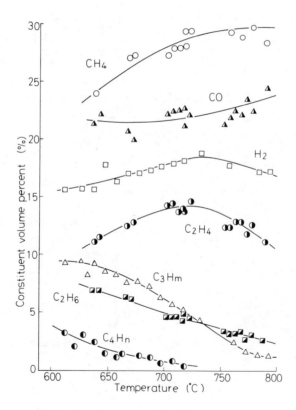

Figure 4.   Gas components vs. temperature

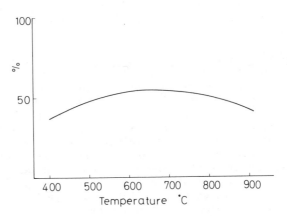

Figure 5.   Energy recovery rate vs. temperature

Table-3   Constituents and volume of Exhaust Gas

Gas Constituent	Two-bed System	Conventional Incinerator
HCl (ppm)	Trace	100 ~ 800
SOx (ppm)	< 20	20 ~ 100
NOx (ppm)	< 80	50 ~ 200
Nm^3/t	2000	5000
Input	plastics-rich refuse (III_G)	general refuse

of gases generated in the two-bed type pyrolysis plant. Generation of nitrogen oxide is negligible because this is essentially a
reducing atmosphere system. Table-3 shows the constituents and
volume of exhaust gas from this system as compared with conventional incinerator.

Treatment of Pyrolysis Waste Water. There is remarkable difference between waste water coming from the pyrolysis line and
that coming from the combustion line. Waste water from the pyrolysis line is a mixture of oil resulting from pyrolysis and water
content in raw material. Oil, about 10 percent of weight component
of waste water, is separated from water and is utilized as fuel
oil for the regenerator to heat fluidized sand. As waste water is
acidic, neutralization is necessary and the facility must be
made of acid-proof material. Since the generated gas is recycled
and no steam is used as fluidized gas, there is a greater reduction in polluted drain water containing condensed oil content, and
thus the drainage treatment system can be kept to a small scale
(Table-4).

Heavy Metals in Pyrolysis Ash. The ash caught by the cyclone
in the combustion line is composed of ash content in raw material
and pulverized fluidized sand. The ash, quantity of which is as
small as about 20g per 1kg of raw material, includes high density
of heavy metals. Ash can be fused together with glass, selectively collected from group I, to make harmless solid matter which may
be recovered as light-weight aggregates. A result of tracing
heavy metals in pyrolysis material shows that copper, lead and
chrome remain mainly in fluidized sand, while mercury and cadmium
remain in waste water.

Elimination of Aluminum Mixed in Municipal Refuse. After a
long period of operation of the pilot plant, some adhesive was
found on the inner wall of the No.1 cyclone for the regenerator,
it being feared that further progress of adhesion may cause possible lowering of dust-collection performance of the cyclone.
Thirty percent or more of this adhesive was aluminum. A protective method has been successfully achieved by the use of an aluminum eliminator composed of a rotary body placed close to the No.1
cyclone and a scraper located on the back of rotary body. The rotary body forces aluminum to adhere to its surface. Meanwhile,
the scraper sweeps out adhesive from the rotary body. Table-5 indicates the components of material collected by the aluminum
eliminator.

Techniques for Stable Operation. For the purpose of obtaining high calorific gas, the both connecting pipes between the
reactor and the regenerator must be packed full with sand to prevent the pyrolysis gas and the exhaust gas from mixing each other.
For packing pipes with sand, both height of the two beds in the
reactors should be controlled in a certain range. The bed height

Table-4   Raw and Treated Water Quality
of Pyrolysis Waste Water (III$_G$)

Component	Raw Water	Treated Water
pH (19°C)	2.1	7.4
N-Hex (mg/ℓ)	416.	2
Phenol	126.	N.D.
CN^-	33.8	0.27
T-P	1.3	N.D.
PCB	0.011	N.D.
Cu	0.3	N.D.
Pb	1.5	N.D.
Cd	0.15	N.D.
T-Hg	0.0006	N.D.
As	0.08	N.D.
T-Cr	1.89	N.D.
Mn	19.9	3.2
F	1.5	N.D.
Zn	13.7	0.9

Table 5   Components of the Powdery Material
Captured by the Aluminum Eliminator

Component	Al	Fe	Zn	Mg	Al_2O_3	SiO_2	Fe_2O_3	C	H_2O
%	32	0.24	0.46	0.04	15	19	4.2	1.5	5.6

can be expressed by the following equation.

$$H_{R12} = \frac{\mp \triangle P_T \pm \{(\,\delta_{R1} - \delta_{R2}) + f(F_S, V_f)\,\} + r_{R12} \cdot \dfrac{W}{r_S \times A_{R21}}}{r_{R12}\left(1 + \dfrac{r_{R21}}{r_{R12}} \cdot \dfrac{A_{R12}}{A_{R21}}\right)}$$

H_R : height of bed
$\triangle P_T$: differential pressure between the reactors
F_S : sand circulating rate
W : sand weight
V_f : superficial velocity in the reactor
A_R : area of the reactor
r_R, δ_R : constant determined by experiments
r_s : bulk density of bed
Suffix-1 : reactor
Suffix-2 : regenerator

Then these two bed-heights are the function of sand weight, sand circulating rate, the superficial velocity in the reactor, and the differential pressure between top of the reactors. As a feature of this system, the air volume used for lifting sand and circulating rate have a linear correlation, which is not influenced by the superficial velocity in the reactor. Accordingly, when the sand circulating rate is given by the balance of refuse feeding rate, moisture content and the temperatures of two beds, the air volume rate is determined as a consequence. Thus, the superficial velocity in the pyrolysis reactor can be determined independently. Also sand weight "W" is set up from the structural factor. Above simple relation can keep operation of this pyrolysis system stable for long period. This system has a feature that the operating point naturally returns into the center of the range against disturbances. Automation of the above relations with some auxiliary equipments, enables this two-bed pyrolysis system to be operated by a few men (Figure 6).

A Preliminary Economic Feasibility of the System. An economic feasibility for a proto-plant of 300 T/D in capacity can be estimated as follows; By pretreating municipal refuse of 300 T/D, the input of the pyrolysis system is estimated to be about 85 T/D, assuming the total system is similar to that of the demonstration plant (100 T/D). The working period of the system is assumed to be 3 months continuously on three shift operation by 21 men, shift detail of which will be as follows;

           plant supervisor 1
           chief operator 1
           plant controller 2
           labourer 2
                 total   5 for each shift

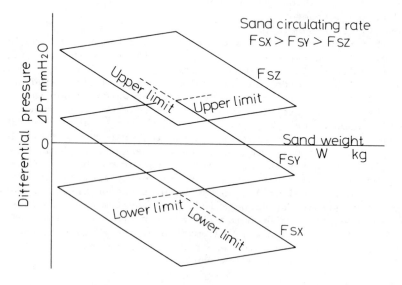

Figure 6.   Operation range

Operation costs for the plant excluding cost of residue disposal, in calender year 1977, is roughly estimated to be about $22.36 per ton as indicated below:

electricity	$10.3
water	$ 1.94
town gas (for pilot burner)	$ 0.945
sand	$ 0.475
chemicals and etc.	$ 8.7
	$22.36 per ton

Assuming the produced fuel gas is sold at the rate of 17.5 cent per 7000 kcal (1 Nm^3), the expected revenue of $53.5/T can make a profit of $31.14/T, offsetting the operation cost of $22.36/T. The residue volume from the system is less than 2%, thus its disposal cost will be below that of the incineration. Consequently, this two-bed system is deemed to be more economical than the conventional incinerator.

Conclusion

The results of the pilot plant of 5t/d in capacity, can be summarized as follows;
I. Comparison with conventional incinerator
   a) High recovery rate of total energy.
   b) Less flue gas that means easier control of air pollution.
   c) Less pollution of NOx.
   d) Easier control of temperature and less damage of reactor despite higher plastic contents.
   e) Less attack of HCl to the reactors (by additives).
II. Features as a pyrolysis system
   a) Less production of tar which is generally difficult to treat.
   b) Clean fuel gas of extremely high calory can be obtained.
   c) Less waste water (due to gas recirculation as fluidization).
III. Constructional features
   a) A new type of refuse feeder enabled stable feeding without gas leakage.
   b) Circulation control of fluidized medium can be easily performed by an air controller specifically developed.
   c) Tangential connection of connecting pipes to the reactors showed better effects on the operation.
   d) The system was comparatively stable against disturbance, and easy to operate.
   e) Aluminum eliminator newly developed solved a hazard of sticking molten aluminum.

Based on the know-how mentioned above, a large pilot plant of the pyrolysis system is now under construction, as a back-end sub-system of a demonstration plant (100 T/D in capacity) at Yokohama City.
The plant shall be totally operated in 1979

Literature Cited

K. Ito and Y. Hirayama "Resource Recovery from Municipal Refuse by Semi-wet Selective Pulverizing System". Conversion of Refuse to Energy, First International Conference and Technical Exhibition (1975) 354-359.

RECEIVED November 16, 1979.

Pyrolysis Process for Scrap Tires

SHIGEO KAWAKAMI, KIMIO INOUE, HIDEKI TANAKA, and
TAMIHARU SAKAI

Kobe Steel, Ltd., Mechanical Engineering Research Laboratory, 1-Chome,
Fukiai-Ku, Kobe, Japan

The number of scrap tires in Japan in 1977 was estimated to
be approximately 47 million with a total weight of 550 thousand
tons. Although some of them are reused as reclaimed rubber, re-
capped tires, etc., it is difficult to recycle all of the scrap
tires by the usual methods, therefore, there is a demand for new
recycling systems for scrap tires.

Kobe Steel, Ltd. formerly developed a new process for producing powdered rubber and constructed an actual plant with a capacity of 7,000 ton per year in Osaka in 1976 (1,2). The powdered
rubber is used as shock absorbing material for railway beds or as
a filler for rubber products.

Several pyrolysis processes using reactors such as a fluidized bed (3), a shaft furnace, an extruder and a rotary kiln have
also been studied in Japan. Pyrolysis using a rotary kiln has
been studied since 1973 (4). A pilot plant test was finished in
1976, and an actual plant with a capacity of 7,000 ton per year
has been constructed at Sumitomo Cement Co., Ltd. in AKO City,
Hyogo Prefecture, in 1979 (1). The plant will recover fuel oil
and carbon black from the scrap tires.

I. Process Flow.

An outline of the process flow from crushing and pyrolysis of
the scrap tires to refining of char into carbon black is shown in
Fig. 1.

The figures in the parentheses in Fig. 1 show the material
balance obtained from the pilot plant tests. The ratio of the recovered products changes according to the pyrolysis temperature as
shown in Fig. 2. The main products are char and oil, which amount
to approximately one third and one half of the total products recovered, respectively. Though the rate of pyrolysis increases at
higher temperatures, the pyrolysis temperature has to be kept
under 600°C in order to get the char from which high quality carbon black can be produced.

0-8412-0565-5/80/47-130-557$05.00/0

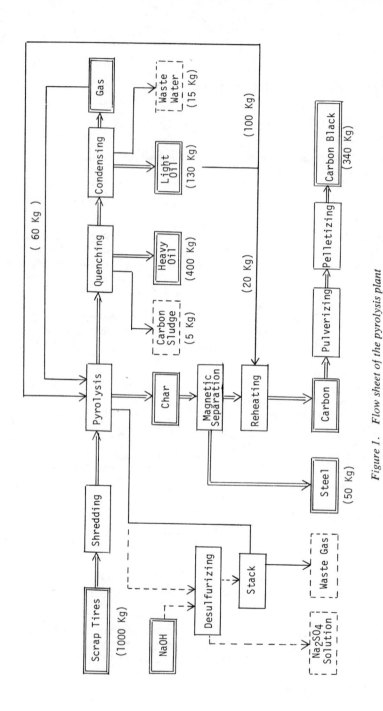

Figure 1.   Flow sheet of the pyrolysis plant

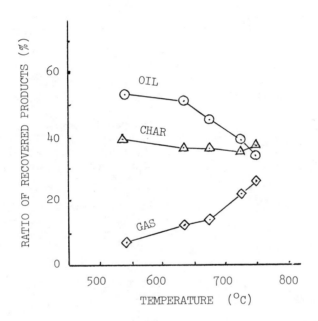

Figure 2. Effect of the pyrolysis temperature on the ratio of recovered products (PP test)

II. Research and Development

To examine the pyrolysis reaction of the scrap tires, pulverized scrap tires were heated in a quartz tube by an electric furnace (4). Then the continuous pyrolysis test plants in Table I were constructed to get the engineering data required for the design of an actual plant.

The pilot plant is shown in Fig. 3.   The longest period of continuous operation of the pilot plant was 120 hrs. In the mini plant tests (MP Test) scrap tires shredded into pieces about 10 mm in size were used. In the pilot plant tests (PP Test) scrap tires shredded into pieces about 50 mm in size as well as the fine shredded tires were used.

Table I. Test plants for the pyrolysis process

	Pyrolysis Bench Test (MP test)	Pilot Plant Test (PP test)
. Dimensions of the kiln (mm)	100^ϕ x 1,200*	425^ϕ x 2,000*
. Heating system	Indirectly heated by electricity	Indirectly heated by oil and gas
. Feeding system	Continuous feeder	Continuous feeder with triple damper
. Discharge system	Batch process	Continuous discharger with triple damper
. Capacity (Kg/H)	~10	~ 120 (Max.250)

* The length of the kiln inside of the furnace.

III. Actual Plant

The actual plant will process 7,000 tons of scrap tires collected each year by Automotive Tire Trading Association. The plant is funded by several public agencies such as Kikai Shinko Kyokai (Japan Society for the Promotion of Machinery Industry), Clean Japan Center, Hyogo Shigen Sairiyo Jigyo-dan (Hyogo Prefecture Resource Recovery Incorporation). **Fig. 4** shows a view of the pyrolysis plant and Fig. 5 is the layout of the plant.

The main apparatus of the plant has the following specifications.

Shredding. The scrap tires piled in the stockyard are carried to the shredding plant by hook conveyers. The scrap tires

Figure 3. View of the pilot plant

Figure 4. View of the actual plant

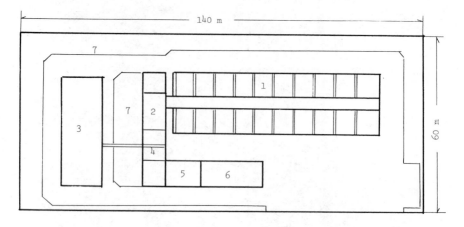

Figure 5. Layout of the pyrolysis plant: (1) tires stock yard; (2) shredding plant; (3) pyrolysis plant; (4) central control room; (5) carbon black plant; (6) storehouse; (7) green zone

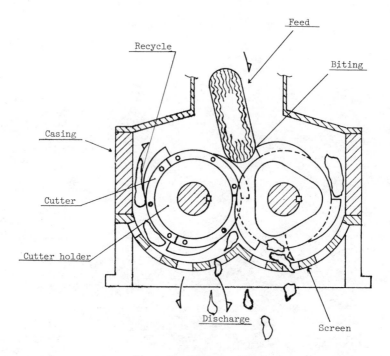

Figure 6. Scrap tire shredder

are shredded into pieces of about 100 mm to 150 mm by two separate shredders. The shredders are twin axis type ones, as shown in Fig. 6, which can shred the scrap truck and bus tires (TB tires) as well as the scrap passenger car tires (PC tires) without removing the beads. The shredded tires are stocked in a feed hopper and are fed at a constant rate into the pyrolysis kiln.

Pyrolysis Kiln. The crushed tires are pyrolized into char, oil, gas and steel by an indirectly heated rotary kiln (Rotary calciner). The rotary motion of the kiln and the action of a specially designed scraper provided in the kiln allow the contents to be heated uniformly. The scraper also removes the carbon deposits that stick to the inside of the kiln. The kiln is sealed by triple dampers at the inlet and at the outlet. The gas pressure in the kiln is kept automatically from 10 to 30 mm Aq. higher than the ambient air pressure in order to prevent air from coming in.

Indirect heating has the advantages of preventing oxidation of the recovered products and dilution of the gas product. By the use of a rotary kiln comparatively large pieces of scrap tires can be fed without removing steel cords or wire, thus decreasing the cost of shredding. The kiln of the actual plant is designed to have a capacity of 1,100 kg per hour.

Quenching and Condensing. The gas produced by pyrolysis is conducted into a quenching tower, into which hot oil is sprayed to condense heavy fraction oil (Heavy oil) and to settle carbon particles in the gas. A thickner and a decanter remove the suspended carbon particles and then the heavy oil is stored in an oil tank.

Light fraction oil (Light oil) with a little water is condensed in the second condenser. After the water has been separated, the light oil is used as fuel for heating the kilns. The remaining uncondensable gas, which is partly returned to the kilns as a purging gas, is also used as fuel for the pyrolysis kiln.

Reheating. After removal of steel by magnetic separators, the char is reheated in another small rotary kiln at temperatures of up to 600°C. to remove volatile matter.

Char Refining. The dried char is pulverized into fine powder (carbon black) by a Jet mill. A small amount of binder is added to the carbon black. The carbon black is compacted into flakes by rolls and is pelletized by a breaker.

Ancillary Apparatus. Flue gas from pyrolysis and reheating furnaces is conducted to the suspension preheater of the adjacent Portland cement plant to remove pollutants.

All equipment is controlled and under observation from the central control room.

IV. Properties and Uses of the Products

Oil. The ratio of the oil recovered is about 50 wt. % of the scrap tires. Table II shows a sample of the properties of the oil (5). Heavy oil can be sold as fuel oil. In the actual plant heavy oil will be blended with petroleum as fuel for cement kilns.

Table II. Properties of oil[*]

		Light oil	Heavy oil
Specific gravity	$15^\circ/4^\circ C$	0.8258	0.9856
Reaction		Alkaline	Neutral
Flash point	$^\circ C$	Room temp.	53
Viscosity	$50^\circ C$, m^2/s	0.730×10^{-6}	10.04×10^{-6}
Pour point	$^\circ C$	$-40 >$	-30
Carbon residue	wt.%	0.53	4.1
Ash	wt.%	$0.01 >$	0.01
Sulfur	wt.%	1.01	1.65
Water	wt.%	0.1	0.1
Copper corrosion	$100^\circ C$, 3 hr.	1 (1b)	1 (1b)
Calorific value	M.J/kg	43.9	42.5

[*] PP test.

Char. Char, most of which comes from carbon black and inorganic additives contained in the scrap tires, can be recovered as carbon black through the refining process.

Table III shows a sample of the properties of carbon black, in which the recovered carbon black is compared with the commercial carbon black (2, 4, 5, 6). The reinforcing properties of the recovered carbon black are higher than those of GPF grade. The recovered carbon black contains a high percentage of ash and a high level of PH (6,7). The properties of the carbon black depend upon the kind of the scrap tires and condition of the pyrolysis, especially the pyrolysis temperature.

Physical properties of the rubber vulcanizates reinforced by the recovered carbon black are shown in Fig.7(a)-(g), (5,6). The reinforcing properties of the recovered carbon black show a maximum at about 600°C. and decrease rapidly at temperatures over 600°C. The decrease over 600°C. is caused by poor dispersion of the carbon black in rubber compounds as shown in Fig. 8, which originates from the increase in cohesion force between the carbon particles of the char. It is thought that the carbon residue which is generated by the secondary decomposition of oil and gas produced, makes the cohesion force so high that pulverizing by a Jet mill is not enough to break the agglomates. This is proven by the fact that the compression force necessary to break char of a particular diameter (about 2 mmϕ) increases with the pyrolysis temperature as shown in Fig. 9.

Table III. Properties of carbon black and rubber vulcanization[*]

		Pyrolysis[**] carbon black	Commercial carbon black	
			GPF	HAF
Heating loss	wt.%	0.5	–	–
Iodine number	mg/g	157	30	94
DBP absorption	ml/100g	95	83	107
Volatile	wt.%	2.0	2.0 >	2.0 >
Benzen discoloration %		90.1	76.7	97.2
pH		8.9	6.3	7.3
Ash	wt.%	8.6	0.1	0.2
Mooney viscosity ML (100°C, 1+4)		68	66	71
Tensile strength $M \cdot N/m^2$	15'	27.0	18.0	28.2
	20'	27.5	18.4	28.3
	30'	27.5	18.9	28.2
	40'	25.9	18.7	28.1
Ultimate Elongation %	15'	640	920	600
	20'	590	740	540
	30'	500	560	450
	40'	440	490	400
300% Modulus $M \cdot N/m^2$	15'	8.6	5.6	12.0
	20'	10.1	7.2	13.5
	30'	12.7	10.2	17.6
	40'	14.7	11.5	19.9
Hardness JIS A	15'	64	61	67
	20'	65	62	68
	30'	66	62	69
	40'	67	64	69
Tear strength $K \cdot N/m$	15'	41.2	52.0	48.1
	20'	41.2	49.0	–
	30'	34.3	44.1	37.3
	40'	34.3	39.2	39.2

[*] Test recipe of rubber compounds   [**] PP test

SBR 1502	100 parts
Zinc oxide	5 "
Stearic acid	2 "
Sulfur	2.2 "
Accelerator MBTS	2.0 "
" TMTD	0.1 "
Carbon black	45 "

Vulcanizing temperature, 140°C

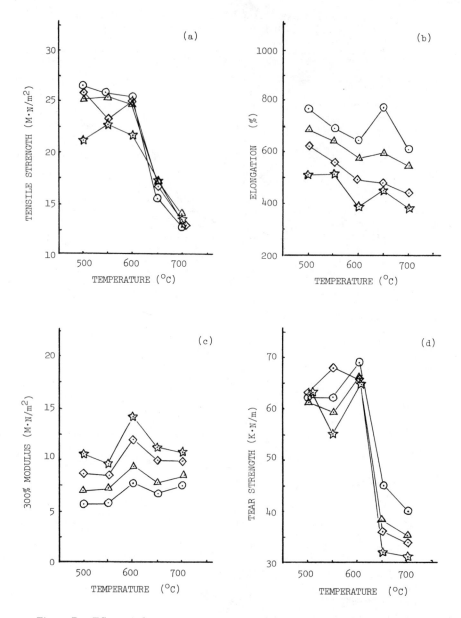

Figure 7.   Effect of the pyrolysis temperature on physical properties (a)−(g) of rubber vulcanizates (MP test): (a) tensile strength; (b) elongation; (c) 300% modulus; (d) tear strength; (e) hardness; (f) cut growth; (g) abrasion. Test recipe of rubber compounds is the same as in Table III; (⊙) 15'; (△) 20'; (◇) 30'; (☆) 40'

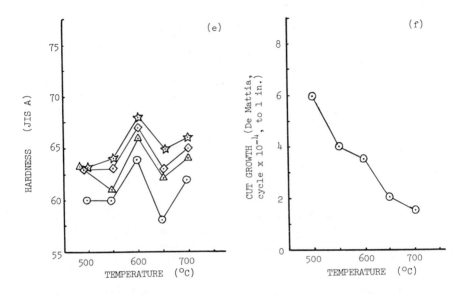

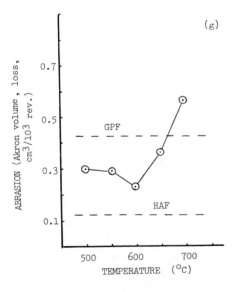

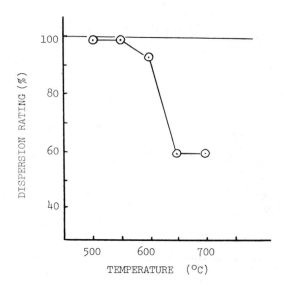

Figure 8.   Effect of the pyrolysis temperature on the dispersion of carbon black (MP test)

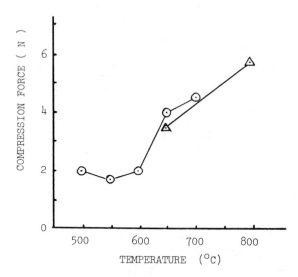

Figure 9.   Effect of the pyrolysis temperature on the compression strength of char ((○) MP test; (△) PP test)

There could be numerous uses for the recovered carbon black for rubber goods such as vehicle tires, footwear, rubber sheets, conveyer belts, dockfenders, etc. Some samples of rubber goods such as bicycle tires and safety shoes have been made with the recovered carbon black and their durability has been tested in comparison with commercial ones. At least in short term tests, for example, three months, the safety shoes made with the recovered carbon black showed higher abrasion resistance than the commercial shoes in which HAF-LS carbon black was used. The test results are shown in Table IV (6).

Flaps for truck tires have also been made with recovered carbon black using 10 to 40% of the total carbon black. These have been tested by JATMA (Japan Automotive Tire Manufacturers Association). The flaps with the recovered carbon black have shown sufficient durability in laboratory tests and actual road tests.

Gas. Gas has a high calorific value of more than 42,000 KJ/ NM^3. It is used as fuel for heating the pyrolysis kiln along with the recovered light oil.

Steel. Steel cords and wires can be recycled as scraps for the steel industry.

V. Economics

In order to evaluate the economics of the process, the plant costs and the total operating costs have been estimated assuming various plant capacities from 4,000 tons to 14,000 tons per year. The revenue from the sale of the recovered products has also been estimated by assuming the quantities and the price they could be sold for. The results are shown in Fig. 10, (5). The larger the plant size, the lower the net operating cost or the disposal cost, which is the difference between the total operating cost and the income from the sale of the recovered products. However, as collection of scrap tires from a wide spread area results in an increase in the transportation cost, the plant capacity should be carefully decided considering such factors as the plant site, the availability of scrap tires, the collection system, etc. Considering these factors, a plant of 7,000 tons per year capacity is thought to be a suitable capacity for Japan.

The disposal cost for a 7,000 tons per year capacity plant is estimated in Fig. 10 to be 2 yen per kg., therefore, it costs 12 yen for an average Japanese PC tire (6 kg). This disposal cost is low in comparison with usual disposal charges presently collected in various districts in Japan, which vary from 50 to 100 yen per scrap PC tire. Three fourths of the income from the sale of the recovered products depends upon the sale of the recovered carbon black, therefore, it is most important that salable carbon black is recovered from the pyrolysis process. In

Table IV. Actual use tests of safety shoes

Test places	Main factors for abrasion	Abrasion Index[*] (%)
Iron and steel works	Heat	96.0
Machine works	Oil, steel chips	91.3
		mean 93.8

[*]   Abrasion Index = (Abrasion of the design on the soles of the trial shoes) / (Abrasion of the design on the soles of the commercial shoes) x 100

[**]   Composition of the shoe sole compound.

        NBR   100 parts
        Carbon black   60   "
        $CaCO_3$ 25   "
        Plasticizer   12   "
        Antioxidants, Accelerators, Sulfur

        Carbon black : Recovered carbon black (PP test)
                         and HAF-LS

[***] The number of shoes tested was ten in each works, and the term of the actual test was three months.

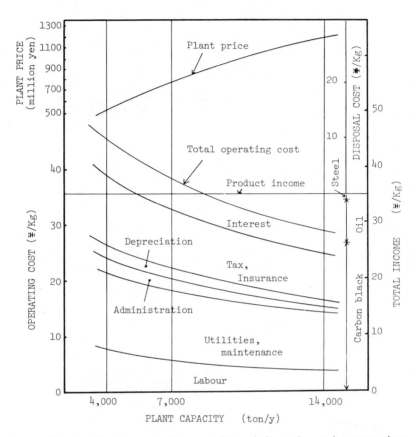

Figure 10.   Relation between plant capacity and disposal cost for scrap tires:

(1) products and selling income (carbon black: 34 wt % yield, 80 yen/Kg, 27.2 yen/Kg tire; oil: 40 wt % yield, 20 yen/Kg, 8.0 yen/Kg tire; steel: 5 wt % yield, 10 yen/Kg, 0.5 yen/Kg tire; total income: 35.7 yen/Kg tire.) (2) depreciation: 9%/year of plant cost; (3) tax, insurance: 2% of plant cost; (4) interest: assuming the plant's funds for building all borrowed at 8%/year from a bank (equal repayment for 10 years, average rate of interest at 4.8%/year); (5) these figures are based on estimations in 1976.

the actual plant the pelletized recovered carbon black, which is named as "REC Carbon" will be recycled for the rubber industry.

VI. Summary

A pyrolysis system has been developed to recover useful material from scrap tires. An actual plant was constructed in 1979. The plant will be on a trial for a year to demonstrate that the equipment of the plant satisfies the test specifications and that the recovered carbon black has the quality demanded by the rubber industry. After passing one year tests, it will be put into commercial operation. This project is a full scale recycling for scrap tires supported by public agencies which also supported the pulverizing plant in Osaka. It is expected that the completion of the project will open a new course for recycling and disposal of waste.

Acknowledgement.

The authors gratefully acknowledge members of the Industrial Research Institute of Hyogo Prefecture for the support of the carbon black and rubber compounding tests.

References

(1) S. Kawakami, K. Sawa, Kobe Steel Engineering Report, 1978, 28, 93.

(2) S. Kawakami, K. Inoue, K. Sawa, Kobe Steel Engineering Report, 1979, 29, 23.

(3) Y. Saeki, G. Suzuki, Rubber Age, 1976, 108, 33.

(4) A. Takamura, K. Inoue, T. Sakai, "Resources Recovery by Pyrolysis of Waste Tires." Conference Refuse to Energy, 1975, 532.

(5) S. Kawakami, K. Inoue PPM. 1977, 8, 20

(6) Y. Hirata, S. Yokoyama, T. Shikada, K. Inoue, T. Sakai, T. Asai, S. Kishi. Journal of the Society of Rubber Industry, Japan. 1978, 51, 283.

(7) K. Fujimoto, T. Miyajima, S. Kijima. Journal of the Society of Rubber Industry, Japan. 1978, 51, 938

RECEIVED November 16, 1979.

PUROX System Demonstration Test on Simulated Japanese Refuse

T. MASUDA

Showa Denko K.K., Environmental Systems Department,27-17, Hamamatsu-cho 1-Chome, Minato-Ku, Tokyo 105, Japan

T. F. FISHER

Union Carbide Corporation, Linde Division, Tonawanda, NY 14150

The PUROX System, developed by Union Carbide Corporation of the U.S., pyrolyzes refuse by use of oxygen. It additionally purifies and recovers the pyrolysis gas and collects the inorganic residue in the form of slag. The wastewater generated during gas purification is discharged after treatment. There are several papers (1, 2, 3, 4) published on the details of the system. Figure 1 shows the schematic flow of a typical PUROX System.

Showa Denko K.K. has imported the technology of the system and modified it to establish a process that meets conditions peculiar to the Japanese market, which are as follows:

Refuse Composition

Since Japanese refuse contains a higher proportion of household garbage, its moisture content is higher than that of its U.S. counterpart. Additionally, its plastic content has been increasing in recent years. A comparison of Japanese and U.S. refuse is given in Table I.

TABLE I.   COMPARISON OF JAPANESE AND U.S. REFUSE

		Japanese Refuse[*]			U.S. Refuse
		H.M.R.	S.M.R.	L.M.R.	
Combustibles	wt.%	25.5	31.9	40.0	49.5
Ash	"	6.0	13.8	20.0	19.5
Moisture	"	68.5	54.3	40.0	31.0
Lower H.V.	kJ/kg	3,050	5,360	7,540	8,080
	(kcal/kg)	(730)	(1,280)	(1,800)	(1,930)
Higher H.V.	kJ/kg	5,190	7,200	9,170	9,630
	(kcal/kg)	(1,240)	(1,720)	(2,190)	(2,300)

Utilization of Recovered Gas

The potential for utilizing the recovered gas is extremely limited due to the distance of plant locations

[*]See abbreviations

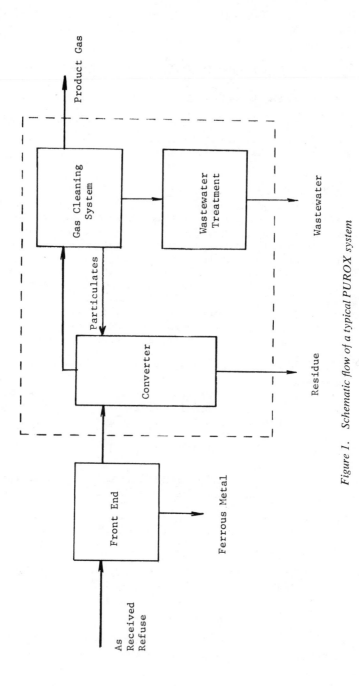

Figure 1.   Schematic flow of a typical PUROX system

from places of possible use and the restraints imposed
by laws and regulations including the gas utility industry law.

Wastewater Treatment

Since the sewer system is poorly developed at many
locations and the competent government agencies for
sewage treatment are different from those for refuse
treatment, the wastewater has to be brought below the
relevant effluent standards within the refuse treatment
plant. The cost for such wastewater treatment is substantial, one reason being the greater amount of wastewater generated due to the higher moisture content of
the refuse.

To meet the abovementioned local conditions, Showa Denko has
established a process whereby the pyrolysis gas is directly combusted and the generated heat is utilized for evaporation and
oxidation of the wastewater from the refuse pit (hereinafter
called the Dry Process). The schematic flow of the Dry Process
PUROX System is as shown in Figure 2.

Simulation Test

Using the PUROX System pilot plant of Union Carbide located
at Tonawanda, New York, U. S. A., Showa Denko conducted a demonstration test on simulated Japanese refuse from June through
August, 1978, after making some modifications of the plant. An
overall view of the pilot plant appears in Figure 3. The purpose
of the demonstration was to achieve the following three objectives.

1) To conduct tests on simulated Japanese refuse and thereby
   collect additional data to supplement those obtained on
   U.S. refuse, in order to obtain the technical approval of
   the Ministry of Health and Welfare (MHW) of Japan for the
   PUROX System,

2) To analyze in accordance with procedures provided in the
   Japanese Industrial Standard or those established by the
   Environment Agency and MHW of Japan, in order to evaluate
   outgoing streams on the basis of Japanese national environmental standards and emission and effluent standards,

3) To obtain engineering data necessary for the design of
   the components of the Dry Process, namely, the gas combustor, waste heat boiler, and electrostatic precipitator.

The refuse used for the test was shredded refuse from St.
Catharines, Canada. Glass, plastics, silage, fruits and vege-
tables, and water were added to them to simulate Japanese refuse.
Refuse of high, intermediate, standard, and low moisture were
prepared to simulate typical Japanese refuse. The refuse mixing
ratio for the four simulated refuse are given in Table II. The
quality of the simulated Japanese refuse tested as well as that
of Japanese refuse are shown in Figure 4.

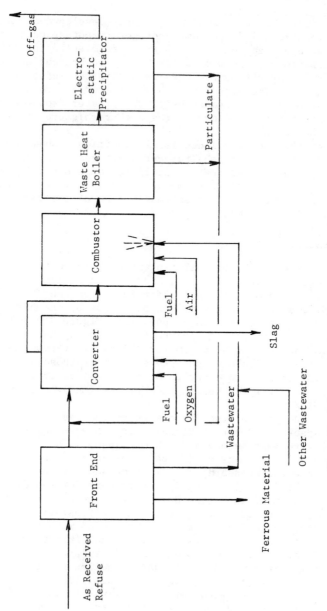

Figure 2.   Schematic flowsheet of the dry process PUROX system

Figure 3.   Overall view of the pilot plant

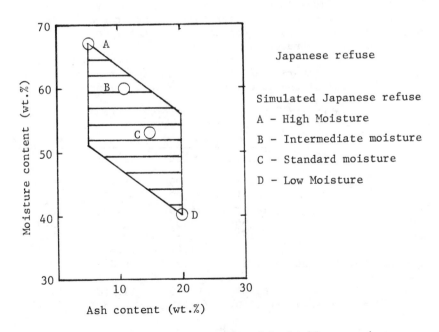

Figure 4.   Quality of Japanese refuse and simulated Japanese refuse

TABLE II. TYPICAL REFUSE MIXING RATIO

	H.M.R.*	S.M.R.*	L.M.R.*
Raw shredded refuse (wt.%)	17.6	37.0 - 38.2	46.8 - 47.6
Silage	0.0	10.4 - 10.6	3.0 - 3.1
Shredded potato	58.0	10.4 - 11.1	9.1
Polyvinylchloride	0.9	1.8 - 1.9	2.9 - 3.0
Polyethylene	1.8	3.8	5.9
Glass	1.6	5.1 - 5.2	9.8 - 10.0
Water	20.1	30.2 - 30.6	21.9 - 22.0

Although the pilot plant had a refuse treatment capacity of 20 Mg/d, it was usually operated at reduced load due to limitations coming from the refuse preparation site. Also schedule constraints prevented replacement of the desuperheater and scrubber employed to clean fuel gas in the pilot plant, by a waste heat boiler and electrostatic precipitator, such as are employed in the Dry Process PUROX System design. Consequently, dust was not recycled.

Refuse prepared at a set mixing ratio was supplied by apron conveyor to the feeder. The feeder pelletized the refuse and charged the pellets into the converter, where it was dried, pyrolyzed, combusted and melted to form pyrolysis gas and molten slag. The slag, or inorganic residue, was continuously tapped from the converter, quenched in the water tank and collected in the slag bunker. The pyrolysis gas was combusted in the combustor and discharged after scrubbing.

The process flow of the pilot plant is shown in Figure 5, together with a material balance for the case of high moisture refuse. A typical energy balance is shown in Figure 6.

TABLE III. ANALYSIS OF PYROLYSIS GAS

		Japanese Refuse*			U.S. Refuse
		H.M.R.	S.M.R.	L.M.R.	
CO	dry gas vol.%	16.00	30.09	28.64	38
CO_2	"	58.91	38.93	36.40	27
H_2	"	18.41	24.16	25.69	23
CH_4	"	2.96	2.52	3.77	5.0
CmHn	"	2.30	2.33	3.85	5.1
Others	"	1.42	1.97	1.65	1.9
H_2O	Nm^3/Nm^3 dry gas	1.3	1.2	0.8	0.7
Lower H.V.					
	kJ/Nm^3 dry gas	6,900	9,420	11,510	14,570
	(kcal/Nm^3 dry gas)	(1,650)	(2,250)	(2,750)	(3,270)

*See abbreviations

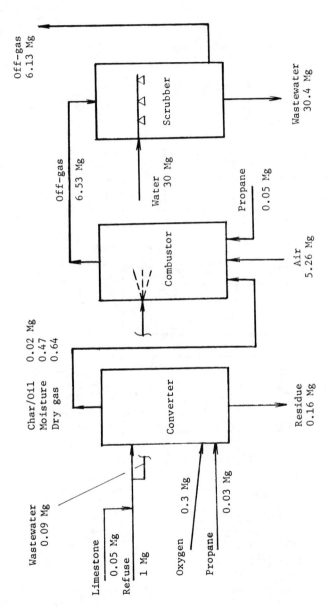

Figure 5. Process flow of the pilot plant and material balance (high moisture refuse)

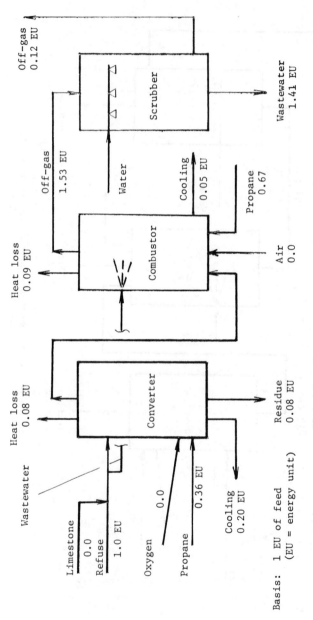

Figure 6.   Energy balance of the pilot plant (high moisture refuse)

A comparison of the pyrolysis gas obtained with Japanese standard refuse and with U.S. refuse (dry gas base) is given in Table III. It is apparent that higher percentage of water and CO_2 in the gas are obtained in the case of Japanese refuse and that the heating value per unit volume of the dry gas obtained is lower.

A comparison of the slag composition is given in Table IV. The ignition weight loss figures, which show the percentages of uncombusted materials, for both the Japanese and the U.S. refuse are substantially lower than the standard Japanese values obtained with the conventional stoker type incinerator, which are below 10 wt.% for a furnace of less than 200 Mg/d and below 7% for a furnace of more than 200 Mg/d.

TABLE IV. SLAG COMPOSITION

		Japanese Refuse			U.S. Refuse
		H.M.R.	S.M.R.	L.M.R.	
SiO_2	(wt.%)	51.6	59.3	56.4	59.7
Al_2O_3		6.7	7.1	4.8	10.5
CaO		21.9	10.2	18.0**	10.3
Na_2O		5.3	7.3	5.9	8.0
FeO		7.4	8.1	10.9	6.2
Ignition loss		0.2	0.05	0.1	0.4

** Limestone added as neutralization agent.

The slag leaching test results are given in Table V. The analyses were conducted in accordance with Notification No. 13 of the Environment Agency of Japan. For purposes of comparison, the leachate standards established by the Prime Minister's Office of Japan are also given.

These results clearly show that the amounts of hazardous substances leached from PUROX System slag are much less than those of the standards. The inert nature of the slag is considered attributables to the pyrolysis of hazardous materials such as organic phosphorus, cyanide, and PCB to form more stable and less environmentally objectionable substances, the temperature being as high as 1,650°C at the bottom of the converter, and trapping of trace metals in the glassy matrix.

Refuse to which PCB and dry cells and Ni-Cd cells had been added was tested. It was not possible, however, to observe the behavior of PCB and heavy metals since their concentrations in the slag did not show any marked difference. This may have been due to addition of cells that had each been cut into three pieces, thereby causing spotty distribution of the heavy metals.

Table VI gives the composition of slag when PCB was and when it was not added. As shown in the Table, the PCB content was below the detection limit in both cases. It is considered that

PCB either decomposes or exists in the slag in such a form as not to leach out and, therefore, does not cause environmental problems.

TABLE V.   SLAG LEACHING TEST RESULTS

(mg/1)	Inland Landfilling			Landfilling on the Sea		
	Analysis	Detection Limit	Standard	Analysis	Detection Limit	Standard
Alkyl mercury	ND	0.0005	ND	ND	0.0005	ND
Total mercury	ND	0.0002	0.005	ND	0.0002	0.005
Cadmium	ND	0.0025	0.3	ND	0.0025	0.1
Lead	0.023	0.005	3	0.015	0.005	1
Organic phosphorus	ND	0.0006	1	ND	0.01	1
Hexavalent chromium	ND	0.01	1.5	ND	0.01	0.5
Arsenic	0.0025	0.0007	1.5	0.0042	0.0007	0.5
Cyanide	ND	0.002	1	ND	0.002	1
PCB	ND	0.0005	0.003	ND	0.0005	0.003

TABLE VI.   EFFECT OF PCB ADDITION ON SLAG COMPOSITION
(Standard Moisture Refuse)

	PCB added	PCB not added
Slag composition (wt.%)		
SiO_2	59.3	48.8
Al_2O_3	7.1	6.2
CaO	10.2	17.6
Na_2O	7.3	5.5
FeO	8.1	17.9
Slag leaching test (ppm)		
PCB - pH 6	<0.001	<0.001
- pH 8	<0.001	<0.001

The combustor off-gas corresponds to the throughput to the gas cooler (e.g., waste heat boiler) of a commercial plant. The gas analysis is given in Table VII. The values for the conventional stoker type incinerator are those in the case of large cities in Japan. They are average values with the exception of HCl, which is slightly higher. The amounts of particulate, SOx and HCl contained in the off-gas of the Dry Process PUROX System are much less than those of the stoker incinerator. The reason is considered to be that the Cl^-, SO_4^{2-}, etc. combine with alkali metals and shift into the slag in the converter.

TABLE VII. COMBUSTOR OFF-GAS ANALYSIS

	Neutralization Agent						
	Not added	Added**					
	S.M.R.	H.M.R.	S.M.R.	I.M.R.	L.M.R.	Stoker Type	Standard
(vol.ppm)							
NOx	108	83	120	64	114	119-204	250
SOx	10	8	8	15	7	41- 79	
HCl	117	54	55	38	26	421-712	430
Parti-culate*	0.65	0.76	0.72	0.38	0.55	1.2-4.4	

* g/Nm^3

** Limestone of about 5 wt.% of the refuse was charged into the feeder together with the refuse.

Furthermore, it was confirmed that addition of limestone to the refuse as the neutralization agent reduces the volume of HCl in the off-gas to about 60%. Further study is required, however, of the relationship between the HCl removal rate and the amount of limestone added and of the method of such addition as well. With the exception of its particulates, off-gas from the Dry Process PUROX System can satisfy Japanese national emission and effluent standards without any special treatment.

An analysis of the collected particulate is given in Table VIII. It is apparent that the particulates contain practically no uncombusted carbon. In other words, it is clear that the char in the gas was almost completely combusted under the operating conditions of the gas combustor used in the present tests.

TABLE VIII. ANALYSIS OF PARTICULATE

C	(wt.%)	0.3	K	(wt.%)	6.4
Cl^-		22.3	Na		17.1
SO_4^{2-}		21.2	Zn		10.1

The percentage of the total input of each trace metal which was shown to exit with the slag is given in Table IX. This result for the pilot plant is based on analyses of the outgoing streams. The gas stream analysis was obtained at the combustor outlet, which would correspond in a commercial plant to the inlet of the boiler. The gas would, however, be subsequently cleaned in an electrostatic precipitator and the collected particulates recycled to the converter to be turned into the slag in the case of a commercial plant. The estimation for a commercial plant is premised on use of an electrostatic precipitator designed for a commercial plant of which the outlet particulate concentration is 0.1 g/Nm^3 dry gas. It is expected that more than 99.7% of the trace metals would be converted to slag.

TABLE IX.   TRACE METAL DISPOSITION IN SLAG

	Pilot Plant	Estimation for Commercial Plant
Cd (wt.%)	98.8	99.8
Pb	98.0	99.7
Cr	99.9	100
Cu	99.9	100
Zn	98.4	99.7
Fe	100	100
Mn	100	100
Ni	99.2	99.9

In comparison with the conventional stoker type incinerator, the Dry Process PUROX System is, as can be seen from the foregoing discussion, a highly pollution-free method for refuse treatment.

Economics of the Dry Process PUROX System

On the basis of the price levels as of spring 1979 and at the translation rate of 200 yen to one dollar, the construction cost of the Dry Process PUROX System in Japan is estimated to be about $13,000,000 for a 200 Mg/d facility ($65,000 for Mg of refuse), exclusive of the land, utility supply facilities down to the battery limits, and fixtures and supplies. This is about 10% higher than the construction cost in Japan of the stoker incinerator, which is estimated to be about $58,500 per Mg of refuse.

Table X shows the running costs of the Dry Process PUROX System as well as the stoker incinerator. In Japan solid refuse treatment plants are operated by local municipalities and taxes and depreciation are not considered components of the running costs of solid refuse treatment plants. They, therefore, comprise electricity, water, chemicals, auxiliary fuel, labor, and maintenance costs. When a comparison is made on the basis of these costs, the Dry Process PUROX System is somewhat higher than the stoker incinerator.

However, the cost difference becomes less if consideration is given to landfilling costs, which include the cost not only of transportation but also of preparation of the site and treatment of the water coming out of it. While the costs are considered to be $5-10 per Mg of refuse for the stoker incinerator, it is about 1/5 that for the Dry Process PUROX System since the volume of slag it generates is much less than that of the incineration ash coming from the stoker incinerator and there is no leaching of the water from the landfilling site.

In view of the fact that available landfilling space is getting scarce in Japan, making it increasingly necessary to

TABLE X.   RUNNING COSTS OF DRY PROCESS PUROX SYSTEM

(Design capacity: 150 Mg/d)

	Unit Price	Dry Process PUROX System		Stoker Incinerator	
Utilities					
Electricity	10¥/KWH	200 KWH	2,000¥	70 KWH	700¥
Water	25¥/m³	1.3 m³	33	1.3 m³	33
Nitrogen	–	–	40	–	–
Fuel	32/kg	20 kg	640	1.2 kg	39
Chemicals					
Ash pelletizing agent	50¥/kg	0.7 kg	35	4.5 kg	225
Others	–	–	25	–	25
Lub. oil and grease	–	–	40	–	20
Sub-total			2,813¥ (14.1$)		1,042 (5.2$)
Maintenance			657¥ (3.3$)		657¥ (3.3$)
Manpower		14 p.	1,130¥ (5.6$)	15 p.	1,200¥ (6.0$)
Landfilling of incineration ash			av. 300¥ (1.5$)		av. 1,500¥ (7.5$)
Total			4,900¥ (24.5$)		4,399¥ (22.0$)

* Consumption and cost figures are as per Mg of as-received refuse.

transport incineration residue long distances, generation of much
less an amount of incineration residue is a particularly attractive feature of the Dry Process PUROX System.

ABBREVIATIONS

H.M.R.	High Moisture Refuse
H.V.	Heating Value
I.M.R.	Intermediate Moisture Refuse
L.M.R.	Low Moisture Refuse
ND	Not Detected
S.M.R.	Standard Moisture Refuse

LITERATURE CITED

1. Anderson, J. E., The Oxygen Refuse Converter,
    Proceedings of the 1974 National ASME Conference
    Incineration Division, April 1974, pg. 337.

2. Fisher, T. F.; Kasbohm, M. L.; Rivero, J. R.,
    A.I.Ch.E. 80th National Meeting, September 1975.

3. Moses, C. T.; Rivero, J. R., Design and Operation of the
    PUROX System Demonstration Plant, Fifth National Congress
    on Waste Management Technology and Resource Recovery,
    Dallas, Texas, December 1976.

4. Moses, C. T.; Young, K. W.; Stern, G.; Farrell, J. B.
    Co-Disposal of Sludge and Refuse in a PUROX Converter
    ACS Symposium on Solid Wastes and Residues at the 175th
    Meeting of ACS, Anaheim, California, March 1978.

RECEIVED November 16, 1979.

Integrated System for Solid Waste Disposal with Energy Recovery and Volumetric Reduction by a New Pyrolysis Furnace

TATSUHIRO FUJII

Technology & Development Headquarters, Hitachi Shipbuilding & Engineering Co., Ltd., 6-14, Edobori 1-chome, Nishi-Ku, Osaka 550, Japan

Rapid progress in incineration technology has been brought about in the field of solid waste disposal. With the aim of reducing solid wastes to render them harmless, great progress has been advanced due to the traditional priority placed on incineration in Japan.

Recent years have seen a great improvement being made in the quality of life, with the population being concentrated in modern cities. However, rising criticism has emerged against pollution problems, calling for greater controls to protect people's health and safety. To meet these needs, administrative measures have been taken to tighten the regulations on pollution-creating factors.

At the same time, a growing problem has developed over the acquisition of incineration plant sites, ash dumping grounds, etc.

Under these circumstances, new needs have arisen for improved technological approaches to meet the diversified requirements of communities, each with its own particular refuse disposal problems. Steady technological progress in this field has brought about alternative possibilities to incineration as methods for waste disposal.

Hitachi Zosen, in its effort to meet the growing needs of modern Japan, started early to develop a new type of "pyrolysis" furnace. Now, with technological difficulties worked out, this product is now ready for the commercial market. The following outline is to acquaint you with this new type of furnace and the technology it represents.

Process Description

As illustrated in Figure 1, the process of the pyrolysis (thermal gasification) system consists primarily of the following steps: receiving and storing refuse; pyrolyzing it;

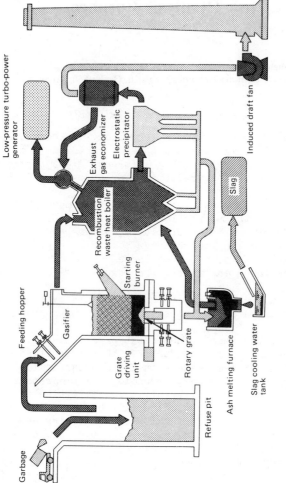

Figure 1. Integrated system for solid waste gasification, ash-melting, and waste heat power generation

burning the gas thereby produced and finally, melting the ash.
The system is planned so that the production of harmful substances in the pyrolyzing process is completely inhibited.
The generation of NO_x in the heat recovery process through
secondary combustion is minimized by controlling the excess air
ratio.

Meanwhile, the pyrolytic ash and the dust resulting from
combustion of the produced gas and then collected by the
electrostatic precipitator are melted in the ash-melting
furnace which is separate from the pyrolysis furnace. These
residues will subsequently be solidified by water cooling.

The produced gas is burnt in the recombustion boiler.
This process converts the refuse into energy in the form of
steam, which is utilized as a source for generation or heating.

Outline of Test Plant and Experimental Results

Construction of Gasifier. The construction of the pyrolysis
furnace, the heart of the pyrolysis-incineration system for
municipal refuse is shown in Figures 2 and 3. It is a shaft
furnace with a rotary grate on the bottom and a refuse feeding
hopper at the top. The air and steam for the water-gas reaction
are supplied through the rotary grate.
The gas produced by pyrolysis is led out through the top of
the furnace, while the ash remover clears out the residues.
The particulars of this test plant are as follows:

Capacity : 20 t/day
Uses : Municipal refuse, industrial waste
Type of furnace : Shaft furnace with rotary grate

Experimental results. The results of experimentation with
the above-mentioned test plant are as follows:

(1) Composition of Municipal Refuse. The experiment was
carried out with municipal refuse, as well as with a simulated
unclassified municipal refuse prepared by admixing the classified
municipal refuse with plastics, glass, metals, etc. The
compositions of the respective refuse are shown in Table I.

(2) Results of Analyses of Pyrolytic Gas and Exhaust Gas.
The compositions of the samples of produced gases from the
pyrolysis furnace are as shown in Table II. The compositions of
pyrolytic gases given in Table II are, as mentioned above,
experimental values when the calorific value of the refuse is in
the range of Hu = 1,265 $\sim$ 1,330 kcal/kg.
The results of the combustion tests carried out with these
produced gases show that when the calorific value of the gas is
as low as approx. 764 kcal/Nm^3 , unaided combustion is difficult,
with assistance from 5 $\sim$ 15 % of the heat content of the
produced gas being essential.

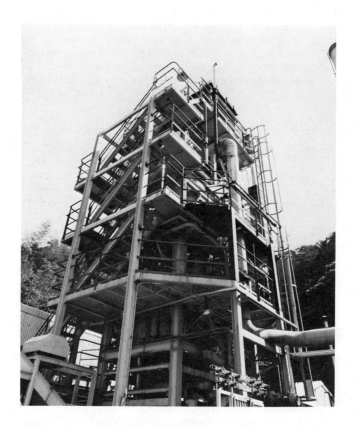

Figure 2.   A 20 TPD test gasifier

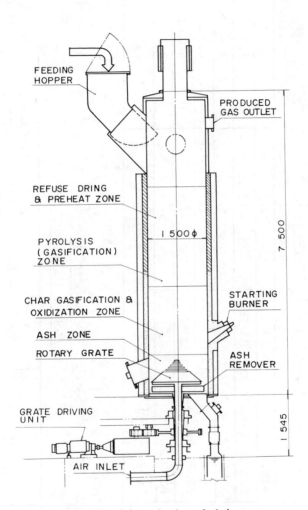

Figure 3.   Schematic of pyrolysis furnace

TABLE I

COMPOSITION OF MUNICIPAL REFUSE

COMPONENTS		WEIGHT PERCENTAGE (%)	
		CLASSFIEDED MUNICIPAL REFUSE	UNCLASSFIEDED MUNICIPAL REFUSE
PAPER		17.3	14.7
GARBAGE & OTHER COMBUSTIBLES		18.9	16.0
PLASTICS		5.2	11.4
INCOMBUSTIBLES	GLASS	1.7	6.9
	METALS	2.4	4.9
MOISTURE		54.4	46.1
CALORIFIC VALUE (NET) (Kcal/Kg)		1265	1329

TABLE II

PRODUCT GAS COMPOSITION

COM-PONENTS	H_2	CO	CO_2	N_2	HYDROCARBON				CALORIFIC VALUE (NET) (Kcal/Nm^3)
					CH_4	C_2H_4	C_2H_6	C_3H_8	
%	6.8~10.9	9.9~13.6	16.8~18.0	51.3~63.6	2.9	~	5.9		764~1,280

Table III shows the results of analyses of harmful gaseous elements etc. in the pyrolytic gas and the exhaust gas.

(3) Properties of Pyrolytic Ash and Slag. Table IV shows the quantities of harmful substances contained in the ash removed from the pyrolysis furnace and also the quantities of harmful substances dissolved out of the slag prepared by melting the residues when immersed in fresh water.

The ignition loss of the pyrolytic ash and the slag and their volumetric reduction ratio are shown in Table V. The tabulated figures show that in the slag solubility-test, the quantities of individual harmful substances liberated are all below the control limits which are set in Japan for land reclaiming work.

It is also apparent that this slag is superior to that from the stoker-type incinerator in ignition loss and volumetric reduction ratio.

The ash removed from the pyrolysis furnace (before melting), however, contains char, hence its ignition loss is slightly greater than that of the ash from the stoker-type incinerator.

(4) Co-disposal of Scrap Tire and Refuse. The experiment was carried out with shreded tires mixed with municipal refuse. The composition of the samples of produced gases from pyrolysis furnace are as shown in Table VI.

The composition of pyrolytic gases given in Table VI are experimental values mixed with 30 % of scrap tires.

System Consideration

Recovery of Resources and Energy. There are basically two systems for pyrolytic disposal of refuse: one consists of pyrolyzing the refuse into fuel; the other consists of directly burning the combustible gas produced by pyrolysis to convert it into energy. (1)

One major problem that is encountered as we attempt pyrolytic disposal of municipal waste in Japan however, is that the refuse has a higher moisture content than that in Europe or America, where calorific values fall in the range of 1,000 ∼ 1,500 kcal/kg. In consequence, the calorific values of the pyrolytic gases produced from them are as low as 700 ∼ 1,300 kcal/Nm^3, as shown in Table II.

For pyrolytic disposal of municipal refuse in Japan, therefore, preference may be given to the pyrolysis-incineration system, which, with its low-pollution feature, allows on-the-spot conversion of the produced gas into thermal energy -- this system is preferred until there should be a radical improvement in the quality of municipal refuse in our country.

TABLE III

PRODUCED PYROLYTIC GAS ANALYSES

	KINDS OF REFUSE	HCl	NO$_x$	SO$_x$	HCN	NH$_3$	H$_2$S	Cl$_2$	R-COOH
PYROLYTIC GAS MEAN $\frac{MAX.}{MIN.}$ (ppm)	CLASSIFIEDED MUNICIPAL REFUSE	478 $\frac{680}{350}$	70 $\frac{95}{105}$	125 $\frac{200}{35}$	—	2320 $\frac{4800}{130}$	27 $\frac{52}{7}$	N.D.	184 $\frac{270}{98}$
	UNCLASSIFIEDED MUNICIPAL REFUSE	2142 $\frac{3340}{1050}$	87 $\frac{105}{64}$	83 $\frac{230}{<10}$	41 $\frac{52}{26}$	900 $\frac{1200}{600}$	398 $\frac{500}{310}$	N.D.	—
EXHAUST GAS MEAN $\frac{MAX.}{MIN.}$ (ppm)	CLASSIFIEDED MUNICIPAL REFUSE	73 $\frac{99}{51}$	74 $\frac{104}{47}$	40 $\frac{46}{31}$	—	—	—	—	5 $\frac{6}{4}$
	UNCLASSIFIEDED MUNICIPAL REFUSE	127 $\frac{177}{48}$	79 $\frac{95}{47}$	21 $\frac{24}{22}$	9 $\frac{17}{4}$	17 $\frac{30}{<5}$	<1	N.D.	—

TABLE IV

METAL CONCENTRATION & SOLUBILITY-TEST RESULT

METAL	CONCENTRATION IN ASH (mg/kg)	SOLUBILITY-TEST OF HEAVY METALS IN SLAG (mg/l)	REGULATED VALUES FOR LANDFILL (mg/l)
T - Hg	0.046	< 0.0005	< 0.005
Al - Hg	< 0.05	< 0.0005	N.D. (< 0.0005)
Cd	2.1	< 0.0005	< 0.3
Pb	140	< 0.05	< 3
Org - P	< 1	< 0.1	< 1
Cr +6	< 1	< 0.05	< 1.5
As	3.5	0.02	< 1.5
T - CN	< 1	< 0.005	< 1
T - Cr	550	< 0.02	-

TABLE V

COMPARISON OF IGNITION LOSS, CAUSTIC-SOLUBLE
CARBON CONTENT & VOLUMETRIC REDUCTION RATIO

	ASH (PYROLYSIS)	SLAG	ASH (INCINERATOR)
IGNITION LOSS (%)	6.5	0	3.0 ~ 10.0
CAUSTIC-SOLUBLE CARBON CONTENT (%)	2.1	0	1.0 ~ 5.0
VOLUMETRIC REDUCTION RATIO	1/20	1/40 ~ 1/50	1/20

TABLE VI

PRODUCED PYROLYTIC GAS ANALYSES
(CO-DISPOSAL OF SCRAP TIRE AND REFUSE)

COMPONENTS	H_2	CO	CO_2	N_2	HYDROCARBON				HCl MEAN$\frac{MAX.}{MIN.}$ (ppm)	SO_x MEAN$\frac{MAX.}{MIN.}$ (ppm)	H_2S MEAN$\frac{MAX.}{MIN.}$ (ppm)	CALORIFIC VALUE (NET) (Kcal/Nm^3)
					CH_4	C_2H_4	C_2H_6	C_3H_8				
%	14.6	19.6	14.0 ~ 16.0	35.8 ~ 41.8	10.0 ~ 14.0				480 $\frac{680}{350}$	403 $\frac{970}{260}$	297 $\frac{610}{64}$	1,920 ~ 2,320

In Table VII, which deals with a system for using the thermal energy obtained through pyrolysis-incineration of municipal refuse to produce steam for power generation, calculated data on the quantity of steam produced, the rate of power generation and the power requirements of the plant are shown.

The power generation rate given in Table VII shows the plant's generating capacity when the condenser vacuum pressure is 700 mmHg, with the assumption that an ample supply of cooling water for the plant is available.

However, depending on the location of the plant, for example, it is possible that another means of cooling, such as air cooling, may be required to compensate for a shortage of water.

Such factors must be taken into due consideration when calculating the power generation rate (capacity), plant's power requirement, etc.

TABLE   VII

COMPARISON OF GENERATING & REQUIRED POWER

ITEMS	PLANT CAPACITY	50t/dx2UNITS	100t/dx2UNITS	150t/dx2UNITS
STEAM GENERATING CAPACITY	STEAM FOR TURBO-POWER GENERATOR (kg/H)	4,298	8,241	12,571
	STEAM FOR PROCESS USE (kg/H)	1,791	4,010	5,807
	TOTAL (kg/H)	6,089	12,251	18,378
GENERATING CAPACITY (KWh)		438	840	1,280
REQUIRED POWER (KWh)		386	730	1,084
EXCESS AND DEFICIENCY OF POWER (KWh)		+ 52	+ 110	+ 196

NOx and HCl Contents of Exhaust Gas. Examples of analyses data showing the NO_x and HCl contents of the exhaust gas from this system are given in Figure 4 and 5. These figures show that the NO_x content of the exhaust gas from this system is approx. one half that of the stoker-type incinerator. This is due to the fact that although the temperature in the pyrolysis furnace can be as high as approx. 1,100°C, the supply of air to

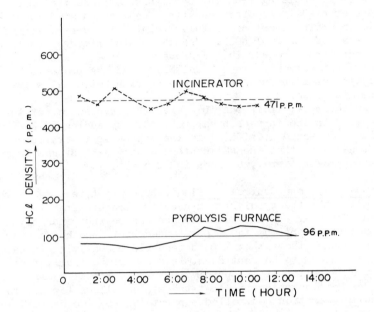

Figure 4.   *HCl density*

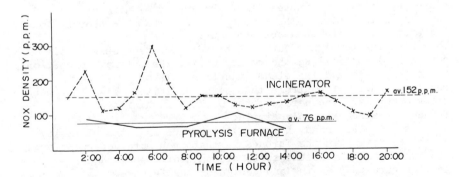

Figure 5.   *No_x density*

the furnace is controlled to a rate just adequate for combustion
of the char resulting from the pyrolysis zone, so that the
furnace as a whole is operated with a deoxidized atmosphere -- a
condition which almost completely inhibits the generation of
the so-called "thermal NO_x". Meanwhile, the nitrogen-containing
compounds in the combustible components of the refuse are
likewise not converted into "fuel NO_x", and the produced
pyrolytic gas is burned in the recombustion boiler at a low
excess air ratio.

Furthermore, it is possible with this system to cut the HCl
production in the furnace by approx. 60 % by adding 50 kg/ton of
calcium hydro-oxide ($Ca(OH)_2$) to the refuse as it is fed into
the furnace, so that the HCl content of the exhaust gas is as
low as approx. 96 ppm., as seen in Figure 4.

Operation and Control of Plant. One great feature of this
system is that the pyrolysis furnace is separate from the ash-
melting furnace, so that each furnace can be independently
designed to be optimum for its intended use. Another feature is
that even when the rather less dependable ash-melting furnace
fails and needs to be shut down, it does not basically interfere
with the operation of the pyrolysis furnace for disposing refuse.

One more advantage is that the system allows for simple
batchwise operation controlled separately from the ash-melting
furnace. Figure 6 shows the increase of temperature inside the
pyrolysis furnace as it proceeds from the shut-down state to
a steady state, as determined through experimentation in the
test plant. As the figure shows, the pyrolysis furnace, after
having been shut down for about 10 hours, proceeds to a steady
state in approx. 30 minutes after the supply of air for
combustion is resumed; the refuse in the furnace is brought to
the shut-down state in about one hour.

With the pyrolysis furnace, proper operation essentially
consists of controlling the rotary grate, the partial combustion
air supply rate and the steam injection rate only -- the
required number of controls thus being definitely less than with
the stocker-type incinerator; the working mechanism of the
furnace is also extremely simple.

On the other hand, however, this means that the stability
of the condition in the furnace is that much more dependent upon
the quality of the refuse to be disposed of, hence requiring
greater care in planning and designing the furnace.

Comparison of Alternative Facilities for Disposal of Municipal
Waste

Now, the economical facets of this system will be compared
with a conventional incinerator of the stoker-type.
The results of a comparison of this system with its capacity of
150 t/d x 2 units, with a conventional stoker type incinerator

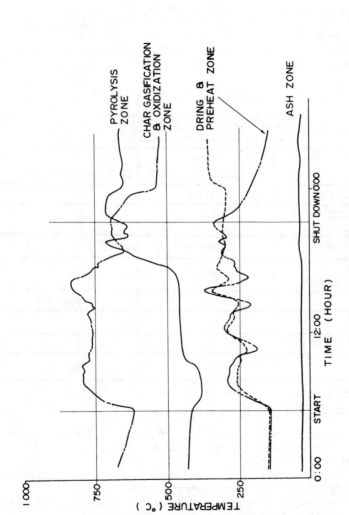

Figure 6. Wall temperature in pyrolysis furnace

are shown in Table VIII in regard to floor space, installation
space, capital cost and operating cost. As the tabulated
figures show, this refuse disposal plant system has a floor
space approx. 70 % that of the stoker type, with 80 % of the
capital costs and 86 % of the operating costs.

Furthermore when there is no requirement for using the
waste heat for power generation or creating steam, the plant
setup can be further simplified by including only an exhaust
gas cooler with water spraying nozzles following the recombustion chamber.

TABLE VIII

COMPARISON OF PLANT SPACE,
CAPITAL COST & OPERATING COST

	TYPE OF SYSTEM	STOKER TYPE INCINERATING PLANT	PYROLYSIS-INCINERATING PLANT
PLANT SPACE	FLOOR SPACE	100	70
	INSTALLATION SPACE	100	64
CAPITAL COST	MECHANICAL	100	85
	CIVIL & CONSTRUCTION	100	73
	TOTAL	100	80
OPERATING COST	UTILITY	100	100
	AMORTIZATION COST	100	80
	TOTAL	100	86

Summary and Conclusion

In this paper, I have tried to acquaint you with our newly
developed pyrolysis-incineration system for the disposal of
municipal refuse in Japan.

In recent years, cities in our country have been confronted
with the growing threat of secondary pollution caused by
incinerating municipal refuse and also by the growing difficulty
of securing dumping areas for residues. All the while
regulations against the discharge of harmful substances become
more and more stringent.
Under such circumstances, need for low-polluting, inexpensive
and energy-saving means of refuse disposal has been steadily
increasing. We are now convinced that the new refuse disposal
system introduced here, with its outstanding features, will be
able to meet modern standards and so be successfully put on the

market in the near future.

Abstract

The system primarily consists of four processes, shaft furnace gasifier with rotary grate, ash melting furnace, combustor of product gases and turbo-power generator for energy recovery.
The aims of this system are to reduce residue volume, assure stabilized and pollution-free operation and enable the recovery of useful resources.
Combustible product gases produced from gasifier are directly burnt in a recombustion boiler with moderate amount of air for power generation. Furthermore, the residues are melted into the slag in the ash melting furnace.
This paper describe such newly developed integrated solid waste disposal system.
The test results applying the gasifier to the processing of municipal refuse and scrap tires with solid waste are presented. The fundamental characteristics of this plant such as installation space, capital cost and operating cost are discussed.

Literature Cited

1. V.L. Hammond, L.K. Mudge, C.H. Allen,
   G.F. Schiefelbein
   Energy from Solid Waste by Pyrolysis-Incineration
   (1972)

RECEIVED November 16, 1979.

Development of a Solid Waste Disposal System with Pyrolysis and Melting

M. ONOZAWA

Solid Waste Melting System Project Office, Plant and Machinery Division, Nippon Steel Corporation, Nakabaru 46-59, Tobata-Ku, Kitakyushu City, Fukuoka Prefecture 050, Japan

1. Introduction

Recently in Japan, the improvement in living standards has given rise to an increase in the quantity and a diversification of the properties of wastes from households. This has become an increasingly important problem for local governments. At present, the greater part of such wastes are being disposed of by burning in stoker incinerators. But an increase in the amounts of difficult-to-burn or incombustible wastes such as plastics, glass bottles and cans, which started in about 1965, has made the present incineration method increasingly ineffective. Hence, ways to overcome the limits of the present method are being investigated.

Taking an interest in this field, Nippon Steel Corporation in 1972 started studying a pyrolysis and melting method using a shaft furnace for the disposing of non-industrial wastes. Since 1974, experiments have been carried out at a 20-ton-per-day pilot plant built in Tobata, Kitakyushu. The object of the fusion-thermal composition method, is to recover energy from inflammable gas; render product ashes into slags at high temperatures, thus making heavy metals, if any, eludible; decrease the volume of the end products; and process the product gas so that less HCl and NO_x are discharged into the atmosphere.

In an attempt to reduce secondary pollution and protect the incinerators, the Tokyo Metropolitan Government, since 1973, has been asking households to separate some items from ordinary wastes before they are put out for collection. The items to be separately collected include those which cannot be burned, such as metals, glass and ceramics, and those which should not be burned, such as plastics, rubber, hides and leather. These items are buried in reclaimed lands, unprocessed.

After inspecting the Tobata pilot plant, the Bureau of Public Cleansing of the Tokyo Metropolitan Government decided to perform experiments to ascertain the applicability of this system to the processing of the classified refuse described above. In response,

Nippon Steel, in conjunction with the Plastic Waste Management
Institute, offered the government use of an experimental plant,
with a capacity of 40 tons per day. The Government, in turn,
subcontracted the performance of the experiments to Nippon Steel.
The experiments were started in January 1978, and completed in
August 1979, yielding almost all the expected results.

Taking advantage of the results obtained from these test
plants, Nippon Steel is constructing a 100-ton-per-day plant (50-
ton-per-day unit x 2) in Kamaishi and a 450-ton-per-day plant
(150-ton-per-day unit x 3) in Ibaragi.

2. Outline of the System

This waste disposal system centers upon a melting furnace.
The system can be divided into two main types, according to how
the gas generated in the furnace is treated.

Figure 1 is a schematic flow sheet of the type of system
that recovers the inflammable gas; this system is used in Tokyo
for processing the classified refuse.

Figure 2 is a schematic flow sheet of the direct combustion
system.

For effective burning of refuse, coke must be added as a
supplementary fuel and limestone as a flux. To melt the slag at
high temperatures, preheated, oxygen-enriched air is blown into
the furnace. Accordingly, iron, glass, ceramics and the like
contained in the refuse are discharged from the furnace in a
molten state.

In the gas recovery system, the dust and tar from the wet
scrubber may be returned to the furnace for further docomposition.
The dust contains some heavy metals as zinc and lead and alkali
chlorides that may vaporize in the furnace, so, not all dust can
be returned. At present in Tokyo, therefore, a certain portion
of the dust and thickener liquid is processed in a rotary kiln.

Figure 3 shows the melting furnace that forms the core of
this system. Refuse is hoisted from a storage pit to the furnace
top, and dumped into the shaft furnace. Two seal valves attached
to the charging hopper open alternately to prevent the leakage of
gas resulting from thermal decomposition. Descending through the
furnace, the charged refuse meets the ascending high-temperature
gas. Any moisture contained in the refuse evaporates and combustibles are gasified. In front of the tuyeres supplying the
oxygen-enriched air, coke and char, which is pyrolysis product,
are burned at high temperatures.
The ash and iron contained in the waste melt into slag and molten
iron. When a certain quantity of these molten materials has
collected in the hearth, they are tapped out, then solidified in
vessels. The molten substances may also be dropped into water
jets to form granulated slags.

In the inflammable gas recovery system, the gas is cleaned
first by collecting coarse dust with a dry dust collector, then

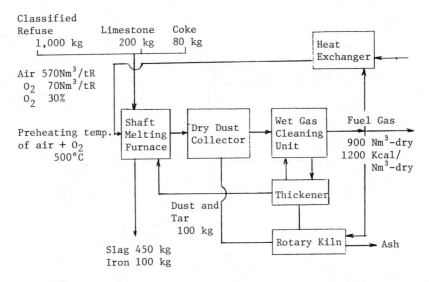

Figure 1. Typical example of gas recovery system for classified refuse

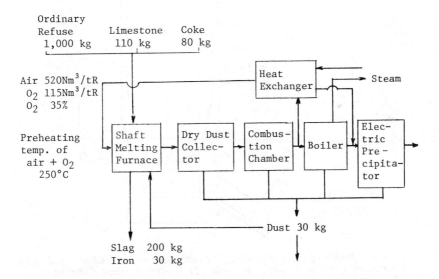

Figure 2. Typical example of direct combustion system for ordinary refuse

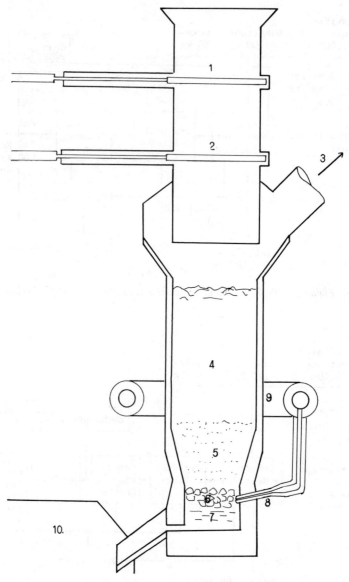

Figure 3.   Profile of melting furnace: (1) upper seal valve; (2) lower seal valve; (3) inflammable gas outlet; (4) drying zone; (5) pyrolyzing zone; (6) melting zone— high temperature; (7) molten slag; (8) oxygen-enriched air inlet—tuyere; (9) blast header; (10) slag granulation

by eliminating the finer dust and a part of tar by using a wet scrubber. Part of the clean gas is consumed in hot stoves for the heating the oxygen enriched air that is blown into the furnace, while the rest is supplied as fuel to boilers and other gas-spending facilities.

In the direct combustion system, the gas heavily laden with steam, dust and tar is burned in the combustion chamber. Part of the gas is used as an energy source for hot stoves, with the remainder being supplied to a boiler for heat recovery. Then, they are both cleaned by an electric precipitator, and discharged into the atmosphere through an induction fan.

3. Experimental Procedures

3.1. Tobata Pilot Plant

The Tobata pilot plant has been conducting experiments since 1974. The gas processing system was of the inflammable gas recovery type originally, but later modified to permit operation on the direct combustion principle, also. At present, the system is operated mostly using the latter method. Using this pilot plant, basic, comparative studies have been made on the furnace profile, effects of the properties of wastes and the oxygen concentration in the blast, the need for supplementary fuel, and how to process the product gas.

(1) Refuse Properties
The refuse used for the experiments were the ordinary urban refuse (all-inclusive) collected from Kitakyushu City. Table I shows their usual properties.

Table I. Properties of Urban Refuse from
Kitakyushu City (On Wet Basis; %)

Physical Properties	Combustibles	Undesirable Combustibles	Incombustibles	Total
	59.6	10.5	29.9	100.0
Chemical Properties	Moisture	Combustibles	Ash	Total
	48.2	33.2	18.6	100.0

(2) Experimental Progress
Eight thousand seven hundred tons of wastes were disposed of in approximately 12,000 hours between November 1974 and June 1979.

(3) Experimental Conditions
1) Blast volume   560 ∿ 1,200 Nm^3/hr

2) Oxygen quantity (in air + enrichment)
                                160 $\sim$ 250 Nm^3/hr
3) Oxygen concentration   20.9% (air) $\sim$ 36%
4) Air preheating temperature   Ambient - 400°C
5) Limestone   80 $\sim$ 120 kg/ton of refuse

    By varying the blast volume, oxygen concentration, preheating temperature and other factors, the optimum quantity of refuse charge, coke requirement and other conditions for stable plant operation have been studied.

3.2. Tokyo Test Plant

    Experiments on the Tokyo Test Plant were started in December 1977 and ended in August 1979. Using mainly the aforementioned classified refuse, this plant was operated continuously for a long time, in order to reconfirm the requirements for stable operation, utilities requirements and pollution control-related figures established by the Tobata pilot plant as well as to study the feasibility of larger-size plants scale-up.

(1) Refuse Properties
    The properties of refuse processed by the Tokyo plant shown in Table II differed widely from those of Kitakyushu.

Table II. Properties of Classified Refuse Treated
at the Tokyo Plant   (On Wet Basis; %)

Physical Properties	Combustibles	Undesirable Combustibles	Incombustibles		Total
			Glass & Ceramics	Metals	
	31.3	19.9	31.0	17.8	100.0
Chemical Properties	Moisture	Combustibles	Ash		Total
	18.6	33.8	47.6		100.0

(2) Experimental Progress
    16,319 tons of refuse were disposed of in 9,400 hours between December 1977 and August 1979.

(3) Experimental Conditions
1) Blast volume   1,280 $\sim$ 1,645 Nm^3/hr
2) Oxygen quantity (in air + enrichment)
                                380 $\sim$ 400 Nm^3/hr
3) Oxygen concentration   23.2 $\sim$ 36%
4) Air preheating temperature   700°C
5) Limestone   Approximately 200 kg/ton of refuse

Experimentation was started with the oxygen concentration of 23.2% and coke consumption of 108 kg per ton of refuse. Tests on varying oxygen enrichment were also made. Nitrogen gas was used for purging the inflammable gas, and cement for solidifying the residue. Substantially the same experimental conditions as at the Tobata pilot plant were used, except for the higher air preheating temperature.

4. Performance

4.1. Experimental Results

Table III. Typical Experimental Results

Item	Tobata Pilot Plant Ordinary Refuse	Tokyo Test Plant Classified Refuse
Air, Nm^3/hr	400 ∿ 550	1,100 ∿ 1,350
Pure oxygen, Nm^3/hr	80 ∿ 100	100 ∿ 140
Oxygen concentration, %	30 ∿ 36	26 ∿ 30
Waste disposal rate, t/hr	0.8 ∿ 1.0	1.8 ∿ 2.1
Coke consumption, kg/t-refuse	70 ∿ 80	65 ∿ 85
Limestone consumption, kg/t-refuse	100 ∿ 120	190 ∿ 210
Tapping rate, kg/t-refuse	200 ∿ 300	430 ∿ 550
Product gas, Nm^3/t-refuse	1,000	1,020
Gas composition, %		
N_2	47.3	48.9
CO	21.3	23.7
CO_2	18.8	14.2
H_2	10.3	9.4
CH_4	1.6	2.2
C_2H_4	0.7	1.6

Table III lists the typical experimental results obtained from the two plants. Within the range of the tests, coke consumption generally decreased with increasing oxygen concentration. For the same oxygen concentration, the classified refuse, despite its markedly larger slag production, consumed less coke than the ordinary urban refuse. Though not fully proved, this is thought to be due to the higher moisture content of the ordinary urban refuse.

The daily disposal capacity was 16 to 22 tons for the Tobata pilot plant, and 45 to 50 tons for the Tokyo test plant.

4.2. Products and Uses

(1) Slag and Iron
        By slowly cooling the molten product in a vessel,
solid rock-like slag and iron are obtainable. When cooled rapidly in water jets, both slag and iron become sandy. The iron can
be made into weights or used for some other applications. The
slag may be used for reclamation. Its use in the civil-engineer-
ing field, at present, is limited to sub-base material. The slag
has the properties shown in Table IV, free from such problems as
the eluting of heavy metals.

Table IV. Slag Composition (%)

SiO_2	CaO	Al_2O_3	MgO	Na_2O	K_2O	T-Fe
46.6	34.5	9.13	1.98	4.87	0.74	1.25

(2) Product Gas
        Within the oxygen concentration range of 28 to 36%
used in the experiments, 800 to 1,100 Nm^3 of gas (dry) were
generated per ton of refuse. This yields 1,100 to 1,400 kcal/Nm^3
of energy (dry). The heating value of the gas increases with
increasing oxygen concentration in the hot blast.

4.3. Environmental Pollution Data

(1) Noxious Gases
        Table V lists the air pollutants contained in the
exhaust gas from the systems. The values vary to some extent
with the properties of the refuse that is processed. The HCl,
NOx and SOx concentrations obtained from the inflammable gas recovery and direct combustion methods are considerably lower than
their concentrations in the conventional incinerator method.
        The concentrations of HCl are especially low in the two NSC
Systems.

Table V. Air Pollutants in Exhaust Gas (ppm)

System	HCl	NOx	SOx
Inflammable gas recovery system (Exhaust gas after burning recovered gas)	15	58	7
Direct combustion system	36	58	16

(After correcting O_2 concentration in exhaust gas to 12%)

(2) Drainage
        If requested by the user, both systems can be designed

so that no water is drained outside the system. The inflammable
gas recovery system cannot dispense with the rotary kiln shown in
Fig. 1, since the recirculated water in it carries organic matter
and heavy metals.

(3) Slag
          The slag is so stable that no heavy metals are eluted.

(4) Dust
          Both systems generate 30 to 50 kg of dust and/or ash
per ton of refuse. To prevent the elution of heavy metals, they
must not be dumped unless they are solidified within cement.
Table VI shows the chemical composition of the dust from the
direct combustion system.

Table VI. Chemical Composition of Dust from
the Direct Combustion System (%)

C	S	Cl	K	Na	Ca	Si	Al	Zn	Pb
2.07	1.23	1.43	1.18	2.71	2.00	8.65	4.32	0.75	0.19

5. Application to Practical Use

The two commercial plants being constructed in Kamaishi and
Ibaragi are of the direct combustion type. Figure 4 shows a
basic flow sheet on which the two plants are designed.
The fuel gas generated in the melting furnace are burned,
together with dust and tar, in the combustion chamber. Although
a small part of the burned high-temperature gas passes through
the air preheater, the greater part flows into the boiler where
waste heat is recovered. The re-joined gas streams are cleaned
by the electric precipitator, then discharged into the atmosphere
through the induction fan. The dust collected by the precipitator, is subjected to the elution-preventing measure, then discharged outside the system.
In the Kamaishi plant, which is rather small, the boiler is
replaced with a gas cooler.
In both plants, slags are water-granulated.
The Kamaishi plant is scheduled to begin operation in August
1979, and the Ibatagi plant in February 1980.

6. Consideration

6.1. Comparison of the Two Gas Processing Methods

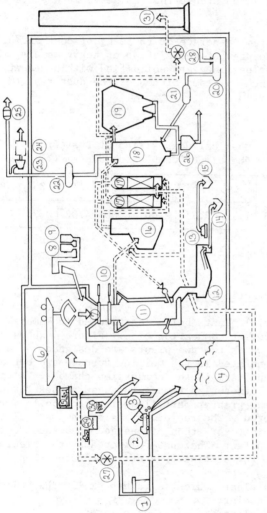

Figure 4.   Process flow sheet

(1) air curtain; (2) collection vehicle; (3) receiving gate; (4) refuse storage bin; (5) crane operating room; (6) refuse crane; (7) refuse charging hopper; (8) coke storage hopper; (9) limestone storage hopper; (10) double seal valves; (11) melting furnace; (12) slag granulation pit; (13) magnet separator; (14) slag storage hopper; (15) iron storage hopper; (16) direct combustion chamber; (17) hot stove; (18) waste heat boiler; (19) electric precipitator; (20) clean water tank; (21) water purifying unit; (22) boiler drum; (23) steam turbine; (24) generator; (25) heat exchanger; (26) concrete solidifying equipment; (27) combustion air fan; (28) induced draft fan; (29) pit for bulky wastes; (30) crusher; (31) stack

Table VII. Comparison of Gas Processing Methods

Description	Inflammable Gas Recovery Method	Direct Combustion Method
Gas Processing	Though dust collector is simple, water and tar processing systems are complex and require large amounts of space.	Gases entering boiler and heat exchanger are heavily laden with dust. Measures to prevent dust trouble are needed. Dust collector takes up much space.
Air Preheating	High temperature is attainable.	High temperature is difficult to attain.
Operation	Complicated by processing of dustladen water.	Simple.
Environmental Pollution	Much lower in SOx and HCl than the incinerators	Lower in HCl and NOx than the incinerators, but a little higher in SOx and HCl than the gas recovery method.

For lack of sufficiently long operational experience, Nippon Steel cannot yet say which of the two methods is more advantageous. The direct combustion system is adopted for the Ibaragi and Kamaishi plants. Generally, the choice between the two calls for very careful consideration of various factors peculiar to each plant.

6.2. Process Economics

In respect to economics, the pyrolysis and melting waste disposal system under discussion has the following advantages over the stoker incinerators that are most widely used in Japan:

(1)   The NSC system is capable of processing refuse that are unsuited for burning or incombustible, and cannot be processed by the stoker incinerators.

(2)   There are few air pollutants emitted in the NSC system, especially low are HCl emissions.

(3)   Whereas the ash from the stoker incinerators involve

the risk of eluting organic matter and heavy metals, the slag from the NSC system is perfectly stable. When buried in the reclaimed land, the volume of slag decreases to half of the stoker incinerator ash, thus making the land serviceable for a longer period.

It cannot be denied that the processing cost of the NSC system is a little higher, because of its need for coke and limestone. Despite that, the pyrolysis and melting system is economical when advantages (2) and (3) are desired or when underirable combustible refuse have to be processed.

RECEIVED November 16, 1979.

BIOMASS ENERGY FOR
DEVELOPING NATIONS

The Potential of Biomass Conversion in Meeting the Energy Needs of the Rural Populations of Developing Countries — An Overview

V. MUBAYI and J. LEE

Brookhaven National Laboratory, Upton, NY 11973

R. CHATTERJEE

Institute for Energy Research, State University of New York, Stony Brook, NY 11794

Biomass, in the form of firewood, dried crop residues and animal dung, is the fuel most extensively utilized in the rural areas of most developing countries. What was true of the United States over a hundred years ago, when wood accounted for over 90% of energy consumption, still holds good for many developing nations, particularly in their rural settlements. Current utilization of these resources generally takes place outside of a market economy (hence the term "noncommercial" usually applied to these resources) and in varying amounts for both energy and nonenergy uses. Crop residues, for example, are used as fuel, building material and as animal feed. Dung is used both as fuel and fertilizer.

Practically all of the biomass use that takes place at present is in the form of direct combustion at very low efficiencies to supply heat for essential human needs such as food preparation. The mechanical power needs of agricultural production are largely met through the metabolic energy input of draft animals and human labor. Constraints on land and the population it can support make the raising of agricultural productivity essential if food production is to expand in step with population growth. Furthermore, the provision of basic human needs and amenities to improve the "physical quality of life" is an important social goal of almost all of the developing countries.

Energy is universally recognized as essential for development. Greater amounts of energy capable of being utilized more efficiently and productively are particularly needed in rural areas of LDCs in order to raise agricultural productivity, provide off-farm employment, essential amenities such as lighting and to improve the physical quality of life. Hence, in addition to low temperature heat necessary for subsistence, mechanical energy is required for agricultural tasks (e.g., pumping water for irrigation), transportation, and small industry, chemical feedstocks whose manufacture requires energy are needed as fertilizer and electricity is needed for lighting.

0-8412-0565-5/80/47-130-617$05.00/0

Historically, in those rural areas where modernization has occurred, petroleum products have provided the necessary energy inputs required for a variety of activities. Kerosene is a widely used illuminant, diesel oil for powering irrigation pumps and other farm equipment, gasoline for transportation and fertilizers are often manufactured from naphtha. However, the world oil situation now makes the "petroleum route" to development an increasingly difficult one for many developing countries to undertake.

Biomass conversion technologies, which can convert already available biomass into a variety of solid, liquid, and gaseous fuels, capable of being utilized more efficiently and productively, will be essential if the future subsistence and developmental energy needs of rural areas in developing countries are to be met. This paper represents a very preliminary attempt at assessing the contribution biomass conversion could make in the context of the rural areas of six developing countries: India, Indonesia, Peru, Sudan, Tanzania, and Thailand. The choice of countries was dictated partly by the availablity of data on resources and consumption and partly by the fact that these countries typify the rural energy situation in a large number of developing countries located in South and South-East Asia, Africa, and South America.

This overview assesses the potential resources of biomass available in the six countries selected. With a few exceptions (such as Peru's forest resources) these resources are already being collected and used, albeit inefficiently. We focus on rural energy end-uses in each of the countries, including a discussion of the current energy consumption patterns and their magnitudes. Current consumption is projected into the future, to the year 2000, with emphasis placed on conceptualizing both subsistence needs (e.g., cooking) and developmental needs (increasing productivity and living standards). The method adopted in our approach is that in the context of rural development an assessment of new technologies has to be anchored within the perspective of future energy requirements to satisfy both the basic subsistence needs of growing populations as well as their developmental needs.

The technologies selected for analysis are: anaerobic digestion of wet biomass to produce methane and pyrolysis of dry biomass to produce charcoal, liquid fuels, and low-Btu gases. Preliminary estimates are made of the amounts of fuels that could be produced in each of the selected countries by a combination of these technologies. We find that in five of the six countries, with the exception of India, implementation of these technologies could potentially meet the future energy needs of their rural populations for both subsistence and development.

Current Sources of Biomass Supply

The principal sources of biomass used for energy purposes are wood, gathered from forests, orchards and farms, agricultural crop residues and animal wastes. Human wastes can also be potentially considered as an energy resource for biomass conversion technologies such as anaerobic digestion if appropriate collection practices are employed (as, for example, in many Chinese villages). We exclude from our purview organic matter which could be potentially grown in "energy plantations" in agriculture, silviculture, or aquaculture. This is not to suggest that such plantations are not potentially important sources of biomass energy in developing countries. However, an evaluation of their contribution would require an analysis of alternative land-use patterns which are specific to each country and falls outside the scope of our assessment.

Table I provides estimates of the biomass produced annually in each of the six countries. However, this biomass is not all available for energy conversion. There are numerous uncertainties involved in attempting to estimate the amounts of biomass that could be used as an energy source such as the fraction of wood, crop residues, and animal wastes that can be collected and the alternative non-energy uses that exist for the amounts that are currently collected. Crop residues, for example, are commonly utilized for animal feed and sometimes as building materials, animal dung is utilized as a fertilizer, and so forth. We have nevertheless attempted preliminary estimates of the biomass available for energy conversion in the six countries based in part on data provided in other studies and in part on reasonable assumptions made in these studies. However crude, for the purposes of an overall assessment of the impact of biomass conversion technology on rural energy needs, these estimates are believed to be realistic approximations of the amounts of biomass that could be made available for energy conversion. Assuming an average energy content of 14 billion joules/dry ton for crop residues, 15 billion joules/dry ton for animal wastes and 16 billion joules/dry ton for fuelwood, the total potential for energy from biomass can then be calculated based on these estimates (Table II).

In order to assess the significance of these estimated energy potentials, we have compiled the best available data on the current rural biomass consumption for energy in the six countries (Table III). It appears that for Sudan the biomass energy potential, estimated at 650×10^{15} joules, is about ten times its current biomass consumption. For Peru, the potential is estimated at 1341×10^{15} joules, which is enough to supply all its rural biomass energy needs 7.5 times. The potential for other countries is about twice the rural consumption for Indonesia, Thailand, and Tanzania, and 1.5 times for India. To fully utilize this potential, however, requires an understanding of a disaggregated pattern of rural energy needs and the biomass conversion technologies that can transform this potential into usable fuels.

TABLE I

Estimated Potential Sources of Biomass in Rural Areas
(per annum)

Source	Peru 10^6 Ton	Peru 10^15 Joules	Thailand 10^6 Ton	Thailand 10^15 Joules	Indonesia 10^6 Ton	Indonesia 10^15 Joules	India 10^6 Ton	India 10^15 Joules	Sudan 10^6 Ton	Sudan 10^15 Joules	Tanzania 10^6 Ton	Tanzania 10^15 Joules
Human Waste[a]	0.2	3.0	1.0	15.0	3.2	48.0	14.3	214.5	0.5	7.3	0.4	6.0
Animal Manure[b]	6.9	103.5	12.8	192.0	10.0	150.0	198.0	2970.0	24.0	360.0	20.6	209.0
Crop Residues[c]	6.0	84.0	37.0	520.0	52.0	728.0	160.0	2240.0	9.3	130.0	2.1	29.4
Fuelwood[d]	510.0	8160.0	198.0	3168.0	720.0	11520.0	300.0	4800.0	132.0	2112.0	152.0	2432.0
Total	523.1	8350.5	248.8	3895.0	785.2	12446.0	632.3	10224.5	165.2	2609.3	175.1	2776.4

[a] Based on UN statistics (1,2) and 33 Kg (dry weight) of waste production per person per year.
[b] Based on livestock population (1) and manure production rate of major livestocks (3,4).
[c] Based on production of major crops production (1) and their respective residue coefficients (5,6).
[d] Based on actual forestry area (6-11) and per unit area increment of wood by type of forest (12).

TABLE II

Estimated Availability of Biomass for Energy Conversion in Rural Areas
(per annum)

Source	Peru 10^6 Ton	Peru 10^15 Joules	Thailand 10^6 Ton	Thailand 10^15 Joules	Indonesia 10^6 Ton	Indonesia 10^15 Joules	India 10^6 Ton	India 10^15 Joules	Sudan 10^6 Ton	Sudan 10^15 Joules	Tanzania 10^6 Ton	Tanzania 10^15 Joules
Human Waste[a]	0.1	1.5	0.5	7.5	1.6	24.0	7.2	107.3	0.3	4.4	0.2	3.0
Animal Manure[b]	5.2	77.6	9.6	144.0	7.5	112.5	148.5	2227.5	18.0	270.0	15.5	231.8
Crop Residues[c]	2.7	38.0	16.7	234.0	23.4	328.0	24.0	560.0	4.2	58.5	1.0	13.2
Fuelwood[d]	76.5	1224.0	29.7	475.0	108.0	1728.0	119.0	1900.0	19.8	316.8	22.8	365.0
Total	84.5	1341.1	56.5	860.5	140.5	2192.5	298.7	4794.8	42.3	649.7	39.5	613.0

[a] Assuming 50% collectibility.
[b] Assuming 75% overall collectibility.
[c] Assuming 50% overall collectibility. It is further assumed that 50% of the collected crop residues is used for animal feed in India, 10% in Tanzania (5) and the remaining countries.
[d] Assuming 15% use of the estimated total annual increment of wood (the percentage may be an overestimate for Peru and Indonesia due to the location of forest resources). For India, the availability has been assumed to equal current consumption.

TABLE III

Estimated Consumption of Biomass as an Energy Source in Rural Areas
(10^15 Joules)

Source	Peru[a]	Thailand[b]	Indonesia[c]	India[d]	Sudan[e]	Tanzania[f]
Human Waste	--	--	--	--	--	--
Animal Manure	25.0	--	--	860.0	--	--
Crop Residue	8.7	54.4	200.0	472.0	2.4	--
Fuelwood	138.0	440.0	800.0	1900.0	60.0	320.0

Sources:

[a] U.S. Dept. of Energy (6).
[b] U.S. Agency for International Development (13), assuming all crop residues are consumed in rural area.
[c] Mubayi et. al. (14).
[d] Mubayi et. al. (15).
[e] Abayazid (11).
[f] Per capita consumption of fuelwood in rural area from Revelle (16). Estimated 1975 rural population from UN (2).

Rural Energy Supply-Demand Pattern

Table IV shows the estimated energy use by type and end use for the six countries studied. It appears that the energy share of cooking and heating, which is primarily by wood, ranges from a high of 98.5% (Tanzania) to a low of 86.4% (India). Lighting (mainly supplied by kerosene) accounts for about 3% or less of the total rural energy consumption. Although the direct fuel use in agriculture (irrigation, soil preparation and harvesting) is proportionately small (from 0.3% in Tanzania to 9.7% in Sudan), the total energy requirements of this sector are in general much greater if draft animal power were to be included. This is perhaps more true in the case of rural transportation which primarily depends on animal draft power. In order to assess the potential increase of fuel demand due to mechanization of agriculture and transportation, we have estimated the amount of useful work contributed by draft animals in the six LDCs based on the number and intensity of utilization of these animals in each of these countries (5,11,16). The estimates show that total animal energy demand varies widely, from a negligible amount in Tanzania to about 94×10^{15} joules in India. Cottage industries which include activities such as brick-making, pottery and metal works mainly use energy to provide process heat and their energy share appears generally small except in India (9.3%).

Technology Overview

Biomass conversion technologies can be classified into three main types: Anaerobic digestion, pyrolysis and alcohol fermentation. In anaerobic digestion, organic materials are broken down by microorganisms in the absence of oxygen to produce methane. Although this process has been used in urban waste disposal for many decades, its utilization for producing energy at the village or individual household level is still relatively new. China and India are two countries where the technology has, perhaps, been most extensively implemented in rural areas (4,21,22). Animal wastes are the most widely used input materials in the digestors currently operating although crop residues or a mixture of crop residues and animal wastes can also be utilized. One great advantage of anaerobic digestion is that the left-over slurry after the digestion process generally preserves the nutrient values of the input feed and can be used as a good organic fertilizer.

Pyrolysis is an irreversible chemical change through heating in an oxygen-free or low-oxygen atmosphere. Depending on the system conditions and input material, the pyrolysis of biomass yields a mixture of solid, liquid and gaseous fuels. Although pyrolytic conversion of wood into charcoal by means of simple earth-covered kilns has been widely used for many centuries in the rural areas of many LDCs the process is very inefficient and does not permit the recovery of the liquid and gaseous energy by-products. How-

TABLE IV

Peru (1977)

	FUELWOOD	CROP RESIDUES	ANIMAL WASTES	COMMERCIAL FUEL	SUBTOTAL	% OF TOTAL
RESIDÉNTIAL						
Cooking/Heating	138.0	--	24.0	6.3	168.3	90.8
Lighting	--	--	--	5.6	5.6	3.0
AGRICULTURE	--	--	--	9.5	9.5	5.7
Irrigation						
Soil Preparation,etc.						
COTTAGE INDUSTRY	--	--	--	1.0	1.0	0.5
TRANSPORTATION	--	--	--	1.0	1.0	0.5
TOTAL	138.0	--	24.0	23.4	185.4	100.0

Estimated Animal work in agriculture and transportation: 7×10^{15} Joules
Source: (6).

Thailand (1978)

	FUELWOOD	CROP RESIDUES	ANIMAL WASTES	COMMERCIAL FUEL	SUBTOTAL	% OF TOTAL
RESIDÉNTIAL						
Cooking/Heating	440.0	6.7	--	--	446.7	86.5
Lighting	--	--	--	13.4	13.4	2.6
AGRICULTURE	--	--	--	46.5	46.5	9.0
Irrigation						
Soil Preparation,etc.						
COTTAGE INDUSTRY	--	6.3	--	1.7	8.0	1.5
TRANSPORTATION	--	--	--	1.7	1.7	0.4
TOTAL	440.0	13.0	--	63.3	516.3	100.0

Estimated Animal work in agriculture and transportation: 8×10^{15} Joules
Sources: (1) and (19).

TABLE IV (cont.)

Tanzania (1977)

	FUELWOOD	CROP RESIDUES	ANIMAL WASTES	COMMERCIAL FUEL	SUBTOTAL	% OF TOTAL
RESIDÉNTIAL*						
Cooking/Heating	320.0	--	--	--	320.0	98.5
Lighting	--	--	--	4.0	4.0	1.2
AGRICULTURE	--	--	--	1.0	1.0	0.3
Irrigation						
Soil Preparation,etc.						
COTTAGE INDUSTRY	--	--	--	--	--	--
TRANSPORTATION	--	--	--	--	--	--
TOTAL	320.0	--	--	5.0	325.0	100.0

Estimated Animal work in agriculture and transportation: Negligible
Sources: (15) and (20).

India (1972)

	FUELWOOD	CROP RESIDUES	ANIMAL WASTES	COMMERCIAL FUEL	SUBTOTAL	% OF TOTAL
RESIDÉNTIAL						
Cooking/Heating	1700.0	460.0	760.0	50.0	292.0	86.4
Lighting	--	--	--	74.0	74.0	2.2
AGRICULTURE						
Irrigation	--	--	--	55.0	55.0	1.6
Soil Preparation,etc.	--	--	--	13.0	13.0	0.4
COTTAGE INDUSTRY	200.0	12.0	100.0	4.0	316.0	9.3
TRANSPORTATION	--	--	--	2.0	2.0	0.1
TOTAL	1900.0	472.0	860.0	198.0	3380.0	100.0

Estimated Animal work in agriculture and transportation: 94×10^{15} Joules
Source: (15).

TABLE IV (cont,)

Indonesia (1972)

	FUELWOOD	CROP RESIDUES	ANIMAL WASTES	COMMERCIAL FUEL	SUBTOTAL	% OF TOTAL
RESIDÉNTIAL						
Cooking/Heating	800.0	200.0	--	--	1000.0	96.4
Lighting	--	--	--	30.0	30.0	2.8
AGRICULTURE	--	--	--	4.0	4.0	0.4
Irrigation						
Soil Preparation,etc.						
COTTAGE INDUSTRY	--	--	--	1.6	1.6	0.2
TRANSPORTATION	--	--	--	1.7	1.7	0.2
TOTAL	800.0	200.0	--	37.3	1035.7	100.0

Estimated Animal work in agriculture and transportation: 67×10^{15} Joules
Source: (14).

Sudan (1974)

	FUELWOOD	CROP RESIDUES	ANIMAL WASTES	COMMERCIAL FUEL	SUBTOTAL	% OF TOTAL
RESIDÉNTIAL						
Cooking/Heating	250.0	2.4	--	--	252.4	96.0
Lighting	--	--	--	3.0	3.0	1.1
AGRICULTURE						
Irrigation	--	--	--	3.0	3.0	1.1
Soil Preparation,etc.	--	--	--	4.0	4.0	1.5
COTTAGE INDUSTRY	--	--	--	0.2	0.2	0.1
TRANSPORTATION	--	--	--	0.4	0.4	0.2
TOTAL	250.0	2.4	--	9.6	263.0	100.0

Estimated Animal work in agriculture and transportation: 1.0×10^{15} Joules
Sources: (10) and (18).

ever, a number of promising pyrolytic processes have been developed to produce charcoal and other high grade fuels: (a) continuous low-temperature pyrolysis of wood and crop residues with various retort designs (22), (b) small scale gasification of wood and crop residues (24,25), (c) medium to large scale distillation of wood to produce methanol (25,26).

Alcohol fermentation is a microbiological process to produce ethanol from a variety of sugar containing materials. Various cellulosic forms of biomass which can be converted to glucose sugar through enzymatic or acid hydrolysis are also being tested as substrates in this process in many on-going research projects (27). Because of geographical, economical and other constraints on producing the fermentable sugar needed in this process on a large scale, Brazil is the only country, at present, which has developed a substantial program to produce ethanol for energy use.

Table V lists eight processes that appear as the most likely candidates for processing organic materials into fuels. It should be noted that the table is intended for descriptive purposes only and not for comparison of these processes. In addition, the determination of the maintenance requirements were highly qualitative and the energy costs of each process were not necessarily derived on the same basis or definitions. In the assessment that follows, we have excluded alcohol fermentation because in its present status of development, it requires grown organic materials such as sugarcane as substrates. In planning a future biomass conversion development strategy, however, the selection of technologies for a country are much more complicated and the following criteria should be taken into consideration: (a) sustainability of substrate (b) maintenance and technology requirements (c) capital costs (d) adaptability to a variety of different environments (e) ability to supply the existing and projected pattern of end use needs (f) acceptability in the cultural and social structure.

Supply-Demand Integration

Based on the current end-use patterns discussed earlier, we have generated a scenario for the year 2000 to examine the future rural energy demands and the potential of biomass conversion to meet these demands. It was assumed that energy for subsistence would grow at the same average growth rate of 2.5% as projected for the rural population in the six LDC's. Direct fuel use for development has been projected to grow at 5% annually. In addition, it was also assumed that 20% of the animal draft energy used in agriculture and transportation would be replaced by mechanization. The aggregate rural energy growth implied by these assumptions are relatively on the high side when compared with other LDC energy projections (32). On the supply end, we have held the biomass availability at the current level to provide a "lower bound" of the future biomass energy potential. The future

TABLE V

Biomass Conversion Technologies

Energy Product	Substrate	Process	Status of Technology	Maintenance Requirement	Sustainability of Substrate	Estimated Energy Cost ($/10⁹J)
Methane	Crop residues Animal wastes	Anaerobic digestion	Well developed	Low	High	2-4[a]
Charcoal	Wood	Carbonization by kilns	Well developed	Low	Country dependent	2-6[b]
Charcoal, pyrolytic oil	Wood Crop residues	Pyrolysis	Available	High	High	1-3[c]
Methanol	Wood	Pyrolysis/ distillation	Available	High	Country dependent	8-10[d]
Ethanol	Sugarcane	Batch fermentation	Available	High	Country dependent	18-20[e]
Ethanol	Crop residues	Batch fermentation	Research stage	High	High	30-50[f]
SNG	Wood	Gasification	Development stage	High	Country dependent	6-8[g]

[a] Based on a 75 m³/day community size plant. Does not include collection cost of wastes (28).
[b] Lower limit based on retail price of charcoal (1977) in Thailand (29) and upper limit based on retail price of charcoal (1977) in Ghana (23).
[c] Lower limit based on production cost of a pyrolytic converter with one ton/day capacity (7).
[d] Upper limit based on production cost of a designed converter with six ton/day capacity in Ghana (23).
[e] Based on the economic feasibility of a plant of 100,000 gallon/day capacity at a feedstock cost of $19/ton dry wood (13).
[f] Calculated selling price based on a feedstock cost of $13.6/ton in Brazil (30).
[f] Based on the economic feasibility of a plant of 75,800 gallon/day capacity at a feedstock cost of $15/ton dry wood (13).
[g] Based on the economic feasibility of a plant of 6.4 x 10⁶ SCF/day capacity at a feedstock cost of $19/ton dry wood (13).

supply of biomass in these countries could be higher because of factors such as increased agricultural production. This scenario can therefore be judged as a conservative one in assessing the potential of biomass conversion.

In matching the projected energy demands by end use and the various fuels that can be produced through the few biomass conversion technologies selected for this study, we have converted the projected biomass energy needed for direct combustion (predominantly cooking at 10%-20% efficiency) into the energy equivalent of converted fuel forms which can be utilized at a much higher end-use efficiency (30% to 50%). In this study we have used a conversion factor of 0.4 to estimate fuel requirements in 2000. This conversion is important because greater end use efficiency requires less energy input to generate a fixed amount of useful energy. Figure 1 is a general network diagram which indicates energy flows from resources available in rural areas of LDCs through current and projected conversion technologies to the disaggregated end uses. This network, which is a subsystem of the Less Developed Countries Energy System Network Simulator (LDC-ESNS) developed at Brookhaven National Laboratory (33), can be utilized to make quantitative assessments of the impacts of new technologies.

Table VI shows the amounts of energy that can be produced by the technologies we have selected matched against the end-use demand for energy in the six countries for the year 2000. Energy for subsistence denotes almost entirely the domestic household demand for cooking and lighting. Energy for development focusses on the energy needs of agricultural production, rural industries, transportation and improving living standards of the rural population (17,32). A very rough attempt has been made to divide the developmental energy needs into the need for heat energy and the need for motive power, which we assume will have to be met almost completely by liquid fuels, and stationary mechanical power, which can be met by gaseous fuels as well (e.g., biogas) or by conversion to electricity. This division, however, is subject to a large number of uncertainties. It is provided here as a purely normative estimate to serve as a quantitative benchmark against which the impacts of biomass conversion technologies can be evaluated.

From this preliminary analysis we see that biomass conversion technologies have the potential to meet the projected rural energy demands in both the subsistence and developmental categories in five out of the six countries. Only in the case of India does there appear to be a relative shortfall and much of that is in the category of liquid fuels. This shortage reflects the fact that the technology we have chosen, methanol from wood, requires an input which is in relatively short supply in India.

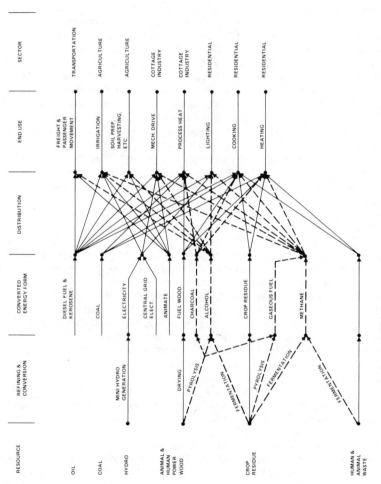

Figure 1. Rural reference energy system

TABLE VI

Projected Rural Energy Demand (2000) and Supply of Bioconverted Fuels
(10^{15} Joules)

Country	Subsistent Energy[a]		Developmental Energy (Heat)[b]		Developmental (Mech. Power)[c]		Supply/Demand (%)	Biomass Surplus[d]
	Demand	Supply	Demand	Supply	Demand	Supply		
Peru	130	130	3	3	40	40	>100	1040
Thailand	330	330	15	15	150	145	99	- 10
Indonesia	960	960	7	7	90	90	>100	500
India	2530	2530	505	400	370	--	86	-1070
Sudan	200	200	1	1	27	27	>100	240
Tanzania	230	230	1	1	4	4	>100	215

[a] In forms of charcoal, biogas, and pyrolytic oil. Assuming 55% conversion efficiency for anaerobic digestion and 66% for pyrolysis of crop residues and fuelwood (15 energy units of feedstock produce 6 units of charcoal and 4 units of pyrolytic oil).

[b] In forms of charcoal, pyrolytic oil, and pyrogas. Assuming 75% efficiency for gasification (10 energy units of feedstock produce 5.8 units of gas and 1.7 units of charcoal).

[c] In form of Methanol. Assuming 50% efficiency for pyrolysis/distillation of wood.

[d] In terms of energy content of biomass wastes.

Conclusion

We have shown, that from an overall technological and resource standpoint, biomass technologies do appear to offer a significant possiblity for meeting the future energy needs of the
rural areas of a number of developing countries. In particular,
biomass conversion technologies can potentially provide the kind
of high-grade energy, especially liquid fuels, necessary for further development and which can serve as an alternative for scarce
petroleum fuels. Biomass resources are, in many cases, locally
available and renewable, if proper resource management, especially
of forests, is practiced. (An exception to this for the countries
studied in this paper are the forest resources of Indonesia and
Peru which are remote from the main areas of rural settlement.)
From a system point of view, since rural energy demand arises from
a large number of dispersed and poorly connected uses, the availability of local resources which can be transformed into useful
energy products on a decentralized level is a definite benefit.
It avoids the necessity of investing in an elaborate centralized
energy distribution system which is susceptible to a number of
problems, as is illustrated, for example, by the experience of
rural electrification programs in many developing countries.
However, the difficulties of implementation of biomass technologies in the rural context should not be underestimated. Current biomass use practices have evolved through many millenia of
social and cultural adaptation and new ways of dealing with the
same resource wil have to overcome a number of institutional, cultural, and social barriers. For example, a community size anaerobic digester operating on the animal waste inputs from privately
owned cattle or other animals presupposes a certain degree of
cooperative arrangements among the owners for effective management
of a collection scheme. Wood that was formerly gathered privately
and used in individual households but now has to be turned over to
a processing plant for manufacture of higher-grade fuels requires
new institutional set-ups to ensure that benefits are properly
distributed. Furthermore, even for technologies that are relatively simple, a certain number of technical skills are necessary
for adequate maintenance and operation. Such skills are not usually available in most rural areas and training and education programs will be needed. Also important is the fact that rural areas
are not a closed system by themselves. Given previous development
patterns it is likely that high-grade fuels would be siphoned to
better-off urban areas leaving the rural areas short of energy as
before. The Brazilian ethanol program whose output largely goes
to fuel urban automobiles is an illustration of this problem.
Most importantly, implementing a biomass conversion program
will require substantial inputs of capital in rural areas. The
low energy consumption and correspondingly low standard of living
of most rural inhabitants reflects their low levels of income and
the economics of biomass conversion has to take this fact into

account. Unlike the situation in developed countries, the collection costs of the basic raw material will be low, but the capital constraints will be high. In this overview, we have not tried to analyze the potential implementation of biomass conversion technologies in terms of a market penetration model. We feel that such an approach would make little sense in analyzing resources and consumption patterns that, broadly speaking, currently occur outside of a market economy. A normative approach that analyzes the energy needs of development and the resources available to supply those needs, appears to us more feasible at present in assessing the impact of conversion technologies. However, a great deal of further analysis aimed at reducing the uncertainties in making quantitative judgements of technology impacts is necessary to better define the potential of biomass conversion in developing countries.

Literature Cited

1. Dept. of Economic and Social Affairs, "1976 Statistical Yearbook," United Nations, New York, 1977.
2. Dept. of Economic and Social Affairs, "Trends and Prospects in Urban and Rural Population, 1950-2000," United Nations, New York, 1975.
3. Mubayi V.; Kahn, P.; Keane, J, "Some Considerations on the Use of Village-Scale Biogas Plants in Developing Countries," (Internal Report), Brookhaven National Laboratory, Upton, New York, 1976.
4. Barnett, A.; Pyle, L.; Subramanian, S. "Biogas Technology in the Third World: A Multidisciplinary Review," International Development Research Center, Ottawa, Canada, 1978.
5. Makhijani, A.; Poole, A., "Energy and Agriculture in the Third World," Ballinger Publishing Company, Cambridge, Mass., 1975.
6. "Joint Peru/United States Report on Peru/United States Cooperative Energy Assessment, Volumes I and V," U.S. Dept. of Energy, Washington, D.C., 1979.
7. Henderson, J.; Barth, H.; Heimann, J.; Moeller, P.; Shinn, R., Soriano, F.; Weaver, J.; White, E., "Area Handbook for Thailand," U.S. Government Printing Office, Washington, D.C., 1970.
8. Vreeland, N.; Just, P.; Martindale, K.; Moeller, P.; Shinn, R.; "Area Handbook for Indonesia," U.S. Government Printing Office, Washington, D.C., 1974.
9. Fuel Policy Committee, "Report of the Fuel Policy Committee,: Government of India, New Delhi, India, 1974.
10. Abayazid, O., "Prospects of Fuel and Energy in the Sudan," University of Khartoum, Sudan, 1975.

11. Herrick, A.; Harrison, S.; John, H.; MacKnight, S.; Skapa B., "Area Handbook for Tanzania," U.S. Government Printing Office, Washington, D.C., 1968.

12. Earl, D., "Forest Energy and Forest Development," Clarendon Press, Oxford, 1975.

13. Agency for International Development, "Thailand-Renewable Non-Commercial Energy" (Draft), USAID/Thailand, 1979.

14. Mubayi, V.; Lee, J., "A Preliminary Analysis of Energy Supply and Consumption in Indonesia," Brookhaven National Laboratory, Upton, N.Y., 1977.

15. Mubayi, V., "Background Information for the Assessment of Energy Technologies for Developing Countries," Brookhaven National Laboratory, Upton, N.Y., 1976.

16. Revelle, R., "Requirements for Energy in Rural Areas," AAAS Selected Symposium 6, Westview Press, Boulder, Colorado, 1978.

17. Reddy, A.; Prasad, K., "Technological Alternatives and the Indian Energy Crisis," Economic and Political Weekly, 1977.

18. Food and Agriculture Organization, "1977 Yearbook of Forest Products," United Nations, Rome, 1979.

19. Lee, J.; Mitchell, G.; Mubayi, V., "A Preliminary Analysis of Energy Supply and Consumption in Thailand," Brookhaven National Laboratory, Upton, N.Y., 1978.

20. "Workshop on Solar Energy for the Villages of Tanzania," Tanzania National Scientific Council and USNAS, Dar Es Salaam, 1977.

21. Food and Agriculture Organization, "China: Recycling of Organic Wastes in Agriculture," United Nations, Rome, 1977.

22. Parikh, J.; Parikh, K., "Mobilization and Impacts of Biogas Technologies," Energy, 1977, Volume 2.

23. Chiang, T.; Tatom, J.; Graft-Johnson, J.; Powell, J., Pyrolytic Conversion of Agricultural and Forestry Wastes in Ghana,: Georgia Institute of Technology, Atlanta, Georgia, 1976.

24. "Symposium Papers: Clean Fuels from Biomass, Sewage, Urban Refuse, Agricultural Wastes," Institute of Gas Technology, Chicago, Illinois, 1976.

25. "A Conference in Capturing the Sun Through Bioconversion: Proceedings," Washington Center for Metropolitan Studies, Washington, D.C., 1976.

26. Jones, J.; Barkhorder P.; Bomberger, D.; Clark, C.; Dickenson, R.; Fong, W.; Johnk, C.; Kohan, S.; Phillips, R.; Semran, K.; Terter, N., "A Comparative Economic Analysis of Alcohol Fuels Production Options," SRI International, Menlo Park, California, 1979.

27. "3rd Annual Biomass Energy Systems Conference Abstracts," Biomass Energy Systems Conference, 1979.

28. Ghate, P., "Biogas: A Pilot Project to Investigate a Decentralised Energy System," State Planning Institute, U.P., Lucknow.

29. "Energy Situation in Thailand-1977" National Energy Adminis-
     tration, Bangkok, Thailand, 1978.
30. Yang, V.; Trindade, S., "The Brasilian Fuel Alcohol Program,"
     Centro de Tecnologia, Promon, Brazil, 1979.
31. Schooley, F.; Dickenson, R.; Kohn, S.; Jones, J.; Meagher,
     P.; Ernest, K.; Crooks, G.; Miller, K.; Fong, W., "Mission
     Analysis - Market Penetration Modeling," SRI International,
     Menlo Park, Calif., 1979.
32. Palmedo, P.; Nathans, R.; Beardsworth, E.; Hale, S. Jr.,
     "Energy Needs, Uses and Resources in Developing Countries,"
     Brookhaven National Laboratory, Upton, N.Y., 1978.
33. Reisman, A.; Malone, R., "Less Developed Countries Energy
     Network Simulation LDC-ESNS: A Brief Description," Brook-
     haven National Laboratory, Upton, N.Y., 1978.

RECEIVED November 16, 1979.

Power Generation from Biomass Residues Using the Gasifier/Dual-Fuel Engine Technique

ANDRE A. DENNETIERE — DUVANT Motors, Valenciennes, France

FRANCOIS LEORAT — RENAULT New Technologies, 92508 Rueil-Malmaison, France

GEORGE F. BONNICI — IDP/RENAULT, One Old Country Road, Carle Place, NY 11514

The increase of the cost of all forms of energy has forced industry to look for the cheapest and most dependable ways of producing and purchasing energy. Potential biomass energy, either under the form of natural resources or by recycling wastes, is rather important throughout the world. For a long time, surface natural combustibles, such as wood and agricultural wastes of some specific products, have been part of energy utilization in plants. They have often been outclassed, in quality as well as in flexibility, by fossile combustibles and so have been abandonned. While waiting for new energy utilization, these surface natural products can be used again. The energy conversion of some wastes may also be considered when they have no other use by reprocessing.

The first technical idea is generally to burn these products and use the direct thermal effect (heat and thermal fluids) or indirect mechanical effect (steam engines). However, this kind of technique is generally costly, inefficient or simply not usable with some processes. As a result, a great deal of biomass waste is poorly converted, lost, or burnt without energy production. Conversely, by gasification of vegetable cellulose, and with the help of well adapted equipment, excellent results may be obtained in converting these products to useful energy, without sacrificing the available energy and flexibility.

Gasification Cycle and Mechanical Power Generation Overall Performances

The cycle consists in converting potential thermal energy of organic wastes into a combustible gas for heavy duty thermal engines. The efficiency of the gasification process is generally about 70 to 80 % (Figure 1).

The thermodynamic efficiency of thermal engines is about 37 %. As we shall see, the best suited engines are of the dual-fuel type, whose output power results from the simultaneous work of two combustibles. A small quantity of liquid combustible

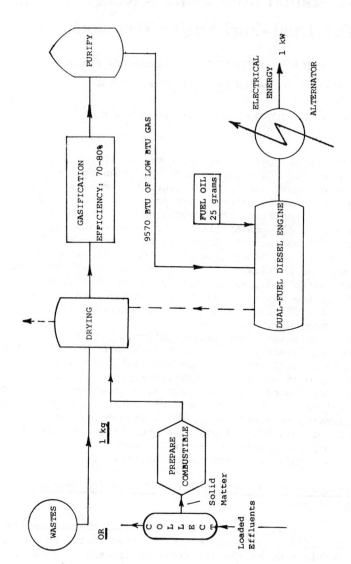

Figure 1.   Operation diagram of a low Btu gas installation

(straight Diesel oil) is injected according to Diesel principles and ensures ignition of the gaseous load entering the cylinder during the admission stroke. In fact, gas has the most important role during the internal cycle working process. As a result, the gas flow rate is controlled depending on the desired power output, while Diesel fuel injection remains constant regardless of engine load. It amounts to a constant reduced Diesel oil consumption per hour.

For utilization with the low BTU gas obtained through gasification of organic waste, the energy brought to the engine by Diesel oil is limited at less than 10 % of the overall fuel consumption at full load. It means that Diesel fuel savings under usual operating conditions is no less than 80 % of the consumption of conventional Diesel engines of identical power. Most of the time, the engine is coupled to an alternator allowing electricity generation with an efficiency of 93 %.

Each kWh (3.6×10^6J) is obtained from:
- 0.055 lb (25g) of Diesel fuel
- 2200 kcal (92.11×10^5J) under gaseous form.

The low BTU gas available from the gasifying equipment is a mixture of:
- combustible elements: CO (14-15 %), H_2 (15-20 %), methane CH_4 (less than 2 %), and some hydrocarbons of a higher order than methane.
- inert elements: CO_2 and N_2.

Its NCV is about 1100 kcal/normal m^3 (46.05×10^5J/m^3). Therefore, the production of 1 kWh requires about 2.2 m^3 of this gas.

Cellulose wastes can be regarded as including about 50 % carbon and 50 % water. Carbon oxydizing conversion means, as a general rule that with a mean gasification efficiency of 75 %, 1 kWh is obtained from 800 to 900 g of anhydrous product.

Good gasification conditions are obtained when the moisture content of the product is low, 10 to 20 % of water, referring to gross weight. So, it is possible to consider, without great error, the following ratio: with 1 kg of so called dry waste, we get 1 kWh at the alternator output. This result is quite satisfactory.

Some Characteristics of Low BTU Gas Engines

These engines are directly derived from standard 4 stroke Diesel engines. They are particularly suited for continuous operation. Patented design features allow a perfectly reliable and high output gas, even with low BTU gas as the main fuel. The Diesel cycle's high compression ratio is continuously maintained, providing high efficiency performance.

Ignition power, obtained by combustion of a small amount of injected Diesel fuel, is much greater than conventional spark ignited engines. This ensures an excellent combustion stability of the combustible gas mixture, even with excess air. This is quite interesting when gas composition may vary, which is precisely the case with gasification or pyrolysis gas. For partial load operation, the ability to operate outside the stocchiometric limits allows it to maintain a high efficiency, as in a pure Diesel cycle (see test results in Table I).

Independent fuel lines provide for the separation of air and gas intake manifolds up to the cylinders, ensuring a good mixture inside them. This feature prevents any risk of explosion outside the engine. Moreover, air and gas intake being separately controlled, burnt gases are efficiently scavenged resulting in reduced internal thermal load and supercharging is possible without any fuel losses.

Supercharging

Supercharging is performed by means of a turbo-compressor, as in a pure Diesel cycle. In the general case, gas is available without any relative pressure. Therefore, our engines are equipped with two turbo-compressors: one for gas, the other for combustive air.

Engine power regulation, i.e. the ability to keep a driven machine's speed constant whatever the load, is automatic.

The engine is started in Diesel cycle, and brought to the desired power level. Then, gas is supplied, and the maximum possible gas quantity replaces Diesel fuel automatically. This is accomplished without diminishing the engine's performance.

In the case of gas production failure (quality and quantity) the lack of energy is automatically compensated for by complementary Diesel fuel (with, of course, a reduction of the savings during this time). If the lack of gas persists, the engine will operate on pure Diesel cycle and will continue delivering the desired power. The regulating functions (transient or working changes) occur inside a narrow statistical range (less than 5%).

These generating sets can accomplish the following programs:
- autonomous generating sets supplying electricity to local consumers under variable load.
- coupled generating sets.
- fixed power generating sets for internal or local electricity, or linked to utilities for cogeneration.
- direct drive of various machines.
- basic elements for "total energy" systems.

A basic quality of these sets is a very high operation flexibility. This essentially results from the dual-fuel concept. Because the engines are able to operate in pure Diesel cycle, a power station can be quite autonomous, provided that a small Diesel fuel storage reserve can be kept.

Table I. RESULTS OF ONE 500 kW GENERATING SET WITH DUAL FUEL
6 CYLINDERS SUPERCHARGED ENGINE (WOOD GAS RUNNING)

TEST	POWER	GAS ADMITTED	SUPERCHARGE PRESSURE		TEMPERATURE AFTER COOLING		CYLINDER EXHAUST TEMP	FUEL OIL QUANTITY	TOTAL EFF.
			GAS	AIR	GAS	AIR			
	HP	ft³/h / Nm³/h	lb/in² / g/cm²	lb/in² / g/cm²	F / C	F / C	F / C	lb/h / kg/h	%
1	375	22,775 / 645	5.9 / 413	4.1 / 290	61 / 16	57 / 14	622 / 328	24.4 / 11.1	31.4
2	545	32,132 / 910	8.2 / 585	6.7 / 475	68 / 20	65 / 19	727 / 386	25.7 / 11.7	34.2
3	600	34,250 / 970	8.5 / 610	7.7 / 550	77 / 25	72 / 22	763 / 406	26.0 / 11.8	36.6
4	715	35,663 / 1,010	12.3 / 930	9.8 / 700	86 / 30	77 / 25	833 / 445	23.2 / 10.55	37.1
*5	540	24,717 / 700	7.4 / 530	5.7 / 410	131 / 55	122 / 50	815 / 435	23.2 / 10.55	39.5

GAS CALORIFIC VALUE   TEST 1 − 990 calories/Nm³; 110 BTU/SCF; 4,145.1 J/m³
ACCORDING TO ANALYSIS   2 − 970   "   103.5   "   4,061.4 "
                                  3 − 955   "   106.9   "   3,998.6 "
                                  4 − 1100   "   123   "   4,605.7 "
                                  5* − 1060   "   118   "   4,438.2 "

*NOTE: CORRESPONDS TO OPERATION WITHOUT WATER SUPPLY TO AIR AND GAS COOLANTS.

Specifications of the Gasification Facility

A) Gasification Apparatus. The above defined generating sets are generally fed with low BTU gas by a down draft gasifier with a fixed carbonization bed.

Gasification is the decomposition of hydrocarbon combustibles, such as cellulose, at high temperature by partial combustion, that is to say with a lack of air.

A fast, but partial, combustion generates carbon dioxide CO_2 with high exothermicity. CO_2 can then be reduced in the presence of the remaining incandescent carbon, yielding CO, a combustible gas. Water from the chemical structure or moisture content also plays a part, generating hydrogen and CO. CO_2 and water reduction are endothermic. Thus, a thermal equilibrium takes place between combustion reaction and gasification reactions: the energy of the first one being used in the best way to ensure the second ones.

With ligno-cellulosic products, we are not dealing with pure carbon. As a result, the needed physical carbon bed must be generated.

Therefore, the first reaction is a distillation or pyrolysis of the combustible that has to pass a first heating phase, where volatile matters are exhausted. These volatile matters would give an important calorific value, but cannot be used directly in engines requiring low intake temperature, because of thermodynamic cycle considerations. The cooling of such gas would generate pyrolignous acids and tar condensations that would cause an important loss in the calorific value and increase the risks of clogging.

Combustible gases are sucked from the lower part of the gas generator. Thus there exists a parallel downward path for combustible air and produced gases, pyrolignous juices, acids and tars. These are cracked on an incandescent coal bed.

The apparatus consists in a vertical tower fitted with refractory materials and including:

- in the upper part, a combustible reserve fed by a continuous or sequential loading mechanism adapted to the working conditions;
- in the central part a combustion and distillation area fed with air by nozzles;
- in the lower part, a reduction area followed by a grille designed for ash removal, generally performed in a hydraulic lock.

In order to improve efficiency, fed-in air is preheated by the combustible gases' sensible heat, by means of integrated heat exchangers.

The complementary cooling and cleaning of these gases is performed in a final scrubbing unit. This scrubber is a vertical tower where the upward flow is mixed with atomized water. The gases, now cleaned and cooled in the scrubber, are driven by a compressor that creates the necessary pressure for piping them to the engine through an intermediate filter.

This chain of equipment allows the gasification of all kinds of solid organic wastes of adequate size. For example, wood wastes including wood bark, coconut shells, corn cobs, etc. can be used without preliminary preparation. More divided combustibles may be added to these, in proportions up to 20 % in weight.

In order to use highly divided wastes such as small seed shells, coffee shells, etc. a combustible has to be prepared by densification. The humidity of these products has to be low. A moisture content less than 20 % is required to get a good steady operation.

For more humid products, which is frequently the case, a drying stage is needed and generally takes place just before loading. Thermal energy necessary for drying is provided by engine thermal losses: cooling circuit water and exhaust gas.

B) Pyrolysis Equipment. The pyrolysis process is also relevent. In that case, the operation aims at utilizing waste by producing a residual char. A specific combustible is obtained (active char or charcoal, refractory ashes...) at the rate of approximately 25 % of the initial weight.

Of course, these combustibles may in turn be gasified in simpler equipment than the ones described in Paragraph A. The gases exhausted during the material heating can be utilized. These gases are often of higher calorific value than the low BTU gas obtained with gas generators, but include higher hydrocarbons. These products must be separated by condensation, in order that the engine burn only the incondensable part.

The condensates may find interesting commercial value or be used as a thermal supply to the pyrolysis process itself. The hydrocarbons can also be dissociated in cracking equipment fed with charcoal burning with air deficiency. This equipment works in nearly the same way as the lower part of the above described down draft gasifier.

Fields of Application

The field of application for the generating set gasifier system that seems to offer the most interesting opportunities is in the power range from 100 kW to a few thousand kW. Single commercially available generating sets cover the 150 to 1000 kVA (approx. 120 to 800 kW) range (Figure 2).

Available gasification equipment is well adapted to this range. Several generating sets can be coupled to one source of gas. Conversely, several gasifiers can be coupled to feed a single set.

Power stations of less than 100 kW are perfectly feasible too. It all depends on local economic conditions, which will define the profitability of the system.

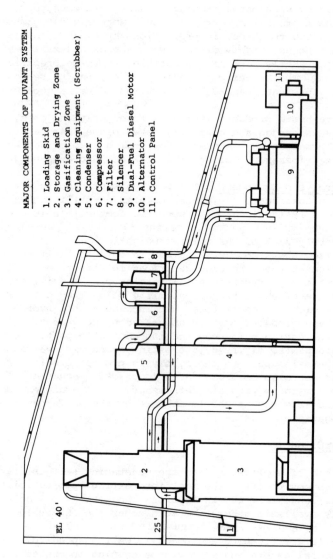

MAJOR COMPONENTS OF DUVANT SYSTEM

1. Loading Skid
2. Storage and Drying Zone
3. Gasification Zone
4. Cleaning Equipment (Scrubber)
5. Condenser
6. Compressor
7. Filter
8. Silencer
9. Dual-Fuel Diesel Motor
10. Alternator
11. Control Panel

Figure 2.   Typical plant layout for Duvant gasifier/dual-fuel diesel motor/alternator system

For more important stations, power units of more than 1000 kVA are being developed. It should be noted that the thermal conversion of solid wastes and biomass is also justified for very big installations. However, this can be done only if a large quantity of waste can be conveniently obtained.

When applied to steam cycle equipment, which has a poor efficiency, it should be used only if there is a need for producing thermal energy (boilers, back-pressure machine systems are thus best adapted).

Moreover, it should be pointed out that the overall mechanical energy needs in factories today is continually increasing, while strictly thermal energy needs are steadily decreasing. Therefore, gas generators and dual-fuel Diesel engine power stations are well adapted to actual and future needs.

Advantages are the following:
- reduced investment cost when compared to heat-power installations
- efficiency and flexibility equal to classic Diesel engines, but with fuel savings of at least 80 %.

Mentioning Diesel stations, it should be noted that actually a good many of them are used as electrical auxiliaries to heat-power systems whose power potential is too low. Other stations also provide the power required for seasonal industries.

In both examples, low BTU gas-Diesel sets replacing conventional Diesel equipment yield savings in money spent for the purchase of petroleum. They can also promote complementary industrialization between seasons, by using local resources producing low value wastes.

Potential for Rural Electrification by Biomass Conversion of Local Wastes. The creation of large generating stations is seldom advantageous. Because of the enormous quantities of fuel required, frequent collection and transportation of waste over long distances is necessary. When one considers the added expense of constructing power lines, it is difficult to regard this as a cost-effective system of biomass conversion.

On the contrary, medium size generating stations provide low cost local electrification for rural areas separated from one another by long distances. Converting local wastes reduces the costs of transporting raw materials and distributing electricity, and greatly alleviates waste disposal problems.

Finally, the developing regions are more interested in local autonomy (self-reliance). Their needs are best satisfied by small or medium sized plants. The answer is found in low BTU gas engines.

Hints for Re-Cycling Waste. Under the form of charcoal and charcoal substitutes, combustible products from wastes can provide interesting commercial opportunities.

The development of manufacturing processes requires a

great deal of mechanical power, because good mechanization is needed to produce high quality briquets or granulates.

Production of this power can be obtained at low cost by means of an engine-gas generator system fed with the same combustible. This enables us to perform several functions with the same equipment.

Better Thermal Conversion of Very Poor Combustibles. Generally the low calorific value of a combustible results from its high moisture content. It should be noted that up to a moisture content of 50 to 60 %, these combustibles can be used in the gasification cycle without efficiency loss, by pre-drying them with the heat obtained from the engine losses.

Very poor combustibles are sometimes incinerated without any gain, because their low calorific value does not justify the investment of boilers with special fire places. However, they can be used in gas-generators, which are less expensive for the identical power.

Surplus Gas Utilization. A gasification plant can feed in parallel a generating set and a thermal production boiler. It can be a classic boiler with rich gas or fuel burners, which means high efficiency.

A burner for low BTU gas allows the utilization of surplus gas during the periods when little or no electricity is needed.

Other Methods of Effective Waste Conversion

The increasing amount of research in waste conversion has led to the development of additional sources of usable fuel. Because of their versatility, durability, and high efficiency, dual-fuel engines serve to maximize the value of these efforts. It may seem to be a contradiction in terms, but "waste" and "garbage" are now being looked upon as "energy resources".

We can consider, for example, the bacteriologic treatment of very diluted wastes allowing methane fermentation. The gases, produced from digesters, are an excellent fuel for these engines.

Some industries generate a large volume of different types of waste, requiring different processes. Solid wastes must be extracted, because they are usually poor combustibles. Nevertheless, they are useful for gasification.

Other wastes, such as effluents, require purification to avoid pollution. The two kinds of produced gases can be combined to feed a gas-Diesel engine station. The dehydration, stabilization and compacting of urban waste, gives a gasifiable combustible if they do not include a great deal of plastic that would generate acid gases.

Other pyrolysis or treatment processes of urban waste provide gases that can be utilized in engines. For these gases, this solution is better than post-combustion which is not always

Figure 3. *Gasifier for a 1000 kVA system*

Figure 4. *View of ash-removal section of gasifier*

Figure 5.   Gas scrubber with condenser mounted on top

Figure 6.   A 1000 kVA gen set with dual-fuel diesel engine (720 rpm)

very well used because of the difficulties encountered in transporting the obtained thermal energy.

Methanol synthesis can be obtained by the reaction of synthesis gases obtained by gasifying ligno-cellulosic wastes with oxygen. Besides the use of methanol in the chemical industry, its utilization as a combustible may be considered too. Indeed, the above described gas-Diesel engines are actually fitted with devices allowing the utilization of methanol instead of gas, together with the usual 10 % Diesel fuel injection.

Some agro-food industry by-products are quite well adapted to alcoholic fermentation. Ethanol obtained by distillation is a good combustible that, as methanol, can be easily stored and transported. Our dual-fuel engines are adapted to its utilization. Fuel savings are actually about 80 %. This ethanol does not demand as strict a rectification as in the case of use in mixture with fuels. The alcoholic degree must be between 90 and 96 Gay Lussac.

Where sufficient low cost thermal sources can be found to create distilleries, waste conversion by ethyl alcohol is justified. The best example is sugar-cane: bagasse surplus allows the preparation of a gasifiable combustible utilized in gas-Diesel engines during normal periods of plant stoppage.

An organization of the plant furnaces fed with bagasse allows the creation of auxiliary distilleries that can convert to fuel, under the form of combustible alcohol, the molasses which cannot be recycled in some places (See Figures 3, 4, 5, and 6).

The engines described are suitable for both applications.

Conclusion

We have tried to show that the economical use of natural forest and agricultural wastes is possible and necessary. The research field for gasification or poor combustibles conversion is wide open.

The engines described are perfectly efficient, reliable and universal machines. As such, they allow the creation of basic or complementary energy sources. Their existance is neither purely theoretical nor in its early developmental stages. Rather, they have been extensively tested and are currently available.

Unlike many laboratory and prototype testing facilities, which are hampered by major problems when commercially installed, these engines were developed and tested in a "working" environment. These operating conditions are typically encountered at site installations.

Therefore, the outstanding performance shown in the accompanying test results can be anticipated for subsequent installations.

RECEIVED January 9, 1980.

46

Studies on the Practical Application of Producer Gas from Agricultural Residues as Supplementary Fuel for Diesel Engines

IBARRA E. CRUZ

College of Engineering, University of the Philippines, Diliman, Zuezon City, Philippines 3004

The Philippines is an agricultural country where crop irrigation has become essential to accelerate food production. Many irrigation systems use diesel engines. In rice production, for instance, a government-supported cooperative-type farm based organization has installed since 1975 a total of 334 diesel engine-driven pumps of sizes ranging from 15 to 150 horsepower (1 horse power is 746 watts) which enabled farmers to harvest two and even three crops in a year. With the continuing increase in price coupled with the scarcity of fuel oils, particularly diesel fuel, it is becoming more and more difficult to continue operating these diesel engine-driven pumps.

The objective of these studies therefore is to assure the continued operation of existing diesel engines, particularly those used in crop irrigation, by converting them to dual-fuel engines with minimum modifications and using producer gas as supplementary fuel. It includes the design of a simple gas producer that can be fabricated inexpensively.

A diesel engine with a little modification, can be operated as a dual-fuel engine, that is, an engine that uses both gaseous fuel and liquid injection fuel. Normally, a diesel engine aspirates air during the intake stroke of the piston and compresses this to a high pressure and temperature. The compression ratio of the diesel engine is high enough so that the temperature of the air inside the engine cylinder after the compression stroke attains a sufficiently high level so as to ignite the diesel fuel that is injected into it. In dual-fuel operation, a mixture of gaseous fuel and air in the proper proportion is aspirated into the engine and compressed during the compression stroke. The gaseous fuel-air mixture is on the lean side so that it does not pre-ignite during the compression stroke. Only the injection of the liquid fuel initiates ignition and final combustion of both gaseous and

0-8412-0565-5/80/47-130-649$05.25/0

liquid fuels. Since combustion of the gas aspirated with air provides power, a much less amount of liquid fuel, compared to
straight diesel operation, need be injected to produce a given
power output. Thus a significant savings in the liquid injection
fuel (diesel oil) is effected.

The economics of dual-fuel operation becomes favorable if
the gaseous fuel can be obtained from indigenous sources. One
such gaseous fuel is producer gas from agricultural residues.
Studies on the production of gas from agricultural wastes have
earlier been reported (1).

Experimental Work With a Single-Cylinder Engine

Initial experimental work on the use of producer gas from
coconut shell charcoal was done with the support of the Philippine Coconut Authority (2). Charcoal as fuel was chosen because
the cleaning of the gas from charcoal was simpler due to less
tar in the gas.

Figure 1 shows the experimental lay-out for studying the performance of a single-cylinder diesel engine when using producer
gas as its main fuel. The engine had a bore of 4.5 inches and a
stroke of 4.25 inches (1 inch is 2.54 centimeters). The engine
was manufactured by Lister-Blackstone.

In Figure 1, it is seen that the producer gas is aspirated
into the engine together with the air. Therefore, the only modification necessary to allow the diesel engine to use producer gas
is a gas pipeline connection to the air intake pipe of the engine
with appropriate control valves for proportioning the air-gas
mixtures.

The gas producer was mounted on a platform scale to allow
weight measurements of the charcoal consumed during a test
run. From derived relations, the weight rate of producer gas
utilized in the engine was calculated. The liquid fuel tank, containing diesel oil was likewise mounted on a weighing scale. A
rotameter or flowmeter was also installed in the liquid fuel line
to serve as a check on the rate of fuel consumption.

The engine was started in the normal way by hand cranking,
with the air intake valve fully open and the producer gas valve
fully closed. Thus the engine was run on diesel fuel alone at the
start.

The engine torque output was measured by a prony brake
mounted on another platform scale, and the engine RPM by hand
tachometer. The brake horsepower output in each run was thus
determined.

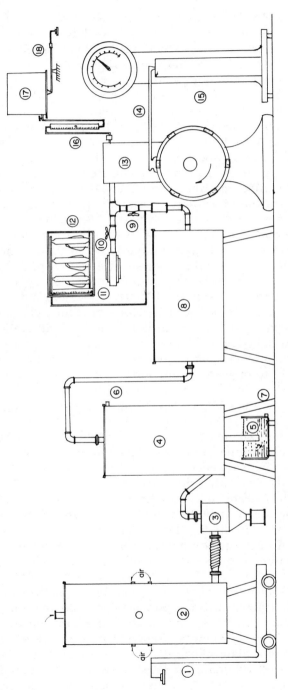

Figure 1. Experimental layout for studying the use of producer gas from solid waste fuels in a diesel engine

(1) platform scale; (2) gas producer; (3) cyclone filter; (4) scrubber; (5) water outlet; (6) water inlet; (7) drain; (8) gas filter (coconut fiber); (9) control valve (gas); (10) control valve (air); (11) air filter; (12) gas analyzer; (13) diesel engine; (14) prony brake; (15) platform scale; (16) flowmeter; (17) liquid fuel; (18) weighing scale

The gas producer was a suction-type, downdraft reactor with 12 air holes around the mid-section of the cylindrical body, and a single gas outlet at the bottom. Connected to the gas outlet was a cyclone separator (3) to remove entrained dust and charcoal fines out of the gas before it went to the engine. The gas scrubber (4) and filter (8) were later additions when fuels with high tar and volatile matter contents were used.

At start-up, the gas producer was initially filled with charcoal crushed to about 1-inch size, up to the level of the air-holes. Feeding of fuel was done by opening the top of the reactor. A burning zone was started on the top of the charcoal bed by igniting small pieces of wood and when the charcoal was burning evenly at all levels of the air-holes (this took about 10 minutes to occur from the time the fire was lighted), the producer was charged with more charcoal until it was full. The top of the reactor was then closed and the gas intake valve to the engine slowly opened. The engine would now aspirate gas from the reactor and speed up. Since the engine was controlled by a speed governor adjusted to about 1000 RPM, the liquid fuel intake would be automatically reduced as more producer gas was aspirated into the engine, until a minimum use of liquid fuel, that which was required only for ignition, was reached.

The initial design of the gas producer was such that it could be converted readily to operate either as a downdraft or as an updraft reactor. Also, the air intake could either be by suction from the engine or by forced draft from a compressed air tank. There was only a single air inlet at the side of the producer when it was operated as a down-draft reactor, so that the combustion zone was concentrated in the vicinity of this single air inlet. When operated as an updraft reactor, the air entered from below the reactor and passed up uniformly through a bar grate. The quality of the gas thus produced was better in the updraft producer. However, downdraft operation produced a cleaner gas, particularly when fuel with high volatile matter was used. Therefore after about half of the experimental runs were finished, the producer was redesigned to operate permanently as a suction, downdraft reactor.

Redesigning the reactor involved providing for additional air inlets so that air distribution to the combustion zone could be more uniform. Also to increase the depth of the combustion zone, the air holes were distributed around three circumferential planes (4 holes to a plane) spaced 3 inches apart thus extending the burning zone to a depth of at least 6 inches. Furthermore, the cross-sectional area of this combustion zone was reduced to

6 inches diameter from the original 10 inches and the longitudinal cross section of the reactor now exhibited a constriction or a "throat" at the combustion zone. The purpose of this throat was to make combustion more intense at this zone. Combustion rate per unit or cross-sectional area would now be higher and hence temperature higher. Thus, the combination of higher temperatures and a deeper combustion zone would lend to more efficient cracking of the volatile and tarry material in the fuel and to the production of more combustible gases.

A mass of data was obtained in evaluating the performance of the engine as a dual-fuel engine using producer gas from charcoal. The parameters used for evaluating performance are the brake-horsepower (BHP) output, the brake thermal efficiency (e_b) of the engine, and the percentage energy from producer gas (EPG) utilized in the engine. EPG is defined as the ratio of the heat released by the combustion of producer gas aspirated into the engine to the total heat released by both liquid injection fuel (diesel) and producer gas, multiplied by 100 to express as a percentage. Derivation of the equation for EPG and brake thermal efficiency is found in the Appendix.

To obtain a more convenient form, the mass of data was reduced to multiple linear regression equations to give the following relationships:

$$BHP = 7.25 - 1.90 \times 10^{-3} (RPM) - 3.22 \times 10^{-3} (CVn) \qquad (1)$$

$$\text{Coefficient of correlation, } R^2 = 0.48$$

$$e_b = 25.63 - 1.34 \times 10^{-2} (RPM) + 8.60 \times 10^{2} (CVn) \qquad (2)$$

$$R^2 = 0.41$$

$$EPG = 22.75 - 13.34 (BHP) + 1.03 (CVn) \qquad (3)$$

$$R^2 = 0.53$$

CVn in the above equations is the net calorific value of the gas in Btu/ft^3 at NTP, i.e., at normal temperature and pressure of 273 K and 1 atmosphere. (One Btu/ft^3 is 37.25913 kJ/m^3).

The range of values of the parameters used in obtaining the above equations are as follows:

(1) Engine speed: $\qquad$ $900 < RPM < 1550$

(2) Brake-horsepower: $\qquad$ $4 < BHP < 6$

(3) Net calorific value
in Btu/ft^3 $\qquad$ $100 < CVn < 135$

(4) Percent energy from
producer gas: $\qquad$ $75 < EPG < 100$

It is interesting to note that when the engine was run on die-sel oil alone, the experimental points indicate that the engine RPM was in the vicinity of 1000 RPM at which point the speed governor was set. However, for short periods of time when conditions were favorable, the engine could run on straight producer gas (EPG = 100%) without any need for the injection of diesel fuel and the maximum speed attained was greater than 1500 RPM, an increase of more than 50 percent. This was understandable since at 100 percent EPG when no liquid fuel was being used, the engine speed was not being controlled by the governor any longer but by the amount of producer gas allowed into the engine by the gas control valve.

That the engine could run on 100 percent producer gas without the need of even a small amount of liquid fuel for ignition purposes was due to the fact that the right combination of producer gas-air mixture, gas calorific value (CVn), and engine load (BHP) led to a condition whereby the combustible charge of air and gas could be ignited by piston compression alone. This condition usually occurred about an hour after start-up of the gas producer, when the calorific value of the gas had improved to about 125 Btu/ft^3. A precaution that had to be observed, however, was to adjust the gas-air mixture to make it leaner as the gas calorific value continued to improve, to prevent severe knocking of the engine.

Thus from equation (3), for an engine load of 4 BHP, CVn must be 127 Btu/ft^3 for EPG to be 100 percent. From equation (1), the engine speed would be 1496 RPM, and from equation (2), brake thermal efficiency e_b is 16.5 percent.

If the load is to be increased to 5 BHP with CVn the same at 127 Btu/ft^3, from equations (1), (2), and (3), the engine speed becomes 966 RPM, the brake thermal efficiency is 22.4 percent, and the energy supplied by producer gas, (EPG) is 87 percent. This means that only 13 percent of the energy is supplied by diesel fuel.

Other fuels such as coconut shells, woodwaste, coal, and rice hulls were tried for gasification in the gas producer and utilization in the 5-BHP diesel engine. Typical results of these trials are shown in Table 1. However, when using these fuels other than charcoal, the problem of cleaning the gas of tar in the present gas cleaning equipment (gas scrubber with a water spray) had not been satisfactorily solved. The engine had to be dismantled after about 50 hours of operation to clean its insides of tar deposits.

Experimental Work With a Six-Cylinder Engine

In a project supported jointly by the Ministry of Energy, the National Science Development Board, the National Irrigation Administration, and the University of the Philippines, experimental work on the performance of a "Fordson" 6-cylinder, 65 brake horsepower diesel engine was conducted at the U. P. College of Engineering (3). The final objective was to use this engine in dual-fuel operation for irrigation of 40 hectares (400, 000 m^2) of riceland in Siniloan, Laguna. Charcoal was used for the gas producer.

The experimental procedures were essentially the same as for the single-cylinder engine experiments except that the brake horsepower output was about ten times more in the larger multi-cylinder engine. For measuring power, a hydraulic dynamometer was used. The set-up is shown in Figure 2 except that the weighing scales on which the fuel tank and the gas producer were mounted during the experiments are not drawn. The dimensions of the gas producer are shown in Figure 3.

Table 2 shows typical data obtained when operating the engine both in dual-fuel mode and in diesel fuel mode. Again, the data are reduced to a more convenient form by multiple regression equations:

(1)   Brake Horsepower.
      (a) Diesel fuel mode:

$$BHP = 4.14 \, (RPM/1000)^{-0.98} \, (F_2)^{1.22} \qquad (4)$$
$$R^2 = 0.97$$

where F_2 is diesel fuel consumption in kilogram per hour (kg/h).
      (b) Dual fuel mode:

$$BHP = 236 \, (RPM/1000)^{2.80} \, (EPG)^{-0.78} \qquad (5)$$
$$R^2 = 0.88$$

Table 1. Comparative Performance of a 5-Brakehorsepower
          (3.7 kW) Diesel Engine Run in (A) Dual-Fuel Mode, and
          (B) Single-Fuel (Diesel) Mode

Fuel	Mode	RPM	Brake horse-power	Pound fuel per Bhp-hour		% Diesel Saved
				Liquid	Solid	
Charcoal	A	1043	4.1	0.148	1.0	83
Diesel	B	1000	4.0	0.892	0	-
Coal	A	1288	4.8	0.246	1.3	80
Diesel	B	1246	4.9	1.261	0	-
Coconut Shell	A	1212	4.7	0.208	2.6	72
Diesel	B	1208	4.2	0.730	0	-
Wastewood	A	1221	4.1	0.357	2.8	62
Diesel	B	1237	4.1	0.950	0	-
Rice Hulls	A	1214	3.4	0.323	5.6	59
Diesel	B	1170	3.3	0.795	0	-
Corn Cobs	A	1287	4.5	0.516	1.0	31
Diesel	B	1222	4.2	0.752	0	-

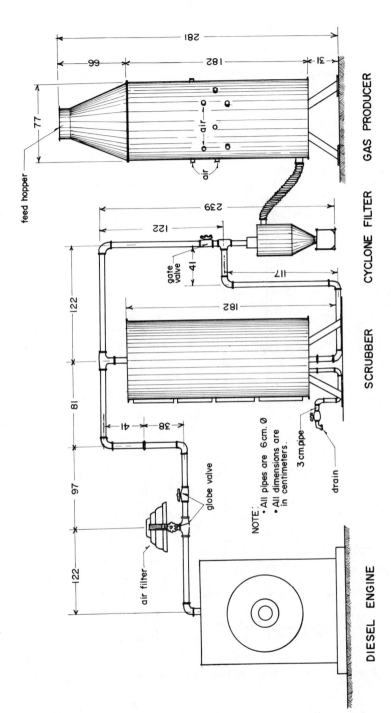

Figure 2. Multicylinder dual-fuel engine set-up

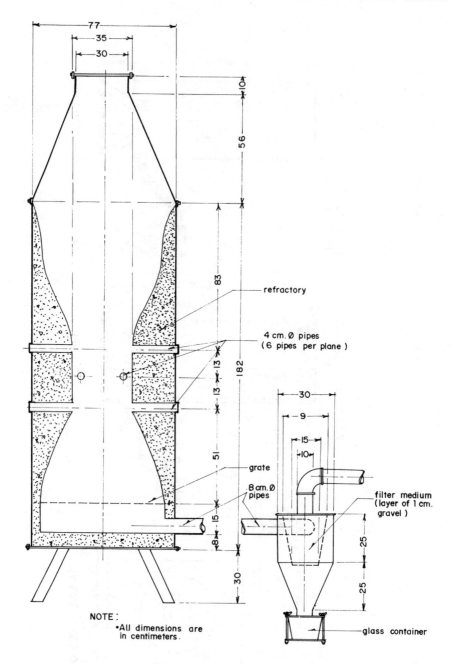

NOTE:
• All dimensions are in centimeters.

Figure 3.   Cross section of down-draft suction-type gas producer

Table 2. Typical Data on Comparative Performance of a Six-Cylinder Compression Ignition Engine Run in (1) Dual-Fuel Mode, and (2) Diesel Fuel Mode

Run No.	Run Duration Minutes	Engine Mode	RPM	BHP*	Fuel Consumption kg/h Diesel	Charcoal	%, e_b
1	120	Dual	1671	39	5.220	15.6	20.3
2	30	Diesel	1631	41	9.860	0	25.2
3	60	Dual	1619	45	6.880	15.8	20.3
4	60	Dual	1568	28	1.818	14.4	22.5
5	60	Dual	1665	31	1.504	19.8	19.7
7	15	Diesel	1590	31	7.384	0	25.2
8	60	Dual	1750	31	1.374	20.4	19.8
9	60	Dual	1747	37	1.847	20.2	22.0
15	15	Diesel	1800	49	11.640	0	25.6
16	60	Dual	1800	42	4.162	21.9	18.1

* 1 BHP = 746 watts

where EPG is percent energy from producer gas.

Also, a good correlation with IFC (Injection Fuel Consumption) is obtained:

$$BHP = 10.56 \ (RPM/1000)^{1.90} \ (IFC)^{0.24} \tag{6}$$
$$R^2 = 0.84$$

(2) Brake Thermal Efficiency:

(a) Diesel fuel mode:

$$e_b = 17.49 \ (RPM/1000)^{-0.93} \ (BHP)^{0.23} \tag{7}$$
$$R^2 = 0.65$$

(b) Dual fuel mode: Poor correlations were obtained.

(3) Charcoal Consumption W_1 in kg/h (with 10 to 20% moisture)

(a) Correlation with RPM and EPG:

$$W_1 = 1.87 \ (RPM/1000)^{2.26} \ (EPG)^{0.25} \tag{8}$$
$$R^2 = 0.70$$

(b) Correlation with RPM and BHP:

$$W_1 = 10.45 \ (RPM/1000)^{3.14} \ (BHP)^{-0.31} \tag{9}$$
$$R^2 = 0.71$$

(c) Correlation with RPM and IFC:

$$W_1 = 4.91 \ (RPM/1000)^{2.57} \ (IFC)^{-0.05} \tag{10}$$
$$R^2 = 0.67$$

(4) Diesel Fuel Consumption F_2 in kg/h During Diesel Fuel Mode:

$$F_2 = 0.349 \ (RPM/1000)^{0.918} \ (BHP)^{0.771} \tag{11}$$
$$R^2 = 0.98$$

The range of values of the parameters in the above equations (4) to (11) are as follows:

(1) Engine speed: $1550 < RPM < 1850$

(2) Brake horsepower: $25 < BHP < 50$

(3) Percent energy from producer gas: $50 < EPG < 90$

(4) Injection fuel consumption in dual fuel mode, kg/h: $1 < IFC < 7$

EPG or percentage energy from producer gas is obtained indirectly from the following relations (See Appendix):

$$EPG = \frac{100\ E_1}{E_1 + E_2} \qquad (17)$$

where E_1 = Energy from producer gas

E_2 = Energy from the liquid fuel

EPG can be measured directly in the field (by means of a graduated burette connected to the fuel tank) by first measuring the injection fuel consumption IFC during dual fuel operation, then the fuel consumption, F_2 during diesel fuel mode at the same engine RPM and BHP.

The percentage diesel fuel saving is calculated as follows:

$$FS = \frac{100\ (F_2 - IFC)}{F_2} \qquad (13)$$

Table 3 compares the values of EPG and FS for the dual fuel experimental runs in Tables 2 and 4. It is seen that FS and EPG are practically the same.

The above equations were useful in estimating the performance of the engine when it was brought to Siniloan, Laguna to power the irrigation pump. In about 60 hours of test spread over 12 days, the engine in dual-fuel mode when driving the pump at 1200 RPM showed an average injection fuel consumption rate of 2.58 kg/h. This was equivalent to a 49 percent diesel fuel saving since during straight diesel operation at the same engine speed, the fuel consumption was 5.09 kg/h. From equations (4), (5) and (6) the power delivered by the engine was estimated to have been 22 brake horsepower. From equations (8), (9) and (10), the charcoal consumption was calculated at about 8 kg/h.

Maximum diesel fuel saved in dual fuel operation in the labo-

Table 3.   Percentages in Diesel Fuel Saving (FS) and Energy
from Producer Gas (EPG)

Run No.	RPM	BHP	Diesel Fuel Consumption kg/h		FS, %	EPG, %
			Dual Mode	Diesel Mode		
1	1671	39	5.22	9.39	44	55
3	1619	45	6.88	10.21	33	49
4	1568	28	1.82	6.92	74	76
5	1665	31	1.50	7.89	81	84
8	1750	31	1.37	8.22	83	86
9	1747	37	1.85	9.39	80	82
12	1795	46	3.02	11.36	73	70
13	1800	40	1.99	10.33	81	80
16	1800	42	4.16	10.67	61	70
17	1800	38	2.83	9.83	71	79
20	1807	44	3.30	11.19	70	73
21	1803	43	2.66	10.86	76	75

ratory was 80 percent at 1800 RPM and 40 brake horsepower. This condition could have been achieved in the field by throttling the pump discharge to prevent overloading the engine at the higher speed of 1800 RPM. At this condition, the injection fuel consumption rate would be 2.4 kg/h and the charcoal consumption about 22 kg/h. The operation, however, would be inefficient due to throttling the pump discharge and the higher frictional losses at higher speeds. It was decided, therefore, to operate the pump at 1200 RPM.

During the dry months of February to May, 1979, the engine was run continuously for 8 hours a day, two days a week on dual-fuel mode. Four days during the week, the engine was run on straight diesel. For precautionary reasons, the engine was not run in dual-fuel mode all the time, since no experience in maintenance during prolonged operation was yet available. Downtime caused by maintenance problems would have disrupted the rice planting season for the farmers in the area. No problems, however, were encountered during the four months and operations were stopped in June, 1979 only because the rainy season had arrived.

Gas Producer Performance

The performance data for the gas producer supplying gas to the multi-cylinder engine is summarized in Table 4. It is interesting to compare this to the performance of the smaller gas producer used with the single cylinder engine as indicated in Table 5. It is evident that the larger gas producer performed significantly better with an average thermal efficiency of 85 per cent and an average gas net calorific value of 128 Btu/nft^3 compared to 70 percent and 113 Btu/nft^3 for the small gas producer.

The reason for the improved performance of the bigger reactor can be partly explained by the higher specific gasification rate at the throat or combustion area (206 kg/h/m^2) compared to that of the smaller reactor (133 kg/h/m^2). Also, the total gasification rate of the larger reactor (20 kg/h) was 8 times that (2.5 kg/h) of the smaller reactor.

Other data on comparative performances are shown in Table 6. It is also noted that even though the calorific value of gas in the smaller reactor is less, the average percentage of energy input from producer gas (EPG) tends to be higher in the single cylinder engine (79 percent) than in the 6-cylinder engine (73 percent). The probable reason for this was the better control of

Table 4.   Typical Data on Gas Producer Performance During
           Dual-Fuel Engine Operation of a Six-Cylinder Engine.

Run No.	1	3	4	5	9	16
Engine	1671	1619	1568	1665	1747	1800
BHP*	39	45	28	31	37	42
Dry Gas Analysis						
% CO_2	3.7	3.3	2.8	4.3	2.9	4.9
% O_2	0.3	0.5	0.3	0.3	0.2	0.6
% CO	27.5	25.7	28.5	28.3	29.5	26.2
% H_2	11.1	11.4	10.0	12.2	11.2	16.7
% CH_4	1.0	0.5	0.7	0.8	0.8	0.8
CV, Btu/nft^3*						
Gross	143	132	139	147	148	155
Net	126	117	124	130	131	136
Gas Temperature, ^{o}F*						
T_1, before scrubber	430	385	378	662	673	555
T_2, after scrubber	76	97	72	90	100	88
Cold Gas Efficiency						
N_{th}, %	84	85	83	84	85	91

---

*Conversion factors
  1 BHP = 746 watts
  1 Btu/ft^3 = 37.25913 kJ/m^3
  ^{o}C = 9/5 (^{o}F - 32)

Table 5. Typical Data on Gas Producer Performance During
Dual-Fuel Engine Operation of a Single-Cylinder Engine.

Run No.	12	26	32	77	81	89
Engine RPM	953	1105	1000	1043	1311	1426
BHP	4.8	4.8	5.3	4.1	4.1	4.0
Dry Gas Analysis						
% CO_2	9.1	5.8	4.0	5.1	5.4	3.7
% O_2	0.4	0.2	0.2	0.5	0.3	0.4
% CO	18.9	24.6	27.6	23.1	26.2	26.2
% H_2	9.0	9.8	10.2	9.4	8.7	10.6
% CH_4	1.0	1.0	1.0	1.8	1.6	2.5
Calorific Value, Net, Btu/nft^3	93	114	124	115	121	134
Gas Temperature (without scrubber) $^\circ$F	205	130	173	114	127	131
Cold Gas Efficiency N_{th}, %	62	70	73	74	71	80
Energy from Producer Gas, EPG, %	44	99	83	75	96	98

Table 6. Comparative Performance Data Between the Multi-Cylinder and Single Cylinder Engine-Gas Producer Systems.

Averages of:	Single-Cylinder	Multi-Cylinder
1. Brake horsepower	4.5	38.6
2. Engine Speed RPM	1048	1735
3. Specific Diesel Fuel Consumption kg/Bhp/h	0.073	0.079
4. Specific Charcoal Consumption kg/Bhp/h	0.535	0.500
5. Percent, e_b	19.5	21.5
6. Percent EPG	79	73
7. Gas Thermal Eff., %	60	85
8. Gasification Rate (based on grate area) kg/m^2/h	48	50
9. Gasification Rate, kg/m^2/h (based on throat area)	133	206
10. Net Calorific Value Btu/nft^3	113	128

proportioning the gas-air mixture in a single cylinder engine than in a multiple cylinder engine.

Abstract

Gasification of various agricultural residues in down-draft, fixed bed gas producers and the utilization of the gas in small diesel engines converted for dual-fuel operation were studied at the College of Engineering, University of the Philippines. Such agricultural residues as coconut shells, wood waste, rice hulls and corn cobs were readily gasified in gas producers of simple design. Cleaning of the gas before its use in diesel engines presented some problems.

Use of charcoal in the gas producers to provide gas to a 5-brake horsepower single cylinder engine and a 65-brake horsepower six cylinder engine proved satisfactory. With charcoal as fuel, the percentage of the total energy from diesel oil replaced by producer gas and utilized in the single cylinder engine was higher (79 percent) compared to that in the six cylinder engine (73 percent). The thermal efficiency of the bigger gas producer, however was significantly better (85 percent) compared to the smaller gas producer (70 percent). The total gasification rate of the bigger reactor (20 kg/h) was 8 times that (2.5 kg/h) of the smaller reactor.

Appendix: Equations for Evaluating Engine and Gas Producer Performance

Brake Thermal Efficiency. The brake thermal efficiency, e_b, is calculated from the following equations:

$$e_b = \frac{(2545)\,(BHP)}{(2.2)(13.76)\,F_1\,(HV_1) + F_2\,(2.2)\,(HV_2)} \tag{12}$$

$$F_1 \quad \frac{(69.3)\,(359)\,(2.2)\,W_1}{(12)(13.76)(\%\,CO_2 + \%\,CO + \%\,CH_4)} \tag{13}$$

where

W_1	=	Dry charcoal consumed in kg/h
F_1	=	Producer gas consumed in kg/h
F_2	=	Liquid fuel consumed in kg/h
BHP	=	brake horsepower output of engine

2545 = conversion factor of 1 BHP to British thermal unit (BTU) per hour

359 = Cubic feet of a mole of gas at NTP

2.2 = Conversion factor of 1 kg to pound (lb)

13.76 = Specific volume of producer gas in ft^3/lb at NTP

69.3 = % C in ultimate analysis of dry charcoal (69.3% C, 5.5% H, 22.2% O, 3.0% Ash)

% CO_2, % CO, % CH_4, % H_2 = percentage analysis of dry producer gas (volumetric)

HV_1 = 3.41 (% CO) + 3.43 (% H_2) + 10.67 (% CH_4) (higher heating value of dry producer gas at NTP in Btu/ft^3)

HV_2 = 19,494 Btu/lb (higher heating value of diesel fuel)

Substituting equation (13) and corresponding HV's into equation (12), the thermal efficiency when the engine is run on producer gas and diesel fuel becomes:

$$e_b = \frac{100\ BHP}{E_1 + E_2} \qquad (14)$$

where

$$E_1 = W_1 \frac{6.17\ (\%\ CO) + 6.15\ (\%\ H_2) + 19.13\ (\%\ CH_4)}{\%\ CO_2 + \%\ CO + \%\ CH_4} \qquad (15)$$

and

$$E_2 = 16.85\ F_2 \qquad (16)$$

To correct for moisture content M (percent) of charcoal, multiply equation (15) by (1 - M/100).

Percentage Energy from Producer Gas (EPG). In equation (14), the denominator represents the total energy supplied to the engine, allocated as follows: E_1 = energy from producer gas and E_2 = energy from the diesel fuel. Therefore,

$$EPG = \frac{100\ E_1}{E_1 + E_2} \qquad (17)$$

Cold Gas Thermal Efficiency (N_{th}). The cold gas thermal efficiency equation is given as follows:

$$N_{th} = \frac{\text{Net Calorific Value of Gas, } CV_n}{\text{Net Heat in Solid Fuel Used}} \times 100$$

$$CV_n = 3.18 \ (\% \ CO) + 2.70 \ (\% \ H_2) + 8.95 \ (\% \ CH_4)$$
$$\text{(in Btu/ft}^3 \text{ at } 60^\circ F \text{ and saturated with moisture)}$$

$$\text{Net Heat in Solid Fuel Used} = \frac{12(HV_n) \ (\% \ CO_2 + \% \ CO + \% \ CH_4)}{379 \ (\% \ C)}$$

For dry charcoal, the net heating value $HV_n = 12,510$ Btu/lb and percentage by weight of carbon, $\% \ C = 69.3$.

Substituting in the cold gas efficiency equation:

$$N_{th} = \frac{55.64 \ (\% \ CO) + 47.24 \ (\% \ H_2) + 156.59 \ (\% \ CH_4)}{\% \ CO_2 + \% \ CO + \% \ CH_4} \tag{18}$$

To correct for moisture content, M (percent) of charcoal, divide equation (18) by (1 - M/100).

Literature Cited

1. Cruz, I.E. Resource Recovery and Conservation, 1979, 2, (3), p. 241 - 256.

2. Cruz, I.E. "Producer Gas as Fuel for The Diesel Engine", U.P. Industrial Research Center Report, 1977 (July), (21 pages mimeographed report).

3. Cruz, I.E. "Studies on the Practical Application of Producer Gas from Agricultural Residues as Alternative Fuel for Diesel Engines", U.P. Engineering Research and Development Foundation Report, 1978 (Dec), (21 pages mimeographed report).

RECEIVED November 20, 1979.

Third World Applications of Pyrolysis of Agricultural and Forestry Wastes

JOHN W. TATOM and HADLEY W. WELLBORN

J. W. Tatom, Consulting Engineers, 4074 Ridge Rd., Smyrna, GA 30080

FILINO HARAHAP and SASWINADI SASMOJO

Development Technology Center, Institute Technology Bandung, Bandung, Indonesia

Third World Countries face extraordinary problems in meeting the energy demands required for development, especially due to their general lack of foreign exchange. They are also confronted with high levels of unemployment, but have the mixed blessing of low wage levels. Therefore, a renewable energy system that could engage many people in its operation would be an especially welcome prospect. Since most Lesser Developed Countries lie in the tropics and already have economies largely based on agriculture and forestry, the potential for biomass as a renewable energy source is therefore great; particularly since it is very labor intensive. In addition, a big supply of biomass is already available in the form of agricultural and forestry residues that, to a very large extent, are currently wasted. These residues are not only scattered about in the fields and forests, but they are also concentrated at processing plants such as rice mills, saw mills, cotton gins, sugar mills etc, and therefore a significant fraction is already available conveniently for use as an energy source.

The main problem, however, with biomass as it is available now is that: at an industrial scale, most current energy conversion equipment is designed to operate on fossil fuels and at a domestic scale, most if not all stoves are for wood or charcoal use and will not operate properly on these residues. Thus there is a need for a means for converting residues into synthetic coal, oil and gas which could be utilized in existing equipment. Pyrolysis, especially low temperature pyrolysis, which favors char and oil production, offers a particularly promising means of conversion, since the char and oil products are storable and easily transportable-an especially attractive characteristic in developing countries.

Morever, recent U.S. experience with the steady-flow, vertical packed bed, partial oxidation pyrolysis process (1,2) has demonstrated that this technology, updated from its earlier froms is a viable means for conversion of forestry and agricultural wastes into synthetic fuels. In addition, the basic simplicity of the process makes it suitable for applications in rural environments because of its few moving parts, low maintenance requirements, and general ruggedness. But the steady-flow, vertical, packed bed,

0-8412-0565-5/80/47-130-671$05.00/0

partial oxidation pyrolytic convertors produced thus far have been
built primarily in developed countries with a high level of automation and capital intensity and are subject to environmental constraints totally inappropriate to developing countries. However,
recent studies (3,4,5,6) of the application of partial oxidation
pyrolysis-in a suitably modified form-to Third World countries have
indicated the technical-economic potential of the process, and
these studies have led to several hardware development programs.

Since the initial fabrication of an appropriate technology pyrolytic convertor dates back only to the spring of 1977, there
still is much to be learned about how to build a truly appropriate system, even though at least five generations of these units
have now been constructed. Thus while it is believed that current
demonstration systems will be shown to be economical, there are no
doubt improvements that will be made in time to upgrade the performance of these early models. Therefore, this presentation can
only be regarded as a progress report on this work. In a later
paper, a more complete presentation will be made.

It should also be noted that due to the often adverse conditions under which the units have been operated that the primary
challenge has been to get the systems properly running, and up until this time, only the most basic data have been recorded. Thus
performance has often had to be estimated using assumed energy
constants, and so until more refined data are available, some re-
servations must be held regarding the exact values presented. But
since the basic process is that of the vertical, packed bed, whose
performance has been studied at considerable length (1,2) and has
been shown to have a char-oil energy conversion efficiency which
is a function only of the air-to-feed ratio, and that seems to be
independent of feedstock, convertor scale, bed depth, and operating technique (see Figure 1), there is good reason to expect that
these appropriate technology designs will have similar performance
characteristics.

The Basic Design Concept

To develop an appropriate technology pyrolytic convertor inevitably means the compromising of many design features, performance,
and environmental constraints to gain economy, simplicity, reliability and maintainability. Thus there is little doubt that
existing, more capital intensive convertors can provide higher performance than described here. But under the conditions for which
they were developed, we believe the characteristics and performance of the designs discussed in this paper are sufficiently advanced to make them at least the basis for a first generation of
economically practical conversion systems.

To illustrate the problems of developing an appropriate technology design; once the basic decision to use the partial oxidation pyrolysis process is made, several questions immediately have
to be dealt with: i.e.

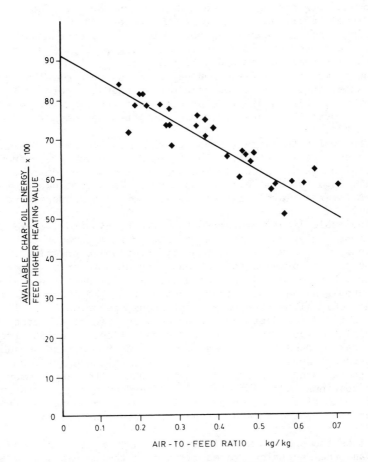

Figure 1. Percent available energy in char–oil mixture composite of all data

(1) How to make the process continuous so that oil and gas also can recovered without the use of expensive controls and input-output equipment?

(2) How to avoid the difficulties experienced in the U.S. with particulates in the off-gas which tend to foul the condenser?

(3) How to avoid corrosion problems using available materials?

(4) How to add the process air without blocking the internal flow and/or burning up the process air delivery system?

(5) How to separate out the oil aerosol and/or condense the oil vapor from the off-gas in a practical condenser?

(6) Using a manual technique, how to maintain proper materials flow within the convertor and avoid the bridging problems associated with conventional packed bed designs?

(7) How to maintain a fixed bed depth to insure a more uniform off-gas temperature; thus allowing proper condenser operation?

(8) How to develop a simple char output system that provides a uniform flow over the convertor cross section-as opposed to other systems that do not?

The design that has evolved to this date to best deal with these questions is presented in Figure 2. The convertor shown can be described as a "batch continuous" system which operates externally in a batch mode, but internally in a continuous mode-so long as the feed level is maintained above the submerged off-gas port.

The basic idea is to store the feed within the reactor itself and thus to allow a continuous flow of feed through the working section of the convertor and thereby provide a practically constant supply of off-gas to the condenser at a nearly fixed temperature, but yet avoid the expensive material handling equipment and controls needed for a continuous system. The submerged off-gas port not only allows the batch continuous operation, but it guarantees a fixed bed depth and provides an effective means for filtering the gas through use of the bed itself, which forms a filter cake around the off-gas port (however there is a substantial pressure drop through the bed using the technique). Moreover, since there is no free surface through which the gases must pass, and no continual addition of feed, as in conventional continuous systems, the fine fraction of the feed has no opportunity to be carried away by the off-gas stream to subsequently clog the condenser.

Since the process air is introduced through the convertor walls and the resulting gases migrate to the entrance of the off-gas port which is located near the center of the convertor, the flow is not vertical and thus the feed generally passes diagonally through the gases. An advantage to this arrangements is that the convertor walls are not exposed over a wide area to the hot off-gases. A disadvantage is that the depth of the bed may be relatively small and occasional pentrations of uncharred feed can occur during rapid processing, especially if sufficient agitation is not employed. However, with proper precautions, such as the in-

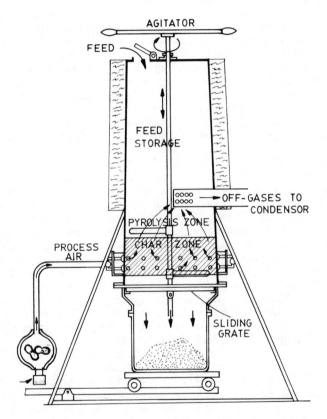

Figure 2.   Appropriate technology pyrolysis convertor

troduction of the air at two levels to insure that the char zone
has a definite thickness-such occasions can be effectively avoided.

Since the maintenance of a free-flowing condition within the
feed/char inside the convertor is essential to successful system
operation, every effort has been made to facilitate the flow. Thus
the manually operated agitator, together with the conic shape of
the convertor and the almost complete absence of any obstructions
to the flow within the reactor effectively guarantee that the
packed bed design will operate properly.

The condenser/demister system chosen, and shown in Figure 3, is
basically a compromise between performance and complexity, since
to effectively remove all the fine aerosol oil mist in the off-gas
leaving the convertor and to condense out the remainder without
collecting an excessive amount of water requires a much more
sophisticated system than believed practical for the application
intended in many LDC's. Thus the condenser/demister design chosen
involves no moving parts and is entirely cooled by natural convection. The off-gas-oil mixture from the convertor enters the unit
and expands through a series of holes in the inside pipe. The resulting jets impinge on the inside surface of the outer natural
convection cooled jacket, thus producing good local heat transfer
and a high degree of turbulence and mixing as the resulting wall
jets expand, collide, roll up, and are reintrained into the imping-
ing jet flows. Thus a single fuild particle, in its complex path
through the condenser, is effectivelly cooled and has many oppor -
tunities to collide with other particles to form larger particles
or droplets and/or be collected on the outer condenser walls. The
demister is of conventional design and uses available fiber to provide a matrix through which the condenser off-gases pass. Vegetable fiber "ijuk" from a species of palm tree has been successfully
used for this purpose. The overall system takes maximum benefit
of the effectiveness of the submerged off-gas filter technique, and
with weekly maintenance operates free of the clogging problems that
have plagued other condenser designs. Thus while its performance
can be improved through further optimization, current oil recovery
levels are acceptable, but not outstanding.

Regarding corrosion; while it is well known that the hot pyrolytic oils are acidic and thus could seriously threaten the condenser, it has been found that the tar in the oil tends to cover
the exposed metal surfaces with an almost enamel-like coating that
effectively protects them from excessive oxidation. Moreover,
while a metal matrix in the demister such as from lathe turnings
is destroyed in a matter of days, the lifetime of ijuk appears to
be indefinite.

Development History and Current Program Status

Several generations of convertors involving the basic batch
continuous, appropriate technology design described above have been
developed during the last two years since the first was built by

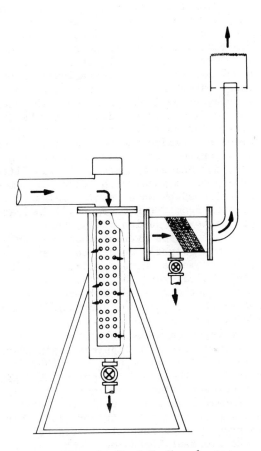

Figure 3. Pyrolysis oil condenser

Tatom and Stone in California out of oil drums and galvanized pipe
(7). This system involved the first use of an annular pebble bed
located near the grate, for process air introduction. A second
convertor, designed and built by Tatom and Wellborn, also of oil
drums, but using manual agitation, and a more advanced off-gas
filtering technique and char removal and storage system served as
the predecessor to the first real prototype. This latter system,
was built in early 1978 at Kumasi, Ghana at the Technology Con-
sultancy Center (TCC) of the University of Science and Technology.
Sponsored under a joint USAID-Bank of Ghana program administered
by the Building and Road Research Institute (BRRI), this system
was designed, fabricated, and tested under subcontract by Georgia
Institute of Technology, Tatom, and Wellborn. This convertor, designed to process 57 kg/hr of hardwood residue, together with three
others and two large batch driers is currently being assembled to
form a system for round-the-clock operation. The char and oil produced will be used to fire a brick kiln at the BRRI, and the gas
will be used to dry the sawdust down to six to eight per cent moisture. These convertors utilize the pebble bed process air intro -
duction technique described previously together with still more
advanced condenser, mechanical agitator, off-gas filter and char
grate designs. The program involves a technical demonstration of
the concept to be followed by an economic evaluation. At the present time, information from the program (8,9) indicates that the
system throughput meets the design goals, that the driers operate
properly, that the char yields average about 23 per cent, but that
the oil yields are only slightly better that seven percent. Thus
clearly, further improvements in the condenser performance are desired. But since tropical hardwoods typically produce only 11 to
12 percent oil/tar from their destructive distillation, the effectiveness of the condenser, while not overwhelming, is still in the
range of 60 to 70 percent.

Another pyrolysis program, which is being conducted by Georgia
Tech and the University of the Philippines Engineering Research and
Development Foundation with the support of UNIDO, uses the basic
Ghana convertor design. The system is primarily designed to pro-
cess rice husks. However, since the program has only recently been
initiated, no performance data are available (10).

Of principal importance to the authors is a joint USAID-Repub-
lic of Indonesia program also recently initiated. This effort
conducted by the Development Technology Center at the Institute of
Technology Bandung, is directed toward the development of an economically viable system for pyrolytic conversion of rice husks at
rice mills, including charcoal briqueting equipment, and other
components such as lamps, stoves, modified diesel engines, gas
burners etc. which can utilize the oil and gas produced. In addition, alternative uses of the oil-tar such as for wood preserving
are being investigated. The Indonesian pyrolysis system incorporates more advanced ideas in process air introduction, grate design

and agitation than in previous programs. This effort to date has
involved both the design, fabrication and test of a 100 kg/hr Technology Development Unit, shown in Figure 4, for research and deve -
lopment and training on the ITB campus, and the design and fabrication of a prototype system for use at rice mills.

   More recently the government of Papua New Guinea has initia -
ted an ambitious program for development of renewable energy
sources through conversion of their wood residues into domestic and
industrial fuels by means of pyrolysis (11). Both appropriate
technology and more capital intensive systems are planned to sup -
ply the various local needs. An initial program at the University
of Technology in Lae to build and operate a small, but automated,
one tonne/day, research and development system using the basic concept described above has just been initiated, and plans are to begin construction of a 25 tonne/day prototype industrial scale system in 1980. Finally, an appropriate technology system using the
design approach discussed before is also planned later this year
or early in 1980 as part of the UNEP Rural Energy Center being
built outside Dakar, Senegal.

System Description

   Since there have been a number of systems developed, each
having significantly different design features, it is important to
distinguish between the various convertors built, yet impractical
to describe more than one. Therefore the system to be described in
this paper is the Technology Development Unit mentioned previously
as part of the Indonesian program and designed primarily to process
rice husks. While many technical improvements have been made using
this system, its principal purpose was to investigate alternative
means of process air introduction and off-gas removal, with the
object of avoiding as many obstacles to the materials flow within
the convertor as possible. Process air configurations investigated
were the pebble bed system used in the Ghana and Philippine con -
vertors and a flush mounted, water jacketed technique, wherely the
air is introduced directly through the convertor walls. The primary objections to the pebble bed design were the very large fraction of the convertor cross section blocked by this component and
the tendency of the design to cause the formation of hot spots resulting from cracks or channels created in the shear layer between
the central, core flow and that above the pebble bed. In addition,
a flush mounted off-gas system having four ports, manifolded together, and the simpler, internally mounted probe system used in the
Ghana design were tested. It was hoped that succes using the four
flush mounted ports would allow removal of the internal probe which
does present some obstruction to the flow. However, from the start
it was also recognized that the resulting internal gas flow pat -
terns might leave a relatively uncharred core in the convertor center, since the shortest path for the gases would be along the convertor walls.

Figure 4.   Indonesian technology development pyrolytic convertor system

But, while there are difference between this system and others described there are also many similarities. Thus the rice husks are added at the top of the convertor through, a quick disconnect inlet feed port and the charcoal removed at the bottom through a simple sliding grate. Process air is introduced near the bottom of the system and passes upward to exit through the off-gas system and then enters the condenser/demister which removes the oil-tar fraction.

The convertor stands 2.35 meters tall, has a base diameter of 0.84 meters and a top diameter of 0.70 meters. It is built of mild steel, typically two to three millimeters thick. Instrumentation includes a process air orifice, and dial thermometers for measuring bed, convertor off-gas and condenser off-gas temperature.

System Operation

The system is started by first introducing cold charcoal from the previous day's operation up to the level of the top layer of air holes. Burning charcoal briquetes are then put into the convertor and are subsequently smothered by the feed which fills up the storage section. The process air is then introduced at about 10 to 20 percent of the full steady-state flow rate and is slowly increased. After several hours the temperature of the off-gas begins to rise rapidly and limited actuation (shaking) of the grate is initiated. Gradually the process air rate is increased as the removal of charcoal accelerates until a desired operating condition is reached.

Perhaps the two most important indices of system performance are the convertor and the condenser off-gas temperatures. If pro - per control of these two is maintained and correct interpretation of variations in these temperatures is made, the system operation can be relatively routine. To illustrate: the average convertor off-gas temperature is not only important so far as oil production is concerned, but rapid increases in it may indicate the formation of cavities in the bed, while decreases may suggest that char is being removed too rapidly. Likewise the condenser off-gas temperature must be maintained above the dewpoint of the mixture to avoid water condensation, but not so high as to reduce the oil yields.

Since each feed contains different ash and moisture contents and because the bed depth for optimum performance varies, it is impractical to specify these temperature too closely, but:
(1) for rice husks with a moisture content of about 11 percent, an ash content of 20 percent, and at a bed depth of about 0.30 meters, the desired convertor off-gas temperature lies between 120°C and 140°C while the condenser off-gas temperature is around 85°C.
(2) for tropical hard-wood, with a moisture content of six to eight percent, an ash content of about one percent and a bed depth of about 0.50 meters, the desired convertor off-gas temperature is in

the range of 130°C to 150°C, while the condenser off-gas temperature is also around 85°C.

Once the desired conditions have been reached and the planned
process rate established, the system can be operated in a steady-
state mode. This may take four or five hours from a cold start or
two to three in a run where hot char from the previous days operation is in the convertor. During the steady mode operation, feed
is added about every 30 minutes, the char grate opened for a few
seconds 30 to 40 times each hour and the char drums changed every
hour. Two char barrels equipped with quick disconnect hardware are
used for rapid replacement of the drums. To prevent burning, the
char is stored in sealed barrels while it cools. The oil is periodically gathered from the condenser and demister. Typically,
the condenser recovers about three quarters of the total oil while
the demister recovers the remainder.

Except for a total of three or four minutes each hour when
feed is added and/or the char drums changed, the system operates
continuously. During these filling/emptying periods the air supply is shut off and operations halt. However, the thermal capacity of the system is sufficient to insure that only minor changes
in temperature occur during these periods. The result-so far as
char and oil production are concerned-is that the operation is effectively constant. However, because the off-gas flow too is interrupted, measures must be taken either to store a short supply
of the gas or to restart the gas utilization system-using a single
convertor. With several convertors, manifolded together such as in
Ghana, the operations can be staggered with no interruptions in
off-gas production.

Since by its nature the system involves a minimum of instru -
mentation, it is vital that maximum use of every other indicator
be made. For example, abrupt changes in capstan torque can suggest cavity formation and indicate the need for more agitation.
Likewise, the average torque is a measure of bed depth and-during
start up-of the degree that charring of the bed has occured. The color
of the off-gas is a measure of the reaction temperature. A very
white smoke indicates localized gasification, while a greyish
brown smoke suggests a more general pyrolysis condition. The color
of the char also is a measure of convertor performance. Hence uncharred material suggests too rapid a throughput, while greyish
white char indicates the process rate is too low. Even the characteristics of the boiling in the water jacket can provide useful information regarding the internal condition of the reactor. Exces -
sive, or film boiling at specific points indicates hot spots, most
likely due to the presence of cavities in the bed and suggests
the need for more agitation of the bed.

Test Results

Since the combination of direct introduction of process air
through the convertor walls and the flush mounted off-gas system is

so attractive, in that it minimizes internal resistance to flow, it was investigated first. Thus a series of tests were conducted to see if this combination would operate satisfactority and with good performance. While the water jacketed process air system worked extremely well, there was no problem in materials handling, the convertor off-gas temperature was in the desired range, and good char yields were achieved, it was found that gas production was exces - sive. While the oil was clean, the yields were low, the maximum throughput was less than half that for which the system is designed, the air-to-feed ratio ran well in excess of unity, the gas heating value was disappointing and excessire heating of the uncooled walls above the water jacket occurred. It was thus concluded that the short circuiting of the off-gas flow that had been feared, did indeed occur, and while some pyrolysis was present, a significant amount of gasification took place near the convertor walls. Therefore since greater throughputs and oil yields were desired, it was decided to abandon this configuration and to test the submerged off-gas port in combination with the water jacketed process air system.

This led to a second, more promising test series. At first, the bed depth was operated at 0.50 meters, which in the Ghana program, using 6-8 per cent moisture hardwood feed, had been satis - factory. But it was found that while improved process rates were obtained, much lower air-to-feed ratios were achieved, and gas production was moderate, the convertor off-gas temperature was too low and oil recovery was still disappointing. The low oil yields have been mainly attributed to the low off-gas temperature obtained, with the assumption that significant oil condensation occured in the bed. Since the current philosophy is to try to avoid the need for a mechanical drier and to use the husks as they are produced, the feed moisture content typically runs at about 11 percent of the wet weight. Thus this relatively high moisture fraction, together with the high silica content (20 percent) of the husks was believed responsible for the low off-gas temperature and oil yields, since in tests with wet feed reported in (2) similar disappointing oil yields had occured. Therefore a series of tests at a bed depth of only 0.23 meters was conducted-in an effort to raise the off-gas temperature. The results of these tests are presented in Table 1 and Figure 5. Study of the results reveals that the char yields are well in excess of those expected from the data correlations in (1) and (2) at the air to-feed-ratios used. This may be due to oil condensation in the bed, but at this time it is not clear why the char yields are so high. On the other hand, recovered oil yields are very low. However, the figure shows that the trends of in - creasing oil yield and decreasing char yield with increasing air-to-feed are similar to those reported in (1) and (2). The consistency of the data, with one exception, is encouraging. It should be noted that the feed throughputs were limited by the capacity of the available air supply, since the pressure drop through the system was unexpectedly high. Therefore with a larger air supply

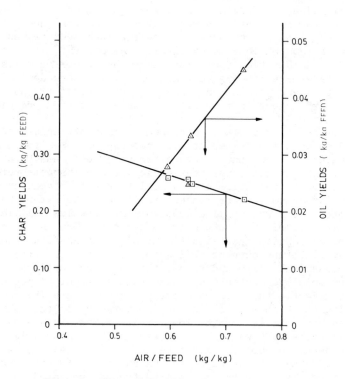

Figure 5.   Char and oil yields as function of air/feed

there is little doubt that the design flow rates can be reached.

Table 1. Preliminary Test Data from Indonesia

Test No	Air/Feed (kg/kg, dry ash free)	Feed Process rate (kg/hr)	Average Convertor off-gas Temp.(°C)	Average Condensor off-gas Temp.(°C)	Char yields (kg/kg, dry ash free)	Recovered oil yields (kg/kg, dry ash free)
1	0.633	64	121	93	0.247	.033
2	0.732	47	160	93	0.218	.045
3	0.596	50	132	85	0.255	.028
4	0.63	47	135	82	0.251	.025

While the recovered oil yields are quite low, it was obser -
ved that significant quantities of oil were lost in the off-gas
stream, apparently due to inadequate condenser/demister perfor -
mance, and thus with proper optimization of this system it is believed that oil yields at least equivalent to those reported from
Ghana (8,9) can be realized. Regarding the products themselves;
the char is of good quality, the oil is very thick and apparently
free of any significant water content, and the gas burns cleanly,
completely, and without odor in both a diffusion burner and a
forced air combustor designed for crop drying applications. Moreover in preliminary tests, the gas, in combination with a very
small quantity of diesel oil, provided the energy to run a small
five horsepower diesel engine. In addition a 45 volume per cent
mixture of the oil-neutralizid with milk of lime - dissolved in methyl alcohol (30 percent) and mixed with coconut oil (25 percent)
also powered the diesel engine in short feasibility tests. However,
while the oils can be used to run a diesel engine, it appears that
their current highest value is as a wood preservative. But perhaps
most important; the charcoal was succesfully briqueted to form
"Fire Balls" by means of a simple agglomeration technique in which
waste cassava pulp from a local starch mill was used as a binder.
Briquet production from the laboratory scale machine utilized has
reached 21 kg/hr at this date. Further improvements are antici -
pated as refinements in the technique are developed.

Economic Analysis

There are many factors which make a complete economic analysis difficult, especially in Indonesia where fossil fuels are heavily subsidized. Thus the question of whether to use the fair
value of the products or their current equivalent market price must
be answered. Moreover because the complete system has not   yet

operated in a real world situation, all the numbers are simply projections. But since a prototype and a briqueting system have now been
built and operated, the required capital investment and the man -
power requirements are fairly well known. Unfortunately the product values are not and must be estimated. Therefore the follow -
ing two brief analyses, while not complete should at least bracket
the economic performance of the system:

Pessimistic outlook-Rice mill runs 150 days/year; unit processes
one tonne/day husks; oil yields remain at 3 percent of input feed;
oil used as coal oil substitute for wood preservation at US$0.185/
liter; char yields are 40 percent of input feed; char as produced
is worth US $40 per tonne wholesale; gas is not used and has no
value; four men run system and are each paid US $2/day, total sys-
tem cost including convertor, air supply, briquetting equipment,
small engine-generator,operating platform, and feed storage is
US $2500; government loan is available at 12½ percent interest
with 10 year payback; annual overhead runs at 15 percent of ini -
tial capital costs. Therefore the following results are obtained:

Annual Capital Costs	453
Annual Labor Cost	1200
Overhead	375
Total Operating Costs US $	2028

Annual Income	
Charcoal	2400
Oil	832
Gas	--
Total	3232

Therefore profit is US $ 1204 and payback is 2.1 years.

Optimistic outlook-mill runs 200 days/year; oil yields are up to 7
percent; oil sells at US $ 0.185/liter, char is worth US $50/tonne;
gas replaces 35 liters/day of un-subsidized diesel fuel (worth
US $ 0.25/liter) needed to run an 8 hp engine powering rice mill.
Therefore the following results are obtained:

Annual Capital Costs	453
Annual Labor Costs	1600
Overhead	375
Total Operating Costs	2428

Annual Income	
Charcoal	4000
Oil	2590
Gas	1750
Total	8340

Thus profit is US $ 5912/year and payback is about five months.

Thus no matter how the analyses are made, the return on investment is certainly attractive, and so regardless of the exact return, the economic outlook is very favorable.

Conclusions and Recommendations

While there is clearly much technical work to be done and improvements to be made and there is an obvious need for extended operating experience using the equipment so far produced, never-the-less, the outlook is promising both technically and economi - cally. But while the prospects are generally good, there is also a pressing need for more detailed studies: of the properties of the char, oil, and gas, of the various applications of these products possible, and of the overall systems involved. This work should start immediately.

From the above discussion it should also be evident that the work in this area is currently in a state of rapid expansion. This rapid acceleration is a vivid indicator of the extent of the energy crises in developing countries and their willingness to inves - tigate promising, evolving technology as a means of solving this crisis, even through all the technical problems have clearly not been resolved. This willingness to risk is to our view a wise philosophy, since the severity and scale of the risks of not finding a workable solution to this problem in the very near future are sufficiently serious to make the investment in alternate energy technologies a trivial consideration. However, one major concern is that an immediate lack of complete success-either technical or economic-will disenchant those who most severely need this technology. We earnestly hope that a proper, fair evaluation of these efforts will be made, with the attitude that if shortcomings arise-and they will-then solutions will simply be bought and found. The promise of this technology is real and exciting; it must be realized now. But only through free and open cooperation among all investigators will its potential be demonstrated.

Acknowledgements

The authors would like to express their appreciation to Mr. Ruben Hardy of Development Technology Center, Indonesia, who made many contribution to the work reported here. Also special thanks is given to Mr. Charles Stone of Sacramento, California, who assisted greatly in the early development of the system described. In addition we wish to acknowledge Mr. Ken Yeboah and Dr. John Powell at University of Science & Technology, Kumasi, Ghana, whose technical suggestions, patience, interest, and diligence greatly facilitated the program in Ghana. The support of Georgia Tech, the U.S. agency for International Development, and the governments of Ghana and Indonesia are acknowledged.

Literature Cited

1. Tatom, J.W., et al., "Utilization of Agricultural, Forestry and Silvicultural Waste for the Production of Clean Fuels", Final Report under EPA Contract 68-02-1485, EES, Georgia Tech, Atlanta, Ga. Sept. 1976.

2. Tatom, J.W., Colcord, A.R., Williams, W.M., Purdy K.R., Demeter, J.J., McCann, C.R., Elkman, J.M., Bienstock, D. "A Prototype Mobile System for Pyrolysis of Agricultural and/or Silvicultural Wastes", Final Report under EPA Grant R-803430, EES, Georgia Tech., Atlanta, Ga. June 1978.

3. Chiang, T.I., Tatom J.W. de Graft Johnson, J.W.S., Powell, J.W., "Pyrolytic Conversion of Agricultural and Forestry Wastes in Ghana-a Feasibility Study" Report under AID Contract AID/ta-C-1290, EES, Georgia Tech, Atlanta, Ga, July 1976.

4. Tatom, J.W., Chiang, T.I., Harahap, Filino, Apandi R.M., Wirjosumarto, Harsono, "Pyrolytic Conversion of Agricultural and Forestry Wastes to alternate Energy Sources in Indonesia-a Feasibility Study" Report under AID Contract AID/ASIA-C-1203, EES, Georgia Tech, Atlanta, Georgia, February 1977.

5. Tatom, J.W., "Demonstration of Alternative Fuel Production through Pyrolysis of Agricultural Wastes at the UNEP Rural Energy Center in Senegal" Special report to UNEP, Atlanta, Ga, November 1977.

6. Tatom, J.W., "Feasibility of Industrial Fuel Production from Pyrolysis of Wood Wastes in Papua New Guinea" report 2 - 79, prepared for Energy Planning Unit, Dept. Minerals and Energy, Konedobu, Papua New Guinea", February 1979.

7. Nauman, Art "Swill to Fuel Project gets off the Ground" The Sacramento Bee, Sunday April 17, 1977.

8. Private Communications, Yeboah to Tatom, Kumasi, Ghana, October 1978 - June 1979.

9. Private Communication, Powell to Tatom, Kumasi, Ghana, June 1979.

10. Anon, "The National Nonconventional Energy Resources Development Program - Progress Report" Republic of the Philippines, Ministry of Energy, January 1979.

11. Anon, "White Paper - Energy Policy and Planning for Papua New Guinea" Ministry of Minerals and Energy, P.O. Box 2352, Konedobu, 1979.

RECEIVED November 27, 1979.

A Preliminary Analysis of the Potential for Developing Nations to Use Biomass Fuels in Vehicle Engines

J. L. JONES and A. K. CHATTERJEE

Chemical Engineering Laboratory, SRI International, 333 Ravenswood Avenue, Menlo Park, CA 94025

Researchers at the Brookhaven National Laboratory recently reported that oil-importing developing nations contain more than 40% of the world's population but consume less than 10% of the world's commercial energy (1a). In two-thirds of those countries, petroleum supplies at least 90% of the commercial energy (commercial energy refers to conventional fuel or energy forms of commerce such as petroleum, coal, natural gas, and electric power.) In 1975, oil imports by all of the developing countries combined was about 2.7 million barrels per day.

As other speakers at this symposium have described, the use of fuel gas produced by biomass or solid waste gasifiers can reduce the use of petroleum fuels in stationary combustion equipment (oil-fired boilers, diesel engines for electric generators or irrigation pumps). Stationary engines and furnaces, however, are not the only big users of petroleum fuels in lesser developed countries (or LDCs, the term we will employ to describe the 88 poorest nations in the world). As is the case for industrialized nations, lesser developed countries need liquid transportation fuels, and probably will for a long time. Brookhaven reports that most LDCs have increased their dependence on highway transport over the last two decades (1b).

Biomass fuels offer a possible source of energy for certain LDCs. Vehicle engines may use biomass by such methods as on-board thermal gasification of solid biomass fuels, or the burning of alcohols produced by either fermentation (ethanol) or by thermal gasification and synthesis (methanol). Vehicle-mounted gasifiers are not new. During World War II, securing fuels for transport, agricultural, irrigation, and stationary combustion engines became an acute problem. The development and adoption of biomass gasifiers using charcoal or wood were directed to the transportation sector. Sweden, Germany, France, and Italy were the first nations to adopt biomass gasifiers for transport and agricultural engines. During 1942, the number of biomass gasifiers used for vehicles rose to 821,604 worldwide (2). During that time, the Italian government decreed that Italy's entire fleet of

0-8412-0565-5/80/47-130-689$05.00/0

68,500 agricultural tractors be fitted with biomass gasifiers
by 1947. Similarly, Germany ordered that all road and farm
tractors of more than 25 horsepower (HP) be fueled with biomass
fuel gas. By the end of 1944, Sweden reported having nearly
90,000 operational vehicles powered by biomass fuel gas (3).

Sweden's statistics showed a preference for charcoal over
wood as a gasifier fuel for private cars, small trucks, and
motorcycles. Charcoal's popularity resulted from its ease of
handling and storage, the small pressure drop it caused in the
gasifier, its good thermal efficiency, and its tendency not to slag.

Sweden's wood- or charcoal-fueled engines supplied between
10 and 200 HP. They were primarily for over-the-road vehicles,
but farm equipment, fishing vessels, road-paving equipment and
stationary engines for driving compressors and electric generators
were also powered by the wood or charcoal gasifiers.

Potential for using alcohol fuels as a substitute or supplement for liquid transportation fuels is a topic of current interest
not only in LDCs but in most areas of the world. The use of
sucrose from sugar cane or by-product molasses (containing sucrose
and various C_6 sugars) has been the subject of discussions at
several recent meetings of the United Nations Industrial Development Organization (UNIDO) and the Organization for Economic Cooperation and Development (OECD)(4,5).

In this paper, we shall present a cursory analysis of the
required capital investment for an LDC to substitute biomass fuels
for petroleum. We have chosen to analyze only investment costs
because, first, alcohol production options are likely to necessi-
tate the construction of relatively large and complex conversion
plants. The plants would require substantial capital investments
(perhaps with equipment purchased abroad) and a sophisticated
engineering and operational staff. Second, the Brookhaven study
suggests that, by the year 2000, LDCs outside OPEC will need to
increase their annual investments in energy resources by 50% to
meet their commercial energy demands. Third, because only a few
countries have substantial access to private international investment, the present lending policies will make it difficult for
many countries to develop their energy supplies (1c). Fourth, any
estimate of operating costs is of questionable value unless the
analysis is based on a specific country or region. Thus, we will
not compare operating costs, or any other costs related to modifications of fuel distribution networks or changes in agricultural
practices that might allow more harvesting of wood or agricultural
residues. For this reason, we will be unable to draw firm conclusions concerning the relative economic desirability of the
options. We believe, however, that the results should be instructive to planners who will be faced with evaluating future energy
options for LDCs.

Bases for the Analysis

For this analysis, we have assumed that the goal for a specific region or country is to replace 1000 barrels/day (42,000 gallons or 159,000 liters) of petroleum fuels. The heating value of this amount of petroleum fuel is ~ 5.7×10^9 Btu/day (~ 6×10^{12} joule/day). The import value of 1000 barrels per day of petroleum fuels is probably close to \$25,000/day or \$9 million/year (assuming a crude oil price of $\geq$ \$20/barrel on the world market).

The types of fuel that will be considered for use in vehicle mounted gasifiers include wood chips ($\leq$ 30 weight percent H_2O) and densified fuel pellets ($\leq$ 10 weight percent H_2O) produced from agricultural residues.

Two approaches may be used to replace petroleum fuels with alcohol fuels. The first is to blend the alcohol with the petroleum fuel in mixtures that contain 5 to 20 volume percent alcohol. Blending alcohols with petroleum fuels at such levels may allow operation of unmodified vehicles with little noticeable change in the volumetric consumption of fuel per unit of distance traveled. (The question of whether or not the volumetric fuel consumption changes for alcohol blends is a topic of current research and debate.)

Ethanol has a volumetric heat content about two-thirds that of a petroleum fuel. Methanol, whose weight percent of chemically bound oxygen is higher than ethanol's, has a volumetric heat content about half that of a petroleum fuel. If 100% alcohol fuels were to replace petroleum fuels, then the volume of fuel required per unit of distance traveled could increase by as much as 50 to 100% depending on the specific alcohol used and the engine design. For this analysis, we will assume that the alcohol fuel is mixed with petroleum fuels in alcohol mixtures of 10 to 20 volume percent and that the alcohol fuel is substituted on an equal-volume basis with petroleum. Therefore, the average alcohol production rates will be

Volume/day	Volume/year
1,000 barrels	365,000 barrels
42,000 gallons	15.3 million gallons
159,000 liters	58 million liters

Alcohol Fuels Production

Ethanol Production from Biomass

Feedstock Options. Ethanol may be produced via fermentation (with yeast) of 6-carbon or 12-carbon sugars from a number of carbohydrate sources including sugar crops, starch crops, or lignocellulosic materials.

With lignocellulosic materials, such as wood or crop residues (wheat straw, rice straw, corn stover), the material must first undergo extensive treatment with mineral acids or enzymes to hydrolyze (or saccharify) the material to soluble sugars for fermentation. Such procedures have proved highly expensive; only the USSR currently has operating commercial plants that produce sugars from wood by acid hydrolysis techniques. We do not consider such hydrolysis techniques to be competitive at this time with other conversion methods for using lignocellulosic materials in most regions of the world.

Starches present in grains or root crops are readily converted to sugars for fermentation to ethanol. With the shortages of food in developing nations, grains would probably not be used for fuel. Therefore, we will consider only sugar crops and molasses to be available for ethanol production. In the initial part of the analysis, we will identify the countries where sufficient sugar is exported or molasses produced to allow production of ethanol to replace 1000 barrels of petroleum fuels.

Many developing nations sustain sufficient sugar production levels to meet their own needs and still export large tonnages. Table I summarizes the world production and consumption of sucrose, by region, for 1977, a year in which production significantly exceeded consumption. The use of cane sugar juice for fuel production may solve a possible future problem related to a decreasing demand for sucrose as certain industrialized nations that import sucrose (such as the United States) increase their production of high-fructose corn syrup to replace sucrose in some applications.

In the developing regions of the world, almost all of the sucrose production is from sugar cane. Production far exceeds consumption in Central and South America and is slightly higher than demand in Africa. Table II lists the 11 major exporters of sucrose that account for ~ 75% of the total exports of world producers. On the basis of the following assumed stoichiometry, ~ 1.63 metric tons of sucrose is sufficient to produce 1000 liters of 100% ethanol.

$$5C_{12}H_{22}O_{11} + 5H_2O \rightarrow 10C_6H_{12}O_6$$
(sucrose)

$$10\ C_6H_{12}O_6 \xrightarrow{\text{Yeast}} 18\ C_2H_6O + 18CO_2 + 6\ (CH_2O)$$
(ethanol)

represents new cells and other products

$$\left(\frac{1710\ \text{kg sucrose}}{828\ \text{kg EtOH}}\right)\left(\frac{796\ \text{kg EtOH}}{1000\ \text{liters EtOH}}\right)\left(\frac{1\ \text{metric ton}}{1000\ \text{kg}}\right) = \frac{1.64\ \text{metric tons sucrose}}{1000\ \text{liters EtOH}}$$

TABLE I   1977 WORLD PRODUCTION AND CONSUMPTION OF SUCROSE

Region	Production (10^6 Metric Tons/Yr)	(% Cane Sugar)[*]	Consumption (10^6 Metric Tons/Yr)
Europe	30.48	2	31.74
North America	4.74	31	11.29
Central American	13.59	100	4.49
South American	13.72	89+	9.07
Asia	18.46	88+	20.98
Africa †	6.07	95+	5.69
Oceania	4.76	100	1.05
	91.83	61	84.31

[*]Sucrose is produced from sugar beets or sugar cane.

†Includes Hawaii, Fiji and Australia.

Source: Reference 6

TABLE II SELECTED MAJOR EXPORTERS OF SUCROSE

Country	1977 Exports (metric tons)
Cuba	6,238,162
Dominican Republic	1,116,587
Argentina	958,310
Brazil	2,486,587
Peru	411,832
Taiwan	644,000
Philippines	2,574,825
Thailand	1,674,540
Mauritius	673,995
South Africa	1,383,867
Australia	2,965,249

Source: Reference 6

TABLE III   LESSER-DEVELOPED COUNTRIES WITH SUCROSE
PRODUCTION LEVELS OF > 600,000 METRIC TON/YEAR
(1977 Data)

Cuba	India	Dominican Repbulic
Mexico	Taiwan	Philippines
Brazil	Argentina	Thailand
China	Indonesia	Colombia
Peru	Pakistan	Mauritius
Iran	Egypt	

Source: Reference 6.

To substitute ethanol for 365,000 barrels/year (85 million liters) petroleum fuels would require ~95,000 metric tons of sucrose annually.

The total exports of sucrose from LDCs (~ 18 x 10^6 metric tons/year) could be converted to roughly 11 billion liters of ethanol, which is equivalent on a volumetric basis to ~ 0.2 million barrels/day of ethanol or 7% of the total volume of petroleum imported by those countries in 1975.

If sucrose were to be used as a feedstock, the sugar juice at 10 to 12 weight percent sugar would go directly to fermentation instead of being used to produce raw sucrose crystals plus molasses.

Even though sucrose cannot now be counted on to provide the major portion of liquid fuels for LDCs, the potential exists in some developing nations for producing significant quantities. In addition to the nine developing nations listed in Table II, 16 other countries now export enough sucrose to produce 1000 barrels/ day of ethanol using the sugar now exported. They are: El Salvador, Guatemala, Jamaica, Trinidad and Tobago, Guyana, India, Mozambique, Swaziland and Fiji, Barbados, Nicaragua, Panama, and Bolivia.

Instead of producing ethanol from cane sugar juice, as mentioned previously, countries could continue to export sucrose and the molasses could be used as the feedstock for ethanol production. Molasses, an alternative to sucrose, is a viscous liquid by-product of raw sucrose production processes. Its composition varies, but, at the time that it comes from the centrifuges, it contains 77 to 84 weight percent total solids. The sucrose level varies from 25 to 40 weight percent and the C_6 sugars from 35 to 12 weight percent, with the sum of the two (total sugars) 50 weight percent or higher. Worldwide, the tonnage of molasses produced is roughly equivalent to about one-third the tonnage of raw sucrose produced. Therefore, about 30 to 31 million metric tons of molasses were produced in 1977 (see Table I for sucrose production data).

Only countries that have total sucrose production levels of ≥ 600,000 metric tons would have enough molasses available to provide the feedstock for the ethanol production capacity being considered here. The developing nations that have large sucrose production levels, and thus large molasses production levels, are listed in Table III.

Relative Value of Raw Sucrose, Molasses, and Petroleum Fuels. If developing nations choose to convert sugar and molasses into fuels, they will forego income from exports, but they will save money on petroleum imports. Figure 1 shows the relationships between import costs, export values, and product prices for the three commodities. We have assumed, in the case of using sucrose originally to be exported, that the process is altered by eliminating the steps of concentration and crystallization. Therefore, no molasses would be produced and about 18% more sugar would be

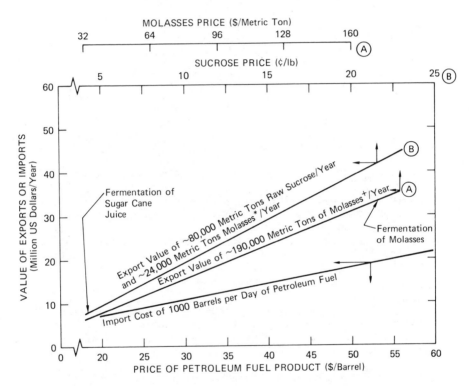

Figure 1. Values of imports and exports as a function of the prices of raw sucrose, molasses, and petroleum fuels (basis: equal volume substitution of 1000 barrels EtOH/day for 1000 barrels petroleum fuels/day). Note: (+) molasses ~ 80 wt % solids, 50 to 55 wt % C_6 and C_{12} sugars; (*) value of molasses based on total sugar content at 2/3 the value of raw sucrose.

contained in the sugar solution than would be available for export
as crystalline sucrose. Two cases are illustrated in Figure 1;
one is of the use of sugar cane juice for fermentation and the
other is of the use of molasses alone.

With the world price of sugar at about 8¢/lb (18¢/kg), and
the world price of molasses above $60/metric ton, the price of
petroleum fuels must exceed $50/barrel before the value of imports
and exports would be in balance. Average prices for sugar have
fluctuated greatly during this decade, as Table IV shows.

TABLE IV   AVERAGE SUCROSE PRICES ON LONDON AND NEW YORK MARKETS
(¢/lb)

	1971	1972	1973	1974	1975	1976	1977
New York	4.5	7.4	9.6	29.9	20.4	11.5	8.2
London	4.5	7.3	9.6	30.1	20.9	11.6	8.2

Sugar price increases have been forecast by U.S. Department of
Agriculture studies completed in 1977 and 1978 (7,8).

For the very large sugar exporters, worldwide demand may
not keep pace with the ability to expand production so that fermentation to ethanol may provide them with a means to reduce petroleum
imports and to keep sugar off the market, thus keeping world sugar
prices high. But, even in this case, the use of ethanol for a
fuel may not be the best economic use for the alcohol. Use of
ethanol as a chemical feedstock may also be economically attractive. This topic, however, is beyond the scope of the current
analysis.

Investment Requirements for Ethanol Fermentation. The actual
investment requirements for a total fermentation capacity of
365,000 barrels/year (~ 15.3 million gallons or ~ 58 million
liters of ethanol) will vary, depending on many factors, such as

- Number of plants required to reach capacity level
  specified.

- Extent that existing utilities and services from adjacent
  sugar mills are used.

- Average days of plant operation per year (daily capacity
  installed to meet annual production level).

- Process design (batch fermentation time, level of
  automation or instrumentation).

- Specific country and sites where plants are to be constructed.

If we assume that the fermentation plants will operate 330
days/year, regardless of the days of operation of the sugar mills,
the investment cost ranges would be roughly as follows:

Feedstock	Factors Considered	Range of Estimates Investment Costs: Million U.S. Dollars (1979)
Sugar cane juice (10 to 12 wt% sucrose)	No. of plants: 1 to 4. Utilities and general services: provided by adjacent sugar mill; installed at fermentation plant	14 to 20
Molasses	No. of plants: 4 to 6. Utilities and general services: provided by adjacent sugar mill;	20

These estimated investment costs were developed by using data
from references 9 and 10 and adjusting the costs on the basis of
current values of construction cost indices and the specific plant
capacities considered.

The range of estimated investment costs are equivalent to
$0.90 to $1.50/gallon of annual capacity ($0.24 to $0.40/liter of
annual capacity). The average investment cost for more than 200
distilleries in Brazil (of an average capacity of 39 million
liters EtOH/year) with roughly a 50/50 split between those using
sugar cane juice and those using molasses is > $1/gal (or
> $0.26/liter) (11).

Methanol Production from Biomass

Feedstock Options. Numerous lignocellulosic materials could
be used for the production of methanol. The conversion technology
entails thermal gasification to produce a synthesis gas stream
(CO and H_2) followed by synthesis to methanol. Agricultural
residues could be used after chopping and perhaps densification
into pellets. The amount of processing required before gasification
will depend on the type of gasifier chosen. The most suitable
feedstock is probably wood, because it will generally be the material
available at the lowest cost in the quantities required. To
produce an amount of methanol equivalent on a volumetric basis to
1000 barrels/day (365,000 barrels/year) of petroleum fuel would
require the processing of ~ 500 metric tons/day of wet wood
(at 50 weight percent moisture) for 330 days/year. This estimate
is based on data presented by Kohan for a wood-to-methanol
plant (12).

Wood Availability. Table V summarizes data for worldwide wood
production in 1974, plus projected production levels for 1985 and
2000. Note that in 1974, 47% of the estimated world wood production was for fuel; 75% of that use was in developing nations. In
1974, 82% of the roundwood production in developing nations went
for fuel use. It has been estimated that the use of wood for fuel
will increase by 10% worldwide by 2000, while the total production
of wood will increase by 50% (13). The use of wood for fuel by
developing nations has been estimated to increase by 43%, or by
374 million cubic meters/year. In terms of the actual percent
increase in production, however, pulpwood production for LDCs has
been estimated to increase the most, 15-fold. These data on
pulpwood production illustrate an important point concerning wood
use. As countries develop, wood is vitally needed for the construction of commercial buildings and residences, and for paper
production. Therefore, even if the fuel demand met by wood remains
at a constant level, either the total harvest must increase to
satisfy other demands for wood or the efficiency of wood utilization for fuel must increase so as to require less wood for fuel
purposes. In certain developing areas of the world, increasing
wood production or even maintaining current production levels may
be difficult. Wood for fuel is now reported to be scarce in large
regions of Southern Asia, African countries bordering the Sahara
from Senegal to Ethiopia, the Andean countries, Central America,
and the Caribbean. Deforestation is already viewed as a serious
problem in numerous areas, and reforestation must be started before
large increases in wood production can be achieved (1d).

TABLE V   WORLDWIDE PRODUCTION OF WOOD PRODUCTS

| | Million Cubic Meters | | |
	1974	1985	2000
Fuelwood and charcoal	1170 (860)*	1200 (1014)	1292 (1236)
Industrial roundwood			
Sawlogs and veneer logs	799 (136)	983 (200)	1191 (272)
Pulpwood	340 (15)	669 (96)	1111 (222)
Other products	202 (40)	208 (51)	206 (62)
Total Roundwood	2511 (1051)	3060 (1361)	3800 (1792)

* Figures in parentheses ( ) are for developing nations.

Source: Adapted from data in reference 13.

In view of current forest resources, it appears South America
offers the greatest opportunity for wood use as fuel. Brazil, for
example, is currently planning a large methanol production program
based on wood as a feedstock to complement present ethanol

production operations that use sugar cane juice or molasses as the main carbohydrate source (14). In some other developing countries, however, in Africa and Southeast Asia, the potential does exist for increased long-term use of forest resources for fuel. But, because of the many varieties of wood produced and the differences in value as a function of wood grade and end use, a discussion of wood value is beyond the scope of this effort.

Investment Requirements for Methanol Production from Wood. According to investment cost estimates prepared by Kohan, a plant that would operate 330 days/year and produce 48,000 gallons/day of methanol (182,000 liters/day) would cost roughly $50 million (12). This investment is equivalent to $2.50/gallon of annual capacity (or $0.66/liter of annual capacity). Because significant economies of scale are achievable in methanol production, a large plant is preferable to minimize the investment requirements per unit of capacity. Single methanol plant capacities for Brazil have been proposed with significantly more than 10 times this capacity, and investment costs per unit of annual capacity have been estimated to be less than $1.25/gallon (14).

Vehicle-Mounted Gasifiers

Feedstock Options

In this analysis, only air-dried wood chips or billets, and pelletized agricultural residues will be considered. The moisture content of these fuels is assumed to be ≤ 20 weight percent. Charcoal briquets are also a highly desirable fuel, but their production requires significant capital investment. The use of charcoal, however, warrants consideration in future studies. (See references 3 and 15 for detailed discussions of biomass fuel use in vehicle engines.)

Investment Requirements

The investment requirements for replacing 1000 barrels/day of petroleum fuel by burning a low-heating-value gas produced using vehicle mounted gasifiers have been estimated based on the calculations and assumptions described in Table VI. Retrofitting of large short-haul trucks or buses in rural areas is the basis for the estimate of an investment cost of from $2 to 4 million. Retrofitting 2000 smaller vehicles (also operating in rural areas) at a cost of ≤ $2000 each might be a more practical initial goal and might be achieved for essentially the same cost.

If five plants were required for the production of densified fuel pellets from agricultural residues, the investment costs for a total capacity of 580 metric tons/day would probably exceed $10 million. Therefore, the investment costs associated with the option of using vehicle-mounted gasifiers may be about $15 million.

TABLE VI BASES AND ASSUMPTIONS FOR THE ANALYSIS OF THE VEHICLE MOUNTED GASIFIER OPTIONS
(Replacement of 1000 Barrels/Day of Petroleum Fuels or ~ 6 x 10^{12} Joules/Day)

• Large short haul trucks and buses with diesel engines are considered for retrofitting.

• Wood billets or pelletized fuels with ≤ 20 weight percent moisture are to be used in downdraft gasifiers; the average heating value of the fuel is 6337 Btu/lb (or 14.7 x 10^6 joules/kg).

• A thermal efficiency of 70% in conversion of solid fuel to a low heating value gas is assumed to calculate the solid fuel requirement.

• [6 x 10^{12} joules/day/ (14.7 x 10^6 joules/kg) (0.7)] ≈ 580,000 kg/day

• ∴ $\dfrac{580,000 \text{ kg solid fuel}}{159,000 \text{ liters of liquid fuel}} = \dfrac{3.65 \text{ kg solid fuel}}{\text{liter of liquid fuel}}$ (or ~ 30 lbs solid fuel/gallon).

• Calculation of specific fuel consumption is based on an engine efficiency of 27% and is equal to 0.95 kg of solid fuel/HP-hr.

• For a 300-HP engine operated 5 hours/day, the solid fuel requirement is
(300 HP x 5 hrs) (0.95 kg solid fuel/HP-hr) = 285 kg/hr.

The solid fuel requirement for 5 hours of operation/day is 1425 kg/vehicle/day

The volume of this fuel is at most [1425 kg/ (300 kg/cubic meter)] = 4.75 cubic meter (168 cubic meter)

The volume could be 50% this amount if densified fuel pellets are used.

• For 1 hour of operation with wood, the gasifier volume must be ~ (285 kg/hr)/(300 kg/m³) ≈ 1 m³

• To replace 1000 barrels of petroleum fuel will require retrofitting 400 large vehicles
(580,000 kg solid fuel/day)/(1425 kg/vehicle/day) ≈ 400 vehicles.

• The average cost to retrofit large vehicle is estimated to range from $5,000 to $10,000;
Therefore, retrofitting 400 vehicles would cost from $2 to $4 million.

Summary and Discussion of Findings

Investment costs to allow replacement of biomass fuels for petroleum fuels used in transportation appear to be significant for all options. To replace 1000 barrels/day of petroleum fuels will require capital investments for conversion facilities as follows:

Option	Estimated Investment (Millions of 1979 U.S. Dollars)
Ethanol production	
From sugar cane juice	14-20
From molasses alone	20
Methanol production	40
Vehicle-mounted Gasifiers	
Retrofitting alone	≤ 4
Retrofitting + densified fuels production	~ 15

For both the cases of methanol production and the use of vehicle-mounted gasifiers, additional investments will be associated with facilities and equipment for harvesting and transport of the biomass to the conversion sites. In the case of ethanol production from currently produced sugar cane juice or molasses, the conversion facilities would be constructed adjacent to sugar mills that already are receiving the biomass raw material. Thus, additional investments for harvesting and transport equipment would not be required.

The use of vehicle-mounted gasifiers would probably present more logistical problems (e.g., the need for frequent refueling) than would the use of alcohol fuel blends. The environmental consequences of using vehicle-mounted gasifiers could also be a problem, as could operating reliability and engine performance (engine derating and response to load). The use of vehicle-mounted gasifiers, however, may represent a simple technological option that offers employment in rural areas and significant savings in imported fuels for transportation.

The use of sugar cane juice or molasses to produce an alcohol fuel may be difficult to justify economically in countries where all of the excess raw sucrose production (and molasses) can be exported at competitive world prices.

Although much enthusiasm currently exists for producing ethanol fuels from local sugar crops in developing nations, the results of this very cursory analysis indicate to us that at least one other option could supplement petroleum fuels used in transportation with biomass fuels at comparable or lower capital investments. This option is the use of vehicle-mounted gasifiers.

Conclusions and Recommendations

On the basis of this cursory analysis, vehicle-mounted biomass gasifiers as well as alcohol fuels appear to us to represent possible means to reduce petroleum fuel use in developing nations. Biomass availability for use as a fuel, however, could restrict the use of the option to only a few LDCs.

We recommend that, in the future, transportation development plans and energy studies for LDCs include consideration of vehicle-mounted gasifiers (especially in rural areas), along with the other options being evaluated. The potential problems with the use of such gasifiers that we have identified in this paper should also be carefully analyzed.

LITERATURE CITED

1. Palmedo, P. F., et al., "Energy, Needs, Uses and Resources in Developing Countries," National Center for Analysis of Energy Systems, Brookhaven National Laboratory, (a) p. XV, (b) p. 95 (c) p. XIX (d) pp 52 to 55, March 1978.

2. Egloft, G., et al., "Motor Vehicles Propelled by Producer Gas" The Petroleum Engineer, 115, December 1943 (pp. 65-73).

3. "Generator Gas--The Swedish Experience from 1935-1945", report translated from Swedish to English by the Solar Energy Research Institute, Golden, Colorado, SERI Publication No. SP-33-140, January 1979, published in Swedish, Stockholm, 1950.

4. O'Sullivan, D. A., "UN Workshop Urges Wider Use of Ethanol," Chemical and Engineering News, April 23, 1979 .

5. Molasses and Industrial Alcohol," Proceedings of the Meeting of Experts Organized by the Development Center of the Organisation for Economic Co-operation and Development, Paris, 1976, published by the OECD, Paris, France, 1978.

6. International Sugar Organization, "Sugar Year Book 1977"; The Whitefriars Press Ltd: London, England, July 1978.

7. "Report on World Sugar Supply and Demand, 1980 and 1985"; Foreign Agricultural Service, United States Department of Agriculture, Washington, D.C., November 1977.

8. "An Update on World Sugar Supply and Demand, 1980 and 1985"; Foreign Agricultural Service, United States Department of Agriculture, Washington, D.C., May 1978.

9. Jones, J. L., Fong, W. S., "Mission Analysis for the Federal Fuels from Biomass Program, Volume V: Biochemical Conversion of Biomass to Fuels and Chemicals"; report prepared by SRI International, Menlo Park, California, for the U.S. Department of Energy, December 1978.

10. Lipinsky, E. S., et al., "Second Quarterly Report on Fuels from Sugar Crops"; report prepared by Battelle Memorial Institute, Columbus, Ohio for the U.S. Department of Energy, October 1977.

11. Yang, V., and Trindade, S. C., "Brazil's Gasohol Program"; Chemical Engineering Progress, <u>75</u>, 4, April 1979 (p. 14).

12. Kohan, S. M., Barkhordar, P. M., "Mission Analysis for the Federal Fuels from Biomass Program, Volume IV: Thermochemical Conversion of Biomass to Fuels and Chemicals"; report prepared by SRI International, Menlo Park, CA, for the U.S. Department of Energy, January 1979.

13. Stone, R. N, Saeman, J. F., "World Demand and Supply of Timber Products to the Year 2000"; <u>Forest Products Journal</u>, 1977, <u>27</u>(10).

14. "Brazil out to Show Methanol Grows on Trees"; Chemical Week, 1979, <u>124</u> (11).

15. Skov, N. A., Papworth, M. L., "The Pegasus Unit". Pegasus Publishers, Inc., Olympia, Washington, 1974.

RECEIVED November 20, 1979.

Social and Economic Aspects of the Introduction of Gasification Technology in Rural Areas of Developing Countries (Tanzania)

MICHIEL J. GROENEVELD and K. R. WESTERTERP

Department of Chemical Engineering, Twente University of Technology,
P.O. Box 217, 7500 AE Enschede, Netherlands

The development of rural areas in the Third World depends a.o. on the availability of energy [1,2]. Makhijani showed that for the improval of the agricultural production an increase in energy consumption is required. This energy in whatever form has to be available in the rural areas in relatively small quantities per user. Consequently energy production by small-scale units scattered throughout the country is required. Nowadays mostly internal combustion engines are used to drive tractors, water pumps, machinery for processing agricultural products or electric generators. It seems attractive to replace the fossil liquid fuels needed for these engines by locally produced fuels. One of the possibilities is the thermal gasification of agricultural waste, which produces a low-caloric combustible gas that can be used directly to drive these engines. In the thermal gasification process the heat of combustion of the waste is transferred to the combustible gas by means of partial oxidation. The same process had been used in World War II with wood blocks as feedstock to drive cars and lorries in countries where petrol was scarce. In our laboratory [3] relatively simple, small scale gasifiers have been built and tested for agricultural and forestry wastes. The fuel gas was used to drive stationary engines.

To identify the social and economic advantages of our process a study has been made of its applicability in rural areas of Tanzania. The evaluation included the selection of a suitable feedstock, the macro- and micro-economical effects and the fitting into the existing social-cultural system.

Description of the process

A gasification process to generate power consists of a gasifier, a gas cooler, a gas-air mixer and a combustion engine and will be described:

```
waste →┌─────────┐    hot   ┌────────┐              air
        │gasifier │→  gas  → │ cooler │ gas → → →  ↓
air   →└─────────┘          └────────┘        →│engine│→ power
                                    condensate ↓
```

Gasifier   From the various possible types the co-current moving bed gasifier was selected because of its tar free gas production and its simple construction [3]. As feedstock agricultural (forestry) waste with a low ash and water content is preferred because of the high caloric value per kg. waste. The gasifier can be divided in four zones (see fig. 1) from top to bottom:
a) bunker. The bunker volume allows for discontinuous filling of waste into the gasifier by hand. In the lower part of the bunker volume drying occurs.
b) pyrolysis zone. Due to the heat transfer from the hotter zones the feedstock pyrolyzes here and thus produces char and (condensable and non-condensable) gases including tar.
c) oxidation zone. In this part air is introduced and a partial combustion takes place, resulting in temperatures of 1200 - 1600 $^{\circ}C$. Due to these high temperatures the tars produced in the pyrolysis zone are cracked.
d) reduction zone, Here the CO_2 and H_2O from the oxidation zone react with the charcoal forming hydrogen and carbon monoxide. The hot product gas is sucked out of the gasifier via the gas cooler by an engine. The ash, which collects in the bottom of the gasifier, should be removed occasionally.
A more detailed description of the gasification of solid waste is presented elsewhere [4]. It is possible to construct the gasifier simply from brickwork and steel, see figure 2, and it has no moving parts.

Gas cooler   To increase the output of the engine the gas should be cooled down to approximately the temperature of the surroundings. This can be done simply with a water or air cooler and with a pebble bed: here the latent heat of the gas is accumulated, so that the bed should be cooled down occasionally. The pebble bed should clean the gas of tar, dust or soot, if by bad operation the gas is dirty.

Gas-air mixer   The gas volume has to be mixed with an about 1.1 times larger, adjustable air flow, which is almost stoichiometric. The total gas-air flow passes through the normal engine filter for final dust removal.

Engine   It is possible to use any internal combustion engine. In a spark ignition 4 stroke engine all the petrol can be re - placed by gas without any adjustment. In case a two-stroke spark engine is used lubricating oil has to be injected separately, in a diesel engine only about 90% of the diesel oil can be replaced by gas: the addition of the same diesel oil is necessary for the ignition. The thermal efficiency of the engine is not affected by the use of gas, but due to the lower caloric value of a gas-air mixture the mechanical output is decreased by ± 30% in comparison to petrol or diesel oil driven engines. Our pilot plant consumed 2,5 kg fuel/kWh (mechanical), but results can be much

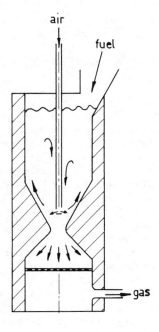

Figure 1. Co-current moving-bed gasifier

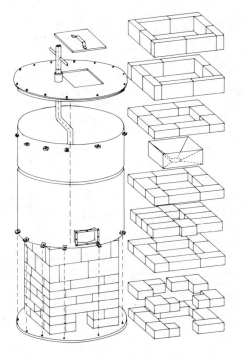

Figure 2. Gasifier appropriate for developing countries

improved with better engines [5,6].

Evaluation criteria

Energy production from agricultural waste by gasification competes with several other possible uses.

The availability of agricultural waste   The production of plant material depends on the solar radiation, the type of vegetation, the climate, the soil and the agricultural system. If and how much of the plant material can be removed from the site for either food or energy production, is determined by the effect on the soil, especially on the humus content. In case of an energy plantation in tropical areas, the solar radiation (yearly average 250 W/m^2) is for 3.8% converted into plant material [7]; assuming an overall efficiency for gasification of maximal about 45%, this results in 4 W/m^2 (electrical). This output can only be obtained in special areas with intensive use of fertilizer in parts of a country specially used for energy production.

In the rural areas of developing countries, however, most of the fertile land has to be used for food production. Consequently only those waste products are available, that can not remain or be returned on the land as a mulch fertilizer. These waste products represent only part of the total crop production, so that energy production depends highly on the organization of the actual agricultural system: only those waste products can be used for energy production, that are already being collected or produced during the processing of the crop.

Gasification should help the development of the country and especially the poorest strata of the society   according to the requirements of the IMF and World Bank. Many poor people live in rural areas, where except in the harvest season a high unemployment exists. The introduction of new sources of energy should create jobs and/or replace unworthy work. New technology used in the rural areas implicates however, that it should be based on unskilled or hardly skilled labour. Local construction or at least local maintenance and repair requires the equipment to be made of well know materials and with simple techniques; so that local technological development is stimulated in an appropriate way.

Finally it is necessary that the introduction of gasification units benefits not only the owner but also the national economy and that it fits in with government policies for unemployment, rural development, trade balance, etc.

Conclusively the gasification of agricultural waste as an alternative method for energy production must be evaluated according to the following criteria:

1. The waste has to be available and usable without damaging the agricultural system or the environment.
2. The technology should fit in the existing sociocultural system.
3. The use has to benefit the development of the poorest strata of the society.

4. The use should be profitable for the owner and the users.
5. The introduction should be of national interest.

Applicability in the rural areas of Tanzania

Selection of a feedstock for the gasification process   In Tanzania
dominate the moist savanne, the dry savanne and the steppe climates. The moist savanne areas show the highest crop production,
and a good infrastructure (roads) exists. The agricultural system
in Tanzania consists of: 1) private farmers 2) large state farms
(or plantations) 3) Ujamaa-villages. The state promotes the
state farms and especially the Ujamaa-villages [8]. In such an
Ujamaa-village the traditional African culture is combined with
a modern locally adapted socialism. These villages are regarded
as the basis and centre for the rural development and as a
communal and cooperative production and marketing unit. In these
villages the land is divided into small individual plots for
each family and a larger communal plot for the production of cash
crops, which enables the Ujamaa to buy tractors, fertilizers and
high quality seeds and to maintain a school, a dispensary, a shop
and water supply. The local availability of a fuel would be very
benificial. The Arusha area has been selected for our field study.
In this area all crops and the associated products have been
studied with special emphasis on:
1. The type and quantity of waste.
2. The need for use as a mulch fertilizer, thus staying on or returned to the soil.
3. The present or planned use.
4. The possibility to gather the waste at the gasification site.
5. The expected technological problems for gasification although
   these might be overcome by further research.
6. The expected power output per unit of area.
7. The most likely use of the power produced, preferably for
   the crop processing. If the waste can be used by the producers
   or processors, costly transport is avoided and no marketing is
   required.
The results of our field study are summarized in table 1 with an
evaluation of different crops. Three crops got a positive
judgement and five have to be studied more in detail. Corn cobs
look a very attractive feedstock for gasification, because they
are reasonably abundant within the villages. Corn is the major
food crop in this region; its production is stimulated by the
National Corn Program of Tanzania, which aims at the food
production for urbanized areas. In this program Ujamaa villages
play an important role. The corn is produced as follows: the
soil is cultivated by a tractor and/or by handwork. All the other
activities are done manually. At harvesting the corn is picked
by hand and transported to the village. There the leaves and the
cobs are removed and the grain either stored or sold to the
National Corn Program. Thus the corn cobs are removed from the
land, but find no use at all for the moment, except sometimes as

Table 1. Evaluation of the use of agricultural waste from Arusha for small scale energy production by gasification.

Crop	Yield	Waste	Dry waste yield kg/ha,j	Recycling to the soil	Present use	Possibility to collect	Possibility to gasify	Output mechanical energy kWh/ha,j	Energy use for crop processing	Recommendation
Coffee	500-1000	branches	?	++	mulch	--	+	?	-	--
		pulp	400-800	++	mulch	++	±	270-530	-	--
		skin	100-200	-	waste	++	+	67-130	industrial	+
Beans	200-1000	straw leaves	?	++	cattle-feed mulch	--	+	?	-	--
Corn hybride	2500-3000	straw	2500-3000	+	cattle-feed mulch	--/+	±	1590-2010	milling	--
		cobs	375-450	-	waste	+/++	++	250-300	shelling drying	++
Sorghum hybride	1700	straw	1700	+	cattle-feed mulch	--	+	11300	milling	further investigation
		plume	750	-	?	+/++	+	500	shelling drying	
Traditional	300-600	chaff ear	900-1800	+	mulch cattle-feed mulch	--	±	600-1200		--
Millet	300-750	straw	300-750	+	cattle-feed mulch	--	±	200-500	milling	further investigation
Finger-millet	300-600	chaff ear	± 10	-	waste	+/++	+	7	shelling	
Pearl-millet	300-750	chaff ear	150-375	-	waste	+/++	+	100-250	drying	
Sugercane	28.000	bagasse	3500	?	fuel	++	+	2300	++	--
Bananas	?	stem	?	+	mulch	--	±	?	-	--
Sisal	750-1600	pulp	3000-6000	-	waste	++	-	2000-4000	industrial	-
Sunflower	500-1500	straw	?	+	mulch	--	+	?	?	--
		receptacle	100-400	+	mulch	--	±	67-270	?	-
Cotton	220-450	straw	?	-	destroyed	+	+	2000	-	-
		seeds	600-1200		oil	++	++	400-800		-
		press residue	?		cattle-feed	++	+		industrial	further investigation
Traditional Rice	800-1200	straw	?	+	mulch	--	±	1070	polishing	-
		chaff	160-240	-	waste	++	±	100-160	drying	
Wheat	1400	straw	?	+	mulch	--	±	?	milling	further investigation
		chaff	280	-	waste	++	?	190	drying	
Ground-nuts	450-700	shells	100-175	?	?	?	++	67-120	?	further investigation
Wood	1000-1500	branches	200-800	+	fuel	+	++	135-200	+	-
		sawmill wastes	100-150	-	fuel board etc.	++	++	67-100	saw mill	+

cattlefeed in very dry periods. Some large scale producers use
them as mulch, but the compostation of the corn cobs takes several
years and returns only a small amount of organic material to the
soil and is 10% of the total plant production, poor in inorganics
like Nitrogen and Phosphate. The corn cobs therefore are already
available in the village without having an alternative use; the
agricultural system is not endangered by burning them. The power
produced by the gasification of corn cobs can be used for water
supply, local electricity or driving a corn mill. Such corn mills
are already operating in many villages. In Arusha corn mills,
driven by an imported diesel engine, are locally constructed and
sold in the rural areas. It is possible to include in this system
a corn sheller and a gasification unit. The sheller replaces the
hand picking of grains from the cob and so diminishes the amount
of work in the busy harvesting period. In this way the cobs are
produced at the gasification site. The sheller can be driven by
the same engine as is used for the corn mill because it consumes
only a small amount of energy (2 kW).
The introduction of gasification of corn cobs in an Ujamaa thus
agrees with the first two evaluation criteria, the other three
criteria will be studied in a social-cost-benefit analysis.

Social cost-benefit analysis

A social cost-benefit analysis gives information on the profit of
a project and further on the influence of the project on employment, production, consumption, savings, trade balance, redistri-
bution of income etc. Here the 'Guidelines for project evaluation' of the Unido [9] are used. Both the investment and operating costs of a project have to be divided into
a) materials, imported or domestically produced
b) labour, skilled or unskilled
c) interest depending on the way of financing
d) taxes paid.
For the various social groups (the government, labourers, tax
payers, etc.) it has to be determined to what extent they consume
the additional income out of a project and how much will be
saved for reinvestment. Further a premium has to be allocated for
the import savings, the increase in the use of unskilled labour
and a penalty for the use of scarce skilled labour. Based on this
information it can be calculated, whether the project is profitable for the various social groups and the country as a whole.

Benefits of corn milling in Tanzania for the owner. In many
Tanzanian villages the corn for the local consumption is brought
once a week to the mill, where people pay 0.15 Tsh/kg (1 Tsh =
0.15 $ 1977) for the grinding. The corn mill has a nominal
capacity of 600 kg/h and is driven by a 20 kW engine. Such a unit
of mill and engine costs around 33000 Tsh, however it is apparently too large for most of the villages. In these villages live
around 400 families (average two adults, four children) each
consuming about 800 kg corn/year. This figure has been obtained

from statements in the villages and is equivalent to 8 kJ/day for adults and 4 kJ/day for children.
The corn mill operates twice a day at sun rise and at sunset for a period of around 2.5 hours/day; it consumes daily around 30 ltr of diesel oil, which costs 2.14 Tsh/ltr. The said diesel oil consumption relates to a motor efficiency of about 15%, which is relatively low.
Based on a depreciation of 5 years and 10% interest rate a cost - benefit analysis is included in table 2. In the same table a cost - benefit analysis is given in case the corn miller would be able to replace the liquid fuel required, by gas from corn cobs. Then the corn mill is enlarged with a suitable gasifier for 2000 Tsh. The economic advantages of the gasifier for the corn miller are evident in view of the profit increase.

Table 2   cost - benefit analysis of a corn mill without (A) and with gasification (B) in Tsh/day.*

	A	B
Daily benefits:		
800 kg/day à 15 Tsh/100 kg	120	120
Costs:		
30 ltr fuel/day ; 3 ltr.	64	6
labour (1 man)	16	16
repair and maintenance	14	16
depreciation (5 yr)	18	19
interest (10%)	5	6
total	117+	63+
Daily profit for the Ujamaa community	3	57

* 1 Tsh = 0.15 $ (1977).

    Social cost-benefit analysis of a gasifier   From the pre-ceding calculation it can be seen that introduction of gasifica-tion is very beneficial for the mill owner, in this case the village community. We now will evaluate the social costs and be-nefits according to the Unido guidelines with their different methods of increasing complexity: 1. at market value; 2. at a national level; 3. per social group.
    Analysis at market value   To make a social cost benefit analysis of the addition of gasification to a corn mill the in-vestment costs have to be on annual basis. We assume that the in-vestments are financed by the government and are repaid by the Ujamaa community in five years including an interest of 10%, in equal installments (annuity of 0.26).
Investment costs for the gasifier (Tsh)

	gasifier	annual costs
—materials imported	800	208
—local unskilled labour	200	52
—materials, domestic	1000	260
total	2000	520

Further it is assumed that 25 Tsh of the fuel cost is the net import saving of oil per day, thus 9125 Tsh/yr. The extra operating
costs for the use of gasification can be seen from table 2: 2 Tsh/
day for repair, thus 730 Tsh/yr in the category of domestic materials. In table 3 an input-output analysis for the gasification
project is given from the national point of view for a period of
5 years of operation.

Table 3   Input-output analysis of the addition of a gasifier to a
corn mill (Tsh/yr).

Annual input : construction costs (imported), annuity	+	208
construction costs (domestic), annuity	+	312
operating costs (domestic)	+	730
Annual output: foreign exchange savings FE	−	9125
Annual national aggregated consumption at market value, MC		7875

The remainder of the savings to the Ujamaa villages of 64-25 Tsh/
day refer to the costs incurred in Tanzania for converting the
imported crude oil into diesel oil in the state owned oil refinery, transport of the diesel oil to the Ujamaa village and local
taxes. These savings to the village, however, do not represent a
saving to the nation as a whole. On the contrary skilled labour
(o.a. lorry drivers) will loose some work and thus wages ($\pm$ 2 Tsh/
day), the rest is lost by the government. On an annual basis the
diesel oil savings per gasifier represent around 78 barrels of
crude oil. In Tanzania easily 4000 gasification units can be installed leading to a potential savings of around 300000 barrels of
imported crude oil per year.

   Analysis on a national level   The market value of the net
aggregate-consumption benefits in a given year of the project as
given in table 3 does not reflect the social opportunity costs on
a national scale. As for the foreign exchange position the heavy
pressure on the balance of payment has resulted in strict quantitative import control and export subsidies to maintain the dollar
value of the Tsh. A dollar of foreign exchange is worth substantially more than the official exchange rate: the opportunity cost
of foreign exchange relative to its official market price is $1 + \phi$,
ϕ representing an premium. Unskilled labour is found in surplus
and the market wage exceeds the opportunity costs of employing
additional workers; the opportunity costs will be denoted $1 + \lambda$,
in which λ represents the unskilled labour premium; λ will be zero
or negative. The social value of skilled labour is measured by the
contribution its services make to aggregate consumption benefits
for the country: usually skilled labour is underpaid. In parallel
with ϕ and λ, χ is defined as the social premium on the market
value of a skilled worker. With respect to the absolute value of
the parameters ϕ, λ and χ the following can be said.
If for example $\phi = 1$, this means that imported goods are put
at the disposal of their consumers at only 50% of their real
price. Also if $\lambda = -1$ (extreme value), unskilled labour has no
social value because otherwise still the primary necessities

of live for unemployed   unskilled labourers has to be provided
for. And also if $\chi = 1$, a skilled worker earns only 50% as wage
of his real contribution to the national economy.
The net aggregate consumption benefits after incorporating the
opportunity cost premiums, SC, now becomes:
$$SC = MC + \phi\ FE + \lambda L + \chi W$$
in which FE are the foreign exchange savings, L the unskilled labour costs and W the skilled labour costs.
With the values of $\phi = 0.5$, $\lambda = -1.0$ and $\chi = 1.0$ we find
$$SC = 7875 + 0.5(9125-208) - 1 * 52 + 1 * 700 = 12982 \text{ Tsh/year.}$$
We see that the benefits for the national economy are much larger
for the introduction of the gasification technology after correcting for the social opportunity costs.

  <u>Analysis on a national level and per social group</u> Another
and more precise approximation to the net aggregate consumption
benefits of the project takes into account the adjustments necessary when the social value of funds devoted to investments exceed
the social value of the same funds devoted to consumption. This
is the case when the government has not been in a position to
raise savings – and investment – to the point where the marginal
rate of return on investment, q, is equal to the social rate of
discount, i. If 100 units consumed now is considered equivalent
to 110 units consumed the next year, the social rate of discount
i is 10%. A unit of money withdrawn from consumption to an investment now should have a rate of return of at least i in order to
increase the consumption in future years to the desired level.
The opportunity cost of investment, p^{inv}, is defined as the ratio
of the social value of investment to the social value of consumption, where "social value" is understood to mean the value of the
relevant time stream of aggregate – consumption benefits discounted back to the present at the social rate of discount. The value
of p^{inv} can be determined according to the following formula [9]:
$$p^{inv} = \frac{(1-s)q}{i-sq}$$
in which s is the economy-wide marginal rate of reinvestments of
profits, expressed as a fraction of total profits; all the parameters are assumed to remain constant over time. This means that
the value of one Tsh for future consumption, if invested now,
equals p^{inv} Tsh's consumed now. Once it is recognized that p^{inv}
does not equal 1, it becomes essential to evaluate the net effect
of a project on the mix of consumption and investment in the
economy. To evaluate the net effect of the gasification project
on the rate of investment, it is necessary to distinguish all the
benefit and cost flows that make up SC as well as the accompanying
cash transfers according to the group that gains or loses and to
estimate the respective marginal consumption and saving propensi-
ties of each group. Three broad groups of gainers and losers have
to be distinguished in the gasification project: the Ujamaa farmers F, skilled labourers W and the Government G. If the construction costs of the project are paid out of government revenues

without any corresponding receipts - so that the governments'
budget is not raised by general taxation - the government is the
loser and the taxed public is not affected. Moreover we assume
that the unskilled labour is provided by the Ujamaa villages for
the local erection of the gasifier. The simplified approach used
before for MC and SC now no more can be applied and cash flows
have to be used for every year of the project. This is done in
table 4.

Table 4   Cashflows in 1000 Tsh/year for one gasifier.

Years	0	1	2	3	4	5
Benefits						
1. Fuel savings						
1 D Domestic currency	–	13.50	13.50	13.50	13.50	13.50
1 F Foreign exchange	–	9.13	9.13	9.13	9.13	9.13
1 W Skilled labour	–	0.73	0.73	0.73	0.73	0.73
Costs						
2. Construction costs						
2 L Unskilled labour	0.2	–	–	–	–	–
2 D Domestic materials	1.0	–	–	–	–	–
2 F Foreign exchange	0.8	–	–	–	–	–
3. Operating costs						
3 D Domestic materials	–	.73	.73	.73	.73	.73
Transfers						
4. Rental and interest paid	–	.52	.52	.52	.52	.52

The government G loses the fuel savings and construction costs.
Skilled labour W loses some work. The Ujamaa farmers F gain the
fuel savings and lose the operating costs. The additional opportunity costs due to the foreign exchange exponents are sustained
by the government, since the government is in effect subsidizing
the use of imported inputs in agriculture by making them available
to the farmer at the official exchange rate. The same reasoning
can be set up for the other social groups. Although the cash
transfer item 4. is not relevant to the evalutation of aggregate
benefits and costs, they are certainly relevant to the distribution of these benefits and costs. Hence, they too must be considered here in assessing the allocation of net benefits among
the three groups G, W and F. Item 4. represents a gain to G and
a loss to F.

The distribution of net consumption benefit SC by group can be
summarized as follows:

$$SC = SC^F + SC^W + SC^G$$
$$SC^F = 1\ D + 1\ F + 1\ W - 3\ D - 4 - \lambda 2L$$
$$SC^W = - (1 + \chi)\ 1\ W$$
$$SC^G = - 1\ D - 2 + 4 + \phi\ (1\ F - 2\ F)$$

where SC^F, etc. represent the value of net consumption benefits
flowing to F, etc. To arrive at the final social value of net
aggregate - consumption benefits C it is necessary to correct SC^F
etc. according to the proportions in which each is divided between
consumption and investment. Thus if the average Ujamaa farmer

saves a proportion s^F of his marginal gains, the social value of
the net consumption benefits flowing to the Ujamaa community is
[9]:
$$C^F = [s^F p^{inv} + (1 - s^F)] SC^F$$
Analoguous formula can be derived for C^G and C^W. Now as objective of the project the net present value V of the total aggregáte
consumption plus the extra aggregate consumption to the Ujamaa
villages is taken
This is done according to $V = C + \Theta C^F$
If no weight is attributed to the additional consumption of the
Ujamaa villages $\Theta = 0$, but otherwise $\Theta > 0$. Θ means that the government is willing to lose $1 + \Theta$ Tsh in other more developped
areas in the country in order that the consumption in the Ujamaa
village raises with 1 Tsh. The social rate of discount i is still
unknown and we will calculate V for several values of i and Θ.
The values of the general parameters we have assumed are given in
table 5. The present values of the cash flows in the year 0 for
the various rates of social discount and the general parameter
values are given in table 6. With these data the various calculations are executed, results are given in table 7 and Figure 3.

Table 5   Value of general parameters

Foreign exchange premium	ϕ	$= + 0.5$
unskilled labour premium	λ	$= - 1.0$
skilled labour premium	χ	$= + 1.0$
marginal rate of return on investment	q	$= 0.20$
marginial rate of reinvestment of profits	s	$= 0.30$
social rate of discount	i	$= 0.075; 0.10; 0.20$
associated social price of investment	p^{inv}	$= 9.33; 3.50; 1.0$
marginal propensity to save:		
farmers	s^F	$= 0.20$
skilled workers	s^W	$= 0.30$
government	s^G	$= 0.90$

Table 6   Present values of cash flows in year 0 (1000 Tsh)

	Social rate of discount		
	7.5%	10%	20%
Fuel savings			
1 D Domestic currency	50.69	46.52	33.19
1 F Foreign exchange	34.28	31.50	22.73
1 W Skilled labour	2.74	2.52	1.82
Construction costs			
2 L Unskilled labour	0.2	0.2	0.2
2 D Domestic materials	1.0	1.0	1.0
2 F Foreign exchange	0.8	0.8	0.8
3 D Operating costs	2.74	2.52	1.82
4 Rental and interest paid	1.95	1.79	1.28

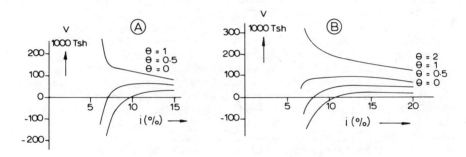

Figure 3.   Socioeconomic present value of the project: (A) one gasifier; (B) one gasifier and sheller

Table 7   Present value of net benefits of one gasifier in year 0
(1000 Tsh)

Item	Social rate of discount		
	7.5%	10%	20%
MC , national aggregate consumption	29.5	26.1	16.1
FE , foreign exchange savings	33.5	30.7	21.9
SC^F, net consumption benefit, farmers	83.2	76.4	54.8
SC^G, net consumption benefit, government	− 34.0	− 31.4	− 23.0
SC^W, net consumption benefit, skilled labour	− 5.5	− 5.0	− 3.6
SC , net consumption benefit, total	43.7	40.0	28.2
C^F , social value of SC^F	222	115	54.8
C^G , social value of SC^G	−289	−102	− 23.0
C^W , social value of S^W	− 19	− 9	− 3.6
C , social value of SC	− 86	4	28
V , Beneficial effects of the project, $\Theta=0$	− 86	4	28
$\Theta=0.5$ weight on redistribution to Ujamaa	25	61	55
$\Theta=1.0$	136	119	83

At small values of i and Θ the net present value is negative
because the government has a much larger propensity to reinvest
then the Ujamaa villagers. But if the government is serious in
trying to increase the net aggregate consumption benefits in the
Ujamaa villages ($\Theta > 0.4$) the project has to be evaluated positive ($V > 0$). The government is the loser in this project because
the state owned refinery produces less diesel oil and the tax
income on the sales of diesel oil is lost. The Ujamaa villagers
lose the operating costs but gain the much higher value of the
fuel savings. The skilled labour loses some employment, but it
seems likely that they can find work somewhere else. The introduction of gasification is thus beneficial for the Ujamaa-villa-
gers and is under almost all circumstances of national interest.
If not only a gasifier is installed but also a sheller at the corn
mill, so that hand picking of the cobs is replaced by a machine,
investment is increased. At the same time unworthy work in the
already very busy harvesting season is avoided. The investment in
a sheller is around 6000 Tsh. The preceding calculations have been
repeated for a project, where a gasifier plus a sheller are installed in a village. Results are given in graph 1B. From this
graph can be read that at $\Theta > 0.5$ this extended project has a positive value with as benefits mainly the replacement of unworthy
work, for which the government has to pay in foreign currency for
the imported parts of a locally constructed sheller. As Θ for the
gasifier and sheller is higher than for the gasifier alone, the
government has to sacrifice more in the cities to benefit the rural areas.

Conclusions

- According to the presented evaluation criteria the gasification
  of corn cobs is acceptable from the economical and agricultural
  point of view in the rural areas around Arusha (Tanzania).

- The gasification system is relatively simple of construction
  and local maintenace is possible. If the sytem is connected to
  the already existing corn mills in the villages it is appropriate to the existing socio cultural system.
- The economic calculations made clear that the use of gasification is attractive for both the owners of the corn mill and the
  government. The advantages for the government are the savings on
  imported oil and the extra income created for the users of the
  corn mill (inhabitants of the rural villages). The government
  loses income from taxes and from the production and transport
  of diesel oil.
- The presented evaluation methods can and should be used for
  gasification projects in other areas as they are undertaken at
  the moment in Nigeria and Indonesia, taking into account the
  differing government policies in their countries.

Aknowledgement

We are grateful to professor Feil, for his continuing stimulating
interest and professor Van Swaaij, who is the research director of
our gasification studies at the Twente University of Technology.
We are also grateful to our colleagues Schaffers, Skolnik and
Hartoungh, who contributed considerably in some part of these
studies and to the Dutch Ministery of Development Cooperation and
the Tanzania Small-Industry Development Organisation, who made the
activities in Tanzania possible.

Literature cited

1. Energy and Power, Scientific American Book, 1971.
2. A. Makhijani, Energy and Agriculture in the Third World, Ford
   Foundation, Energy Policy Project Report, 1974.
3. M.J. Groeneveld, W.P.M. van Swaaij, The design of co-current
   moving bed gasifiers fuelled by biomass, this symposium.
4. M.J. Groeneveld, W.P.M. van Swaaij, Gasification of char particles with CO_2 and H_2O, Sixth International Symposium on
   chemical reaction engineering, Nice, 1980.
5. I.E. Cruz, Producer gas as a fuel for the diesel engine,
   University of the Philippines, 1977.
6. P. Schläpfer, J. Tobler, Theoretische und praktische Untersuchungen über den Betrieb von Motorfahrzeuge mit Holzgas,
   Bern, 1937.
7. C.K. Dietz, Ammonia production from Brava Cane, IGT symposium
   Energy from biomass and wastes, Washington, D.C., 1978.
8. J. Nyerere, Freedom and development, Oxford University Press,
   1973.
9. P. Dasgupta, Guidelines for project evaluation United Nations,
   New York, 1972.

RECEIVED December 7, 1979.

INDEX

INDEX

A

Acetaldehyde 218
Acetone 218
  decomposition products 221
Acid, acetic 218
Acid, formic 218
Activation 303–306
  reaction, pyrolysis 303t
  thermodynamics of 307
African iron ore 110
Afterburning 60, 405
  by the multiple combustion method,
    NO$_x$ prevention at 516t
Agricultural
  and forestry wastes, third world
    applications of pyrolysis of 671–688
  residues 86, 317, 337, 691, 697
    crop 618
    densification systems for 179–193
    pelleted 187
  in a rotary kiln, pyrolysis of 337–350
  as supplementary fuel for diesel
    engines, producer gas
    from 649–669
  waste, availability of 708
Air 292, 307
  classification 4, 129, 352
  classifier light fraction yield,
    research 131f
  classifier test configuration,
    vertical 130f
  equilibria, carbon–steam– 308f, 309f, 310f
    –CH$_4$/stoichiometric 311f
  feed system 383
  for gasification 286
  heaters, process 295
  introduction technique, pebble
    bed process 678
  pollution control 65–66
    devices 67
  pressure requirements 286
  secondary (overfire) 64
  soot blowing, compressed 64
Alcohol(s)
  biological conversion to 209
  fermentation 621, 647
  fuels production 691–699
Alkali
  carbonates 370
  concentrations 119
  silica ratio 102

Alternator system, Duvant gasifier/
  dual-fuel diesel motor 642f
Aluminum 63, 97
  eliminator 552t
  mixed in municipal refuse,
    elimination of 550
Ammonia 43, 358, 443
  capacity, requirement for additional 351
  case investment 38
  from wood 38
Anaerobic digestion 619, 621
  process 7
  of wet biomass 618
Animal(s) 153
  dung 617
  wastes 618, 621
Anthracene start up oil 25
Apparatus, bench-scale test 229
Aromatic(s)
  by-products 209
  highly condensed 432
  hydrocarbons, condensed 484
  oils, highly 484
  products 218
  polynuclear 357
Ash 134, 198, 200, 229, 534, 589
  chute quenching system 66
  content 135, 262, 454
    char, high 351
  -free combustibles
    fuel 144
  fusion temperatures 139
    average 139t
  leaching of heavy metals from 493
  heavy metal in 535t
    pyrolysis 550
  –melting furnace 589
  –melting and waste heat power
    generation, integrated system
    for solid waste gasification
    and 588f
  removal 46, 385, 466
    continuous 71
    section of gasifier 645f
    system 383
  residue, inert solid 291
  and slag, properties of pyrolytic 593
  and total carbon, analysis for 122t
  zone 46
Assessment conclusions,
  technology 327, 328t
Athabasca oil sands 317

B

Bacteria ... 153
Baking ... 303
Bale, rectangular ... 186
Barbados ... 694
Bark ... 86, 355
  Ponderosa pine ... 102
  white fir ... 102
Bed unit, auger ... 67
Benzene ... 424, 428, 433, 480, 484
  homologs, $C_8/9$ ... 484
  production ... 435f
  yield of ethylene and ... 435f
Biological
  activity ... 160
  process ... 7
  stability ... 155, 160
Biomass
  anaerobic digestion of wet ... 618
  conversion ... 22
    in meeting the energy needs of the rural populations of developing countries, potential of ... 617–633
    routes ... 330f
    table of supplementary sources for ... 332
    technology of R&D program, Canada's ... 317–333
  –derived activated carbon, thermodynamics of pyrolysis and activation in the production of waste- or ... 301–314
  design of co-current moving-bed gasifiers fueled by ... 463–477
  Energy Systems Branch ... 13
  ethanol production from ... 691–697
  to fuels and feedstocks, thermochemical conversion of ... 13–26
  methanol production from ... 697–699
  replacement of natural gas, payback from ... 90f
  residues using the gasifier/dual fuel engine techniques, power generation from ... 635–647
  resources at various cost levels, availability of ... 320f
  supply, current sources of ... 618–620
  technology and resource assessment chart ... 319f
Birds ... 153
Boiler(s) ... 6, 7, 38, 64, 148, 196
  application ... 103–104
  cleaning methods ... 64
  dacha ... 64
  electrical generation ... 295
  emissions grate burning pelleted fuels, stoker-fired ... 202f

Boiler(s) (continued)
  energy recovery in ... 65
  fuel ... 15, 38
  heat recovery ... 511
  installation, FBC ... 105f
  oil/gas fired ... 51
  rertofit system, capital and operating costs for a wood gasification/ ... 52t
  shutdown ... 296
  stoker ... 135
  tube corrosion ... 60
  water-tube wall furnace/ ... 65
Boiling behavior of the condensate from the pyrolysis of old tires ... 448f
Bolivia ... 694
Brazil ... 625
Brick-making ... 621
Briquetting ... 205
Bulk densities ... 134
Bunker ... 706
Bureau of Mines ... 291
Burner ... 386
  emission rate, demonstration unit Roemmc ... 202f
  system ... 384
Burning ... 424
  of municipal solid waste, mass ... 50–66
  rate per person ... 60
  rate of the refuse ... 93
Butadiene ... 434
Butane ... 365
Butter ... 63
Butylene ... 209

C

$C_8/9$-benzene homologs ... 484
CH_4/stoichiometric air equilibria, carbon–steam– ... 311f
Cables ... 480
Calcium oxide ... 374, 428
Calorific value(s) ... 341, 348
  gas, low ... 351
  product gas, typical yield and ... 401t
Canada's biomass conversion technology R&D program ... 317–333
Canadian hybrid poplar ... 370
Carbon ... 302, 307
  activated ... 301
  analysis for ash and total ... 122t
  black ... 428, 557, 564
  dispersion of ... 568f
  and rubber vulcanization, properties of ... 565t
  content and volumetric reduction ratio, comparison of ignition loss, caustic-soluble ... 595t
  conversion efficiency ... 238
  dioxide ... 11, 294, 301, 329, 365, 640

Carbon (*continued*)
disulphide 442
fixed 303
monoxide 120, 211, 261, 294,
329, 364, 365
and hydrogen synthesis gas
Fischer–Tropsch liquids via 209
oxides 209, 352
oxysulphide 442
–steam–air equilibria 308*f*, 309*f*, 310*f*
–steam–methane/stoichiometric
air equilibria 311*f*
–steam equilibria 311*f*, 312*f*
–steam–oxygen equilibria 312*f*
thermodynamic of pyrolysis and
activation in the production of
waste- or biomass-derived
activated 301–314
weight of 329
Carbonates 119
alkali 370
Carbonization bed, fixed 640
Carbonization gas 444
Carcinogens 36
Cardboard 295
Cash flows in year 0, present
values of 716*t*
Cash flows in 1000 Tsh/year for
one gasifier 715*t*
Catalysis in wood gasification,
role of 369–378
Catalysts, direct liquefaction of wood
using nickel 363–368
Catalysts, inorganic salts as gasification 370
Catalytic gasification 46
Caustic soda 499
Cellulose 363
carbon conversion efficiency 239*f*
cotton 240
prior research on 251*t*
fuel pellets 287
gasification 238
physical properties for 248
pure 304
pyrolysis 240, 244, 247, 410
equilibria 306*f*
volatilization of 238
Cellulosic materials, pyrolysis of 216
Ceramics 603
Cigarette papers and filters 295
Cleaning
methods, boiler 64
shot 64
system, wood gas 384
Chain cleavage 216
Channeling 403
Char 36, 38, 210, 216, 238, 240,
251, 267, 303, 369, 526,
541, 564, 598, 678, 682

Char (*continued*)
analysis for pyrolysis of straw and
stover 343*t*
compression strength of 568*f*
formation 93, 220*f*, 364
gasification 237
high ash content 351
oil 275
mixture composite, percent
available energy in 673*f*
yields as a function of air/feed 684*f*
oxidation 93
and reduction zones 287
particle diameter 467*f*
by pyrolysis of wood, production
of oil and 35*f*
reaction zone 294
refining of 557, 563
residue 240
selling price of 40*f*
yield(s) 343
from biomass by pyrolysis,
typical 305*f*
Charcoal 275, 618, 643, 650,
655, 681, 690
briquettes 190
consumption 660
Charring 303
Chemicals, low molecular gaseous
and liquid 479
Chemical production 38–41
from wood—estimated investment .. 39*t*
from wood—product revenue
requirements 41*t*
Chicago 59
China 621
Chloride(s) 119
corrosion 59
metallic 301
Chlorine 64
Chopping 185
Coal 17, 86, 301, 371, 654
anthracite 287
cleaning 109
conversion 22
gas producer, low volatile 287
gasification technology 369
particle density of 135
supplementary firing and co-firing
of 65
Coconut shell 301
Cogenerated products 30–32
costs 32
Cogeneration (production of elec-
tricity and steam) 360*t*
based on wood feedstock 15
facilities 32
by wood combustion 34*f*
Coke 287, 604

Combustibles, better thermal conversion of very poor 644
Combustion 14, 109
  chamber 511
  complete 64
  development activities, direct 16
  energy recovery from wood
    residues by fluidized-bed 85
  equipment, stationary 689
  facilities—estimated investment,
    wood 31
  /gasification processes 302
  heat of 330f
  of liquid wastes 143
  method, NO_x prevention at after-
    burning by the multiple 516t
  of pine sawdust and pellets, heats of 171t
  reaction, gas phase 94
  residues 181–183
  of solid wastes, fluid bed 93–108
  system(s) development of modular
    biomass gasification and 285–290
  systems, direct 14–15
    for ordinary refuse 605f
  tests 99
  wood 30
    facilities—product revenue
      requirements 33t
    production of electricity by 35f
      and steam (cogeneration) by .. 34f
  zone 46
Combustor(s) 110, 114
  gas 285
  gasifier/reactor/ 287
  general arrangement of a
    gasifier/reactor 289f
  high heat release 93
  industrial fluid bed 94
  MS–FBC 113f
  off-gas analysis 583t
  pressurized fluid bed 94
  two-stage 67
  waterwall 6
  wood-fiueled 15
Commercial
  activities 354, 358
  application 48
  facilities 89–91
Comparison of ignition loss, caustic-
  soluble carbon content, and
  volumetric reduction ratio 595t
Composition 167
Compost plant 60
Composting 61
Compression
  direct 169
  experiments 171t
  strength of char 568f
Compressors, stacks, and bales 186

Condensate from the pyrolysis of
  old tires 450t–451t
  boiling behavior of 448f
  viscosity of................................ 448f
Condensation 486
Condenser mounted on top, gas
  scrubber with 646f
Condensing 563
Conservation and Recover Act,
  Resource (RCRA) 66
Controlled-air incinerators 67
Conversion
  to alcohols, biological 209
  efficiency, carbon 238
    cellulose 239f
  to methanol, catalytic chemical 209
  other methods of effective waste 644
  process evaluations 317
  processes, economic assessment
    for bioenergy 324t–325t
  thermochemical
    projected fiscal year 1979
      budget— 24f
    projects, biomass 25
    projects, geographical distribu-
      tion of 24f
  systems using wood feedstocks,
    economic overview of large-
    scale thermal 29–42
  technology in biomass, the role of .. 317
  technology R&D program,
    Canada's biomass 317–333
  temperature 467f
Convertor
  appropriate technology pyrolysis 675
  design, Ghana 678
  system, Indonesian technology
    development pyrolytic 680f
Conveyor belts 569
Conveyor, screw 405
Corn
  cob(s) 711
    pyrolysis powdered 245t
  mill, cost–benefit analysis of 712t
  mill, input–output analysis of the
    addition of a gasifier to 713t
  milling in Tanzania for the owner,
    benefits of 711
  stover pyrolysis, gas composi-
    tion from 345f
Corrosion 6, 62, 71, 99
  boiler tube 60
  causes 64
  chloride 59
  tube erosion– 64
Cost(s)
  –benefit analysis of a corn mill 712t
  –benefit analysis, social 711
  of a gasifier 712

Cost(s) (*continued*)
  capital
    estimate 323
    of a gasification system for
      small communities 271
    of wood gasification systems 48
  cogenerated products 32
  considerations, production 203
  for conventional and densified
    waste collection 163f
  disposal 11
  of dry process PUROX system,
    running 585t
  electricity 32
  energy 204
  feedstock 48
  of large modular refuse
    incinerator systems 80t
    with heat recovery, investment 79t
  LBG gas 50t
  MBG gas 50t
  net disposal 62
  operating 160, 501
    comparison of plant space,
      capitol cost and 600t
    effect of steam value on 77
    for the installations 77
  for the pyrolysis case 38
  for scrap tires, disposal 571f
  of small modular refuse inciner-
    ation system 78t
  solid waste collection 161
  of solid waste management
    unit operations 152f
  for the source separation of
    solid waste 271
  steam 32
  for the wood biomass feedstock 53
  wood gas production 48
  for a wood gasification/boiler
    retrofit system, capital and
    operating 51t
  utilities and operating 504t
Cotton cellulose 240
Countercurrent operation 406
Countries
  gasifier appropriate for 707
  potential of biomass conversion in
    meeting the energy needs of
    the rural populations of
    developing 617–633
  social and economical aspects of the
    introduction of gasification
    technology in the rural areas
    in developing 705–719
  with sucrose production levels,
    lesser-developed 693t
Cracking, naphtha 408f
Cracking/reforming of volatile matter   237

Crop(s) 317
  irrigation 649
  residues 619
Cubes 188

D

Data acquisition system 264
Decomposition 140, 410
  of polymers, thermal 409
  ratio of combustibles 512f
Defense, Department of 69
Deflection 172
Dehydration 216
Demand integration, supply 625
  facility, actual 526–529
  plant, flow sheet of 528f
  plant test, operation data of 529
Densification 197
  apparatus, energy requirements
    for commercial 175t
  balance sheet 192
  device, experimental ram 165f
  energy requirements, biomass 169–177
  plant processes 197
  power requirements for 191
  process(es) 158, 164, 175
  sequence, typical 183
  of solid waste 258–261
  system(s) 258
    for agricultural residues 179–193
    available today 183–184
    solid waste shredding and 264
  test configuration 170f
  work vs. density 170f
Densified biomass fuels, benefits of
  the use of 196
Densifier 132–134
Density(ies) bulk 179
  HCl 597f
  NO_x 97f
  for various products, values and 182f
Department of Defense 69
Department of Energy (DOE) 13–26, 29, 94
Design(s)
  of co-current moving-bed gasifiers
    fueled by biomass 463–477
  concept, basic 672–676
  criteria, gasifier 381
  of a 100 t/d gasifier 474
  modular unit 68
  plant 29–30
Destrugas process 405
Desulfurization, flue gas **109**
Devolatization of wood 46
Diesel
  engine, producer gas from solid
    waste fuels in 651f

Diesel (continued)
  generator system 384
  for power generation, Northern
    Saskatchewan using 391t
Differential thermal analysis 410, 416
Digestion process, anaerobic 7
Disposal costs 11
Disposal of municipal refuse by the
  two-bed pyrolysis system 541–555
Distillation................................... 640
  techniques 223
Dockfenders 569
DOE (Department of Energy) 13–26, 29, 94
Drainage 610
Drop-shatter test(s) 136
  results 138t
Dryer, steam 509
Drying .... 197'
  application 104
  –devolatization zone 46
  installation, FBC hog fuel 106f
  –pyrolysis process 509
    behavior of nitrogen oxide in
      exhaust gas of 516t
    direct pyrolysis process and
      incineration process, com-
      parison of 517
    performance of 513
    pilot plant 510f
    zone 44
DSS 115
  feed system 114
  feedstocks, experiments with
    MSW and 118–120
Dual fuel engine set-up, multicylinder 657f
Dulong diagram 330f
Dulong's formula 329, 341
Dust 611
  from the direct combustion system,
    chemical composition of 611t

E

Eco-fuel II process 143–149
  description of the Bridgeport 146–149
Eco-fuel II properties of 145t
Economic(s) 161–164, 569
  analysis 72–81, 348, 685–687
  incineration plant capacities
    considered for economic
    analysis 73t
  assessment for bioenergy conver-
    sion processes 324t–325t
  bases 75t
  comparison model 321–328
  considerations 295–299
  of the dry process PUROX system .. 584
  evaluation 48–55
  feasibility of the system, a pre-
    liminary 553

Economic(s) (continued)
  of gasification system, tentative ..267–271
  of operating large modular refuse
    incinerators with heat recovery
    and particulate control 81f
  of operating an 18.2 metric
    tons/day refuse incinerator 79f
  overview of large-scale thermal
    conversion systems using wood
    feedstocks 29–42
  process 358, 538, 613
  system 89
Economical aspects of the introduc-
  tion of gasification technology
  in the rural areas in developing
  countries, social and 705–719
Efficiency of the steam system 89
Effluence and recovery compari-
  son of 506t
EHE 110, 113f, 114, 115
El Salvador 694
Electric power plant 32
Electrical energy, production of
  heat and 455f
Electricity 30–32, 360
  costs 32
  hydro/nuclear 331
  selling price of 34f
  and steam (cogeneration) by wood
    combustion 34f
  steam production and cogenera-
    tion of 360
  by wood combustion, production of 35f
Electrification by biomass conversion
  of local wastes, potential for rural 643
Electrostatic precipitator 65
Eluates of pyrolysis slag from experiments with municipal refuse 459t
Emission data pilot plant 98t
Energy 662t
  balance 462, 580f
  mass and 476t
  of municipal refuse pyrolysis unit 532f
  for the pyrolysis of municipal
    refuse in Kiener System 461f
  in char-oil mixture composite,
    percent available 673f
  conversion, small scale 262–264
  costs 204
  data 346
  Department of (DOE) 3–12, 94
  and materials, recovery of 3–12
  for the MMC, replenishable
    organic 195–205
  needs of the rural populations of
    developing countries, potential
    of biomass conversion in
    meeting 617–633
  plants, refuse-fired 61

Energy (continued)
from producer gas, percentage 668
product/feed ratios 346, 347f
production of heat and electrical 455f
recovery
in boilers 65
from MSW and sewage sludge
using MS-FBC tech-
nology 109–123
rate vs. temperature 549
of resource and 593
from solid wastes, modular
incinerators for 67–82
and volumetric reduction by new
pyrolysis furnace,
integrated system for
solid waste disposal
with 587–601
from wood residues by fluidized-
bed combustion 85
requirements, biomass densification 169–177
requirements for commercial densification apparatus 175t
revenue 322
supply–demand pattern, rural 621
systems, fluid flame 90f
system, rural reference 628f
utilization, solid waste gasification and 291–299
yield, operation with emphasis on .. 454
Engine(s) 386, 706
characteristics of low Btu gas 637–638
diesel 379
dual-fuel engine operation of
single cylinder 665t
dual-fuel engine operation of a
six-cylinder 664t
experimental work with a
single-cylinder 650–655
experimental work with a
six-cylinder 655–663
and gas producer performance,
equations for evaluating .,...... 667
–gas producer systems, multi-
cylinder and single cylinder 666t
1000kVA gen set with dual- diesel 646f
generating set with dual-fuel six
cylinders supercharged 639t
–generator 258
set-up, multicylinder dual-fuel 657f
performance data of Deutz 390t
potential for developing nations to
use biomass fuels in vehicle 689–703
producer gas from agricultural residues as supplementary fuel for
diesel 649–669
producer gas from solid waste fuels
in a diesel 651f

Engine(s) (continued)
technique, power generation from
biomass residues using the
gasifier/dual fuel 635–647
Engineering model of the reduction
zone, reaction 466
Entrained flow, reactors characterized
by 44
Environmental pollution data 610
Environmental Protection Agency
(EPA) 4, 36, 64, 94, 210
Erosion 6, 99
–corrosion, tube 64
Ethane 294, 365
Ethanol 625, 691
fermentation, investment require-
ments for 696
production from biomass 691–697
Ether, triethylene glycol dimethyl 218
Ethyl formate 218
Ethyl hydrosulphide 442
Ethylene 209, 211, 434
and benzene, yield of 435f
Europe 60
European
processes, review of 397–421
PTGL-plants, status of 403
PTGL-technology, survey of 401
Eutectics, low melting 119
Evaporation of domestic sewage
sludge, multisolid fluidized-bed
combustion process with contact 111f
Exhaust gas(es) 515
air pollutants in 610t
combustion 341
of drying-pyrolysis process, behavior of nitrogen oxide in 516t
NO_x and HCl contents of 596
results of analyses of pyrolytic gas and 589
Experimental
procedure 339–341, 509–523
results and discussion 513
results, typical 609t
Experiments, high-flux 246
Experiments, low-flux 243
Explosives 38
Exports as a function of the prices of
raw sucrose, molasses, and petroleum fuels, imports and 695f
Extruded logs 188
Extruded solid waste, samples of
ram and screw 157f
Extrusion
modified mil-pac system, ram 154f
process 169
schematic of 154f
screw 154f
processing, distortion of die wall
during 166f

Extrusion (*continued*)
ram 153
rate vs. pressure 174*f*
of RDF, work of 172*t*
screw 153
temperature vs. pressure 174*f*

F

Farm heating 348
FBC
boiler installation 105*f*
hog fuel drying installation 106*f*
specifications, longview 104*t*
Feed
material 371
physical nature of 355
system, air 383
system, wood 367, 381
Feeder 545
Feedstock(s) 285
characteristics 341
cost(s) 48
for the wood biomass 53
experiments with MSW and DSS 118–120
experiments with MSW and
water 117–118
for the gasification process,
selection of 709
methanol from biomass 22
options 691–694, 697, 699
oxygenated 209
pyrolysis products comparison
using various 222*f*
refuse as a substitute 351–354
stover 339
straw 339
thermochemical conversion of bio-
mass to fuels and 13–26
wood 22
economic overview of large-scale
thermal conversion systems
using 29–42
Fermentation, alcohol 621, 647, 689
Fertilizers 618
synthetic nitrogen 38
Fibers 186
synthetic 38
wet 186
Fiji 694
Film, gasification of waste 235
Fines, content of 134
Fire hazard 151
Firing and co-firing, supplementary
of coal 65
of fuel-oil 65
of solvents 65
of waste oil 65
Firewood 617

Fischer–Tropsch liquids via a carbon
monoxide and hydrogen synthe-
sis gas 209
Fixed-bed
pyrolysis 38
reactors characterized by 44
wood gasifier 47*t*
single stage 45*f*
Flare system 383
Flies 151
Florida 69
Flow
models, cold 469
patterns, double circular 470*f*
process 557
sheet 612*f*
rates, recycle 117
Fluid bed(s)
combustion of solid wastes 93–107
combustors, industrial 94
combustors, pressurized 94
Municipal Solid Waste fired 94
wood waste fired 94
Fluidized-bed 262, 557
combustion
energy recovery from wood
residues by 85
process with contact evaporation
of domestic sewage sludge,
multisolid **111*f***
process with steam generation
in external heat exchanger,
multisolid 112*f*
dense-phase 110
entrained 110
external 117
gasification 46, 354
process, pyrolysis of plastic waste
and scrap tires using 423–439
pyrolysis 425*f*, 427*f*
of plastic waste and scrap tires,
products of 426*t*
reactors 424, 527*f*
gasification of solid waste in
dual 525–540
single 543
system for wood gasification 356*f*
technology 85
Footwear 569
Forest biomass 355
Forestry 317
wastes, third world applications of
pyrolysis of agricultural and 671–688
Forintek fluid-bed gasifier 373*f*
Forintek, laboratory studies at 371
Fossil fuel 146
France 689
Fruits 575

Fuel(s) 267
  aluminum-free 97
  analysis distribution by size,
    peanut hills 198t
  ash 144
  automotive 38
  benefits of the use of densified
    biomass 196
  biomass 317
    gasification at the focus of the
      Odeillo (France) 1 Mw$_{th}$
      solar 237–255
  boiler 38
  carbonaceous 261
  cells 38
  for co-combustion with oil, produc-
    ing a storable transportable 143–149
  consumption, auxiliary 519f
  consumption, diesel 660
  debris 104
  for diesel engines, producer gas
    from agricultural residues as
    supplementary 649–669
  densified biomass 198
  engine operation of single-cylinder
    engine, dual- 665t
  engine operation of six-cylinder
    engine, dual- 664t
  and feedstocks, thermochemical
    conversion of biomass to 13–26
  form, choice of biomass derived 329–331
  fossil 146
  gas(es) 46, 302
    analysis of 354t
    by-product 210
    effect of oxygen enrichment on
      heating value of 359f
    effect of wood moisture on
      heating value of 356f
    from wood, program for the
      production of 355
  gaseous 291
  high moisture 144
  imports and exports as a function
    of the prices of raw sucrose,
    molasses, and petroleum 695f
  mode, diesel 659t
    single 656t
  mode, dual- 656t, 659t
  oil 36
    supplementary firing and
      co-firing of 65
  pellet(s)
    analyses 199f
    cellulose 287
    sale of roemmc 204t
  product end use 299
  production
    alcohol 691–699

Fuel(s) (continued)
  production (continued)
    liquid 36–38
    for wastes
      process for 227–236
      molten salt 228f
    from wood—estimated invest-
      ment, liquid 37t
    from wood—product revenue
      requirements, liquid 39t
  properties 138–139
  refuse-derived (RDF) 4, 63
  saving, diesel 662t
  shredded and air classified solid
    waste 99
  solid, liquid, and gaseous **621**
  source separation system, proposed
    combustible 260f
  specification summary, longview 104t
  stoker fired boiler emissions grate
    burning pelleted 202f
  substitutes, fossil 143
  user experience with densified
    biomass 201
  in vehicle engines, potential for
    developing nations to use
    biomass 689–703
Furnace(s) 6, 509
  application of theoretical data to
    real world 313
  ash-melting 589
  /boiler, water-tube wall 65
  electric 560
  firing, industrial 295
  high efficiency wood 379
  integrated system for solid waste
    disposal with energy recovery
    and volumetric reduction by
    new pyrolysis 587–601
  melting 604
    profile of 606f
  rotary kiln 480
  pyrolysis 589
  schematic of 591f
  shaft 486, 557, 604
  vertical 292
  stoker-fired 201
  wall 63
    temperature in pyrolysis 599f
Funds, Urban Waste Technology 11

G

Garbage 351
Gas(es) 351, 419, 447, 460, 569
  -air mixer 706
  air pollutants in exhaust 610t
  analysis 372
    combustor off- 583t

Gas(es) (continued)
  analysis (continued)
    off- 230
    produced pyrolytic 594t, 595t
  carbonization 444
  cleanup system 258
  combustible 533
    produced from the mixed waste .. 538t
    from pulp sludge 535t
  combustor 285
  components vs. temperature 549f
  composition, product 592t
  converter, drawing of 458f
  converter, two-stage solid waste
    pyrolysis with drum reactor
    and 441–462
  cooler 706
  cost, LBG and MBG 50t
  emission, noxious 94
  exhaust 515
    gas, results of analyses of
      pyrolytic 589
    NO_x and HCl contents of 596
  -fired boiler, oil/ 51
  flue 533
    analysis of incineration 537t
    incineration from pulp
      sludge 534t, 535t
  fuel 46, 302
    effect of oxygen enrichment on
      heating valve of 359f
    effect of wood moisture on
      heating valve of 356f
  generation 358–360t
  generator and location of reaction
    zones, schematic arrangement
    of a pyrolysis-based 288f
  hazardous 160
  hydrogen chloride 428
  intermediate-Btu 32
  lean 401
  low-Btu 15, 262, 618, 637
    engines, some characteristics
      of 637–638
    installation, operation diagram of 636f
  low calorific value 351
  low heating value 227
  medium-BTU 17
  natural 17, 53, 331
  noxious 610
  obtained through the pyrolysis of
    old tires, process 449t
  -phase combustion reaction 94
  power generation, wood gasifica-
    tion for 379–394
  processing methods, comparison of 613t
  producer
    cross-section of down-draft
      suction-type 658f

Gas(es) (continued)
  producer (continued)
    percentage energy from 668
    performance 663
    equations for evaluating
      engine and 667
    system, multicylinder and single
      cylinder engine- 666t
  product 488, 610
  purification 486
    high pressure 223
  pyrolysis 280
    analysis of 578t
    components, produced 547t
    composition, from corn stover 345f
    composition, from straw 344f
    harmful substances in 442–444
  recovery system for classified refuse 605f
  sampling system 383–384
  scrubber with condenser mounted
    on top 646f
  from solid waste fuels in a diesel
    engine, producer 651f
  stream, high temperature 294
  supply, threatened curtailment
    of natural 351
  synthesis 15, 32
  turbines 38, 360
  typical yield and calorific value
    of product 401t
  utilization of recovered 573
  utilization, surplus 644
  velocity and reaction rate 413f
  wood 53, 276
    cleaning system 384
    production costs 48
    reforming of 375t
  yields 343
    vs. temperature 548f
Gaseous substances, harmful 454
Gasifier(s) 46, 230, 386, 706
  appropriate for developing
    countries 707f
  ash removal section of 645f
  batch fed downdraft 264
  bench-scale molten salt 228f
  biomass 287
  cashflows in 100 Tsh/year for one 715t
  coal 22
  co-current moving-bed 465f, 475f, 707f
    description of 463–464
    fueled by biomass, design of ..463–477
  construction of 589
  to a corn mill, input–out analysis
    of the addition of 713t
  design criteria 381
  design of a 100 t/d 474
  downdraft 262
    or updraft 304

Gasifier(s) (continued)
/dual-fuel diesel motor alternator
    system, Duvant 642f
/dual-fuel engine technique, power
    generation from biomass resi-
    dues using 635–647
fluidized-bed 406
Forintek fluid-bed 373f
fueled with source separated solid
    waste, operation of a down-
    draft 257–274
for a 1000 kVA system 645f
laboratory-scale solid waste 265f
in operation, solid waste 266f
options, analysis of the vehicle
    mounted 700t
oxygen blown 17
/reactor/combustor 287
    general arrangement of 289f
social cost–benefit analysis of 712
solid waste 293f
temperature profile 389f
20 TPD test 590f
vehicle-mounted 699–700
vertical shaft 295, 403
wood
    analysis of co-current moving-bed ... 464
    fixed-bed 47t
    single-stage fixed-bed 45f
in year 0, present value of net
    benefits of one 718t
Gasification 14, 261–262, 291, 296,
                          364, 409, 411–414, 486
accomplishment of 292–294
air for 286
apparatus 640
of acid pit sludge 231
of biomass 22
    and combustion systems, devel-
        opment of modular 285–290
    at the focus of the Odeillo
        (France) 1Mw_th solar
        furnace 237–255
    –indirect liquefaction process
        development units 21t
catalytic 370
catalysts, inorganic salts as 370
catalytic 46
cellulose 238
char 237
cycle and mechanical power gen-
    eration overall performances .. 635
downdraft 263f
environment, resource and social
    impact of 386
facility, specification of 640–641
fluidized-bed 46, 412
fundamentals of pyrolysis and 402
future of solid waste 299

Gasification (continued)
graphite 370
hybrid popular 374
–indirect liquefaction systems 15–22
of leather offal 235
of organic waste 637
plant, V.U.B. pyrolysis and 417f
present industrial application of 294–299
process 705
    combustion/ 302
    description 43–46
    selection of a feedstock for 709
    solid waste 487f
    thermal 257
reactions 216, 380
results of hybrid poplar 375t, 376t
of rubber tires 231–235
of solid waste 262
    in accordance with the SFW–
        FUNK–PROCESS 485–489
    ash-melting and waste heat power
        generation integrated system
        for 588f
    in dual fluidized-bed reactors 525–540
    and energy utilization 291–299
    future of 299
    system 297f
    for small communities 259f
suspended particle 412
systems 7
    for small communities, capital
        costs of 271
    tentative economics of 267–271
technology
    low-BTU 15
    in the rural areas in developing
        countries, social and eco-
        nomical aspects of the
        introduction of 705–719
of waste film 235
wood 32–36, 43–55, 235, 379
    /boiler retrofit system, capital
        and operating costs for 51t
    facilities—estimated investment .. 33t
    facilities—product revenue
        requirements 37t
    fluidized-bed system for 356f
    for gas and power generation 379–394
    oxygen blown 331
    plant, schematic of 382f
    program 355
    role of catalysis in 369–378
    status 46–48
    systems, capital costs of 48
    technology 369
    tests, operating characteristics
        during 355t
zone 46
    and oxidation 44

Gasoline 327, 618
  polymerization to 225
  pyrolysis gas purification and
    polymerization to 224f
  from solid wastes by noncatalytic,
    thermal process 209–226
Generation curve income 298f
Generator 285
  refuse-fired steam 60
  system, diesel 384
Geographic distribution of thermo-
  chemical conversion projects 24f
Geometries, tar production at different 469
Germany 61, 689, 690
Ghana convertor design 678
Glass 97, 352, 575, 693
Goals, Canadian PRGL R&D 331–332
Grain drying 348
Grain size 262
Graphite gasification 370
Grate system, mass burning 63
Grate units, basket 67
Grinding 185
Guatamala 694
Gumz model 464
Guyana 694

                        H

HCl removal, scrubbers for 66
Halogens 229
Hamburg 59
Harmful gaseous substances 454
Harmful substances in pyrolysis
  gases 442–444
Hazard, explosion 401
Hazardous substances 581
Health 587
Hearth, multiple 261
Heat
  balance 531
    of mixed waste pyrolysis 537t
  and electrical energy, production
    of 455f
  exchanger, multisolid fluidized-bed
    combustion process with steam
    generation in external 112f
  penetration 473, 474t
  power generation, integrated system
    for solid waste gasification,
    ash-melting and waste 588f
  of reaction 407–409
  recovery
    boiler 511
    estimated investment costs for
      large modular refuse incin-
      eration systems with 79t
    mass and energy balance for an
      incinerator with 73f

Heat (continued)
  recovery (continued)
    and particulate control, econom-
      ics of operating large modu-
      lar refuse incinerators with   81f
    system from MR 495
    source considerations 307–313
    transfer 99, 403, 473
      equations 466
Heating
  direct 406
  indirect 406
  internal 403
  value of fuel gas, effect of oxygen
    enrichment on 359f
  value of fuel gas, effect of wood
    moisture on 356f
Hemacellulose 363, 410
  hydrolysis of 368
Hides 603
Homologs, naphthalene and its 480
Horsepower, brake 655
Horsepower distribution 204t
Hudson Bay plant details 381–385
Human scavengers 153
Human wastes 618
Hydrocarbon(s) 358
  condensed aromatic 484
  destroying pathogens and other 65
  low molecular weight 352
  resins, thermoplastic 484
Hydrochloric acid 443
Hydrofluoric acid 443
Hydrogen 261, 302, 352, 364, 428
  chloride 352, 499
  gas 428
  cyanide 443
  sulphide 352
  synthesis gas, Fisher–Tropsch
    liquids via a carbon monoxide
    and 209
  thiocyanate 442
  weight of 329
Hydrolysis 424
Hydropulping technology 6

                        I

Ignition loss, caustic-soluble carbon
  content and volumetric reduction ratio, comparison of 595t
Ignition loss of residue 512f
Illuminants 303
Imports and exports as a function of
  the prices of raw sucrose, molasses, and petroleum fuels 695f
Incineration 493
  flue gas 534t
    analysis of 537t
    from pulp sludge 535t

Incineration (*continued*)
option, flow diagram of 500*f*
plant capacities for economic
analysis 73*t*
process, comparison of drying-
pyrolysis process, direct
pyrolysis process and 517
system(s) 498, 499
costs of small modular refuse 77*t*
estimated investment for 18.2-
metric ton modular 76*t*
with heat recovery, estimated investment costs for large
modular refuse 79*t*
for treatment of MR, application
of PTGL processes as an
alternative to 495
Incinerator(s) 6, 60, 65, 485
controlled-air 67
economics of operating an 18.2
metric tons/day refuse 79*f*
for energy recovery from solid
wastes, modular 67–82
extent of use of modular 68
with heat recovery, mass and
energy balance for 73*f*
with heat recovery and particulate
control, economics of operat-
ing large modular refuse 81*f*
18.2-metric-ton 72
plants 397
surface-rich fume 296
starved-air 67
systems 227
cost of large modular refuse 80*t*
India 618, 619, 621, 627, 694
Indonesia(n) 618, 619
preliminary test data from 685*t*
technology development pyrolytic
convertor system 680*f*
Inerts 97
Infectious material 229
Installation, large modular 77
Investment
ammonia case 38
capital 61–62
costs for large modular refuse
incineration systems with
heat recovery........................ 79*t*
estimated
chemicals production from wood 39*t*
costs for 18.2-metric ton modular incineration system 76*t*
liquid fuels production from
wood 37*t*
wood combustion facilities 31*t*
wood gasification facilities 33*t*
methanol case 38
plant facilities 30

Investment (*continued*)
requirements 699
for ethanol fermentation 696
for methanol production from
wood 699
Iron 610
Italy 689

J

Jamaica 694
Jamming phenomena 164
Japan 59, 65, 525
characteristics of municipal refuse
and sewage sludge (SSL) in 495–498
overview of PTGL processes in 493–507
pyrolysis developing in 494*t*
Japanese refuse 577*f*
PUROX system demonstration test
on simulated 573–586
Japanese and U.S. refuse, compari-
son of 573*t*
Jet, penetration depth of 472*f*
John Deere/Papakube Cuber,
schematic of 157*f*

K

Kiener System
energy balance for the pyrolysis
of municipal refuse in 461*f*
material balance for the pyrolysis
of municipal refuse in 461*f*
two-stage refuse plant 445*f*
Kiener waste pyrolysis system 444–462
Kerosene 618, 621
Kiln
pyrolysis 563
rotary 67, 262, 424, 557
furnace 480
indirectly heated 434
pyrolysis of agricultural
residues in 337–350
pyrolysis process 338*f*
screw- 424
Kinetics and mechanism 409–415
Kunii–Kunugi process 525

L

Labor charges 77
Laboratory-scale experiments 424–428
Laboratory studies at Forintek 371
Laguna 655
Landfill(s) 9, 60, 227, 295, 296
life 161
sanitary 257
concept 61
Landfilling, direct 493
Leachate 160

Leaching of heavy metals from ash 493
Leather 603
  offal, gasification of 235
  scraps 227
Leaves 185
Levoglucosan 218
  tars 216
Lignin 363
Lignite 301
Lignocellulosic materials 240, 692, 697
Lime 499
Limestone 119, 583, 604
Liquefaction
  catalytic 36
  direct 14
    systems 22–25
  of wood using nickel catalysts 363–368
  processes 257
    development units, biomass
      gasification–indirect 21t
  studies, thermal 364
  systems (midterm), gasification–
    indirect 15–22, 20f
  technology, medium BTU gasi-
    fication (MBG) 18f
Liquid fuels 618
  production 36–38
    from wood—estimated invest-
      ment 37t
    from wood—product revenue
      requirements 39t
Log chips, whole 86
Log yard wastes 86
Lubrication, die 135

M

Magnesium 119
  chips 63
  oxide 371, 428
Magnet 148
Manure 180, 371
Market value, analysis at 712
Mass and energy balances 476t
Mass transfer equations 466
Material
  balance 460, 531
    for the pyrolysis of municipal
      refuse in Kiener System 461f
  composition of input 546t
  considerations 294–295
  recovery
    operation with emphasis on 444
    and pyrolysis option, flow
      diagram of 503f
    and pyrolysis system 498, 501
Mechanical rapping 64
Melting
  development of solid waste disposal
    system with pyrolysis and . 603–614

Melting (continued)
  furnace 604
    profile of 606f
  system, pyrolysis- 498, 501
  vessels 424
Metal(s) 352, 454, 603, 621
  concentration and solubility
    test result 594t
  disposition in slag, trace 584t
  heavy 513
    in ash 535t
    pyrolysis 550
  molten 403
  and other constituents
    in MSW, typical concentrations
      of 116t
    in scale, typical concentrations
      of 116t
    in sewage sludge, typical con-
      centrations of 116t
Methanation reactors 17
Methane 7, 211, 294, 307, 364,
                         365, 380, 428, 618, 621
  additions 331
  injection 468–469
  tracer 470f
  and wood hybrid 332
Methanol 36, 38, 43, 332, 691, 697
  from biomass feedstocks 22
  case investment 38
  catalytic chemical conversion to 209
  production 361, 361t
    from biomass 697–699
    from wood ,investment
      requirements for 699
  synthesis 647
    gas to 358
Methyl acetate 218
Methyl ethyl ketone 218
Methylnaphthalene 432
Military bases, potential use of
  modular incinerators 69
Mill, hammer-type 529
Mines, Bureau of 291
Model
  calculations, thermodynamic 466t
  Gumz 464
  of the reduction zone, reaction
    engineering 166
  Schläpfer 464
Modular incinerators on military
  bases, potential use of 69
Moisture 97, 134, 198
  content 167, 355
  of the refuse 155
Molasses and petroleum fuels, relative
  value of raw sucrose and 694–696
Motor/alternator system, Duvant
  gasifier/dual-fuel diesel 642f

Moving bed gasifier,
   co-current 465*f*, 475*f*, 707*f*
   analysis of 464
   description of 463–464
Mozambique 694
Municipal refuse (MR) 442, 529
   analysis 530*t*
   composition of 497*t*, 589, 592*t*
   elimination of aluminum mixed in .. 550
   eluates of pyrolysis slag from ex-
     periments with 459*t*
   general characteristics of 496*t*
   handling options, operating costs
     for 505*f*
   in Kiener System, energy and mate-
     rial balance for the pyrolysis
     of 461*f*
   pulverized 530*t*
   pyrolysis 531*t*
     unit, energy balance of 532*f*
   raw material and the reaction prod-
     ucts in experiments with .456*t*–457*t*–
   and sewage sludge (SSL) in Japan,
     characteristics of 495–498
   by the two-bed pyrolysis system,
     disposal of 541–555
Municipal solid waste (*see* MSW)
Municipal waste, comparison of alternative facilities for disposal of 598
Muskeg 317
MS–FBC
   combustor 113*f*
   operating conditions 121*f*
   technology, background on 109–110
   technology, energy recovery from
     MSW and sewage sludge
     using 109–128
MSW (municipal solid
     waste) 127, 145, 221, 317
   analysis, typical 115*t*
   and DDS feedstocks, experiments
     with 118–120
   feed system 114
   pelletized 115
   processing facility materials flow
     diagram 100*f*
   properties of 145*t*
   and sewage sludge using MS–FBC
     technology, energy recovery
     from 109–123
   typical concentration of metals and
     other constituents in 116*t*
   and water feedstocks, experi-
     ments with 117–118

N

Naphtha 221
   cracking 408*f*

Naphthalene 424, 432
   and its homologs 480
Nations to use biomass fuels in
   vehicle engines, potential for
   developing 689–703
Nature, fibrous or nonfibrous 179
New York 59
   markets, average sucrose prices
     on London and 696*t*
Nicaragua 694
Nickel
   carbonate 367
   catalysts, direct liquefaction of
     wood using 363–368
   Raney 364, 367
Nitrogen 280, 294, 329
   compounds 443
   content 15
   fertilizers, synthetic 38
   oxide(s) 229
   in exhaust gas of drying-pyrolysis
     process, behavior of 516*t*
NOₓ prevention at afterburning by
   the multiple combustion method 516*t*
Nuclear fission (breeder) 332

O

Oat straw 341
Odors 160
Oil 327, 369, 401, 447, 486, 557, 678
   anthracene start up 25
   char
     mixture composite, percent
       available energy in 673*f*
   and char by pyrolysis of wood,
     production of 35*f*
   diesel 618, 637
   distillate 53
   drums 678
   fuel 36
   /gas fired boiler 51
   heavy 317
   highly aromatic 484
   properties of 564*t*
   producing a storable, transportable
     fuel for co-combustion with 143–149
   pyrolysis 432, 480
     condenser 677*f*
     processing 483*f*
     refinery 483*f*
     from special wastes 482*f*
   residual 53
   sands 317, 363
     Athabasca 317
   selling price of pyrolytic 40*f*
   shale 221
   yields as function of air/feed,
     char and 684*f*

Olefin(s) 223
 mixture, polymerization of 210
 selective pyrolysis to 210–223
Organic(s)
 compounds, trace 301
 industrial wastes, special 444
 liquid 351
  disposal of 46
 oxygenated 223
Oxidation zone 706
 gasification and 44
 and pyrolysis zone, analysis of 468
Oxidizing agent 292
 steam as an 292
Oxygen 302, 307, 363, 364
 blowing 313
 -blown wood gasification 331
 content 290
 enrichment experiments 358
 enrichment on heating value of
  fuel gas, effect of 359f
 equilibria, carbon–steam– 312f
 weight of 329

P

Pallets, transport 295
Panama 694
Paper 371
 packaging 295
Paraffin waxes 430
Paris 59
Particulate(s)
 analysis of 583t
 control, economics of operating
  large modular refuse incinerators with heat recovery and ... 81f
 fine 360
Pathogens and other hydrocarbons,
 destroying 65
PCB addition on slag composition,
 effect of 582t
Peanut hulls, fuel analysis distribution by size 198t
Peat 301, 317
 conversion 337
 pyrolysis, comparison 348
  of straw, stover, and 349t
 pyrolysis system 337
Pebble-bed process air introduction
 technique 678
Pellet(s) 188
 cohesion 135
 densified 153
 densities 134, 135
 length 134
  distribution 137f
 quality 201
Pelleted byproduct 187

Pelletizer, animal feed 132
Peru 168, 619
Petroleum fuels, relative value of raw
 sucrose, molasses, and 694–696
Phenol(s) 218, 357
 formation 366
Philippines 649
Pilot plant 509–512, 544f
 experiments 428–439
Pipe, galvanized 678
Plant
 actual 560
 description 541
 design, basis for 499
 facilities investment 30
 life 321
 operating requirements 72, 74t, 598
Plastic(s) 38, 442, 444, 480, 575, 603
 film 63
 pyrolysis reactor with a feed
  screw for 431f
 waste and scrap tires, pyrolysis of .. 429f
  products of fluidized-bed 426t
Plumes, particulate stack 71
Plywood trim 86
Pollutants, atmospheric 196
Pollutants in exhaust gas, air 610t
Pollution
 air and water 493
 control 161
  air 65–66
  devices 67
 data, environmental 610
 regulations, air 397
Polyethylene 402, 419, 432
 pyrolysis of 432, 433t
Polymers, thermal decomposition of .. 409
Polypropylene 432
Polystyrene 419
 pyrolysis 418f, 419
Poplar hemicellulose 366
Poplars, hybrid 363
 Canadian 370
 gasification 374
 results of 375t, 376t
Potassium carbonate 370, 374
Pottery 621
Powder, high density 145
Power
 comparison of generating and
  required 596t
 consumption 134
 generation
  from biomass residues using the
  gasifier/dual fuel engine
  technique 635–647
  integrated system for solid waste
  gasification, ash-melting and
  waste heat 588f

Power (*continued*)
    Northern Saskatchewan using
        diesel for 391*t*
    overall performances, gasification
        cycle and mechanical 635
    wood gasification for gas and 379–394
    requirements for densification 191
Precipitator, electrostatic 65, 583
Pressure 189–191
    application 175
    to bale straw or fiber 190*f*
    tap locations, thermocouple and ... 113*f*
Price
    of char, selling 40*f*
    of electricity, selling 34*f*
    of pyrolytic oil, selling 40*f*
    of steam, selling 34*f*
Process
    analysis 517–520
    development units, biomass gasifi-
        cation–indirect liquefaction 21*t*
    equipment capacities 200
    material streams, three-way split ... 198*t*
    options in resource recovery,
        technical 4
Processes, pyrolysis 406
Product(s)
    cogenerated 30–32
    feed ratios, mass 342*f*
    ligno-cellulosic 640
    sampling of raw material and 489
    uses 346–348
    yield and composition 341–345
Programming, linear 19
Propane 341, 365
Propene 434
Propylene 209
Prototype unit, description of 352
PTGL processes
    as an alternative to incineration for
        treatment of MR, application of 495
    evaluation of 498–507
    to industrial wastes, application of . 498
    in Japan, overview 493–507
PTGL-project of the University of
    Brussels 416
Public Health Service 61
Pulp sludge, combustible gas from 535*t*
Pump, peristaltic 230
PUROX system
    demonstration test on simulated
        Japanese refuse 573–586
    economics of the dry process 584
    running costs of dry process 585*t*
    schematic flow of a typical 574*f*
    schematic flowsheet of the dry
        process 576*f*
Pyrolysis 14, 237, 275–283, 292,
        303–306, 407, 409–411, 621, 640

Pyrolysis (*continued*)
    and activation in the production
        of waste- or biomass-derived
        activated carbon, thermody-
        namics of 301–314
    and activation reactions 303*t*
    of agricultural and forestry wastes,
        third world applications
        of 671–688
    of agricultural residues in a rotary
        kiln 337–350
    ash, heavy metals in 550
    -based gas generator, and location
        of reaction zones, schematic
        arrangement of 288*f*
    basic principles of waste 397–421
    of cellulose 216, 244, 247, 410
    flash 240
        prior research on 251*t*
    characteristic times for solar flash .. 248*t*
    comparison with peat 348
    comparison of straw, stover,
        and peat 349*t*
    convertor, appropriate technology .. 675*f*
    convertor system, Indonesian
        technology development 680*f*
    data, experimental 213*t*
    of differing special wastes 480
    with drum reactor and gas con-
        verter, two-stage solid
        waste 441–462
    of dry biomass 618
    energy balance of municipal refuse 532*f*
    equilibria, cellulose 306*f*
    equipment 641
    fixed-bed 38
    fluidized-bed 425*f*
    furnace 589
        integrated system for solid waste
        disposal with energy recov-
        ery and volumetric reduc-
        tion by new 587–601
        schematic of 591*f*
        wall temperature in 599*f*
    gas(es) 280
        analysis of 578*t*
        chromatography 410, 416
        components 547*t*
        composition from corn stover 345*f*
    composition from straw 344*f*
        harmful substances in 442–444
        purification and polymerization
        to gasoline 224*f*
    and gasification, fundamentals of .. 402
    and gasification plant, V.U.B. 417*f*
    heat balance of mixed waste 537*t*
    kiln 563
    of agricultural residues in a
        rotary 337–350

Pyrolysis (*continued*)
mechanism 238
and melting, development of
solid waste disposal system
with 603–614
-melting option, flow diagram of 502*f*
-melting system 498, 501
of mixture of sludge and plastic
waste 536
municipal refuse 531*t*
in Kiener System, energy and
material balance for 461*f*
oil(s) 36, 480
condenser 677*f*
processing 483*f*
refinery 483*f*
from special wastes 482*f*
of old tires
boiling behavior of the condensate from 448*f*
condensate obtained
through 450*t*–451*t*
process gas obtained through 449*t*
solid residues obtained
through 452*t*–453*t*
viscosity of the condensate from 448*f*
to olefins, selective 210–223
option, flow diagram of material
recovery and 503*f*
physical phenomena occurring
during 411
pilot plant study on sewage
sludge 509–523
plant
flow sheet of 558*f*
layout of 562*f*
rotary drum 481*f*
plastic waste and scrap tires 429*f*
products of fluidized-bed 426*t*
using a fluidized-bed process 423–439
of polyethylene 432, 433*t*
polystyrene 418*f*, 419
process(es) 257, 404, 406
behavior of nitrogen oxide in
exhaust gas of drying- 516*t*
developing in Japan 494*t*
drying- 509
direct and incineration proc-
ess, comparison of 517
performance of 513
pilot plant 510*f*
partial oxidation 671
rotary kiln 338*f*
for scrap tires 557–572
test plants for 560*t*
products
comparison using various
feedstocks 222*f*
formation of primary 219*f*

Pyrolysis (*continued*)
products (*continued*)
formation of secondary 220*f*
nitrogenous 280
prototype for whole-tire 434–439
of pulp and paper sludge 535
rate of 93
reaction overview, high tempera-
ture 217*f*, 219*f*
reactor(s) 430
the biomass flash 241
with a feed screw for plastic 431*f*
recovery of raw materials by 481*f*
schematic 212*f*
slag from experiments with munici-
pal refuse, eluates of 459*t*
of the solid waste 291
of special wastes 482*f*
of straw and stover, char
analysis for 343*t*
system(s) 7
detailed schematic of 340*f*
Kiener waste 444–462
material recovery and 498, 501
peat 337
two-bed
disposal of municipal refuse
by 541–555
flow diagram of 542*f*
pilot plant of 545
temperature 558*f*, 566*f*–567*f*, 568*f*
thermochemical formation of gases
of oxygenated liquids by 209
typical char yield from biomass by 305*f*
unit 279*f*
basic principles of 399–421
of differing special 480
with drum reactor and gas con-
verter, two-stage solid 441–462
heat balance of mixed 537*t*
water, treatment of 550
of wood, production of oil and
char by 35*f*
zone 44, 706
analysis of the oxidation and 468
Pyrolytic
gas and exhaust gas, results of
analyses of 589
oil, selling price of 40*f*
recovery of raw materials from
special wastes 479–484
Pyrolyzer 277*f*

Q

Quenching 563
system, ash chute 66

R

Radiation, solar 240
Raney nickel 364
Raw material(s) 203t
  heterogeneous biomass 198
  preparation 486
  and product, sampling of 489
  by pyrolysis, recovery of 481f
  and the reaction products in ex-
    periments with municipal
    refuse 456t–457t
  recovery of 466f
  from special wastes, pyrolytic
    recovery of 479–484
d-RDF
  process flow 127–134
  storage 140–141
    effects 140t
  physical properties of 134–138
Reaction
  exothermic 291
  gasification 380
  of heat 407–409
  low temperature endothermic 291
  products in experiments with
    municipal refuse raw mate-
    rial and 456t–457t
  pyrolyzer 291
  watershift 464
  zones 44
Reactor(s)
  assembly 372
  biomass flash pyrolysis 241
  characterized
    by entrained flow 44
    by fixed-bed 44
    by fluidized-bed 44
    by stirred moving-bed 44
  combustor, gasifier 287
    general arrangement of 289f
  continuous 480
  cracking 526
  engineering aspects, chemical ... 414–415
  fluidized-bed 374, 403,415, 527f, 543
    bench-scale 416
  and gas converter, two-stage solid
    waste pyrolysis with drum 441–462
  gasification 258
  of solid waste in dual fluidized-
    bed 525–540
  prototype 438f
    for whole-tire pyrolysis 437f
  pyrolysis 430
    with a feed screw for plastic ... 431f
  quartz 242f
  spout-fluid bed 402
  stirred-tank 402
  types 261–262
  vertical shaft 415

Recovery
  comparison of effluence and 506t
  of energy and materials 3–12
  operation with emphasis on
    material 444
  resource
    nontechnical issues in 9
    process system developed by
      national project 498
    technologies, developmental
      studies of 8t
    technical process options in 4
Recycling 257
Reduction zone 706
  reaction engineering model of 466
Refinery, pyrolysis oil 483f
Refining, char 563
Refractory damage 71
Refuse
  burning rate of 93
  co-disposal of scrap tire and 593
  comparison of Japanese and U.S. ... 573t
  composition of 60, 573
  direct combustion system for
    ordinary 605f
  derived fuel (RDF) 4, 63, 144, 181
    low-grade 146
    preparation and proprieties of
      densified 127–142
    processing plants 148
    production tests, summary of 354t
    properties of 139t
    work of extrusion of 172t
  during storage tests, daily tem-
    peratures for cubed and
    shredded 159f
  -fired energy plants 61
  -fired steam generator 60
  gas recovery system for classified .. 605f
  generation per person 60
  handling 62–63, 286
  high moisture 579f
  incineration system, costs of small
    modular 77t
  incineration systems with heat
    recovery, estimated invest-
    ment costs for large modular .. 79t
  incinerator(s)
    economics of operating an 18.2
      metric tons/day 79f
    with heat recovery and particu-
      late control, economics of
      operating large modular 81f
    systems, cost of large modular 80t
  Japanese 577f
  PUROX system demonstration
    test on simulated 573–586
  from Kitakyushu City, properties
    of urban 607t

Refuse (*continued*)
mixing ratio, typical 578*t*
moisture content of 155
municipal 68, 371
    solid fuel (RDF) from mixed 351
plant, Kiener System two-stage 445*f*
properties 607, 608
as a substitute feedstock 351–354
treated at the Tokyo plant, prop-
    erties of classified 608*t*
Regulated utility 30
Residue(s)
agricultural 691, 697
    conversion 46
    crop 618
    as supplementary fuel for diesel
        engines, producer gas
        from 649–669
characteristics of 180–181
combustion 181–183
composition of 189
conversion, wood 46
crop 180
    dried 617
    forest 180
    ignition loss of 512*f*
    logging 181
    pelleted agricultural 187
    sawmill 181, 192
    solid 447, 460
    obtained through the pyrolysis of
        old tires 542*t*–453*t*
Resource
assessment chart, biomass tech-
    nology and 319*f*
Conservation and Recovery Act
    (RCRA) 66
and energy, recovery of 593
recovery 257
    pilot plant 353*f*
Revenue requirements, product
chemicals production from wood— 41*t*
liquid fuels production from
    wood— 39*t*
wood combustion facilities 33*t*
wood gasification facilities 37
Rice hulls 654
Rice husks 678, 679, 681
Rodents 151
Rubber 227, 432, 603
compounds 566*f*, 567*f*
sheets 569
tires, gasification of 231–235
Rugs 444
Rural energy supply–demand pattern 621
Rural populations of developing
    countries, potential of biomass
    conversion in meeting the
    energy needs of 617–633

S

Safety 587
shoes, actual use tests of 570*t*
Salt(s) 403
as gasification catalysts, inorganic .. 370
molten 402
    fuel production from wastes
        using 227–236
    gasifier, bench-scale 228*f*
    process for fuel production
        for wastes 228*f*
    test facility 230–231
        Santa Susana 233*f*
sodium halide 229
test facility, schematic diagram of .. 232*f*
Sampling system, gas 383–384
Sand(s)
oil 317, 363
disengager 110, 114
recycle rate 117
Sander dust 86
Sanitary landfill concept 61
Sawdust 185, 278, 355
analysis of pine 171*t*
and pellets, heats of combustion
    of pine 171*t*
Scale, typical concentration of metals
    and other constituents in 116*t*
Scandinavia 59
Schläpfer model 464
Screening 129
Scrubber(s) 65, 339
with condenser mounted on top,
    gas 646*f*
for HCl removal 66
venturi 88
water 460
Scrubbing 485
SFW–FUNK–PROCESS, gasification
    of solid waste in accordance
    with 485–489
Silage 575
Silica sand 110
Slag(s) 403, 610, 611
composition 581*t*
    effect of PCB addition on 582*t*
    formation 71
    leaching test results 582*t*
    properties of pyrolytic ash and ... 593
    removal 71
        and soot 447
    trace metal disposition in 584*t*
Sludge 371
acid 480
and plastic waste, pyrolysis of
    mixture of 536
pulp 537*t*
    incineration flue gas from 535*t*

Sludge (*continued*)
    pyrolysis of pulp and paper 535
    sodium-based pulp mill secondary .. 103
Slugs 153
Separation
    of the product, cryogenic 404
    system, conventional source 260*f*
    system, proposed combustible fuel
        source 260*f*
    sludge (SSL) 59
        general characteristics of 496*t*
        in Japan, characteristics of
            municipal refuse and 495–498
        multisolid fluidized-bed combus-
            tion process with contact
            evaporation of domestic 111*f*
        pyrolysis, pilot plant study on 509–523
        typical concentration of metals
            and other constituents in 116*t*
        using MS–FBC technology,
            energy recovery from
            MSW and 109–123
Shavings 278, 355
Shells, coconut 654
Shoes, actual use tests of safety 570*t*
Shredded refuse 153
Shredded solids 143
Shredder 258, 354
    scrap tire 562*f*
    waste 480
Shredding 4, 352, 560
    and densification system, solid
        waste 264
    process 144
    secondary 129–132
SNG 32
Social and economical aspects of the
    introduction of gasification tech-
    nology in the rural areas in
    developing countries 705–719
Social group, analysis on a national
    level and per 714
Socio–economic present value of
    the project 717*f*
Sodium 119
    -based pulp mill secondary sludge .. 103
    carbonate 370
    melt 229
Solar flash pyrolysis, characteristic
    times for 248*t*
Solar radiation 240
Solid fuel (RDF) from mixed
    municipal refuse 351
Solid waste 292
    commercial 68
    costs for the source separation of 271
    cubes with wood chips, comparison
        of 267

Solid waste (*continued*)
    densification of 258–261
    disposal with energy recovery and
        volumetric reduction by new
        pyrolysis furnace, integrated
        system for 587–601
    disposal system with pyrolysis and
        melting, development of ...603–614
    fluid-bed combustion of 93–108
    fuel, shredded and air classified 99
    fuels in a diesel engine, producer
        gas from 651*f*
    gasifier 293*f*
        laboratory-scale 265*f*
        in operation 266*f*
    gasification 262
        in accordance with the SFW–
            FUNK–PROCESS 485–489
        ash-melting and waste heat power
            generation, integrated
            system for 588*f*
        in dual fluidized-bed reactors 525–540
        and energy utilization 291–299
        process 487*f*
        gasification system 297*f*
        for small communities 259*f*
    modular incinerators for energy
        recovery from 67–82
    municipal 67, 69, 181, 209, 304
        experimental results from pyro-
            lyzing material derived from   211
        fired fluid beds 94
        programs 95–102
    by a noncatalytic, thermal process,
        gasoline from 209–226
    operation of a downdraft gasifier
        fueled with source
        separated 257–274
    problem 257
    pyrolysis with drum reactor and gas
        converter, two-stage 441–462
    shredding and densification system   264
    source separation of 258
    urban 3–12
Solubility test result, metal concen-
    tration and 594*t*
Solvents, supplementary firing and
    co-firing of 65
Soot
    blowers, high-pressure steam 64
    blowing, compressed air 64
    corrosion and buildup of 71
    formation 246
    removal, slag and 447
SSWEP 99
    parametric test results 101*t*, 102*t*
Starch content 179
Steam 30–32, 360, 499

Steam (continued)
  carbon–
    –CH₄/stoichiometric air
      equilibria 311f
    –air equilibria 308f, 309f, 310f
    equilibria 312f
    –oxygen equilibria 312f
  costs 32
  dryer 509
  generation 109
    in external heat exchanger, multi-
      solid fluidized-bed combus-
      tion process with 112f
  generator(s) 196
    pulverized coal-fired 203
    refuse fired 60
  as an oxidizing agent 292
  –producing facilities 32
  production
    and cogeneration of electricity 360
    typical system for 86
    of process 295
  selling price of 34f
  soot blowers, high-pressure 64
  system, efficiency of 89
  value on the operating costs,
    effect of 77
  (cogeneration) by wood combustion,
    production of electricity and .. 34f
Steel 569
Stirred moving-bed, reactors
  characterized by 44
Stover, char analysis for pyrolysis
  of straw and 343t
Stover and peat pyrolysis, comparison
  of straw and 349t
Straw(s) 185, 343
  pyrolysis, gas composition from 344f
  and stover, char analysis for
    pyrolysis of 343t
  stover and peat pyrolysis, com-
    parison of 349t
Strawdust 185
Styrene 424
Sucrose
  molasses, and petroleum fuels,
    imports and exports as a func-
    tion of the prices of raw 695f
  molasses, and petroleum fuels,
    relative value of raw 694–696
  prices in London and New York
    markets, average 696t
  production levels, lesser-developed
    countries with 693t
  selected major exporters of 693t
  1977 world production and con-
    sumption of 693t
Sudan 618, 619, 621
Sulfates 119

Sulfite mill liquor 301
Sulfur 64, 229, 329
  compounds, organic .... 442 '
  dioxide emission 196
  oxides 499
Supercharging 638
Supply–demand integration 625
Suspension burner, cyclonic 201
Suspension fired systems 63
Swaziland 694
Sweden 689, 690
Switzerland 63
Syncrudes 317
Synthesis gas 32
  to methanol 358
  production 333, 357
Synthetic fibers 38

                        T

Tanzania 618, 619, 621
  applicability in the rural areas of 709
  for the owner, benefits of corn
    milling in 711
Tar(s) 46, 468, 486, 604
  aromatic 223
  condensed 287
  production at different geometries .. 469
  yields 343
Technical processes 5f
Technologies, bridge 332
Technology overview 621
Temperature(s)
  ash fusion 139
    average 139t
  bed 355
  constraints 117
  conversion 467f
  for cubed and shredded refuse
    during storage tests, daily 159f
  energy recovery rate vs. 549f
  gas components vs. 549f
  gas yields vs. 548f
  profile, gasifier 389f
  profile, typical 118
  pyrolysis 558f, 566f, 567f, 568f
  in pyrolysis furnace, wall 599f
  vs. elapse time, reduction zone 269f
Thailand 618, 619
Thermal
  conversion systems using wood
    feedstocks, economic overview
    of large-scale 29–42
  decomposition 479
  efficiency
    cold gas 668
    brake 660, 667
    maximum 302

Thermal (*continued*)
  gasification and synthesis
    (methanol) 689
  gasoline from solid wastes by a
    noncatalytic 209–226
Thermochemical conversion
  of biomass to fuels and feedstocks ..13–26
  catalysis in 22
  process 14
  projected fiscal year 1979 budget— 24*f*
  projects, biomass 25
  projects, geographic distribution of 24*f*
Thermochemical formation of gases
  or oxygenated liquids by pyrolysis 209
Thermocouple and pressure tap
  locations 113*f*
Thermodynamic
  fundamentals 302–303
  model(s) 464
    calculations 466*t*
  of pyrolysis and activation in the
    production of waste- or
    biomass-derived activated
    carbon 301–314
Thermogravimetric analysis 410, 416
Thermoplastic hydrocarbon resins 484
Third world applications of pyrolysis
  of agricultural and forestry
    wastes 671–688
Tires 444, 480, 569
  gasification of rubber 231–235
  granulated rubber 427*f*
  pyrolysis of old
    boiling behavior of the
      condensate from 448*f*
    condensate obtained through 450*t*–451*t*
    process gas obtained through 449*t*
    solid residues obtained
      through 452*t*–453*t*
    viscosity of the condensate from.. 448*f*
  pyrolysis, prototype reactor for
    whole 437*f*
  removal or reuse of old 423*t*
  scrap
    disposal cost for 571*f*
    and refuse, co-disposal of 593
    products of fluidized-bed
      pyrolysis of plastic waste
      and 426*t*
    pyrolysis of plastic waste and 429*f*
    pyrolysis process for 557–572
    shredder 562*f*
  shredded 402
  using a fluidized-bed process,
    pyrolysis of plastic waste
    and 423–439
Tobacco containers, used 295
Tobago 694
Tobata pilot plant 607

Tokyo test plant 608
Toluene 424, 480, 484
Trash, industrial plant 68
Trinidad 694
Turbines, gas 38, 360

U

University of Brussels, PTGL-project
  of the 416
Urban Waste and Municipal Systems
  Branch 3
U.S. refuse, comparison of Japanese
  and 573*t*

V

Vegetable(s) 575
  fiber "ijuk" 676
Viruses 153
Viscosity of the condensate from the
  pyrolysis of old tires 448*f*
Volatiles, combustible 216
Volatilization of cellulose 238
Vulcanizates, rubber 566*f*–567*f*
Vulcanization, properties of carbon
  black and rubber 565*t*

W

Waste
  availability of agricultural 708
  or biomass-derived activated car-
    bon, thermodynamics of py-
    rolysis and activation in the
    production of 301–314
  cellulose 637
  collection 162
    comparison of costs for con-
      ventional and densified 163*f*
  combustible gas produced from
    the mixed 538*t*
  combustion of liquid 143
  control problem 9
  conversion, other method of
    effective 644
  disposal
    company, acquisition of 351
    Facility, Municipal 143
    municipal solid 46
  -to-energy systems, worldwide
    inventory of 59
  film, gasification of 235
  gas shift reaction 294
  gasification of organic 637
  high-sulfur oil refinery 227
  hints for recycling 643
  molten salt process for fuel
    production for 228*f*

Waste (*continued*)
  oil, supplementary firing and
    co-firing of 65
  plastic 537t
    and scrap tires, pyrolysis of 429f
    products of fluidized-bed 426t
    using a fluidized-bed
      process 423–439
  potential for rural electrification
    by biomass conversion of local 643
  process for fuel production from 227
  product application 299
  programs, wood 102–103
  pyrolysis
    basic principles of 397
    heat balance of mixed 537t
    of mixture of sludge and plastic .. 536
    of special 482f
     differing 480
     oils from 482f
    system, Kiener 444–462
  pyrolytic recovery of raw mate-
    rials from special 479–484
  shredder 480
  slugs 155
  solid
    collection costs 161
    management unit operations,
     cost of 152f
    mass burning of municipal 59–66
    municipal 172
    pyrolysis of 291
    small scale source densification
     of Navy 151–168
  third world applications of pyrolysis
    of agricultural and forestry ..671–688
  urban 644
    technology funds 11
  using molten salts, fuel produc-
    tion from 227–236
  water 499, 529
    raw and treated water quality
     of pyrolysis 551t
    treatment of pyrolysis 550
  wood 641
Water 575
  absorption 160
  drinking 301
  feedstocks, experiments with
    MSW and 117–118
  gas shift 17, 303
    shift reaction 370
  purification 38
  raw and treated water quality of
    pyrolysis waste 551t
  scrubber 460
  treatment 493, 575
    of pyrolysis waste 550
  vapor 216

Watershift reaction 464
Wax(es) 185, 419
  paraffin 430
West Germany 59, 65
Wood 227, 301, 371, 444, 697
  ammonia from 38
  availability 698–699
  biomass feedstock, costs for 53
  char 187
  chips 46, 355, 691
    comparison of solid waste cubes
     with 267
    typical analysis of spruce 387t
  combustion 30
    facilities—estimated investment.. 31t
    facilities—product revenue
     requirements 33t
    production of electricity by 35f
  content 102
  devolatilization of 46
  —estimated investment, chemicals
    production from 39t
  —estimated investment, liquid
    fuels production from 37t
  feed system 367, 381
  feedstock(s) 22
    cogeneration based on 15
    economic overview of large-scale
     thermal conversion sys-
     tems using 29–42
  fines, dried 103
  furnaces, high efficiency 379
  gas
    production costs 48, 53, 276
    cleaning system 384
    reforming of 357t
  gasification 32–36, 43–55, 235
    boiler retrofit system, capital
     and operating costs for 51t
    facilities—estimated investment.. 33t
    facilities—product revenue
     requirements 37t
    fluidized-bed system for 356f
    for gas and power generation 379–394
    plant, schematic of 382f
    program 355
    role of catalysis in 369–378
    systems, capital costs of 48
    tests, operating characteristics
     during 355t
  gasifier
    analysis of co-current
     moving-bed 464
    fixed-bed 47t
     single stage 45f
  hogged 278
  hybrid, methane and 332
  investment requirements for metha-
    nol production from 699

Wood (*continued*)
  moisture on heating value of fuel
    gas, effect of 356*f*
  pellets 287
  —product revenue requirements,
    chemicals production from 41*t*
  —product revenue requirements,
    liquid fuels production from .. 39*t*
  products, worldwide production of 698*t*
  production of oil and char by
    pyrolysis of 35*f*
  program for the production of
    fuel gas from 355
  residues by fluidized-bed combus-
    tion, energy recovery from 85
  shavings 86

Wood (*continued*)
  tropical hard- 681
  using nickel catalysts, direct
    liquifaction of 363–368
  waste 97, 641, 654
    fired fluid beds 94
    programs 102–103

X

X-ray film, waste 227
Xylene 480

Z

Zinc oxide 430
Zurich 59